INTRODUCTION

SIMON TAYLOR

Of all books written on the subject of Scottish place-names, none more deserves to be kept in print than this one, *The History of the Celtic Place-Names of Scotland* (*CPNS*). It underpins the study of place-names (toponymics) in Scotland, and remains an essential reference work; one could almost say *the* essential reference work. There are two reasons for this, one positive, one negative. To start with the negative one: toponymics in Scotland has advanced relatively little since the publication of *CPNS* in 1926. The greatest advance is without doubt W.F.H. Nicolaisen's book *Scottish Place-Names*, first published in 1976: it embraced all the place-names of Scotland, not only those of Celtic origin; it constructed a clear and solid methodological framework for the discipline, and extended and tested the boundaries of what place-names can tell us about history, settlement and language. However, on the basic level of linguistic interpretation, i.e. how you translate a Celtic place-name into modern English, Nicolaisen offers little that is new, taking as his Celtic data-set chiefly the names in Watson's book. The next great advance in Scottish toponymics must be in the production of in-depth local place-names surveys, such as those covering most English counties (produced by the English Place-Name Survey) and about a fifth of Northern Ireland (produced by the Northern Ireland Place-Names Project). Scotland has only two comparable published county surveys, and significantly, and with typical foresight, the first one was produced by W.J. Watson himself, *The Place-Names of Ross and Cromarty*, which appeared exactly a hundred years ago (1904).[1] The publishers of this new reprint of *CPNS* considered a revised edition, but to be frank there is not enough new material to warrant this. The next great book on Scottish place-names, to rank with Watson and Nicolaisen, must await the completion of at least one county survey, ideally one from each of the nine linguistic zones of Scotland. These are the negative reasons why Watson's *CPNS* should be reprinted. The positive reason is, quite simply, that no matter how far the subject develops and advances, nothing will take away from this book its timeless excellence, its breath-taking scholarship, its encyclopaedic breadth and its good sense.

William John Watson was born on 17 February 1865, the son of Hugh Watson, a blacksmith, in Milton (Milntown of New Tarbat or Baile a' Mhuilinn Anndra), Kilmuir Easter, Easter Ross, on the original main road between Invergordon and Tain.[2] A native Gaelic-speaker, he was initially educated in Easter Ross by his uncle James Watson, himself an accomplished Gaelic and Latin scholar, then continued his education in Aberdeen and Oxford. In 1894 he became rector of Inverness Royal Academy, then in 1909 rector of the Royal High School, Edinburgh. In 1914 he was appointed professor of Celtic at the University of Edinburgh, a position he held until 1938, when he was succeeded by his son James C. Watson, who was killed in action in April 1942. W.J. Watson died on 9 March 1948, aged 83.[3]

Shortly after his appointment as professor of Celtic at Edinburgh, he was invited to give the prestigious Rhind Lectures on Archaeology for the Society of Antiquaries of Scotland. He gave six lectures in November 1916, and these, as Watson himself tells us in his Preface, form the nucleus of *CPNS*. Out of these six lectures developed the fifteen chapters of the book, in addition to an Introduction ('Introductory') and nine pages of Additional Notes.[4]

The slow, and no doubt much interrupted, evolution from lectures to book over ten years may explain why there is some repetition and overlap, as well as a rather haphazard system of cross-referencing. For example on

WILLIAM J. WATSON

THE HISTORY OF THE CELTIC PLACE-NAMES OF SCOTLAND

Birlinn

This edition published in 2004 by
Birlinn Limited
West Newington House
10 Newington Road
Edinburgh
EH9 1QS

Reprinted 2005

www.birlinn.co.uk

First published in Edinburgh and London, 1926

ISBN10: 1 84158 323 5
ISBN13: 978 1 84158 323 5

Printed and bound by Antony Rowe Limited, Chippenham

To
MY WIFE

CONTENTS

CHAPTER IX

CHAPTER X

CHAPTER XI

CHAPTER XII

CHAPTER XIII

CHAPTER XIV

CHAPTER XV

p. 321, in his discussion of Ernán dedications, he cross-references to his previous discussion of this saint, but not to another discussion relating to Tannadice and Ernán on p. 271, which adds important new information; or in his list of names containing British **lanarch* 'clearing, glade', he does not mention Caerlanrig ROX, which he discusses on p. 368; or on the same page his discussion of the element **perth* 'wood, copse' does not mention Panbart ELO, which he discusses on p. 374.

It is no exaggeration to say that everything Watson says about a name or an element deserves serious consideration. However, because of the structural constraints of the book, it is easy to miss important material. This situation is not improved by the absence of an elements index.[5] The first attempt to address this shortcoming was made by the entomologist (sic) Eric Basden, who compiled an Elements and Subject Index to *CPNS* in 1978. This was printed by the Scottish Place-Name Society in 1997, the Elements Index running to 73 pages of A4, in double columns, the Subject Index to six pages. The linguist and etymologist Dr Alan James has made a full revision of this Elements Index, linking together different forms of an element under one preferred headword, excising items not place-name related, supplying accents (length-marks), and adding anglicised versions of Celtic elements. This is available in digital form on the Scottish Place-Name Society's website.[6]

Several of Watson's etymologies stated with various degrees of certainty need to be revisited and reassessed, and such reassessments can have important consequences for our understanding of Scottish language- and settlement-history. One of Watson's great strengths is his intimate knowledge of his native tongue, Scottish Gaelic. Already in his Preface Watson states that, besides early forms, it is 'absolutely necessary to ascertain the traditional Gaelic forms of the names in all cases where that is possible' (p. xxxiii (xi)). While fully endorsing this, I would observe that sometimes Watson lets the modern Gaelic form override the evidence of early forms, even when these forms are many hundreds of years old. In other words, he does not always allow for processes of change such as assimilation and re-interpretation to apply to Gaelic as he does to Scots or Scottish Standard English. The important and frequently occurring Ruthven is a good example of this. By way of the later Gaelic form of this place-name, ***Ruadhainn***, he suggested a derivation from Gaelic *ruadh-mhaighin* 'red spot, red place'. Early forms such as ***Rothuan***, ***Roth(e)uen*** and ***Rothfan*** belie this, and there can be no doubt that this name needs to be completely rethought.

Any work as complex, far-ranging and ambitious as *CPNS* can be improved upon, and Watson himself saw the book as standing at the beginning of a process, not in any way a culmination.[7] Even with the relative lack of progress in Scottish toponymics since its publication, there is much that can be added and corrected. For example Watson regards the frequently occurring *-as*/*-es*/*-os*/*-us* ending on places-names, such as Dallas, Rothes, Duffus, as deriving from Gaelic *fas* 'stance, station (i.e. place for stopping)' (*CPNS* 498–9). However, the Irish scholar T. S. Ó Máille has argued convincingly that these endings *-as*/*-es*/*-os*/*-us* are in fact originally Gaelic abstract endings which, when used in a place-name context, mean simply 'place of' or 'place at'.[8]

It is somehow reassuring, as well as salutary, to see even Watson draw a blank despite lengthy and deep consideration. Take for example Fyrish, Alness parish ROS (***Fyrehisch*** 1479, ***Feris*** 1539): in the main text of his *Place-Names of Ross and Cromarty* Watson gives the Gaelic as ***Foireis*** and states 'probably from Norse "fura" or "fyri" [*fúri*], pine-tree' (1904, 77). However, in the Notes to the same book, he improves on his own rendering of the Gaelic, giving rather ***Faoighris***, and adding, somewhat plaintively, 'I fear that the name is Pictish' (277). This betrays an unusual moment of either terror or apology in the face of the realisation that anyone studying Scottish place-names in the north had to grapple with the Pictish dimension – little enough known today, but how much less so in 1904![9] Watson never flinched from tackling this issue head-on, which makes this little remark all the more endearing. In 1909 (149), he gives as the Gaelic for Fyrish both *Foighris* and *Faoighris*, suggesting that it may be for *fo-iris* 'under-roost' or 'small roost', referring to a 'remarkable projection or spur of considerable size, surrounded by a deep gully'. It would seem that he finally gave up on this intractable place-name, as he mentions neither Fyrish nor the element *iris* in his *magnum opus* of 1926.

CPNS has become a kind of Bible not only for those who study the place-names of Scotland, but also for anyone with even a passing interest in the early history of the language, culture and settlement of this country; while its author has been accorded a status as near to divine as our secular age will allow. Having lived and worked closely with this book for over a decade, I can safely say that both book and author are fully worthy of their status.[10/11]

BIBLIOGRAPHY OF WILLIAM J. WATSON 1865–1948

This is fullest bibliography of W.J. Watson yet attempted in print. While it cannot claim to be complete, it does include to the best of my knowledge all place-name related material. I am very grateful to Peadar Morgan, who has been compiling a full, annotated bibliography over several years, for allowing me to include his work here.

W.J. Watson was very ready to share his toponymic expertise with fellow Scottish historians and linguists, and his name appears in footnotes or acknowledgements in a wide variety of publications. For example he contributed extensively with revisions and additions to the list of Proper Names in E. Dwelly's *Illustrated Gaelic-English Dictionary* (1911), 1003–30. In the *Charters, Bulls and other Documents relating to the Abbey of Inchaffray* (Scottish History Society 1908) the writer of Appendix III ('Notes on the Place-Names in the Inchaffray Charters', 323–30), Donald MacKinnon professor of Celtic at the University of Edinburgh, 'desires to associate with himself in contributing these Notes Mr W.J. Watson, Rector of Inverness

Academy' (footnote, p. 323). Thirty years later in a similar publication Watson's contribution to the elucidation of place-names in Appendix VII of *Charters of the Abbey of Inchcolm* (Scottish History Society, 1938, 249–53) is gratefully acknowledged by the editors D.E. Easson and Angus Macdonald. And A. and E. Ritchie's *Map of Iona with a Sketch Historical and Geological of the Island*, 1928, contains a place-names appendix with suggestions by W.J. Watson (29–35). One of his last contributions of this nature was his foreword to Seton Gordon's *Highways and Byways in the Central Highlands* (London), vii–viii, and his 'Hints on Gaelic Pronunciation' followed by a list of Gaelic place-names, with their meanings, 415–9. This was published in 1948, the year of his death.

1904, *Place-Names of Ross and Cromarty* (Inverness; reprinted 1976 and 1996)

1904a, '*Place Names of Scotland*: A Review' *Inverness Courier* [Review of the new and enlarged edition of J.B. Johnston *Place-Names of Scotland* 1892, second edition 1903)] [also Watson 2002, 33–43]

1905, 'The Study of Highland Place-Names', *Celtic Review* 1 (1904–05), 22-31 [also Watson 2002, 44–53]

1905a, 'Tara', *Celtic Review* 1 (1904–05), 286 [also Watson 2002, 55]

1905b, 'Paisley', *Celtic Review* 1, 288 (1904–05) [also Watson 2002, 54]

1906, 'Some Sutherland names of places', *Celtic Review* 2 (1905–06), 232–42, 360–8 [also Watson 2002, 56–66, 66–75]

1906a, 'The Celtic Church in Ross', *Transactions of the Inverness Scientific Society and Field Club* 6 (1899–1906), 1–14

1906b, 'Study of Scottish Place Names', *Transactions of the Inverness Scientific Society and Field Club* 6 (1899–1906), 279–80

1906c 'Faclair Gaidhlig' (review of E. Dwelly *The Illustrated Gaelic-English Dictionary*, pp. 1–280), *Celtic Review* 2 (1905–06), 383-4

1907, '*Innis* in Place-Names', *Celtic Review* 3, 239–42 [also Watson 2002, 76–8]
1908, 'Note on Maolrithe', *Celtic Review* 4 (1907–08), 96 [also Watson 2002, 79]
1908a, 'Cliar Sheanchain', *Celtic Review* 4 (1907–08), 80–8
1908b Review of P. Power, *The Place-Names of Decies*, *Celtic Review* 4 (1907–08), 373–5
1908c 'Ancient Celtic Cavalry Terms', *Celtic Review* 4 (1907–08), 383–4
1909, 'Topographical Varia [I]: *fo; lòch; ialo-s; coll, call, calltuinn; Ibert and Offerance*', *Celtic Review* 5 (1908–9), 148–54 [also Watson 2002, 80–6]
1909a, 'Topographical Varia [II]: *tros; esc; benn; mion; gàg; ith, iodh, ithir*', *Celtic Review* 5 (1908–9), 337–42 [also Watson 2002, 87–92]
1909b, Reviews and Notes *Celtic Review* 5 (1908–9), 288–90
1909c, *Prints of the Past Around Inverness*, published by *Northern Chronicle*, Inverness; this includes 'Names of Places around Inverness' [Revised edition published 1925, for which see 1925b]
1910, 'Topographical Varia -III: *fortair, gwerthyr, verterae; céith, keith,céto-n; eag; air: ur*', *Celtic Review* 5 (1909–10), 236–41 [also Watson 2002, 93–8]
1910a, Review of A. Holder, *Alt-Celtischer Sprachschatz* 19, *Celtic Review* 6, 383
1911 Contribution to discussion following William Mackay 'Saints associated with the valley of the Ness', *TGSI* 27 (1908–11), 160–1 (Mackay's paper, with discussion, 145–62)
1912, 'Topographical Varia -IV: *ath, ate; eadar, *enter, inter; fonn; Brannradh; comraich, tearmann, teagarmachd; Connel, Congal; fas, foss; fasadh, fossad; Invernahyle*', *Celtic Review* 7 (1911–12), 68–81 [also Watson 2002, 99–112]
1912a, 'Topographical Varia -V: *dubron dobhar; mig; Baile Bhaodan; Dùn Bhallaire*', *Celtic Review* 7 (1911–12), 361–71 [also Watson 2002, 113–22]
1912b, Review of *Fianaigecht*, *Celtic Review* 7 (1911–12), 95–6

1912c Review of E. Hogan's *Onomasticon Goedelicum, Celtic Review* 7 (1911–12), 379–84

1913, 'Topographical Varia-VI: *-nt terminal; braon*; Prefixed Nouns used as Adjectives', *Celtic Review* 8 (1912–13), 235–45 [also Watson 2002, 123–32]

1913a, 'The Circular Forts of North Perthshire', *PSAS* 47 (1912–13), 30–60 [paper delivered December 1912]

1913b Review of Orain Ghaidhealach le Donnchadh Macantsaoir [sic], *Celtic Review* 8 (1912–13), 255–61

1913c Review of *Revue Celtique* Vol. XXXIII, No 1, (review) *Celtic Review* 8 (1912–13), 265–6

1913d 'Breisleach', *Celtic Review* 8 (1912–13), 288

1914, 'Aoibhinn an Obair an t-Sealg', *Celtic Review* 9 (1913–14), 156–68 [article in English about deer hunting and associated place-names]

1914a 'Circular Forts in Perthshire', *TGSI* 28 (1912–14), 151–5 [paper delivered December 1912]

1914b, 'Ciuthach', *Celtic Review* 9 (1913–14), 193–209

1914 Reviews of various publications including works by C.J.S. Marstrander, K. Meyer and A. Holder (*Alt-Celtischer Sprachschatz* 21), *Celtic Review* 9 (1913–14), 174–7, 259–61

1915, *Rosg Gàidhlig* (Inverness)

1915a, Circular Forts in Lorn and North Perthshire', *PSAS* 49, 17–32

1916, 'Some Place-Names in the Cairngorm Region', *Cairngorm Club Journal* 8, 133–6 [also Watson 2002, 133–6]

1916a, The Position of Gaelic in Scotland', *Celtic Review* 10 (1914–16), 69–84

1916b, 'The Celtic Church in its Relation with Paganism', *Celtic Review* 10 (1914–16), 263–79

1916c, Review of J. B. Johnston, *The Place-Names of England and Wales*, *Celtic Review* (1914–16) 10, 280–4

1916d, Elrick (note) *Celtic Review* 10 (1914–16), 287

1916e, 'The Death of Diarmid', *Celtic Review* 10, 350–7

1918, *Bàrdachd Ghàidhlig* (Inverness)

1918a, 'Classic Gaelic poetry of Panegyric', *Proceedings of the Royal Philosophical Society of Glasgow* 49, 134–56

1919, 'Classic Gaelic poetry of Panegyric', *TGSI* 29 (1914–19), 194–235 [paper delivered 1918]
1921, *The Picts: their Original Position in Scotland* (reprinted from *The Inverness Courier*, Inverness)
1922, 'Place-names of Strathdearn', *TGSI* 30 (1919–22), 101–21 [also Watson 2002, 137–54] [paper delivered March 1920]
1922a, 'The Picts: their original position in Scotland', *TGSI* 30 (1919–22), 240–61 [paper delivered April 1921].
1922b, Alexander MacBain, *Place Names of the Highlands and Islands of Scotland*, edited with Introduction and Notes by W.J. Watson (Stirling)
1924 *Ross and Cromarty* (Cambridge County Geographies, Cambridge)
1925, 'The Celts (British and Gael) in Dumfriesshire and Galloway', *Transactions of the Dumfriesshire and Galloway Natural History and Antiquarian Society*, Third Series 11, 119–48
1925a, 'Personal Names: The Influence of the Saints', *TGSI* 32 (1924–25), 220–47
1925b *Prints of the Past Around Inverness*, published by *Northern Chronicle*, Inverness; this includes 'Names of Places around Inverness', which is now Watson 2002, 155–62 [Revised edition of 1909c]
1926, *The History of the Celtic Place-Names of Scotland* (Edinburgh and London; reprinted several times, most recently Edinburgh 1994)
1927 'Saint Cadoc', *SGS* 2 part 1, 1–12
1928, 'The Place-Names of Breadalbane', *TGSI* 34 (1927–28), 248–79 [also Watson 2002, 163–92]
1929, *Rosg Gàidhlig, Specimens of Gaelic Prose* (Glasgow) [2nd edition; 1st edition 1915]
1930 'Early Irish Influences in Scotland', *TGSI* 35 (1929–30), 178–202 [printed in 1939; paper delivered April 1929]
1930a, 'Place-Names of Perthshire: The Lyon Basin', *TGSI* 35 (1929–30), 277–96 [printed in 1939; also Watson 2002, 193–210]
1930b, 'Some place-names of the North', *Northern Chronicle*

(Inverness) [Highland Exhibition; also Watson 2002, 211–35]
1931, 'Varia: Reply to a Review', *SGS* 3, 203–8 [the review replied to was by Dr E.G. Gwynn of *The History of the Celtic Place-Names of Scotland* [also Watson 2002, 236–41]
1933, '*Annaid*', TGSI 36 (1931–33), 399–400 [also Watson 2002, 242]
1933, 'The Macdonald Bardic Poetry', *TGSI* 36 (1931–33), 138–58
1933a 'The Celts in Britain', *TGSI* 36 (1931–33), 241–61
1936, 'The History of Gaelic in Scotland', *TGSI* 37 (1934–36), 115–35 [published 1946]
1937, *Scottish Verse from the Book of the Dean of Lismore*, Scottish Gaelic Texts Society I (Edinburgh)
2002, *Scottish Place-Name Papers* (London and Edinburgh)

CORRIGENDA and ADDENDA

In a work ranging over such a wide geographical area, and dealing with such a large number of place-names, errors have inevitably crept in. Rather than revise the text, it was deemed more appropriate to include a list of corrigenda. It cannot be regarded as exhaustive. The list of addenda must be seen as even less exhaustive, reflecting the interests of those who have contributed to it. A full addenda of this book would turn into a major review and update of the whole study of Celtic toponymics in Scotland, which is not within the purview of this re-issue.

22–3 Watson states that the true form of the name of the tribe occupying east central Scotland north of the Forth known variously as Venicómes, Vennicónes, Vernicónes, Vennicones, Venicones, Venicontes) is too uncertain for satisfactory explanation. However, it would seem, from the place-name Maen Gwynngwn in the early Welsh poem *Gododdin,* that the true form is *Venicones* (*Wenikones*), meaning 'The Kindred Hounds', more figuratively 'The Noble Kindred' (Koch 1980).
32, note 2 Add 'twenty' before 'seven'.
48, note 1 Watson asks: 'Is "Forne", which is given as an

old name for the Beauly river, a misreading of "Forire", and merely a ghost name?' The answer to this question is definitely no: Forn(e) is a genuine name for this river, attested in a wide range of medieval and early modern texts.

105 'In 678 (AU) Domnall Breac, king of Dàl Riata, was defeated at Calathros.' While the Annals of Ulster (AU) do ascribe the defeat of Domnall Brecc (AU spelling of the name) to the year 678 it was in fact much earlier, probably c. 634.

119 'Buchquhane in Strathore, Fife, 1530 (*RMS*), now apparently obsolete;' this is in fact Mountquhanie, Kilmany FIF.

136 'Balantrodach, now Arnieston, in Temple parish [MLO] . . . for *Baile nan Trodach*, Stead of the Warriors . . . And there can be little or no doubt that the name was given with reference to the Knights' Templar, who 'had a chapel there in the time of David I'. In fact the earliest form is *Plent[r]idoc* 1175 x 1199 (*Glas. Reg.* i no. 41), and, along with the *Blantrodoc* form from *Kelso Lib.* (i, no. 223) dated 1287 (quoted as *Blantrodoch* and undated by Watson), it is clearly a British place-name containing the element **blain*, cognate with Welsh *blaen* 'end, summit, point; upland; source or upper reaches of a water-course'. It cannot originally have had anything to do with the Knights Templar.[12]

138 Watson suggests that Ballindean (Balmerino parish FIF) and Ballindean (Inchture parish PER) both contain Gaelic *deadhan* '(ecclesiastical) dean'. In fact they derive from Gaelic *baile an t-fhàin* 'stead of the (lower) slope', containing Gaelic *fàn* 'slope, hollow', and are equivalent to Scots *Netherton*.

'Deuchar occurs also in Peebles and in Tannadyce [sic] parish, Forfar.' For 'Tannadyce parish' read 'Fern parish'. Forfar(shire) is the old designation of Angus.

141 Barbauchlaw (now Barbachlaw) is in Edinburghshire (now MLO) not Haddington(shire) (now ELO). It is in Inveresk parish near Musselburgh (as Watson correctly states p. 266).

143 For 'Balnoon, Inverkeithing [FIF]' read 'Balnoon, Inverkeithny [BNF]'.

144 Bolgyne (which belonged to the son of Torfyn and which Macbeth gave to the church of St Serf, Loch Leven) is Bogie, now on the north-west edge of Kirkcaldy FIF.

Torsappie, now Tarsappie, is in Perth parish south-east of the town. Its earliest form is *Torsoppin* 1157 x 1160 *RRS* i no. 157.

147 'Glenpuitty, near Dalmeny' WLO. This is the form of the name on the Ordnance Survey 1-inch 2nd edition map (1898). However, on other editions of the Ordnance Survey maps, both earlier and later, the name appears as 'Glenpunty (Wood)' (NT17 76), which is the historically more correct form (see Harris 1996 under 'Glenpuntie'; curiously it is not discussed in MacDonald 1941). This means that the derivation proposed by Watson cannot be correct.

165 For 'Gilendonrut' read 'Gilcudbricht' (see Black 1946, 300).

175–77 Watson is quite right to want to re-situate the *Niduari* of Bede's 'Life of St Cuthbert' from south-west Scotland to north of the Forth. Later research has fully vindicated him in this, going so far as to connect the name with the Fife settlement– and parish-name Newburn (earlier *Nithbren*), which Watson himself discusses in a different context *CPNS* 54–5. The best summary of this later research can be found in Duncan 1975, 69, and note p. 78.

182 For 'Arnmannoch Kirkpatrick Irongray' read 'Arnmannoch Lochrutton' KCB. However, it is very near the Kirkpatrick Irongray parish boundary.

'Balgeuery . . . from the places that go with it on record it appears to be now Balwearie'. In a footnote Watson shows that he is not happy with this equation, and justifiably so. Balgeuery is a now obsolete name in Kinghorn parish, while Balwearie is in Kirkcaldy parish FIF.

187 For 'St Ciarán's servant's dale' read 'St Ciarán's servant's water-meadow'. This is Watson's suggested etymology of Dailly AYR. However, a much earlier form of Dailly (*Dalmakeran* 1236 *Pais. Reg.* 427) suggests that it means simply 'St Ciarán's water-meadow'.

189 For 'Culbirnie' read 'Culburnie'.

195 For 'Lintheamine' read 'Lintheamina'. The form quoted by Watson is a Latin genitive singular of a first declension noun, the nominative being 'Lintheamina'.

St Madoes PER is not dedicated to St Cadoc but to St Aedoc or Aedan (Aidan). There is also some doubt as to whether Kilmadock contains this name. [see Corrigenda and Addenda below under p. 327]

199 For 'Drumquhassil [now Drumquhassle] (in Dunbartonshire)' read 'in Stirlingshire' (Drymen parish).

204 For 'Villa mineschedin' read 'Villa mineschadin' and in the same line for '1172' read '1173'; in the next line for 'villa Inienschedin' read 'villa Inienschadin'.

242 Early forms of Kingoldrum ANG (e.g. *Kingoueldrum* 1178 *RRS* ii no. 197) show that it cannot contain *coll* 'hazel'; more likely *gobhal* 'fork'.

267 Dulbachlach etc survives as the farm-name Dunballoch (near Lovat Bridge over the Beauly River INV). It was the old name of Wardlaw parish, which together with the old parish of Farnway or Farnua, makes up the parish of Kirkhill. Frequent early forms such as *Dulbatelauch* (1221) rule out a derivation from *bachall*.

'Pittincleroch 1489 (*RMS*) . . . was in the earldom of Strathearn'. It survives as Pittencleroch, Fowlis Wester PER.

268 For 'Pitlour in Kinross-shire' read 'Pitlour in Fife'.

270 Ardeonaig, a medieval parish, on the south side of Loch Tay PER, has a much earlier form, *Ardoueny* 1275 Bagimond's Roll.

271 For 'Furvie' read 'Forvie'.

For 'Sanct Eunendi's Seit' read 'Sanct Eunandis Seit'; there is some doubt as to whether this does in fact refer to the hill now known as St Arnold's Seat (see Taylor 1999, 63–4).

The church of Inch (for which read Insh) in Badenoch is an Adomnan dedication, and contains a famous early bell associated with him.

289 'A piece of land near Pethnick was "Sanct Malrubus stryp", 1576 (*RMS*).' Pethnick, now Paithnick, is in Grange

parish by Keith BNF. The form in the source (*RMS* iv no. 2644) in fact reads *Sanct Mulrubus stryp*. A *stryp* can refer to both a strip of land and a small burn. The context shows that in this case it refers to the latter.

293 'Gillemelooc was an Aberdeenshire name, *c.* 1200'. In fact Gillemelooc (at the reference given, RPSA [= *St A. Lib.*] 290–91) is a witness to a charter relating to north-east Fife.

307 'with the staff of St. Munde called in Gaelic (*Scotice*) *Deowray*' (1497 *RMS* ii no. 2385). This should read 'the staff of St. Munda', as the male St Mundu has become female in this charter ('cum baculo Sancte Munde, Scotice vocato *Deowray*').

309 Senchán . . . 'may be Senchán of Imlech Ibair, Dec. 11, called Mo-Shenóc by Oengus'; this is a reference to the early 9th century Old Irish verse martyrology by Oengus. In fact Oengus, and the notes, which date to around the 11th century, link Mo-Shenóc celebrated on 11 Dec. with Belach Mugnae, now Ballaghmoon, County Kildare (see Stokes 1905, 251, 259).

314 'He (Coeddi) seems to appear also in Inchcad, conjoined with Clony (Clunie in Stormont), 1275 (Theiner)'. In fact Inchcad is now Inchadney, and is in Kenmore parish PER, near Taymouth Castle. See also Watson's own notes to pp. 273, 308 (n. 2) on pp.517–18

Kilconquhar FIF: from early forms it is clear it does not derive from Conchobar, but rather from Dunchad (Duncan). See Taylor 1996, 100, 106.

323 'The church of Strageath in Strathearn and the neighbouring churches of Blackford and Dolpatrick are dedicated to Patrick'. In fact we are dealing with just one Patrick dedication here. The medieval parish kirk of Strageath was dedicated to Patrick. Blackford parish is the modern successor to much of the parish of Strageath, which explains its Patrick dedication. Dolpatrick, now Dalpatrick, was also part of Strageath, and is now in Crieff parish.

324 For 'St. Cyrus in Forfarshire' read 'St. Cyrus in Kincardineshire'. Note also *Lungyrg* KCD (*Arbroath Lib.* i no. 127), which contains this saint's name.

Kilgraston, Dunbarney parish PER, contains the personal name Gillegirg, and is probably named after the Kilegirge (Gillegirg) son of Malis (*c.* 1197) mentioned here by Watson.

Kentigern and Glengairn ABD: for the suggestion that Mo-Thatha is in fact Kentigern, see Ó Baoill 1993.

324–5 St Lolan is probably also commemorated in the name Bonhill DNB (*Buthelulle* 1247 x 1259 *Glas. Reg.* no. 177; *Bothlul* 1274 *Pais. Reg.* 216).

327 For the suggestion that Cadog is probably not the saint of Kilmadock PER, but rather St Doc or Docgwin, see Brooke 1963, 298.

St Madoes, Carse of Gowrie PER does not contain Cadog. The forms 'ecclesia de San[c]to Maghot' (1274 Bagimond Roll, 54) and 'ecclesia de Sancto Mathoco' (1275 Bagimond Roll, 71) show that the saint in question is Aedan (Aidan), from a hypocorism (i.e. pet-name form) of Mo Aedoc ('my little Aed').

331 Eglismenythok etc. by Monymusk ABD is now Abersnithock, Monymusk. See Alexander 1952, 136, and Corrigenda and Addenda below for p. 465.

341 'It is not unlikely that we have the same word [the first element of Edinburgh] in Etin's Ha', the name of a broch on Cockburnlaw in Berwickshire . . . "The Reid Etin" is mentioned ...as a popular story of a giant with three heads . . .' In fact in both these cases we are dealing with the Scots *etin* 'giant', deriving ultimately from Old English *eoten* 'giant'. It can have nothing to do with the first element of Edinburgh, which is Celtic.

350 For 'Trostrie, Wigtown' read 'Trostrie, Kirkcudbrightshire'. To names containing this element **tros* can be added Trustach, Banchory-Ternan KCD and probably Trusta, Fern ANG.

352–3 An important recent contribution on the element **carden* has been made by Andrew Breeze, in which he points out that the evidence for the Welsh cognate *cardden* meaning 'thicket, brake' etc. is dangerously flimsy, and suggests a meaning such as 'encampment, enclosure' instead (Breeze 1999).

'But the Urquharts and Leden Urquhart of Strathmiglo in Fife, and the Urquhart near Dunfermline are of quite different origin [from the Urquharts further north]'. In fact, early forms of the Fife Urquharts show that they are all of the same origin as the northern ones, containing **carden*.

353 'Kincardine . . . in Menteith; in Kincardine-on-Forth, Perthshire': Kincardine-on-Forth, where the Kincardine Bridge is today, is in Tulliallan parish FIF (although it was in a detached part of Perthshire till 1891).

353 'I have no sure instance of *pen* [p-Celtic 'head, end'] north of the isthmus of Forth and Clyde, nor from Galloway or Ayrshire'. For north of the Forth–Clyde line, Watson was certainly right to be cautious, though the following should be considered: Pennan, Aberdour parish ABD (*Pennand* 1587); Pandewan [hill], Lochlee parish ANG; Pinderachy, Tannadice/Fern parishes ANG, and Pinnel Hill, Dalgety FIF (*Pin-hill* 1756).

It is more surprising that he failed to mention a significant group of names in Galloway and Ayrshire which probably contain this element, such as: Pencloe, New Cumnock parish AYR, Penderry Hill, Ballantrae parish AYR, Penneilly Cairn, Balmaclellan parish KCB, Pen Hill, Sorbie parish KCB, Pennan Hill, Kelton parish KCB, Penwhaile [hill], Girthon parish KCB, Penwhapple, Dailly parish AYR and Penwhirn, Inch parish WIG. There is also Panbreck Hill on the AYR-LAN border, and only 2 km to the south (in AYR) is Penbreck Rig.

Another surprising omission from names containing p-Celtic *pen* is Pentland MLO, an early parish name (*Pentland* mid 13th century *St A. Lib.* 28), probably 'end-enclosure or end-church'. (This must not be confused with Pentland of Pentland Firth, from Old Norse *Pettalandfjörðr* 'Pictland firth').

359 'There are on record Treverman in Cumberland . . .' This is now Triermain near Birdoswald, Cumberland.

373 For 'Parbroath in Forfarshire' read 'Parbroath in Fife' (Creich parish). He was unaware of the earliest forms with *parte-* or *porte-* (*Partebrothoc* 1315 *Scottish Historical Review* 2, 173; *Portebrothok*[13] 1335 x 1337 NLS Adv MS 34.6.24, p.

409) so assumes that the first element is the obscure and poorly attested **par*.

374 'Panlaurig 1509 (*RMS*) "in the territory of Duns", Berwickshire, is probably for "Panlanrig", in which case it would mean [in British] "hollow in the glade".' However later forms show that *Panlaurig* is the correct reading (*Panlawrig* 1535 *RMS*; *Panlawrig* 1574 *RMS*; *Pannalrig* 1595 x 1609 *RMS*; *Panlawrig* 1621 *Retours*), and the final element is probably Scots *rig*.

378 'In Linlithgow there is on record Okelfas, Ogelfas etc'. This has survived as Ogilface Castle, Torphichen parish WLO.

Ogilvie BNF, which is mentioned in 1655, took its name from the family of Ogilvie, and has not survived.

380 Craighorn, a hill behind Alva CLA, is *Craigharr* on Stobie's map of 1783, and is unlikely to contain the element **gronn* 'myre'.

Chingothe (*RRS* i no. 123) is now Kingoodie, Longforgan parish PER. It is on the coast, and well illustrates the more specific meaning of Gaelic *gaoth* '(tidal) inlet' (beside the meaning 'marsh, bog' discussed by Watson).

381 For 'Knockcoid in Kirkcudbrightshire' read 'Knockcoid in Wigtownshire' (Kirkcolm parish).

382 The early forms of Balkeith near Tain (*Balmathothe* 1533 *RMS* iii no. 1304, probably for *Balmachothe* or *Balnachothe*; and *Ballecuthe* 1539 *RMS* iii no. 2043), make it very unlikely that it contains p-Celtic **cét-* 'a wood'.

The early form of Balmakeith near Nairn, *Balnecath* (1238 *Moray Reg*. no. 40) militates against it having anything to do with a personal name.

The second element in Inverkeithing FIF and Inverkeithny BNF is in fact very likely to be a burn- or river-name containing p-Celtic **cét-* 'a wood'.

383 To Manor PEB can be added Manor, Logie parish STL.

383–4 Castle Lyon, which Watson suggests may be a very early name containing a British loan-word from Latin *legio*, *legionis* 'legion', in fact contains the family-name 'Lyon' (see MacDonald 1941, 34).

386 The most likely derivation of Partick in Glasgow is from a diminutive (or locational) form of p-Celtic **pert(h)*

'grove, wood', for which see *CPNS* 356-7.

387–8 Ruthven. See Introduction above, p. x.

399 The chapel of *Munmaban* has no particular geographical or ecclesiastical link to the church of Kirkurd (*Horda*); it simply follows it in a long list of churches and chapels belonging to the church of Glasgow (*Glas. Reg.* no. 62).

402 Kilbrackmont, Kilconquhar parish FIF, has forms going back to the 13th century, making it clear that the first element is Gaelic *ceann* 'head, end'.

For 'Montrave in Largo parish' read 'Montrave in Scoonie parish'.

403 Watson lists three places in Fife as obsolete: Monthquoy, Montripple and Munquhany etc. These are in fact early forms of the modern names Montquey, Aberdour parish, Monturpie, Largo parish, and Mountquhanie, Kilmany parish.

407 Pitalmit and Pitchalman in Glenelg INV represent only one place, the former being an early (and poorly transcribed) form of the latter.

Watson's approximate numbers for place-names containing the element *pett (Pit-)* have to be completely revised. For example in Fife and Kinross, instead of Watson's 57, there are at least 77.

408 Blairfetty, Blair Atholl parish PER: Watson analyses this as 'field of the place of *petts*', from *pett* 'holding, estate'. However, the early form *Blairquhatti* (1515 *RMS* iii no. 32) shows that the later *f* is probably from *ch*, from *chat-*, a lenited form of Gaelic *cat* 'cat' (in place-names mainly referring to the wild cat). The final element is probably the locational suffix *-in* 'place of'.

408–9 Pitmiclardie, which Watson locates in Fife, is a mystery. I have been unable to locate this. It is certainly not in Fife.

410 Pitarrow, which Watson places in Forfar (ANG) and FIF, exists only as such in KCD. There is no place of this name in FIF; while the Angus example may be Pitarris, south-west of Montrose.

411 'Blato-bulgio(n), a place in Britain'. This has now been identified with Birrens DMF.

Pethmolin in Crail parish FIF is now Pitmilly, Kingsbarns parish (formerly part of Crail parish).

412 For 'Pitcowden in Aberdeenshire' read 'Pitcowden in Kincardineshire'.

'Pitfoskie [New Deer parish ABD] is from *fosgadh* (now with us *fasgadh*), shelter'. Early forms (such as *Badorosky* c.1300 *Aberdeen-Banff Coll.* 189, probably for *Badcrosky*; *Badforsky* 1587 *RMS* v no.1309) show that a) the first element is Gaelic *bad* 'clump, spot', not *pett*; and b) the second element is Gaelic *crasg, croisg* 'a crossing' (an element discussed by Watson *CPNS* 485).

413 Pitfoules in Fife is a ghost-name, arising from a miscopying in *Retours* Fife no. 1399 (1698) of an original Pitsoulie, Torryburn FIF.

Watson equates Petnaurcha, which he calls one of the oldest names on record in Fife, with Urquhart near Dunfermline. In fact Petnaurcha is the older name for the Dunfermline suburb of Blacklaw (see Taylor 1994, 11, Note 2).

456 For 'Baldowrie in Kettins parish, Fife' read 'Baldowrie in Kettins parish, Forfarshire (ANG)'.

465 Abersnithock, Monymusk parish ABD is not in fact an *Aber*-name. Early forms such as *Eglismenythok* show that the first element was originally **egles* 'church'. Early forms can be found in *CPNS* 331. Watson does not make the connection between the *Aber-* and *Eglis-* forms. See also Corrigenda and Addenda above for p. 331.

478 '. . . Belladrum, "ford-mouth ridge", near Beauly.' Belladrum (Kiltarlity and Convinth parish INV) is more likely to contain *baile* 'farm' as its first element (*Beldrum* 1496 *RMS* ii no. 2320; *Balladrum* 1512 *RMS* ii no. 3730).

479 'Kinnoull near Perth is also *Cinn Alla*, "at head of crag" '. In fact early forms such as *Kynwl* 1250s show that the second element cannot be *all*, genitive *alla* 'cliff, crag'.

500 Magask (and Magus) near St Andrews FIF does not contain the element *magh* 'plain'. Early forms such as *Malgaskis* 1196 x 1204 *RRS* ii no. 411 or *Ovirmalgask* 1438 *St A. Lib.* 430, show that the first element is Gaelic *maol* 'bare'.

502 All the early forms which Watson quotes under 'Alloa' are in fact of Alva, Clackmannanshire.
508 Dunnone 1493, Dunnoyne 1494 in Forfarshire is now Denoon, Glamis parish ANG.
521 For Additional Notes: 'p. 432' read 'p. 431'.

INDEX OF PLACES AND TRIBES

523 For 'Affric 450' read 'Affric 451'.
535 For 'Eunág 444' read 'Eunág 448'.

NOTE ON ORTHOGRAPHY

W.J. Watson's use of *á* in unstressed syllables (e.g. in the *-á(i)n* and *-á(i)g* endings on Gaelic words and names) is not part of Standard Scottish Gaelic orthography.

REFERENCES AND ABBREVIATIONS OF SOURCES

Aberdeen-Banff Coll. Collections for a History of the Shires of Aberdeen and Banff (Spalding Club 1843)

Alexander, William M., 1952, *The Place-Names of Aberdeenshire* (Third Spalding Club)

Arb. Lib. Liber S.Thome de Aberbrothoc (2 volumes, Bannatyne Club, 1848–56)

Bagimond's Roll 'Bagimond's Roll: Statement of the Tenths of the Kingdom of Scotland' ed. A.I. Dunlop, *SHS Misc.* vi (1939), 3–77

Barrow, G.W.S., 1998, 'The Uses of Place-names and Scottish History–Pointers and Pitfalls', in S. Taylor (ed.) *The Uses of Place-Names* (Edinburgh), 54–74

Black, George F., 1946, *The Surnames of Scotland* (New York; reprinted 1993, Edinburgh)

Breeze, Andrew, 1999, 'Some Celtic Place-Names of Scotland, including *Dalriada, Kincarden, Abercorn, Coldingham* and *Girvan*', *Scottish Language* 18, 34–51

Brooke, Christopher, 1963, 'St Peter and St Cadoc' in N.K. Chadwick (ed.), *Celt and Saxon*, 258–322

Duncan, A.A.M., 1975, *Scotland, The Making of the Kingdom* (Edinburgh)

Glas. Reg. Registrum Episcopatus Glasguensis, Bannatyne & Maitland Clubs, 1843.

Harris, Stuart, 1996, *The Place Names of Edinburgh* (reprinted 2002, London and Edinburgh)

Kelso Lib. Liber S. Marie de Calchou, Bannatyne Club, 1846.

Koch, John T., 1980, 'The Stone of the *Weni-kones*', *Bulletin of the Board of Celtic Studies* 29, 87–9

MacDonald, Angus, 1941, *The Place-Names of West Lothian* (Edinburgh and London).

Moray Reg. Registrum Episcopatus Moraviensis, Bannatyne Club 1837

Nicolaisen, W.F.H., 2001, *Scottish Place-Names* (revised edition, Edinburgh; first published London; 1976)

Ó Baoill, Colm, 1993, 'St Machar – Some Linguistic Light?', *Innes Review* 44, 1–13

Pais. Reg. Registrum Monasterii de Passelet, Maitland Club 1832; New Club 1877

PSAS Proceedings of the Society of Antiquaries of Scotland

Retours Inquisitionum ad capellam domini regis retornatarum . . . abbreviatio, ed. T. Thomson (3 vols., 1811–16)

RMS Registrum Magni Sigilli Regum Scottorum (*Register of the Great Seal*), ed. J.M. Thomson & others, Edinburgh 1882–1914 (reprinted 1984)

RRS i Regesta Regum Scottorum vol.i, (*Acts of Malcolm IV*) ed. G.W.S. Barrow, Edinburgh 1960

RRS ii *Regesta Regum Scottorum* vol.ii, (*Acts of William I*) ed. G.W.S. Barrow, Edinburgh 1971

SGS Scottish Gaelic Studies

St A. Lib. Liber Cartarum Prioratus Sancti Andree in Scotia, Bannatyne Club 1841

Stokes, W. 1905 (ed.), *The Martyrology of Oengus*, Henry Bradshaw Society 29 (London; repr. Dublin 1984)

Taylor, Simon, 1994 'Some Early Scottish Place-Names and Queen Margaret', *Scottish Language* 13, 1–17

Taylor, Simon, 1996, 'Place-names and the Early Church in Eastern Scotland', in B.E. Crawford (ed.) *Scotland in Dark Age Britain* (Aberdeen), 93–110

Taylor, Simon, 1999, 'Seventh-century Iona abbots in Scottish place-names', in D. Broun and T.O. Clancy (edd.) *Spes Scotorum Hope of the Scots* (Edinburgh), 35–70

TGSI Transactions of the Gaelic Society of Inverness

OTHER ABBREVIATIONS, INCLUDING COUNTY ABBREVIATIONS

The county abbreviations are those used by W.F.H. Nicolaisen in Scottish Place-Names (see Nicolaisen 2001, xxi–ii), and refer to the pre-1975 counties.

ABD Aberdeenshire
ANG Angus (Forfarshire)
AYR Ayrshire
BNF Banffshire
CLA Clackmannanshire
DMF Dumfriesshire
DNB Dunbartonshire
ELO East Lothian (Haddingtonshire)
FIF Fife
INV Inverness-shire
KCB Kirkcudbrightshire
KCD Kincardineshire
KNR Kinross-shire
LAN Lanarkshire
MLO Midlothian (Edinburghshire)
NAI Nairnshire
PEB Peebleshire
PER Perthshire
ROS Ross and Cromarty
ROX Roxburghshire
STL Stirlingshire
SUT Sutherland
WIG Wigtownshire
WLO West Lothian (Linlithgowshire)

NOTES

1. The other is *The Place-Names of West Lothian*, Angus MacDonald, 1941. A survey of Midlothian exists as a PhD by Norman Dixon ('Place-Names of Midlothian', Edinburgh University, 1947), but remains unpublished.
2. Watson 2002, 10. His obituary in *SGS* 6 (1949) by John MacDonald states that he was born at Kindeace (also Kilmuir Easter) (215).
3. For full biographical details, see W.F.H. Nicolaisen's article 'In Praise of William J. Watson (1865–1948): Celtic Place-Name Scholar', first published in *Scottish Language* 14/15 (1995–96), then in an extended version as the introduction to Watson 2002, 9–25. Biographies are also given in *TGSI* 33, v–x (1932) and *SGS* 6 (1948–49), 215–16.

4. For more details of the content of these lectures, see Nicolaisen 'In Praise' in Watson 2002, 13.
5. It should be pointed out that even *CPNS*'s two indexes, 'Index of Places and Tribes' and 'Index of Personal Names' are by no means comprehensive.
6. www.st-and.ac.uk/institutes/sassi/spns. Also much of the text of *CPNS* is on this website in searchable form.
7. W.J. Watson writes in his Preface to *CPNS*: 'To deal exhaustively with our Celtic names of places is beyond the power of any one man, and I have not attempted anything of that sort' (p. xxxi (ix)).
8. T. S. Ó Máille, 'Irish Place-Names in *-as, -es, -os, -us*', *Ainm* 4 (1990), 125-43. They are remarkably common in Scotland. Other examples quoted by Watson are Altas SUT, Clunes and Fleenas NAI. To these can be added Phoineas by Beauly INV, Leuchars, Wemyss and Ceres, all FIF, and Happas and Kellas ANG to name but a few.
9. It will be clear even to the most casual reader of *CPNS* that Watson carefully avoids 'Pictish' as a linguistic term, using rather the more neutral 'British'. This in no way takes away from what he clearly considered an important Pictish dimension to Scottish place-names north of the Forth.
10. In the context of his towering achievement, his humanness shows itself in a comment in a letter of July 1924 to the early medieval Scottish historian A.O. Anderson: 'I am still trying to wind up the book on place names [*CPNS*]. I am now in the unhappy stage when I almost would like to see the whole thing burned.' (A.O. and M.O. Anderson Archive, St Andrews University Library, St Andrews).
11. I would like to thank Dauvit Broun, Alan James, Gilbert Márkus and Peadar Morgan for reading and commenting on this Introduction and list of corrigenda.
12. As pointed out by Geoffrey Barrow (1998, 73).
13. This is from an 18th-century transcript; the first syllable may read *Parte-*.

ST
St Andrews
June 2004

PREFACE

THE nucleus of this book is the six Rhind Lectures delivered in 1916. Since then I have continued work on the subject and have rewritten and expanded the original part and have added much fresh matter. To deal exhaustively with our Celtic names of places is beyond the power of any one man, and I have not attempted anything of that sort. At the same time, certain elements have been treated of with some approach to completeness, and in the regional surveys I have tried to give a fair idea of the general character of the names found in the various districts.

The names, of course, reflect the language or languages spoken at one time or another in the regions where they occur, and thus compel inquiry into the historical circumstances under which they arose. The study, in fact, touches the very roots of our history, involving points that have occasioned keen controversy. Here I have tried in all cases to go back to the original authorities, and as a consequence some of the conclusions are so old as perhaps to have the appearance of novelty. This applies especially to the treatment of the problems connected with the position, geographical and political, of the Picts, and the extent and nature of early Gaelic influence.

As to the linguistic and racial position in the early part of the Christian era and before that time, I would hold with the Irish Nennius that 'the Britons at first filled the whole island with their children, from the sea of Icht to the sea of Orcs,' *i.e.* from the English Channel to the far north. The ruling families throughout were of the same Celtic stock; under them were the pre-Celtic inhabitants. Conversely I would hold with Kuno Meyer that 'no Gael ever set his foot on British soil save from a vessel that had put out from Ireland.' Regarding the Picts, it is important to keep in view that while all Picts were Cruithnigh, *i.e.* Britons, all Cruithnigh were not Picts.

Settlement from Ireland began on the west and behind the Wall of Antonine during the Roman occupation, and it was probably on a considerable scale. These warrior communities would hold land under terms of military service, and it was not till the sixth century that the Scots of the west became independent. Even when Columba came to Iona, he received the grant of that island by the donation of the Picts, as well as from the king of Dál Riata. The position in the east to the north of the Wall offers some interesting problems. The evidence for early settlement from Munster in that region seems to me conclusive, while the relations of the early kings of Dál Riata with the midlands deserve reconsideration. When the sovereignty of the east passed to the Scots in the middle of the ninth century, Pictland must have been already largely Gaelic-speaking.

The energy and the courage displayed by the early Gael in settlement and conquest were equally conspicuous in missionary enterprise, and that enterprise was vigorous in the east as well as in the west. The early Irish Church was independent of the civil power; so too were its daughter churches in Britain. In Northumbria, however, at the synod held at Whitby in 664 an Anglic king took upon himself the position of arbiter between the Gaelic and the Roman clerics, and in consequence of his decision the former had to leave Northumbria. This precedent was followed in Pictland when in 717 the Pictish king expelled the communities of Iona westwards across Druim Alban. Such interference in ecclesiastical matters was probably part of the 'servitude' to which the Church was subjected among the Picts, and from which it was freed by Giric, who reigned from 878 to 889. The Irish Church must have exercised a great influence in spreading Gaelic among the Picts.

For explaining the meaning of names the first requisite is proper data; in other words, it is necessary to ascertain as accurately as possible the real forms of the names, otherwise we shall be dealing with what may be termed ghost names. For this purpose it is unnecessary to stress the importance of old record forms, especially such as occur in the ancient literature of Ireland and of Scotland. In addition to these it is absolutely necessary to ascertain the traditional Gaelic forms of the names in all cases where that is possible. How deceptive the anglicized map forms can be may be illustrated by a single example. A little cape in the isle of Gigha, spelled on maps Ardaily, is explained by H.C. Gillies in his *Place-Names of Argyll* as *àird àillidh*, 'beautiful cape.' The real name, however, is *àird èaláidh*, 'cape of the boat-passage'; *èaládh*, Ir. *éalódh*, means 'creeping stealthily,' and secondarily 'a passage for boats between two rocks or between a rock and the mainland,' and the Rev. Kenneth MacLeod of Gigha tells me that off Ardaily there is a passage for boats between an outside rock and the shore. The area over which this method can be applied has contracted since the end of last century: it would, I fear, be impossible now to obtain the Gaelic forms of the names of Easter Ross which I collected about 1900, or those of Nairnshire, or of the Callander and Trossachs district, which I collected later, for the fine old men who gave them are all dead. But, speaking generally, it may be said of the names of great part of Caithness, and of the whole of Sutherland, Ross, Inverness-shire, Argyll, and the north and west of Perthshire, that they have been recorded by the late Dr. A. Macbain or the Rev. C.M.

Robertson or myself, or that they are still available. Mr. F.C. Diack has recorded many names of Aberdeenshire and some other neighbouring districts from the remnant of the Gaelic speakers of Braemar. Certain names outside the present Gaelic area are known either generally or in particular districts that are still Gealic-speaking; *e.g. Obar-bhrotháig*, Arbroath, *Dun-dèagh*, Dundee, Eilginn, Elgin, and other such, are known widely; in Perthshire one may get names of Fife and of Forfarshire; in Argyll and Arran one hears some names of Ayrshire and Renfrewshire. It is right to add that in districts where Gaelic is greatly decayed or moribund, as, *e.g.*, in the Cromdale part of Strathspey, the Gaelic forms of local names are unreliable; they are apt to be simply anglicized forms taken over into Gaelic: the tradition has been broken. For the parts of Scotland that have ceased to be Gaelic-speaking the data—apart from sporadic instances mentioned—consist of the modern anglicized forms and the forms on record, so that it is vain to expect absolute certainty in all cases: if Ardaily occurred in Galloway instead of in Gigha, its meaning would be doubtful to the end of time. In the text I have distinguished genuine traditional forms (1) by placing them before their anglicized forms, and (2) by using the phrase 'in Gaelic' or some such expression. All the Gaelic forms given for places within the Gaelic area, as indicated above, are the real forms, not hypothetical reconstructions.

My grateful thanks are due to those, all over the country, who have helped in the work, both the host of intelligent Gaelic speakers who supplied information on the spot and correspondents who kindly answered inquiries. My regret is that the plan of the book and limits of space have prevented me from using all the information collected. In investigating the names of north Perthshire I was assisted by a grant from the Carnegie Trustees. The Royal Celtic

Society has generously defrayed the heavy expense of producing this volume, taking the risk of repayment from sales. I have also to thank the Society, and especially its President, Dr W.B. Blaikie, for much personal encouragement.

Early Celtic names and terms are from Holder's *Altceltischer Sprachschatz* when the source is not otherwise specified. For names occurring in Irish literature Hogan's *Onomasticon Goedelicum* is most useful, but I have verified his references and have supplemented them from my own reading of manuscripts and printed texts. Illustrations of Welsh phonetics are for the most part from *A Welsh Grammar*, by Sir J. Morris-Jones.

NOTES

Throughout the text open *a* (i.e. *a* as in Latin) in final syllables is distinguished by an acute accent—*á*.

In the Index of Places and Tribes stressed syllables are indicated as explained in the note prefixed.

On p. 211 it is to be added that in the Gaelic of Skye, Aberdeen is *Obar (Dh)athan*, where *Dathan* (for *Daan*, two syllables) corresponds to the British form *Doen*. In Edderton parish, Ross-shire, there is a stream *Dathan*, i.e. *Daan*.

On p. 431, to 'thundering' streams add *Allt Tairrneachán* (O.S.M. Allt Tarruinchoin), near the church of Foss, Strath Tummel.

On p. 503 it may be added that the Struy Hill is a landmark for the Moray Firth, called *Giltrax* (stress on *Gil*, with *g* hard) by the fishermen of Cromarty.

ABBREVIATIONS

Aberd. Brev.: the Breviary of Aberdeen, ed. William, bishop of Aberdeen, 1509, 1510; Bannatyne Club and Spalding Club.

Acall.: Acallamh na Senórach, ed. Whitley Stokes; Irische Texte, vol. iv. part i.

Adv. Libr.: Advocates' Library Manuscripts; now the National Library of Scotland.

Anecdota: Anecdota from Irish Manuscripts, ed. O. J. Bergin, R. I. Best, Kuno Meyer, J. G. O'Keeffe; Halle and Dublin.

Ann. Camb.: Annales Cambriae in Y Cymmrodor, vol. ix.

Ant. A. and B.: Illustrations of the Topography and Antiquities of the Shires of Aberdeen and Banff; Spalding Club.

Arch. f. celt. Lexik.: Archiv für celtische Lexikographie, ed. Whitley Stokes and Kuno Meyer.

AU: Annals of Ulster from A.D. 431 to A.D. 1540, ed. W. M. Hennessey and B. Macarthy.

Bain's Cal.: Calendar of Documents relating to Scotland, ed. Robert Bain (A.D. 1108 to 1509).

BB: the Book of Ballymote; photographic facsimile by the Royal Irish Academy; with Introduction by Robert Atkinson.

Bede, Hist. Eccl.: Bede's Ecclesiastical History, ed. C. Plummer.

Blaeu: Jan Blaeu, Geographia, vol. vi. (Blaeu's Atlas of Scotland).

B. of Llandaf: see L.Land.

Cal. of Scottish Saints: Kalendars of Scottish Saints, ed. Alexander Penrose Forbes.

Celt. Rev.: Celtic Review, ed. E. C. Carmichael (Mrs. W. J. Watson); nine vols.; Edinburgh.

Celt. Scot.: Celtic Scotland, a History of Ancient Alban, by W. F. Skene; three vols.; Edinburgh.

Celt. Zeit.: Zeitschrift für celtische Philologie, ed. Kuno Meyer and L. C. Stern, and Julius Pokorny in collaboration with Rudolf Thurneysen.

Chart. Hol.: Liber Cartarum Sanctae Crucis (Charters of Holyrood); Bannatyne Club.

Chart. Inch.: Charters of Inchaffray Abbey, ed. William A. Lindsay, John Dowden, and J. Maitland Thompson; Scottish History Society.

Chart. Lind.: Chartulary of Lindores Abbey (1195–1479), ed. John Dowden; Scottish History Society.

Chart. Monial. de North Berw.: Carte Monialium de Northberwic; Bannatyne Club.

Chart. Pasl.: Registrum Monasterii de Passelet; Maitland Club.

Cóir Anmann (Fitness of Names); ed. Whitley Stokes; Irische Texte, vol. iii. part ii.

Contrib.: Kuno Meyer, Contributions to Irish Lexicography (A—Dno.).

Cormac's Glossary: Sanas Cormaic; Anecdota, vol. iv.

Early Sources: Early Sources of Scottish History, A.D. 500 to 1286; Alan Orr Anderson.

Ex. Rolls: Exchequer Rolls (Rotuli Scaccarii Regum Scotorum).

Fél Gorm.: Félire Hui Gormain; The Martyrology of Gorman, ed. Whitley Stokes: Henry Bradshaw Society.

Fél Oeng.: Félire Oengusa; the Martyrology of Oengus the Culdee; ed. Whitly Stokes; Henry Bradshaw Society.

Fleadh Dúin na nGéadh: the Banquet of Dún na nGéadh and the Battle of Magh Rath; ed. John O'Donovan; Irish Archaeological Society.

FM: Annals of the Four Masters.

Geog. Coll.: Geographical Collections relating to Scotland made by Walter Macfarlane; Scottish History Society.

Hogan: Onomasticon Goedelicum, an Index, with Identifications, to the Gaelic Names of Places and Tribes, by Edmund Hogan; Dublin.

Holder: Alt-celtischer Sprachschatz, by Alfred Holder; Leipzig (a Lexicon of Early Celtic, including names of places, persons, and tribes).

Irish Lives: Bethada Naem nErenn (Irish Lives of the Saints of Ireland); ed. C. Plummer.

Ir. Nenn.: the Irish Version of the Historia Britonum of Nennius; ed. J. H. Todd; Irish Archaeological Society.

Ir. Texte: Irische Texte. ed. Whitley Stokes and Ernest Windisch; Leipzig.

Johnston: Place-Names of Scotland, by J. B. Johnston; Edinburgh.

LBr.: Leabhar Breac, published in facsimile from J. O'Longan's copy; Royal Irish Academy.

LL: Leabhar Laighean; the Book of Leinster, published in facsimile from J. O'Longan's copy; Royal Irish Academy (about A.D. 1150).

L.Land. L: Liber Landavensis; the Book of Llandaff, ed. J. Gwenogvryn Evans and John Rhys; Oxford.

LU: Leabhar na h-Uidhri, 'the Book of the Dun Cow,' published in facsimile by the Royal Irish Academy (about A.D. 1100).

Latin Lives of the Irish Saints: see Vitae SS. Hib.

Lawrie: Early Scottish Charters, prior to A.D. 1153, collected, with Notes and an Index, by Sir Archibald C. Lawrie; Edinburgh.

Leabhar Gabhála: the Book of the Conquests of Ireland, ed. R. A. MacAlister and John MacNeill; Dublin.

Lib. Calch.: Liber S. Marie de Calchou (Charters of Kelso Abbey); Bannatyne Club.

Lib. Dryb.: Liber S. Marie de Dryburgh (Charters of Dryburgh); Bannatyne Club.

Lib. Melr.: Liber S. Marie de Melros (Charters of Melrose); Bannatyne Club.

Lives of the the Saints from the Book of Lismore, ed. Whitley Stokes; Oxford.

Macdonald: Place-Names of West Aberdeenshire, by James Macdonald; Spalding Club.

Mackinnon, Catalogue: a Descriptive Catalogue of Gaelic Manuscripts in the Advocates' Library, Edinburgh, and elsewhere in Scotland, by Donald Mackinnon; Edinburgh.

Mart. Don.: Martyrology of Donegal, ed. J. H. Todd; Irish Archaeological Society.

Mart. Gorm: see Fel. Gorm.

Mart. Taml.: Martyrology of Tamlacht (Tallaght) as contained in the Book of Leinster, pp. 355–365.

O'Curry, M. & C.: On the Manners and Customs of the Ancient Irish, by Eugene O'Curry.

Ord. Gaz.: Ordnance Gazetteer of Scotland, ed. Francis H. Groome.

Orig. Paroch.: Origines Parochiales Scotiae; Bannatyne Club.

Orkn. Saga: Orkneyinga Saga, ed. Joseph Anderson.

RM: Registrum Episcopatus Moraviensis (Register of the Bishopric of Moray), ed. Cosmo Innes; Edinburgh.

RMS: Registrum Magni Sigilli Regum Scotorum (Register of the Great Seal of Scotland); references from 1306–1424 are to the new edition by J. Maitland Thompson.

RPSA: Liber Cartarum Prioratus S. Andree in Scotia (the Register of St. Andrews).

Rawl.B. 502: a Collection of Pieces in Prose and Verse in the Irish Language, compiled in the eleventh and twelfth centuries, published in fascimile with an Introduction and Indexes by Kuno Meyer; Oxford.

Reg. Arbr.: Liber Sancti Thomae de Aberbrothoc (Charters of Arbroath Abbey); Bannatyne Club.

Reg. Brech.: Registrum Episcopatus Brechinensis (Register of the Bishopric of Brechin): Bannatyne Club.

Reg. Cup.: Rental Book of the Cistercian Abbey of Cupar-Angus, ed. C. Rogers; Grampian Club.

Reg. Dunf.: Registrum de Dunfermelyn; Bannatyne Club.

Reg. Ep. Aberd.: Registrum Episcopatus Aberdonensis (Register of the Bishopric of Aberdeen).

Reg. Glas.: Registrum Episcopatus Glasguensis (Register of the Bishopric of Glasgow); Bannatyne Club.

Reg. Lenn.: Cartularium Comitatus de Levenax (Chartulary of Lennox); Maitland Club.

Reg. Neub.: Registrum S. Marie de Neubotle (Charters of Newbattle Abbey); Bannatyne Club.

Rel. Celt.: Reliquiae Celticae: Texts, Papers, and Studies in Gaelic Literature by Rev. Alexander Cameron; ed. A. Macbain and J. Kennedy; Inverness.

Ret.: Retours: Inquisitionum ad Capellam Domini Regis Retornatarum quae in Publicis Archivis Scotiae adhuc servantur Abbreviatio.

Rev. Celt.: Revue Celtique: Paris.

Sanas Cormaic: see Cormac's Glossary.

Silv. Gad.: Silva Gadelica, a Collection of Tales in Irish, edited and translated by Standish H. O'Grady.

Theiner: Vetra Monumenta Hibernorum et Scotorum (1216–1547). A collection of papal letters.

Thes. Pal.: Thesaurus Palaeohibernicus; a Collection of Old Irish Glosses. Scholia, Prose, and Verse; ed. Whitley Stokes and John Strachan; Cambridge.

Tighernach: the Annals of Tighernach, ed. in Revue Celtique, vols. 16–18, by Whitely Stokes.

Vergl. Gram.: Vergleichende Grammatik der keltischen Sprachen (Comparative Grammar of the Celtic Languages): Holger Pedersen.

Vision of Mac Conglinne: Aislinge Meic Conglinne, a Middle Irish Tale, edited and translated by Kuno Meyer; London.

Vita S. Columbae, or Life of Columba, by Adamnan, ed. (text and notes) by W. Reeves, and (text, translation, and notes) by W. F. Skene, based on Reeves' edition. References are to Skene's edition, unless otherwise indicated.

Voyage of Bran: the Voyage of Bran, son of Febal, to the Land of the Living, an Old Irish Saga, edited with translation by Kuno Meyer, and an Essay by Alfred Nutt; Grimm Library.

Wardlaw MS.: Chronicles of the Frasers, by James Fraser, Minister of the Parish of Wardlaw (Kirkhill in Inverness-shire), ed. William Mackay; Scottish History Society.

YBL: the Yellow Book of Lecan, edited in facsimile for the Royal Irish Academy by Robert Atkinson. A collection of pieces, prose and verse, in the Irish language, compiled in part at the end of the fourteenth century.

CELTIC PLACE-NAMES OF SCOTLAND

INTRODUCTORY

THE place-names of Scotland fall into two great divisions, Celtic and Teutonic, representing the types of languages which have been spoken over the whole or a part of the country within historic times. Of these the Celtic division is the older and the larger. Teutonic names were introduced first by the Angles who settled in the north-east of what is now England, and after long struggles got a firm footing between Tweed and Forth in the early part of the seventh century. Thence English names have spread with the growing influence of the language, and are still spreading. A second important source of Teutonic influence was the great Norse occupation of the north and west which began about two hundred years later. On the east coast, the southern limit of Norse occupation, as indicated by the place-names, was the Beauly valley, where we have Eskadale, O.N. *eskidalr*, ash-dale. On the west, the islands are full of Norse names, and on the western mainland they appear with varying frequency from Cape Wrath to the Firth of Clyde. So far there has been no systematic study by any competent scholar of the English names, but a good deal of sound work has been done on the Norse element.

I have said that the Celtic names are older than the English or the Norse, but even they do not necessarily form the very oldest stratum. Scotland was inhabited long before any man of Celtic speech set foot in Britain, and that the pre-Celtic population was by no means wiped out is proved by the fact that their descendants are still plentiful. The approximate date of the Celtic conquest has been much discussed; the present tendency is to put it later than was once the fashion, and it would be certainly rash to ascribe the conquest of Scotland to a period much earlier than the fourth century B.C.[1] When we consider further that the Celts, when they did come, formed rather a military aristocracy than the staple of the population, it need cause no surprise if we find some ancient names difficult to explain from Celtic sources. It is well to note that the people of these islands never called themselves Celts, nor are they ever so called by the classical writers. The term was unknown in native literature till it was introduced in quite recent times.

As applied to language the term Celtic, like the term Teutonic, covers more than one group. There are now, and probably were in prehistoric times, many dialects, but all fall under one or other of two great groups, differentiated most readily (but by no means wholly) by their treatment of the primitive Indo-European *qu* sound, the sound heard approximately in *equal*, or in Latin *equus*. One group retained this sound, making it later into *c* hard or *c* aspirated (*ch*). The other turned it into *p*. The former, called for convenience the Q-group, is represented now by Gaelic in its varieties of Irish, Scottish, and Manx. Traces of *qu* in continental Celtic are very slight; a possible instance is Sequana, the Seine. The other, or P-group, in early times included Gaulish and Old British; it is now represented by Welsh, Cornish (extinct), and Breton. The following examples illustrate the difference :—

GAULISH and OLD BRITISH	WELSH	GAELIC
maponos, a youth (*Apollini Mapono*)	*mab*; O.W. *map*	*mac*; Ogham *maquos*
pennos, head	*pen*	*ceann*; O.Ir. *cend*
petor, four (*petor-ritum*, a four-wheeled vehicle)	*pedwar*	*ceithir*; O.Ir. *cethir*
epos, horse; *Epona*, the goddess of horses	*ebol*, colt	*each*; O.Ir. *ech*
pempe-dula, a five-leaved plant	*pump*, five	*cóig*; O.Ir. *cóic* Lat. *quinque*

[1] The comparative lateness of the Celtic conquest is confirmed by the fact that the remains of Celtic art in Great Britain and Ireland belong almost entirely to the later or La Tène period, which is reckoned to begin about B.C. 400. Objects assigned to the later part of the preceding or Hallstatt period have been found in England.

This difference of treatment is not confined to the Celtic languages. It is seen to some extent as between Latin and Greek, *e.g.* Lat. *sequ-or*; Gr. *ἕπομαι*. It is a feature of dialects within Latin, *e.g.* Lat., *Quintus*, *Quinctius*; Samnite, *Pontius*; Umbrian and Oscan, *Pompeius* (compare Welsh *pump*). It is also found dialectically within Greek, *e.g.* Attic *πῶς*, Ionic *κῶς*.

Celtic of both groups has the important characteristic that it has dropped Indo-European *p* initially and between vowels; in other positions this original *p* has been modified. Thus we find:—

LATIN AND GREEK	GAELIC	WELSH
somnus (*sopn-us*), *ὕπνος*	*suain*; Ir. *suan*	*hun*
palma, *παλάμη*	*làmh*; O. Ir. *lâm*	*llaw*
plēnus, *πλήρης*	*làn*	*llawn*
pro, *πρό*	*ro-*, very	*rhy-*, Gaul. *Ro-smerta*
s-ub, *ὑπό*	*fo*	*go-*, O.W. *guo*
septem, *ἑπτά*,	*seachd*; Ir. *seacht*	*saith*
tep-idus,	*teth*; Ir. *te*	

Similarly, Lat. *piscis*, G. *iasg*, fish; Lat. *parcus*, G. *airc*, penury. It follows that no genuine native Gaelic word contains *p* unless as the result of later changes. The *p* of Gaulish, Welsh, etc., represents an original *qu*.

Our knowledge of British is derived from inscriptions on coins and from names of persons and places recorded by the classical writers. Goidelic—the usual term for proto-Gaelic—is known from the Ogham inscriptions on stones, of which there are about 300 in Ireland, about 30 in Wales, 5 in Devon and Cornwall, 1 in Hampshire, and 17 in Scotland. Both British and Goidelic were fully inflected languages, with stems and case-endings similar to those of Latin. By degrees certain changes took place, by which the ancient forms were converted into the earliest stage of their respective modern forms, that is to say into Old Welsh and into Old Irish. This change was of course gradual, and in its main features was probably completed by about A.D. 550, though learned men no doubt retained a knowledge of the old forms long after they had become obsolete in speech.

Of these changes the most important are the dropping

or modification of the old case-endings and of the stem-endings in compound words. Thus Early Celtic or British *magos*, a plain (stem *mages-*), gives Welsh *ma*, a place, Gaelic *magh*, a plain; *vernos*, alder, gives W. *gwern*, Ir. *fearn*, Scot.G. *fearna*. *Verno-magos* gives *Fernmag*, now *Fearnmhagh*, alder-plain. Gaelic still preserves inflection in a modified form, and distinct traces of the old declensions. Welsh has plural forms, but otherwise it has discarded inflection. It may be convenient to note here a few of the changes undergone by vowels and consonants.

E.Celt. *ū* is modified to *ī* in Welsh, but remains *ū* in Gaelic: *dūnon*, a fort, W. *dīn*, G. *dūn*.

E.Celt. *oi* before a consonant gives *u* in Welsh, *ōi*, *ōe* in O.Ir., *ao* in Mod.G., O.Lat. *oinos*, Lat. *ūnus*, one, W. *ūn*, O.Ir. *ōen*, Mod.G. *aon*.

E.Celt. *ei* (usually written *ē*) gives *wy* in Welsh before a consonant, in Gaelic it becomes *ia* if followed originally by a broad vowel in the next syllable, but before a slender vowel it gives *ēi*: *lētos*, grey, W. *llwyd*, G. *liath*, but its genitive *lētī* gives G. *lēith*.

E.Celt. *ou* before a consonant (except *s*) gives *u* in W., *ua* in G.: *boud-*, victory (*Boud-icca*), W. *budd*, profit, Ir. *buadh*, triumph; Gaulish, *Roud-ios*, W. *rhudd*, G. *ruadh*, red.

Initial *s* before a vowel gives *h* in Welsh usually, sometimes *s*; in Gaelic it remains: W. *hafal*, G. *samhail*, Lat. *similis*, like; W. *haf*, summer, G. *samh*, but W. *saith*, seven, G. *seacht*, Lat. *septem*.

Between vowels, in the body of a word, *s* becomes *h*, and then vanishes: Lat. *esox*, salmon (borrowed from Gaulish), Mid.W. *ehawc*, W. *eog*, Ir. *eo*, genitive *iach*.

Initial *sm-*, *sn-*, *sl-*, *sr-*, remain in Gaelic; in Welsh they become *m-*, *n-*, *ll-*, *rh-*, respectively: G. *smear*, marrow, W. *mer*; G. *sleamhain*, slippery, W. *llyfn*; G. *sruth*, stream, W. *rhwd*.

Before *r* or *l*, *st* remains in Welsh; in Gaelic it becomes *s* (as always): W. *ystrad*, strath, G. *srath*; W. *ystlys*, side, G. *slios*.

-lm- and *-rm-* remain in Gaelic ; in Welsh the *m* becomes *f* or *w* : Gaul. *kourmi*, G. *cuirm*, W. *cwrf*, *cwrw*.

Initial *v* gives *gw* in Welsh, *f* in Gaelic : *vernos*, alder, W. *gwern*, Ir. *fearn* ; Lat. *vīnum*, wine, W. *gwīn*, O.Ir. *fīn*, G. *fīon*.

When stressed or following a stress, *-ijos*, *-ijon* give *-ydd* in Welsh ; *-ija* gives *-edd* ; in Gaelic all yield *-e* : E.Celt. *novios*, W. *newydd*, O.Ir. *núe*, new.

The consonants *p*, *t*, *k*, *b*, *d*, *g*, *m* become in Welsh between vowels, *b*, *d*, *g*, *f* (*dd*), *f*, respectively : *Caratācos*, W. *Caradawg* ; *Cunotamos*, W. *Cyndaf*. This is called the soft mutation, otherwise lenition.

In Gaelic these consonants between vowels become aspirated : *Caratācos*, Ir. *Carthach* ; aspiration is also called lenition.

In the body of a word, between vowels, soft or lenited *g* came to vanish in Welsh : Lat. *legion-* ; *Caer-leon*.

Under the nasal mutation in Welsh, initial *b*, *d*, *g* become *m*, *nn* (*n*), *ng*, respectively ; the tenues (*c*, *t*, *p*) become mediae (*g*, *d*, *b*) : *in Bangor* becomes *ym Mangor*. The corresponding change in Irish is called eclipsis : *in Breatnaibh* becomes *i mBreatnaibh*, pronounced *i Mreatnaibh*. Eclipsis has long been dropped in Scottish Gaelic, but the place-names show many cases of its former influence.

In Welsh early *pp* (British or Latin) gives *ff* ; *tt* gives *th* ; *cc* (*kk*) gives *ch* ; in Gaelic they give *p*, *t*, *c* (*k*), respectively : *Britton-es*, W. *Brython*, G. *Breatan* ; Lat. *peccat-um*, sin, W. *pechod*, G. *peacadh*.

British or Latin *act*, *oct*, *uct*, *ect*, *ict* give respectively in W. *aeth*, *oeth*, *wyth*, *eith* (*aith*), *ith*. In G. the *ct* in each case becomes *cht* (Scot.G. *chd*) : W. *caeth*, Ir. *cacht*, slave ; W. *noeth*, Ir. *nocht* ; W. *wyth*, Ir. *ocht*, Lat. *octo* ; W. *rhaith*, Ir. *reacht*, Lat. *rect-um* ; W. *rhith*, O.Ir. *richt*, G. *riocht*, *riochd*, form.

E.Celt. *ks*, Lat. *x* give in Welsh *is*, in Gaelic *s* : *laxus*, W. *llaes*, G. *las* ; *pexa*, a tunic, W. *peis* ; *coxa* (*koksa*) ; W. *coes*, Ir. *cos*, Sc.G. *cas*, a foot ; *Saxo*, W. *Sais* ; *Saxones*, W. *Sacson*, Ir. *Sasan-ach*, an Englishman.

Long before the Romans actually entered North Britain the country was not wholly unknown by report. Albion, the ancient name of Great Britain as a whole, appears to go back to Himilco the Carthaginian, who explored the coasts of the North Sea about B.C. 500. Nearly two hundred years later (*c.* B.C. 320) came the famous expedition of Pytheas, organized by the traders of Massilia, to the same parts, and though Pytheas' own account of his voyages is lost, fragments of it have been preserved. Pytheas had mentioned Thule. It was six days' sail north of Britain, near the frozen sea, and the region about it was neither firm land nor sea nor air, but a mixture of all three resembling a jellyfish in consistency. Strabo, referring to this description, calls Pytheas an utter liar; as for himself, he does not know whether Thule is an island or whether the region near it is habitable. Orcas, first mentioned by Diodorus as one of the three chief capes of Britain, is also from Pytheas; the other capes are Belerion and Cantion, Land's End and South Foreland. The Roman geographer Mela (*fl.* A.D. 45) places Thule off the coast of the Belcae, 'a Scythian tribe.' He is the first to mention the Orcades which, he says, number thirty, all close together; and as the number of inhabited islands in the group is now twenty-nine, it would appear that Mela wrote on good authority. The poet Lucan and his contemporaries (A.D. 39-65) have heard of the Caledonian Britons. Pliny (A.D. 23-79) names as sources of his information Pytheas, Timaeus (*c.* B.C. 352-256), and Isidorus of Charax, who was probably an elder contemporary of his own; there were others also whom he does not name. He states that there are forty Orcades, thirty Hebudes, seven Acmodae. He also mentions Mona, now Anglesey, between Britannia and Hibernia; Monapia, now Man; Riginia, equated with Rechrann or Rathlin; Vectis, the Isle of Wight; Silumnus; Andros or Adros, perhaps Ireland's Eye; Dumna, the Long Island; and he has heard of Caledonia Silva, the Caledonian Forest. Such are the indications of the knowledge of North Britain possessed by the Romans before Agricola's campaigns (A.D. 80-85). It relates chiefly to islands, and it is just the sort of information that might be got from seafarers who knew little about the interior. Probably most of it is referable ultimately to Pytheas.

Agricola's campaigns form a distinct epoch, and it is unfortunate for us that Tacitus, his son-in-law and biographer, did not include in his *Life of Agricola* a systematic account of the country and tribes among which his father-in-law operated, as he might so easily have done. From Tacitus we hear for the first time of the rivers Clōta, the Clyde, Bodotria, the Forth, Tanaus or Taus, the Tay. The part north of Forth and Clyde is Caledonia; its inhabitants are Britanni. He names only one tribe, the Boresti, but he implies the existence of other tribes, and gives the important information that they joined together under one leader to make common cause against the Romans. The champion of liberty—the first native of Scotland whose name appears on record—was Calgăcus, 'Swordsman,' the most distinguished among all the chiefs for courage and for lineage. Historians have done him scant justice; he was of the type and race of Vercingetorix, the hero of Alesia, and, one might add, of Wallace; but though they are both commemorated by statues, he is not. The position of Mons Graupius, where he gave battle to the invaders, is still uncertain. Agricola showed his desire for further knowledge of the North by ordering his fleet to sail round the north coast. Three results of this cruise are claimed by Tacitus: Britain was proved to be an island; the Orcades, hitherto unknown, were discovered and subdued; the mysterious Thule, their northmost goal, was seen in the distance. The first of these results would be attained by rounding the north coast and sailing southward along the west coast to a point already known, such as the Firth of Clyde. As to the Orkneys, they were known by report, as we have seen, long before Agricola's time, as Tacitus was doubtless well aware. What he meant was, we may suppose, that the Romans had now for the first time direct and accurate knowledge of these islands. By Thule he means Shetland. The position of the Trucculensis Portus, the port from which the fleet set out and to which it returned, cannot be settled;

but it must have been Montrose or some place either on the Firth of Forth or on the Firth of Tay. There can be no doubt that the voyage brought much new information about the North and West.

In the first half of the second century the famous mathematician, astronomer, and geographer, Claudius Ptolemy of Alexandria, embodied in his great *Introduction to Geography* the facts relating to North Britain as known in his time. He fixes his places by latitude and longitude, and following Marinus of Tyre, his older contemporary, he takes as his furthest north fixed point Thule, whose position had been fixed by Agricola's survey. Ptolemy's measurements of latitude and longitude show that he turned all Britain north of Solway and Tyne through a right angle : his furthest north point is the Mull of Galloway ; he makes the west coast face north, the north coast face east, and the east coast face south. Apart from this extraordinary error, which I am quite incompetent to discuss and which really does not concern our present purpose, the outline of the map constructed from the data which Ptolemy supplies is very creditable, and we shall see reason to believe that his names of tribes and places deserve great respect. He records and locates in what is now Scotland 16 or 18 tribes, 17 rivers, 16 towns, 10 islands, 7 capes, 3 bays, and 4 other names. Here is a great advance in method and in knowledge of detail. When, however, it comes to fixing the position of these on a modern map, the matter is one of much difficulty. The difficulty is least in the case of names that still survive, and greatest in the case of inhabited sites or 'towns.'

Latin and Greek writers after Ptolemy's time give little topographical information bearing on our subject. Dio Cassius' contemporary account of the doings of the Emperor Severus in Scotland in the beginning of the third century is lost; but from the epitome of his history by Xiphilinus we learn that the two leading tribes then were the Caledonii and the Maeatae, and that the names of the others had practically been absorbed in these. 'The Maeatae dwell close by the wall that divides the island into two parts, the Caledonii beyond them.' The reference here is probably to Hadrian's Wall, between Tyne and Solway ; this would put the Maeatae south of Forth, between the Walls.

The orator Eumenius in A.D. 297 speaks of the Britons of the province as accustomed to Picti and Hiberni (Picts and Irish) as enemies. In A.D. 310 another orator says : 'I do not mention the woods and marshes of the Caledonians, the Picts, and others,' or, according to another reading, 'of the Caledonians and other Picts.' In or about A.D. 364, 'the Picts divided into two tribes (*gentes*), the Dicalydones and the Verturiones, and also the Atecotti, a warlike nation, and the Scotti, ranged far and wide (in the Roman province) and made great ravaging.' The Scotti were, of course, the Irish ; the Atecotti are styled by Jerome 'a British tribe' (*gentem Britannicam*),[1] but nothing certain is known as to their position. It is important to observe that in the fourth century the Caledonians have come to be a division of the Picts, and are not heard of subsequently. The inference is that that powerful tribe, which held the hegemony at Mons Graupius and after whom the North was called Caledonia, had for some reason lost their position of leadership, and that the Picts, who are first mentioned in A.D. 297, had taken their place. Thereafter the North came to be called Pictland (Pictavia), till the hegemony passed from them to the Scots. Thus we have the succession—Caledonia, Pictavia, Scotland.

[1] This does not exclude the possibility of the Atecotti having been Irish. Professor John MacNeill says, 'The names Scotti and Atecotti . . . are probably of a general application, not designative of special groups.'—*Early Irish Population Groups*, § 3.

CHAPTER I

EARLY NAMES

WE have noted that the old distinctive name of what is now Great Britain was Albion. The account of Himilco's voyage, as turned into Latin verse by Rufus Festus Avienus, who flourished in A.D. 366, from the Greek translation of Eratosthenes (b. B.C. 276), states that at a distance of two days' voyage from the Oestrymnides islands there lies the Sacred Isle, peopled widely by the race (*gens*) of the Hierni, and that near them stretches the isle of the Albiones. The two islands meant are Great Britain and Ireland. A Greek geographical tract, once ascribed to Aristotle, says that in the Ocean beyond the Pillars of Hercules (the Straits of Gibraltar) lie two very large islands, Albion and Ierne, called the Bretannic Isles, beyond the land of the Celts. Diodorus, probably after Posidonius, speaks of 'those of the Pretanni who inhabit Iris,' *i.e.* Ireland.[1] Isidore of Charax makes Albion the largest of the Bretannic Isles. Pliny says that Britannia was once called Albion, while the islands as a group were called 'Britanniae,' 'the Britains.' Ptolemy writes of 'Albion, a Prettanic isle.' From these references we note (1) that in the view of the ancients 'the Britannic Isles' was a group term, including Albion and Ierne; (2) that to Himilco, about B.C. 500, this group name was apparently unknown.

Bede, who died in A.D. 735, repeats the statement that Britain was once called Albion, but as a matter of fact the Irish form of Albion; was current in that sense in Bede's own time and long after it. The oldest form in Irish is

[1] φασί τινας ἀνθρώπους ἐσθίειν, ὥσπερ καὶ τῶν Πρεττανῶν τοὺς κατοικοῦντας τὴν ὀνομαζομένην Ἶριν. 'They say that some (of the northern peoples) eat men, as also do those of the Britons who inhabit Iris'; v. 32, 3.

Alpe, Albe (nom.), exactly as it ought to be. Somewhat later it is found as nom. *Alba, Albu*; gen. *Alban*; dat. *Albain*—on the analogy of *Muma, Mumu* (Munster); gen. *Muman*; dat. *Mumain*, as Kuno Meyer suggests. In Sc.G. it is now nom. *Alba*; gen. *Albann*; dat. *Albainn*, Scotland. The usage in earlier Irish literature needs only a few illustrations. In the glossary ascribed to Cormac mac Cuilenan, king of Munster, who died in 908, Alpa or Alba means regularly Britain, *e.g.* Glastonbury is in Alba. The antiquary Duald mac Firbis, murdered in 1670, quotes from an ancient poem :—

Fairenn Alban co muir nIcht
Gaoidhil, Cruithnig, Saix, Saxo-Brit.

'The population of Alba to the sea of Icht (the English Channel) consists of Gael, Cruithnigh, Saxons, and Saxo-Britons.'[1] In the *Book of Leinster*, written about 1150, the men of Alba are defined as Saxons, Britons, and Cruithnigh;[2] the writer takes the Gael for granted. In short, in the earlier Irish literature Alba is used as the native term for Britannia. After the establishment of the Gaelic kingdom of Dál Riata in the west, its people are referred to as *fir Alban*, 'the men of Britain,' or, as Adamnan calls them, in Latin *Scoti Britanniae*, 'the Scots (or Gael) of Britain,' as distinguished from the Scots or Gael of Ireland. When Aedán mac Gabráin, king of Dál Riata, is styled *rí Alban*, 'king of Alba,' the term Alba appears to be used in a restricted sense equivalent to Gaelic Britain, as opposed to *Cruithentuath*, 'Briton-land,' 'Pictland,' or the eastern side of Scotland north of Forth. At a later period, Alba came to be used to denote the Gaelic kingdom of Scone, as Skene has shown.

As to meaning, Albion has been referred with some probability to the root seen in Lat. *albus*, white, whence also *Alpes*, older *Albes*, the Alps. The inference is that the

[1] Reeves, *Vita S. Columbae*, 276. The second line contains a syllable too many, and may be amended by reading 'Saxain, Brit,' as the last two terms.

[2] 'Albanaig . i . Saxain 7 Bretnaig 7 Cruithnig'; LL 29a.

name 'White-land' was given by the Celts of Gaul, with reference to the chalk cliffs of the south coast.

Adamnan says that the Scots of Britain are separated from the Picts by 'the mountains of the ridge of Britain' (*montes dorsi Britannici*). This is the range which was called in Gaelic *Druim Alban*, the oldest mention of which occurs in an ancient prophecy regarding Niall Nói-giallach, king of Ireland towards the end of the fourth century: *biet ile a gluinn ar Druim nAlpuind*, 'many shall be his deeds on Drum Alban.'[1] There is another Drumalban in Lanarkshire, south-east of Tinto, which may mean 'the ridge of the men of Alba,' for *Albain* is sometimes used in this sense.[2] The great upland of the head waters of the Tay has long been known as *Bràghaid Alban*, Breadalbane, the Upland of Alba. It is described as thirty miles (Scots) long from east to west, and ten miles broad between Glen-Lyon and Lairig Ìlidh (head of Glen Ogle).[3] It thus included Glen Lòcha, whose river is still called *Lòchá Albannach*, 'Locha of Alba,' to distinguish it from the other stream of the same name which flows westwards from near Tyndrum to join the Orchy, and which is therefore called *Lòchá Urchaidh*, 'Locha of Orchy.' It is quite possible that in these two latter names we have a reminiscence of a period when Alba was used to denote the 'Kingdom of Scone.' In Strath Dinard, Sutherland, there is *Allt an Albannaich*, 'the Albanian's Burn.' In Easter Ross we have *Allt nan Albannach*, called on record 'Scottismennis burne,' now Scotsburn, with tradition of a battle and a cairn 'called cairnne na marrow alias Deidmanniscairne.'[4] Two places called Scotsburn occur in Morayshire, and in Ayrshire there is Altan Albany or Albany Burn, a tributary of Stinchar, representing Mid.G. *Allt an Albanaigh*, 'the Scot's burn.' It would appear that the people who gave these names

---

[1] *Zeit. f. Celt. Phil.*, iii. p. 463.

[2] 'co nAlbain 7 Bretnu 7 Saxanu,' 'with men of Alba and Britons and Saxons'; Rawl. B 502, 81 b 46; 'Albain gan chaomh re chéile,' 'men of Alba have no love for each other'; Adv. Lib. MS. LII, 34a.

[3] Macfarlane's *Geog. Coll.*, ii. p. 563-5.

[4] *Place-Names of Ross and Cromarty*, p. 60.

regarded the Scots as different from themselves, and that they date from a period when Alba was not understood to apply north of Spey, or south of the isthmus between Forth and Clyde. ‘Fons Scotiae,’ Scotlandwell, in Kinross-shire, may have been held the chief fountain of this Alba.

We have seen that in early usage the term Britannic Isles included Great Britain and Ireland, and that the larger island was called Albion. By Caesar's time and thenceforward it is regularly called in Latin Britannia. Classical spellings of Britain and Britons vary greatly. We have first a set with initial P, preserved by Strabo and Diodorus, probably on the authority of Pytheas. They write *Πρεττανοί* with *tt*, but Stephanus of Byzantium (*c.* A.D. 500) states that Ptolemy wrote *Πρετανίδες νῆσοι* with one *t*. Secondly we have the form *Βρεττανοί* used, if the texts can be trusted, by Polybius in the second century B.C., and later by Plutarch, Pausanias, Appian, and other Greek writers. Lastly there are *Britannia*, *Britanni*, used by Horace, Catullus, Juvenal, and Latin writers generally; Lucretius, however, has *Brittanidis* (gen. of *Brittanis*), VI., 1106.[1] The modern forms corresponding in Welsh and Gaelic certainly indicate that the spellings with *P* and single *t* are correct. The Welsh *Prydain*, older *Prydein*, Britain, implies an origin from an early Pritan(n)ia. With this goes *Prydyn*, commonly rendered ‘Picts,’ but obviously meaning originally Britons; it implies, according to Kuno Meyer, a singular *Pryden*. Parallel to *Pryden*, *Prydyn*, are the Old Irish *Cruthen*, plural *Cruthin*; compare W. *pryd*, G. *cruth*, form; W. *pryf*, G. *cruimh*, a worm. The whole group has been referred to a root *qrt*, cut, whence Lat. *curtus*, G. *cruth*, form, shape, giving to Pretani the meaning of ‘Figured Folk,’ with reference to the practice of painting the skin mentioned by Caesar. From other references it appears that the ancient pre-Celtic Britons tattooed as well as painted themselves. This is in substance the traditional explanation, and it makes the country, Britain, get its name from its inhabitants, the Britons, as Germani gives Germania; Galli, Gallia;

---

[1] Greek initial *p* becomes *b* in Latin in *Πύρρος*: ***Burrus*** (Ennius), ‘Pyrrhus’; *πύξος*: *buxus*, ‘box.’

Graeci, Graecia, etc. On this theory the Celts, who, probably before they crossed the Channel, called the land opposite them Albion, named its inhabitants Pretani. The latinized country name, Britannia, appears to have been a purely Latin formation. It is significant that the ancient Albion has remained as the name of Scotland in Gaelic.

From *Cruthen* are formed the compounds *Cruthen-tuath*, and *Cruthen-clár* (the latter poetic), meaning the land of the Cruithne, 'Pictavia' in Scotland; also applied to the Cruithnean territory in Ireland. Further formations are *Cruithne*, a collective term, and *Cruithnech*, a Cruithnean, plural *Cruithnigh*, *Cruithnich*. Hence come *Clais nan Cruithneach*, the Cruithneans' Hollow, in Stoer, Sutherland; *Airigh nan Cruithneach*, the Cruithneans' Shieling, in Applecross; *an Carnán Cruithneach*, the Cruithneans' Cairnie, a hill in Kintail. These names, like those involving *Breatan* and *Albannach*, imply that the namers considered the Cruithne to be different from themselves.[1]

Alongside of the ancient terms Pryden, Cruthen, we have another group: W. *Brython*, a Briton, a Welshman; G. *Breatan*, a Briton of Wales, Strathclyde, or Cornwall. These come from *Britto*, stem *Britton-*, a Briton, which is a shortened or hypocoristic form of *Britannus*, dating from the latter half of the first century and common in mediaeval Latin. In Gaelic there is from *Britto* also a noun or adjective *Britt*, 'British,' as in *Fergna Britt*, 'Fergna of Britain'; *an Britt a Cluaid*, 'The Briton from Clyde.—Skene, *P.S.*, 87; compare the surname Galbraith, *i.e.* 'foreign Briton.' Procopius (*c.* A.D. 500-565) has a strange account of an island which he calls Brittia, and which was thought to be Jutland, but must be really Britain. From

[1] Kuno Meyer (*Zur keltischen Wortkunde*, 39) derives *Cruithne* from *Cruthen*, with addition of the suffix *-ne* (early *-inion*). That it was neuter is seen from its dative *Cruithniu*. Many similar formations are collected by Professor John MacNeill in his paper on *Early Irish Population Groups* (Proceedings of the Royal Irish Academy, 1911), including such people-names as *Cuircne*, *Delbna*, *Luaigne*, *Semaine*, etc. If Delbna is based on *delb*, form, it would be parallel to Cruithne, *i.e.* if the latter is to be referred to *cruth*, form.

Brittia comes *Breiz*, Brittany, but it has left no descendant in Welsh or Gaelic.

The plural of *Breatan* is *Breatain*, dat. *Breatnaibh*. In our names of places it is found both south and north of the Forth and Clyde isthmus. Hence *Dùn Breatann*, Dumbarton, 'Fortress of the Britons,' once the acropolis of the Strathclyde Britons. *Clach nam Breatann*, the Stone of the Britons, on the western side of Glen Falloch, at the head of Loch Lomond, probably commemorates either a boundary or a battle. Balbarton, in Fife, is 'the Stead of the Britons,' and in Aberdeenshire there is Drumbarton, 'the Britons' Ridge.' There is also Drumbretton, in Dumfriesshire. In Ayrshire we find Balbrethan, showing Irish influence.[1]

In modern Sc.G. *Breatann* means Britain; a Briton is *Breatnach*, *Breatannach*.

The ancient tribal names recorded by Ptolemy and others are all of the same type as the tribal names of Gaul—plural in form. The Caerēni, Cornavii, Lūgi, and Smertae occupied what is now Sutherland and Caithness. Two may be assigned to Ross-shire, the Decantae in the east and the Carnonācae in the west. In mid-Scotland were the Caledonii, the Vacomagi, Taexali, Venicōnes; and on the west the Creōnes, Cerōnes (if indeed these were distinct from the former), and the Epidii. South of Forth and Clyde were the Damnonii, who reached, however, into Stirlingshire; the Novantae and Selgovae were in Galloway and Dumfriesshire, and in Lothian the Otadini and the Gadeni or Gadini, unless, as seems likely, the latter name is merely a blundering repetition of the former. It is worth noting that the geographer Marcian, who is supposed to have lived in the fifth century, states that Ptolemy mentioned thirty-three tribes in Britain. As seventeen of these belong to England, this leaves only sixteen for Scotland, a number which excludes both the Gadeni and either the Creōnes or the Cerōnes. Tacitus' Boresti adds one more to the list of first-century tribes.

---

[1] In Irish *Breatnaibh*, *Breatnach*, become *Breathnaibh*, *Breathnach*, through assimilation of *t* to *n*. Similarly *d* is assimilated to *n* in *céadna* (*Céadhna*), *Séadna* (*Séadhna*), etc.—v. Kuno Meyer, Contrib. s.v. Bretnach.

The Caerēni are placed in the north-west of Sutherland, and the name means 'Sheep-folk': O.Ir. *caera*, G. *caora*, a sheep. The suffix is seen in Gaulish *ep-ēnos*, knight, based on *epos*, a horse. From the same base we have the Caeracates and the Caerosi on the Continent. Many names of ancient Irish septs are formed from names of animals, and in the case of one of these, the Gamanrad or 'Stirk-folk' of Connacht, there is close connection with a wondrous cow from the Sídh or fairyland, who was apparently regarded as a tribal divinity. Caerēni, however, is probably merely an occupation name.[1]

The Cornavii occupied the district east of the Caerēni, now Caithness, and their name is based on *corn-*, a horn; O.Ir. *corn*, W. *corn*, Lat. *cornu*, with the ending *-āvios*, meaning 'Folk of the Horn,' *i.e.* of the promontory. Ptolemy mentions another tribe of the same name in England, who reached apparently from about the Worcestershire Avon to the mouths of the Dee and Mersey; the horn in this case may have been the cape between the two rivers. Later the Britons of the Dumnonian peninsula are called in Irish *Bretain Cornn*, in A.S. *Cornwealas*, 'Strangers (Welsh) of the Horn,' whence Cornwall, in Latin Cornubia. On a horn-like bend of the Danube stood *Cornācon*, Horn-place, now Sotin; compare also Aber-corn, Bede's Aebbercurnig. The ending *-āvios* is seen in Litavia, Letavia, M.Ir. *Letha*, W. *Llydaw*, Armorica or Brittany; Nant-avia, Nem-avia, etc.

The Lūgi (Lougoi) were south of the Cornavii and east of the Smertae, and as the Smertae occupied the basin of the Oykel, the Lūgi must have occupied the south-east part of

---

[1] Other tribal or sept names with the collective suffix *-rad*, fem., in combination with animal names are *Moltrad*, 'wedder-folk,' *Torcrad*, 'boar-folk'; with the suffix *-raige*, older *-rige*, there are *Dartraige*, from *dart*, a year-old bull or heifer, *Cattraige*, from *catt*, a wild cat, *Luchraige*, from *luch*, a mouse. See John MacNeil's *Early Irish Population Groups*; Micheál Ó Briain, *Irischen Völkernamen* in *Celt. Zeit.*, xv. p. 222. We may compare the Hyatæ (Ὑᾶται), Oneatæ (Ὀνεᾶται), and Choereatæ (Χοιρεᾶται), 'Pig-folk,' 'Ass-folk,' and 'Swine-folk,' tribes of Sicyon, instituted according to Herodotus, by Cleisthenes.—Hdts., v. 68.

Sutherland or Sutherland proper. The name may, as Macbain considered, be connected with G. *luach*, O.Ir. *lóg*, worth, value. But a Gaulish word *lougos*, a raven, is mentioned as explaining the name Lugudunon, and though the alleged connection with Lugudunon is wrong, the word *lougos*, raven, may well be authentic. The Lougoi may have been a dark pre-Celtic people, like the Silures. The people of Lochcarron, in Ross-shire, are still called *Fithich dhubha Loch Carrann*, the black Ravens of Lochcarron, with reference to their swarthy colouring.

The Smertae lived in the basin of the Oykel and probably also in the adjoining basin of the Carron to the south of it. Most MSS. of Ptolemy read Mertae, and the true form has been rather doubtful, but the name, as I discovered many years ago, is still extant in *Carn Smeart*, the name of a hill on the north side of Strathcarron, in the ridge between it and Strathoykel. Smertae is a participial formation from the base *smer-*, smear, cognate with G. *smior*, marrow; Ir. *smior*, E.Ir. *smir*, gen. *smera*; the meaning is 'Smeared Folk.' In Gaulish inscriptions a goddess named Cantismerta appears once; Rosmerta appears twenty-one times, regularly associated with the god Mercurius, who, according to Caesar, was most worshipped of all the Gaulish gods. Mercurius, says Caesar, was held to be the inventor of all arts, and guided men on journeys; he was also the god of gain and of merchandise. But the Gaulish Mercurius was also a ferocious war-god: 'In the language of the Gauls Mercurius is called Teutates, who was wont to be worshipped among them with human blood. Among the Gauls, Teutates is placated by thrusting a man head first into a full vat, to be suffocated there.'[1] Mercurius was also equated with the war-god Esus. Of these divinities Lucan says: 'Those (tribes) by whom pitiless Teutates is appeased by means of horrid blood (of men), and grisly Esus with savage sacrifices, and Taranis, no gentler than the altar of Scythian Diana.'[2] Further inscriptions occur to 'Mars Smert-atius,' to 'Mercurius Ad-smer-ius' (a god of the Pictones), and to a god Ate-smer-ius. Ro-smerta means 'the greatly smeared

[1] Holder, 1805, 48.

[2] *Pharsalia*, i. 444.

goddess'; the other terms are similar in meaning. There can be little doubt that in each case the reference is to smearing with blood, either that of victims slain to placate the divinity or of men slain in battle. We may compare the ritual in connection with the idol Cromm Cruaich on Magh Sleacht in Ireland. 'Milk and corn they would ask of him speedily, in return for one-third of their healthy (or whole) issue: great was the horror and the scare of him.'[1] Here we have a deity, similar to Mercurius in some of his aspects, placated with human sacrifices. *Smertu-*, *Smerto-*, appears in the personal names Zmerton (for Smerton), on an inscription in Galatia; Smerto-māra, a woman's name; Smertullus, a man's name, and possibly also an epithet of the war-god Esus; Smerto-rix, a man's name. In Ireland there were *Sliab Smertain* in Co. Limerick, Magh Smerthach or Magh Smertrach in Offaly, and Magh Smerthuin in King's County. The Irish hero Cuchulainn, who is beardless, provides himself on occasion with a magic *ulcha smerthain* (or *smertha*), 'smeared beard,' by pronouncing a spell over a handful of grass.[2]

It is difficult to be certain as to the exact connotation of the name Smertae, but it can hardly have been given from the practice of dyeing or painting which was in use among the ancient Britons, for if the custom was general, there would be no point in naming a particular tribe after it. Moreover it does not apply to the Gaulish instances, for the Gauls did not dye themselves. On the whole the most likely explanation is that the 'smearing' in question had reference to being reddened with the blood of their enemies; Solinus says of the ancient Irish, 'the victors drink the blood of the slain, and then smear their own faces therewith' (c. 22, 3).

The name of the Decantae in Easter Ross shows the common suffix *-anto-*, participial in form, seen in Nov-antae, etc. The base is most likely *dek-*, good, noble, seen in O.Ir. *dech*, best, noblest; Lat. *dec-or*, *dec-us*, *dec-et*, *decent-is*, giving the meaning of 'nobles,' 'aristocrats,' 'optimates.'

---

[1] *The Voyage of Bran*, ii. p. 304.

[2] Windisch, *Táin Bó Cualnge*, pp. 309, *n.* 5; 310, *n.* 1; LU 74 b 38.

In Wales there was Decantorum arx, 'the citadel of the Decanti,' now Deganwy from *Decantovio-*. Irish Oghams have *Deceddas*, genitive of a presumed nominative *Deces*, and inscriptions in Devon and Anglesey—both once held by the Gael—have *Decheti, Decceti, Decceti*. All the Gaelic forms, it is to be noted, drop *n* before *t*, a characteristic of Gaelic even in the early inflexional stage; in British and Welsh it is retained; compare O.W. *bryeint*, Mid.W. *breint*, Mod.W. *braint*, privilege, from an early *briganti*—(cf. Brigantes).

Carnonācae—Wester Ross—is probably to be analyzed Carn-on-ācae, from *carno-*, meaning 'trumpet,' as in Gaul, or 'congeries lapidum,' a heap of stones, as in Gaelic, or 'rupes,' a rock, a cliff, as *carn* is explained in O.Welsh. This last agrees with the use of *carn* in the north of Scotland, where it is often applied to a high rocky hill. The suffixes are *-on-*, as in the personal name *Carn-ono-s*, and *ācus*. The meaning, therefore, is either 'Folk of Trumpets' or 'Folk of the Cairns,' *i.e.* rocky hills, and the latter is certainly appropriate to the district occupied by the tribe. The Gaulish tribal name Carnūtes is thought to mean 'trumpet folk.'

The Calēdonii extended from the 'Lemannonios Kolpos' or Lemannonian Gulf to the Varar estuary, and above them (*i.e.* to the west of them) was the 'Kalēdonios Drumos.' The Varar estuary is undoubtedly the Beauly Firth, for the old name of the Beauly river was Farrar. On the north, therefore, they marched with the Decantae, as indeed Ptolemy states. The identification of the Lemannonian Gulf—evidently a sea inlet—is difficult. Etymologically it goes with Lennox, *Leamhnacht*, and its river, the Leven, *Leamhain*, and one would naturally think of Loch Long, which bounds Lennox on the west. This was Skene's view. Other authorities made it Loch Fyne, and their view was accepted by Macbain, mainly on the ground that he required Loch Long to represent Ptolemy's river Longos. He admits, however, that 'Loch Long is technically more correct,' and in this most people would probably agree.

With regard to the 'Kaledonios Drūmos,' Skene held that *drūmos*, which in Greek means 'an oak-coppice,' was really the native term seen in *Druim* Alban, Adamnan's Dorsum Britanniae, 'converted by his (Ptolemy's) Latin translator . . . into 'Caledonius Saltus' or Caledonian Wood.'[1] It must be pointed out, however, that the historian L. Annaeus Florus, who wrote some time before Ptolemy, uses this very term in comparing the Ciminius Saltus in Etruria with the Caledonius (saltus) and the Hercynius (saltus) in Germany, where *saltus* is applied to a wooded range of mountains.[2] Further, both Ptolemy[3] and other Greek writers apply the term *drūmos* to the Hercynian ranges.[4] The Latin translation is therefore correct, and the 'Kaledonios Drūmos' is identical with Pliny's 'silva Calidonia' and the Welsh *Coed Celyddon*, the Caledonian Forest, though its position—west of the Caledonians and along Drum Alban—is strange. It is probable, however, that the reference is to the oak copses and pine woods which then clothed the slopes and filled the passes of that mountain range, and that here again *drūmos* applies to a wooded hilly country. In any case the Caledonians appear to have been east of Drum Alban, and to have occupied west Perthshire, marching on the south with the Damnonians. It was from their position astride the Grampians that they were styled Di-calydones, 'the double Caledonians,' and the term 'Dvē-caledonios,' applied to the western sea, may indicate a time when their territory included the western seaboard.

Ptolemy calls the Caledonians *Calēdŏnii*, and so do most other writers, with variants *Calidonii*, *Calydonii*, whence the latinized *Calēdonia* as the name of their country. In

---

[1] *Celtic Scotland*, i. p. 75.

[2] Ciminius interim saltus in medio, ante invius plane quasi Calidonius vel Hercynius, adeo tunc terrori erat ut senatus consuli denuntiaret ne tantum periculi ingredi auderet.—*Flor.*, i. 17.

[3] *Geographia*, ii. 2.

[4] In view of these facts it is unnecessary to stress another consideration—that it is very doubtful whether the first-century form of *druim* would be readily confused with *drūmos*.

Florus and Eumenius they are called *Caledones*, and that this more correctly represents the native form is indicated by the Colchester inscription: 'Donum Lossio Veda de suo posuit nepos Vepogeni Caledo'; 'Lossio Veda, a Caledonian, *nepos* (grandson, descendant) of Vepogenos, made this offering out of his own means.' This is our only example of the name in the nominative singular. It can hardly be separated from the name *Caledu*,[1] which appears on a bronze of the Arverni and on silver coins of the Caletes; in fact this latter is the true Celtic form, *Caledo* being latinized.[2] With Caledonii, Caledones, we may compare *Cingonios* from *Cingu*, gen. *Cingonos*; *Damnonii* from *Dumnu*, gen. *Dumnonos*; the ending *-ios* denotes 'belonging to'; it is tribal or gentilic. For declension *Caledu* may be compared with *Cinoth* or *Ciniod*, gen. *Cinadon*; Adamnan's *Nemaidon* (gen.), and the genitives *Lugedon*, *Dovvaidonas*, etc., cited by Professor John MacNeill.[3]

The Old Welsh form of Caledones is seen in *coit Celidon*, 'the Wood of the Caledonians,' later *coed Celyddon* as already mentioned. This, however, cannot represent an early *Calēdon-*, for *ē* (*ei*) yields *wy* in Welsh, *i.e.* *Calēdon-* would become *Calwyddon*. Further, the change of *cal-* to *cel-* indicates that the vowel of the next syllable must have been *i*, *i.e.* *Calidon-*,[4] which is the form in Solinus.

The tribal name is found with us in three names: *Dùn Chailleann*, Dunkeld, 'Fort of the Caledonians'; (princeps Dúin Chaillden AU 873; abb Dúine Caillen AU 965; Dún Callden in Book of Deer); Ro-hallion near Dunkeld, 'Rath of the Caledonians'; *Sìdh Chailleann*, Schiehallion, 'Fairy Hill of the Caledonians.' In these the vowel of the second

---

[1] Holder, s.v. Caledones.

[2] The Gaulish ending *-ū* corresponds to Lat. *ō*, Gr. *-ων*, in *n*-stems. The true nom. sing. of Pictones, Senones, Redones, and such are Pictu, Senu, Redu.

[3] *Notes on Irish Ogham Inscriptions*, § 18 (Proceedings of the Royal Irish Academy, 1909).

[4] See Morris-Jones, *A Welsh Grammar*, p. 92, § 70.

syllable vanishes in accordance with the law of syncope in Gaelic, the resulting *ld* becoming *ll* by assimilation. The English form 'Dunkeld' appears to show Welsh influence. As to derivation, the old explanation from W. *celli*, grove, bower, G. *coille*, wood, has been long given up. The base is probably either *kal*, hard,[1] whence O.Ir. *calath*, later *caladh*, W. *caled*, hard; or *kal*, cry, as in Lat. *cal-are*, Gr. *καλ-έω*, call, *κέλαδος*, din, clamour.[2]

The Vacomagi were 'below' the Caledonians, *i.e.* east of them, and like the Caledonians appear to have occupied both sides of the Grampians, including Speyside and East Perthshire. Sir John Rhys suggested that their name means 'Men of the Open Plains,' connecting with W. *gwag*, empty, and E.Celt. *magos*, a plain, an explanation which, as he states, would require not Vaco-magoi but Vaco-magioi. Apart from this and the rather unsatisfactory meaning, W. *gwag* is borrowed from a Low Lat. *vacus*, Lat. *vacuus*, and could not well form part of the name of a northern tribe at this period. For the element *vacos* may be compared *Bello-vaci*, a Belgian tribe; Vaco-caburius, a Gaulish god (compare the personal name Caburus; O.Ir. *cobair*, help), Vacontion, and the man's name Sego-vax; it is evidently Celtic, but of unknown meaning. *Magoi* is here most probably the plural of *magos*, great; O.Ir. *mag*, great, potent; *mag-lo-s*, a chief; O.Ir. *mál*, a prince; Lat. *magnus*. The name is on the model of *Sego-māros*, great in strength, or great by reason of strength, and it means 'great in, or by reason of *vacos*'; compare *Esu-magios*, 'mighty by reason of the god Esus,' or perhaps 'mighty as Esus.'

Eastwards of the Vacomagi were the Taexali (variants are Taezali, Texali) in Aberdeenshire, and south of them again the Venicōnes (Venicōmes, Vennicōnes, Vernicōnes, Venni-

---

[1] Compare Καλυδῶνά τε πετρήεσσαν, 'rocky Calydon,' in Aetolia; Homer, *Il.*, 2, 640.

[2] The explanation of the long *e* of Calēdonia may be that it was due to the poets, who are our earliest authorities for it, and that it arose from the requirements of metre: *Călĕdŏnĭus* does not fit into an hexameter line; *Călēdŏnĭus* does.

cones, Venicones, Venicontes) extended apparently to the Firth of Forth. The former name is obscure; it is not, as Macbain suggested, connected with Tough, the name of a parish in Aberdeenshire (on record as Toulch, G. *tulach*, a hillock, pl. *tulcha*). The true form of Venicōnes is too uncertain for satisfactory explanation: Vernicōnes would mean 'Swamp Hounds,' 'Alder Hounds'; Vennicones suggests comparison with the Irish tribal name Vennicnioi, 'descendants of Vennos,' mentioned by Ptolemy.

After the battle of Mons Graupius, Agricola led his troops to winter quarters among the Boresti. This tribe, which is mentioned only by Tacitus, appears to have been near the estuary of the Forth. There is no subsequent trace of the name, and its meaning is obscure.

On the West Coast, southward of the Carnonacae, were the Creōnes, in the district corresponding to *na Garbh Chrìochan* or 'Rough Bounds' of later times, long part of the patrimony of Clan Ranald. South of them came the Cerōnes, perhaps between Loch Leven and the river Add at Crinan. South of the Cerōnes, were the Epidii, reaching to the Mull of Kintyre. We have no direct information about the tribal names of the inhabitants of the western islands. Creōnes and Cerōnes show the ending *-ōn-es* common in Celtic tribal names, as Eburōnes, Suessiōnes. With the former may be compared *Crich Cera* (dat. Ceru: *i Ceru i Connachtaib*, LL 359 b) in Mayo, and *Cerna* (gen. Cernai; dat. Cernu; acc. Cernai) in Bregia. As some MSS. omit Cerōnes, others Creōnes, it has been supposed that the two names are really one, but the extent of territory covered is in favour of two tribes.[1]

The name Epidii is from E.Celt. *epos*, a horse, with the adjectival ending *-idios* as in *Lug-idius*, *Mag-idius*. The singular occurs as a personal name in North Italy; Vergil's tutor in rhetoric was Epidius, a Gaul by name at least. A name exactly parallel is *Marc-idius*, Horseman, from E.Celt. *marcos*, G. *marc*, W. *march*, a horse. The Epidii,

[1] These names can have no connection with *an Crianán*, Crinan, between Knapdale and Lorne.

then, were probably horse-breeders and horse-breakers, like the Gauls, an occupation for which their country was and is well suited. Ptolemy says that the Epidii stretch eastwards (*i.e.* northwards) from the Epidion Akron, the Epidion Promontory, *i.e.* the Mull of Kintyre. In the Old Irish saga, *Aided Chonrói*,[1] 'the Death of Cúrói,' mention is made of a heroic character named Echde, who lived in *Aird Echdi i Cinn Tíre*. Here, as Kuno Meyer points out,[2] *Echdi* may be taken either as genitive of a personal name *Echde*, in which case the meaning would be 'Echde's Cape,' or as the dative of an adjective *echde*. In the latter case, as *echde* is the exact form which E.Celt. Epidion would assume in Gaelic, *Ard Echde* is simply *Epidion Akron* taken over into Gaelic. Kuno Meyer held that Echde as a personal name was merely a fiction of the saga-makers; he is certainly a very shadowy character. The tale mentions also *Tor Echde*, 'the Epidian Tower.'[3] Here, then, we have clear proof that the Epidii were a British tribe, and an instructive example of the gaelicizing of a British name. It is not without significance that Kintyre is the home of the MacEacherns, whose name is an 'anglicization' of *Mac Each-thighearna*, 'Son of Horse-lord.'

Of the four southern tribes, the Damnonii occupied the counties of Ayr, Renfrew, and Lanark, and spread round by Dumbarton into Stirlingshire, marching on the north with the Caledonians, on the south with the Novantae, and on the east with the Votadini. There were Dumnonii in the south-west of England, whose name survives in Devon. In Ireland they appear as *Fir Domnann*. A people of this name lived in Connacht and gave their name to the peninsula of Irrus Domnann, now Erris. *Domnann* was also a name

---

[1] *Ériu*, ii. p. 32.

[2] For Kuno Meyer's identification of Ard Echde *v.* his *Zur keltischen Wortkunde*, 41.

[3] Tor Echde may have been a broch. A very old poem says of Labhraidh Loingseach that he razed eight towers in Tir Iath (Tiree), and in Tiree there are a number of ruined brochs.

for the men of Leinster.[1] The three warrior septs (*laech-aicme*) of Ireland were the Gamanrad of Irrus Domnann, the Clann Dedad in Temair Luachra of Kerry, and the Clann Rudraige of Ulster. The Fir Domnann were reckoned to be descended from Semion of the race of Nemed,[2] which means that they were not, in the opinion of the Irish learned men, of Milesian—*i.e.* Gaelic or Scotic—origin, but of the same race as the Britons. Some references to the great Ultonian hero Cuchulainn throw a very interesting sidelight on this. Cuchulainn's first name was Setanta, which represents an earlier form, Setantios. The Setantii were an ancient British tribe near Liverpool: Ptolemy places 'the harbour of the Setantii' at or near the mouth of the Ribble. The inference is that Setanta means 'a Setantian,' and that Cuchulainn was of British origin. In the single combats which he fights in connection with the *Táin Bó Cualnge*, 'the cattle drive of Cualnge,' Cuchulainn ultimately encounters Ferdiad, the mighty hero of the Damnonian Gamanrad and his own fellow-pupil and friend. He is extremely unwilling to fight Ferdiad, and even expostulates with him, reminding him that they are of the same race and of the same blood. 'Tu m'aicme, tu m'fine,'[3] are the words ascribed to him; 'Thou art my own race; thou art my own kin.' Again, in the account of Cuchulainn's last battle, a certain satirist demands his spear on pain of satirizing his tribe (*cenél*) in case of refusal. Cuchulainn replies, 'Truly tidings of my shaming shall not reach a land that I have never visited, ere I reach it myself.'[4] This puts Cuchulainn's foreign origin very clearly, while the former reference shows that he and the Damnonian Ferdiad were considered to be of the same stock. The Clann

[1] 'Galion tra 7 Domnann, anmann sin do Laignib'; 'Galion, now, and Domnann are names for Leinstermen'; LL 311 a.

[2] 'Cland Semioin dano Galeoin 7 Fir Domnand uile'; 'Now the Galeoin and Fir Domnann are all of them the children of Semion'; LL 8 b 47. 'Sil Semioin . . . Galion . . . Fir Bolg 7 Fir Domnand'; 'the Galion, the Fir Bolg, and the Fir Domnann are the children of Semion'; LL 7 a 52.

[3] *Táin Bó Cualnge*, l. 3470 (Windisch).

[4] 'Tír ém nad ránacsa riam, ni ricfat scéla m'écnaig remum'; LL 121 a 44.

Rudraige, too, were of the race of Nemed, and so were the Cruithne.[1] The Fir Domnann, we are informed, were so called from 'digging the earth'; they were 'the men who used to deepen the earth.' 'To Rudraige and to Genann with their people was the name applied. And it was at Inber Domnann that they took harbour.'[2] Along with them go the Fir Bolg, 'Bag-men,' 'for it is they who were carrying the earth in bags (*bolg*).'

The tradition, therefore, is that the Fir Domnann were of the same race as the British, and apparently that they had come into Ireland from a mining district. As to the latter point, we must keep in mind the readiness with which the learned Irish invented historical accounts merely on a basis of philological speculation. Allowing for this tendency, we must still admit the probability that the Irish Fir Domnann really came from the Devonian peninsula to Ireland, as their fellow countrymen migrated later, in the fifth century, to Brittany, whence the name Domnonia applied to a wide district there in mediaeval times. It may be more than a coincidence that the Scottish Damnonii occupied a tract rich in iron.

The name is based on E.Celt. *dubnos*, *dumnos*, deep, world; as in *Dumno-rix*, World-king; G. *domhan*, W. *dwfn*, deep, the deep, world; whence apparently was formed *Domnu*, 'goddess of the deep'; compare *Tuatha Dé Danand*, the tribes of the goddess Danu. Fir Domnann may mean 'Men of Domnu,' *i.e.* under the care of Domnu, goddess of the deep, again suggesting a reference to mining.[3] Damnon-ii is formed from *Dumnu*, gen. *Dumnon-os*, with the gentilic ending *-ios*, as in Caledon-ius.

About five furlongs west of the church of Yarrow, in a field on the right-hand side of the road as one goes up the valley, stands a stone with an inscription in Latin, in part

[1] *Leabhar Gabhála*, 117. 'Is do síl Neimid sin tra/Cruithne is Bretnaigh Cluada'; 'now of the seed of this Nemed are the Cruithne and the Britons of Clyde'; BB 229 e.

[2] *Leabhar Gabhála*, 119. Inber Domnann is identified with Malahide Bay, a little to the north of Dublin.

[3] For another goddess of the same name *v.* p. 412.

very hard to read under ordinary conditions. The following is the reading given by Sir John Rhys :—

HIC MEMORIAE ET
BELLO INSIGNISIMI PRINCI
PES NUDI
DUMNOGENI HIC IACENT
IN TUMULO DUO FILII
LIBERALIS.[1]

He was of opinion that the monument commemorates two sons of Nudd Hael, 'the liberal,' of the ruling house of Strathclyde, and that it was inscribed in the latter part of the sixth century. For our immediate purpose it is of great interest as showing that the name of the Dumnonians survived in the time of Columba and Aedán mac Gabráin, and that there were then princes who claimed to be sprung from their royal line.

The name of the Novantae of Galloway is formed like Decantae, the base being *nov-*, new, fresh, seen in the river name *Novios* ; O.Ir. *núe*, *nói-* (*nóicrothach*, fresh of form) ; W. *newydd*, new, fresh. They are usually supposed to have been named after the River Novios, the Nith, which may have formed their eastern boundary, but this is not necessarily the case. The meaning may be 'lively, vigorous folk,' and we must compare *Tri-nov-antes*, 'the thrice-vigorous folk,' around Middlesex. The closely related name Noviantum appears very often in France as Nogent, 'Fresh or Green Place' ; we might compare *Noid*, Nude, in Badenoch (p. 445).

Selgovae, the name of the tribe who occupied Dumfriesshire, and probably a good deal more between the Cheviots

---

[1] Rhys translates freely—

Here Nudos' princely offspring rest,
Dear to fame, in battle brave,
Two sons of a bounteous sire,
Dumnonians, in their grave.

He considers the inscription to consist of two Latin hexameters, of a kind in vogue in Britain at that period ; *Y Cymmrodor*, xviii. 5. For other readings, see *Proceedings of Society of Antiquaries of Scotland*, 1912/13, p. 375.

and Tweed, is from the base seen in O.Ir. *selg*, a hunt ; W. *hela*, to hunt ; O.W. *helghati*, hunt thou ; the ending *-ovo-*, *-ova-*, appears in Gaulish *Co-med-ovae*, and elsewhere.

The name which appears in the MSS. variously as Otalinoi, Otadinoi, Otadēni, was corrected by Rhys to Votadinoi, rightly, if they are to be equated with the Guotodin of early Welsh literature, who lived in the region which they occupied, and his correction is usually accepted, though Holder does not mention it. The Gaelic form is preserved in the *Duan Albanach*, 'Poem of Alba,' an historical poem of the eleventh century which states that Alba got its name from a certain Albanus, who was driven out of it by his brother Briutus, and proceeds—the poet is explaining in his own way the change from the name Alba to Britain :—

> 'ro-gabh Briutus Albain āin
> go rinn fiadhnach Fotudāin.'[1]

'Briutus took possession of noble Alba up to the far-seen point of Fotudán.' Here *Fotudāin* corresponds to W. *Guotodin* (later *Guododdin*) ; for the ending compare Lat. *vespertīn-us*, Ir. *espartán*. The conspicuous point in question is, I think, Berwick Law. If the name is to be divided as *Votad-inoi*, it may be compared with E.Ir. *fothad*, 'support,' and the personal name *Fothad*.

Though Ptolemy does not mention by name the tribe who inhabited the Orkneys, their name may be inferred from the names Orcas, Orcades. Orcas is an adjective formed in the Greek manner (sing. m. f. *orcas*, pl. *orcades*) from a noun *orcos*, which is the Celtic cognate of Lat. *porcus*, a pig : O.Ir. *orc*, a pigling, a young boar. Diodorus' (*ἀκροτήριον*) *Ὀρκάν* shows the nom. sing. neut. ; Ptolemy's *Ὀρκὰς ἄκρα*, the nom. sing. fem., and his *αἱ Ὀρκάδες νῆσοι* the nom. pl., all used adjectivally. Orcades, with the word for islands understood, is used as a group name like Cyclades, Echinades, Sporades, etc. The usual explanation is 'Swine Isles' (Stokes), or 'Whale Isles' (Macbain) ; the latter might be compared, though Macbain does not do so, with Lat. *orca*, a whale. In Irish *orc* does not seem to mean

---

[1] Skene, *P.S.*, p. 57.

'whale,' though it is given as meaning 'egg' and 'salmon.' But the true significance of the ancient term appears clearly to lie in the fact that it was formed from a tribal name, *Orcoi*, Boars. The Celts admired the boar as they did the war-hound (*árchú*), for his strength and ferocity, and they used his figure as a fitting emblem to adorn their shields and helmets. Cormac's Glossary has 'orc tréith . i . nomen do mac ríg,' 'a lord's boar is a name for a king's son.' We have seen that tribal names were often taken from animals, among the examples being *Torcrad*, Boar Folk. Another, which survives to the present day, is the Cat tribe with which we shall deal immediately.

In Ross-shire we find now, 'Daimh mhóra Radharaidh, Buic Srath Ghairbh, Fithich dhubha Loch Carrann, Clamhanan Loch Bhraoin,' 'the Big Oxen of Raddery, the Bucks of Strathgarve, the Black Ravens of Lochcarron, the Kites of Lochbroom.' [1]

When we consider the usage in literature, we find that the term has always a tribal suggestion: 'indse Orc' (BB 210 a), 'the isles of Orcs'; 'fecht Orc' (AU 579), 'an expedition against the Orcs'; 'Cairnech fri secht mbliadhna immór rígi Bretan ⁊ Cat ⁊ Orc ⁊ Saxan'; 'Cairnech was for seven years high king over Britons and Cats and Orcs and Saxons' (BB 208 b 14 a); 'for firu Bolc ⁊ for Orcca,' 'on the Fir Bolg and on the Orcs' (Rawl. 512 84 a); 'bellum for Orcaib,' 'a battle against the Orcs' (AU 708). Orkney is in modern Gaelic *Arcaibh*, representing an older *i nOrcaib*, 'among the Orcs,' just as *Cataibh*, Sutherland, represents *i Cattaib*, 'among the Cats,' and *Gallaibh*, Caithness, represents *i Gallaib*, 'among the Strangers.' [2] I think, therefore, that Inse Orc, 'the Isles of Boars,' contains a tribal name. As Orcas goes back to Pytheas, it follows that already before B.C. 300 a people of Celtic speech were established

---

[1] Much information on this subject is collected by the late William Mackenzie, Secretary to the Crofter Commission, in *The Old Highlands* (Gaelic Soc. of Glasgow), pp. 82-9.

[2] Orc was used as a personal name: 'ben Oirc meic Ingais'; *Silv. Gad.*, ii. p. 476; 'Orc allaid na Ruadhacain' (wild boar), AU 1039; 'Colman aue Oirc ab Cluana hIraird,' AU 701.

in these parts. 'Dun-orc auiel' and another Dunorc formed part of lands in Fife which were set apart on behalf of the poor scholars of St. Andrews,[1] now Denork, south-west of St. Andrews.

Coupled with *Inse Orc* in early Irish literature we find *Inse Catt*, 'the Isles of Cats,' a tribal name analogous to the Irish *Cat-raige*. Examination of the passages in which *Inse Catt* occurs proves that here we have the pre-Norse name of the Shetlands. At some period antecedent to the Norse invasion this tribe of Cats had occupied part of the mainland, for the Norsemen called the north-eastern extremity of our island *Katanes*, Cat-cape, now Caithness. The Cats held more than Caithness: they extended well into Sutherland. In fact our name for Sutherland is *Cataibh*, from the old *i Cattaib*, among the Cats. *Machair Chat* 'Lowland of Cats' was the coast district between the Ord of Caithness and Dunrobin. *Bráigh Chat* (Braechat), 'Upland of Cats,' is described in an old account as twenty or twenty-two miles in length, and divided into two parts by the River Shin. *Dìthreabh Chat*, 'Wilderness of Cats,' is described as belonging to the parish of Kildonan, and as marching with Strath Ullie (Helmsdale) and Strath Brora.[2] These names help to fix the position of the tribe in Sutherland: they apparently entered from Caithness, and occupied the eastern and south-eastern part of the modern county—Sutherland proper, as opposed to Strathnaver and Assynt. It may be added that at the present day a Sutherland man is a *Catach*; the Earl of Sutherland was *Morair Chat*; the Duke is *Diuc Chat*; the Kyle of Sutherland is *an Caol Catach*. That the Norsemen regarded the mainland Cat tribe as Picts appears from the fact that they called the narrow sea north of their territory *Pettaland-fjörðr*, Pictland Firth, now Pentland Firth.

We may next note the 'towns' which Ptolemy assigns to certain tribes. All the southern tribes have 'towns,' but of the northern ones only the Vacomagi, Taexali, and

[1] *R.P.S.A.*, p. 316.

[2] Macfarlane, *Geog. Coll.*, under Braechat, Diriechat.

Venicōnes—all between the Moray Firth and the Firth of Forth. The indications of position are so indefinite that attempts at identifying sites are for the most part mere guesswork.

The Vacomagi have four, Pterōton Stratopedon, Tuesis, Bannatia, Tamia. The first appears to be a translation of the Latin Alata (or rather Pinnata) Castra, 'Winged Camp,' *i.e.* camp provided with *pinnae*, artificial shelters on the walls. It has been thought to be Burghead, which has remains of an important fortification. For our purpose the point of interest is that the name was given by the Romans. Tuesis was probably on the river of that name, identified with Spey. Bannatia seems to be based on *bann*, a horn ; W. *ban, bann*, horn, peak ; O.Ir. *benn*, horn, peak ; Gaul. *Canto-benn-icus*, 'white-peaked' ; compare the Mid. Irish *Dún dá Benn*, Fort of two Horns. The same element is seen in Banna-venta, 'Horn-marketplace,' St. Patrick's birthplace. Tamia may be compared with E.Ir. *tám*, meaning (1) rest, repose ; (2) death, pestilence, and the Irish place-name *Táimleacht*, 'memorial of people who have died of plague.' Probably Tamia means simply 'station.' It has been supposed to indicate a station on the river *Teimheil* (Tummel), O.Ir. *temel*, darkness, but the vowel phonetics are not satisfactory. Bannatia and Tamia were probably south of the Grampians.

The Taexali have only one town, Dēvana, which undoubtedly means 'Dee-town,' from Dēva, the River Dee of Aberdeenshire. Its position on Dee is unknown.[1]

Orrea, the town of the Venicōnes, I take to be Latin Horrea, 'barns, granaries, corn-stores,' which Tacitus mentions as part of the Roman organization in Britain.[2]

---

[1] Skene thought that the name of Dēvana is preserved in Loch Daven, near Dinnet (*Celtic Scotland*, i. 74) ; this is impossible, for as I am informed by Mr. F. C. Diack, the Gaelic of which Loch Daven is an anglicization is Loch Dabhain (? Damhain).

[2] Compare Ὄρρεα, a town of Moesia Superior (Serbia) mentioned by Ptolemy, in many of the MSS. Ὀρρέα. For other instances of Horrea as a place-name, v. Dr. George Macdonald's Paper on 'The Agricolan Occupation of Scotland' (*Journal of Roman Studies*, 1919).

Towns of the Damnonians were Colanica, Vindogara, Coria, Alauna, Lindon, Victoria.

Colanica, in most MSS. Colania, may be the station on the Roman Wall called Colanica by the anonymous geographer of Ravenna.[1] The name may be compared with E.Ir. *colainn*, flesh, body; W. *celan*, with the common adjectival suffix *-ico-s*, meaning 'Place of Bodies'; compare *Corpach*, 'Place of Bodies,' near Fort William, and in Jura; also 'Creagnecolon' (? *Creag nan Colainn*, 'Rock of the Bodies') on the east side of Loch Doon. Coria probably means 'Hosting-place'; it would become in Gaelic *cuire*, which is common in E. and M.Ir. in the sense of 'band, host'; 'the battle of Cuire in Alba' is mentioned once at least in Irish literature.[2] The same element is probably seen in *Petru-corii*, 'Four Hosts'; *Tri-corii*, 'Three Hosts,' both tribal names. In Vindo-gara (otherwise Vandogara, Vandovara) the first part is E.Celt. *vindos*, white, O.Ir. *find*, G. *fionn*, W. *gwyn*; the second part may perhaps be compared with G. *gar*, *garan*, a thicket. The place appears to have been near Girvan, older Garvane.

Alauna of the Damnonians is to be compared with Alauna of the Votadini. The position of the former approximates to that of Dumbarton; the site of the latter, as Macbain pointed out, agrees with Inchkeith in the Firth of Forth, but it must have been on the mainland and on the south side of the Firth. There can be little doubt, as it seems to me, that the western Alauna was actually the Rock of Dumbarton, which in the first part of the fifth century was the seat of the kings of Strathclyde and must

[1] He says in his *Cosmographia*, compiled about the seventh century A.D., that there are ten 'cities' (*civitates*) in a straight line from sea to sea in the narrowest part of Britain, connected by a road, namely, Velaunia, Volitanio, Pexa, Begesse, Colanica, Medio Nemeton (leg. Medionemeton), Subdobiadon, Litana, Cibra, Credigone; v. *The Roman Wall in Scotland* (Dr. George Macdonald), p. 153.

[2] 'Ocus do bí (Donn) secht mbliadan fichot i rígféinnidecht Éirenn ocus Alban nó gur marb Dubh mac Dolair i cath Cuire thall i nAlbain é'; 'and Donn was seven years chief over the Fiana of Eire and of Alba, till Dubh son of Dolar slew him in the battle of Cuire over in Alba'; *Silv. Gad.*, i. 145.

have been a place of strength from very early times. The British king Coroticus, to whom Patrick wrote an epistle, is styled in Irish *rex Aloo*, 'king of the Rock,' that is of *Ail Cluade*, in British *Alclut*, 'the Rock of Clyde.' This suggests that here, in O.Ir. *ail*, gen. *alo* (later *ailech*), British *al*, we have the base of Alauna, which is formed by the addition of suffixes as in Gaulish *acaunos*, *acounos*, stone, rock, and other words, and means 'rock-place.' Similarly, the eastern Alauna may be identified with the Rock of Edinburgh, later in British *Din Eidyn*, in Gaelic *Dún Éideann*. Ptolemy mentions two rivers called Alaunos, identified with the Alne of Northumberland and the Axe of Devonshire. The Ravenna geographer has the Alauna, considered to be the Alun, a tributary of the Dee near Chester. On the Continent there are Alauna near Cherbourg, now Alleaume-les-Vallognes; two places of that name, now Allonnes, in the department of Sarthe, besides five other places in France now called Allonne or Allonnes. North-east of Marseilles was Alaunium, near a place now called Alaun. The local god of Alaunium was Alaunios, and local divinities called Alauniae are commemorated in inscriptions found at Chieming, Traunstein, in Bavaria. It does not, however, follow that all these terms are of the same origin as the British names.

Lindon is simply O. British or E. Celtic for 'pool, lake,' to be compared with M.Ir. *lind*. neut., gen. *lindi*, a pool, lake, W. *llyn*. Skene places it at Ardoch; in Holder it is stated to be Dalginross at Comrie, in Perthshire, but there is nothing really to substantiate either position. The most likely site is on a lake, and the most likely lake is *Lin Lumonoy* of Nennius, now Loch Lomond. It may have been at Balloch, at the lower end of the loch, where the early Gaelic rulers of Lennox had their seat.

Victoria is Latin and most probably a Roman name, not a translation of a native name.

The towns of the Novantae are Loukopibia and Rerigonion. A variant of the former is Loukopiabia, and Müller suggests that both are due to a 'conflation' of old variants Loukopia and Loukobia. In any case the first part is

E.Celt. *leucos*, *loucos*, white, cognate with Lat. *lūc-eo*, I shine; Gr. λευκός, white. The Leuci, 'Whites,' so called, perhaps, from their dress, were a tribe of Belgic Gaul; Lucetia, 'White Town,' is now Paris (Lucetia Parisiorum). There is also Leucetios, the name or epithet of a Gaulish war-god, to be compared with O.Ir. *lóche*, gen. *lóchet*, lightning; Ir. *luach-te*, white-hot. The name has been equated, not unreasonably, with Candida Casa, 'White House,' Whithorn, in Galloway; certainly the meaning of both is, approximately at least, the same. There can be no connection with Luce river and bay, for 'Luce' represents G. *lus*, an herb, a plant.

Rerigonion stood on the Rerigonios gulf or bay, which is agreed to be by position and name Loch Ryan. The three national thrones of Britain, according to the Welsh triads, were Gelliwig in Cornwall, Caerlleon upon Usk, and Penrhyn Rhionydd in the north. Another triad has 'Arthur, the chief lord at Penrionyd in the north, and Cyndern Garthwys, the chief bishop, and Gurthmwl Guledic, the chief elder.' Here *Rionyd*, later *Rhionydd*, is exactly the form that ought to be assumed by *Rigonio-* (for *Rigonijo-*) in Welsh, and there can be no doubt that Penrhyn Rhionydd is the old Welsh name of the northern end of the Rhinns of Galloway, that is to say, of the promontory on the west side of Loch Ryan. Somewhere on this peninsula was the third 'national throne' of Britain, and it may well be that this once famous site was also the site of Rerigonion, the town of the Novantae. In the light of *Rionyd*, *Rhionydd*, the name must be analyzed *Re-rīg-on-ion*, where *rīg-on-* is to be compared with W. *rhion*, lord, chief, formed from the base seen in Gaulish *Bitu-rīg-es*, 'world kings'; O.Ir. *rí*, gen. *ríg*, king; Lat. *rex*, gen. *rēgis*. The first part is the same as in *Re-gulbium*, now Reculver, from *re-* and the base seen in O.Ir. *gulban*, a beak; W. *gylf*, a beak; *re-* is evidently intensive, and is probably cognate with Lat. *prae*, before, in front of.[1] Rerīgonion thus means 'very

[1] Compare 'ni chuir formsa remthus rerig' (Windisch, *Ir. Texte*, i. 270), which means 'to go in advance of princes does not disquiet me.' Here *rerig* is gen. pl. of *rere* for *reri*, formed from *re* and *rī*, like *rure* from *ro* and *rī*. Compare McClure, *British Place-Names*, p. 189.

royal place'; it was, in fact, the royal seat of the Novantae, and the name bears out the tradition embodied in the Welsh triad just quoted. Connected with it is the name *Port Rìg*, 'King's Haven,' which will be mentioned later.

Towns of the Selgovae are Carbantorigon, Uxellon, Corda, Trimontion. The first part of Carbantorigon is E.Celtic and British *carbanton*, a chariot, borrowed into Latin as *carpentum*; E.Ir. *carpat*, palate, gum, jaw; chariot—the double meaning is due to the resemblance in shape between the lower jaw and the old wicker chariots. The second part should probably be compared with W. *rhiw*, an acclivity, slope, in which case the name means 'chariot-slope'; compare *Fán na Carpat*, 'Slope of the Chariots,' at Tara in Ireland.

Uxellon means 'high place,' the neuter of *uxellos*, high, W. *uchel*, G. *uasal*, noble; in our place-names now Ochil. The name does not seem to be preserved in the country of the Selgovae, but there is Ochiltree (W. *Ucheldref*, High-stead) in Wigtownshire.

Corda appears also in the Ravenna geographer; its meaning is obscure to me.

Trimontion, 'Three-hill Place,' Trimuntium of the Ravenna geographer, is thought to have been at the Eildon Hills, whose three peaks are so far seen and impressive. It must be admitted, however, that the Eildons would more naturally fall within the territory of the Votadini, which extended from the Firth of Forth southwards well into Northumberland. The name is Latin as it stands, and there was another place so called in Trachonitis (Palestine), better known as Philippopolis, but it may represent a similar native name with the same meaning.

The Votadini are assigned three towns, Coria, Alauna, and Bremenion, of which the two first have been discussed already. Bremenion is known, on the evidence of an inscription, to have been at High Rochester in Redesdale, in Northumberland. It is to be compared with *Breamhainn*, the river Braan in Perthshire; Gr. *βρέμω*, I roar; W. *brefu*, to low, bleat, bray, roar, and means 'rumbling or roaring place.'

It appears that the bounds of the Votadini reached well south of Tweed.

Ptolemy names six capes or promontories. 'The peninsula of the Novantae' is the Rhinns of Galloway. 'The cape of the same name' is made by Müller Corsill Point (Corsewall Point) on the northern extremity, but had Ptolemy meant this, he would have called it 'the Rerigonian Cape'; what is meant is the Mull of Galloway.

The Cape of the Epidii we have seen to be the Mull of Kintyre. The Cape of the Taexali is identified with Kinnaird Head in the north-east corner of Aberdeenshire.

On the north there are three capes. Cape Orcas, as we have seen, was mentioned by Diodorus, who got it from Pytheas, either directly or through Timaeus. Ptolemy gives 'Orcas or Tarvedum,' which latter is given by Marcian as Tarvedunum, the more correct form doubtless, being for E.Celt. *Tarvo-dūnon*, 'Bull-fort'; compare Duntarvie in Lothian. Macbain identified it with either Holborn Head or Dunnet Head, the two headlands that guard Thurso Bay, noting that Thurso is Norse *Thjórsá*, 'Bull's Water.' Of the two, Holborn Head is much nearer the river; Dunnet Head is next to the Orkneys, and perhaps the former corresponds to Tarvedūnum, the latter to Orcas. The latinized ending *-um* (for Gr. *-ον*, Celt. *-on*) indicates that the name came to Ptolemy through Latin.

Virvedrum, also latinized in termination, is agreed to be Duncansbay Head. The name is probably to be divided *Vir-vedron*, where *vir* is latinized from *ver*, the common intensive prefix, cognate with Gr. ὑπέρ, Lat. *super*, G. *for*, W. *guor*; *vedron* is the neuter of E.Celt. **vedros*, whose feminine is seen in *Vedra* (W. *Gueir*), Ptolemy's name for the river Wear, meaning probably 'clear.' Thus *Ver-vedron* means 'very clear (cape)'; compare for meaning *rinn fiadnach Fotudáin*, 'the conspicuous cape of Fothudán.'

South of Virvedrum and corresponding to Noss Head, near Wick, is Cape Verubium, another latinized form, representing E.Celt. *Ver-ub-ion*, where *ver* is intensive, and *ub* is to be compared with Ir. *ubh*, point, in *ubh claidhimh*, point of a sword (Stokes): the meaning is 'pointed cape.'

The plural of *ubios* is seen in *Ubii*, a tribe on the German side of the Rhine, and with *ubios* may be compared *Calgācos*. The same idea gave rise to *Rinn a' Chuilg*, 'Sword-point Cape,' on Loch Tay near Morenish.

Of the Ebudae (whence, by a mis-reading of the form Hebudes, the ghost-name Hebrides) Ptolemy says: 'Above Ivernia (*i.e.* north of Ireland) are the islands called Eboudai (var. Aiboudai), five in number. The furthest west is called Ebouda (var. Aibouda), that next to it on the east is also Ebouda; then comes Rhicina (var. Eggarikenna), then Malaios, then Epidion.' Marcian, according to Stephanus of Byzantium, wrote Aiboudai; Pliny has Hebudes. Solinus, who lived in the third century, says that the Ebudes, five in number, are two days' sail from 'the promontory of Calidonia,' and a sail of seven days and seven nights from the Orchades, but as we do not know what promontory is referred to, the former datum is valueless. The one thing of which we can be quite sure is that Malaios is Mull; then there is a reasonable probability that Epidion represents Islay; the island of Islay is famed for horse raising; the saying goes that an Islay man will carry a bridle and saddle for a mile in order to ride half a mile. The identification of the others is uncertain. Rhikina is in Pliny Riginia (var. Ricnea, Rignea), and has been equated with Adamnan's 'insula quae vocatur Rechru,' 'Rechrea insula,' which again is probably Rathlin, 6½ miles long, off the coast of Antrim, but the connection between Rhikina and Rechru is not obvious. The curious variant *Eggarikenna* (*Enga-*?) would suggest 'Eigg and Rhikenna,' were it not that this would make up six islands instead of five. The two isles called Ebouda were thought by Captain Thomas to be North and South Uist, but Ptolemy puts them with Ivernia (Ireland), not with Albion. Bute, in Gaelic *Bód*, can have no philological connection with Ebouda. It may be worth mentioning that we have in Gaelic the combination *Bód is Ìle is Arainn*, Bute and Islay and Arran —I have heard it in the west—and Irish tradition joins *Ara ⁊ Íla ⁊ Rechra*, Arran and Islay and Rathlin; *Íle is Ara* are coupled too in St. Berchan's 'Prophecy.' It may be that here we have an indication of an old group corresponding to the Eboudai. The meaning of Eboudae is unknown, and the word is probably pre-Celtic. Professor John MacNeill has pointed out that the ancient tribal name *Ibdaig* in Ireland represents exactly an earlier *Ebudāci*, 'men of the Ebudae'; they are said to be of the Ulaid, *i.e.* of the non-Gaelic race known as Dal nAraide.[1] He also refers to an early **Ebudagnos* the personal name Iubdán, borne by the king of an oversea country of dwarfs, whose adventures in Ireland are told in the tale of 'the Death of Fergus' in *Silva Gadelica.*

Malaios is now *Muile*, anglicized into Mull. It shows the common suffix *-aios,* which becomes in Gaelic *-e*; compare Gaul. *Bed-aios*, corresponding to *Bede* in the Book of Deer, and, for the change in the first syllable, Lat. *badius*, G. *buidhe*, yellow. The intermediate stage appears in Adamnan's *Malea insula*, 'the Malean isle,' or, as we say now, 'an t-Eilean Muileach,' where *Malea* is an adjective fem. from *Maile*, the form which would be assumed quite regularly in Adamnan's time by E.Celt. *Malaios*; compare E.Celt. *magos*, neut., a plain (stem *mages-*); O.Ir. *mag*, gen. *maige* (for *mage(s)os*); G. *muighe*. The first part, *mal-*, may perhaps be compared with G. *mol-adh*: W. *mawl*, praise; G. *muileach*, dear, beloved. If these are connected with Ch.Slav. *iz-moleti*, 'eminere,' 'to stand prominent,' the E.Celt. Malaios might have the very satisfactory meaning of 'Lofty Isle,' foreshadowing our *Muile nam Mór-bheann*, 'Mull of the Great Peaks.'

The Eboudai are placed by Ptolemy under the head of Ireland. Later on he says, 'Off the Orcadian Cape (near Thurso) there lie adjacent to Albion the islands of Skitis (var. Skētis—one MS.; Okitis; Okētis), and Dumna; above which (*i.e.* north of which) are the Orcades, about thirty in number, and still further above these the island of Thule.'

As there are no islands between the mainland and Orkney, it is evident that this statement is very confused. Skitis

[1] *Early Irish Population Groups*, § 135.

or Skētis—the Ravenna geographer has Scetis—is usually, and no doubt rightly, identified with Skye. The two readings may not be really different, for E.Celt. *ē*, representing an older *ei*, is sometimes written *i*, as in *devo-*, *divo-* (Dēvona, Dīvona). If the name is Celtic, it appears to be an *i-* stem, either *Skī-ti-s* (*Skē-ti-s*) or *Skīt-i-s* (*Skēt-i-s*); the former would be like E.Celt. *vla-ti-s* (root of *val-eo*, am strong); O.Ir. *flaith*, sovereignty, sovereign; O.W. *gulat*, lordship; W. *gwledig*, lord, king; the latter would be like E.Celt. *vāt-i-s*; O.Ir. *fáith*, prophet. The root would be ultimately the same in both cases, as in *sgian*, knife; *sgiath*, wing; Lat. *scindo*; Gr. *σχίζω*, I cut; Ger. *scheiden*, divide, and the meaning would be 'Divided Isle,' or, as Stokes puts it, 'Winged Isle.'

The old forms which occur in Irish literature are as follows :—

Nom.—Scia insula (Adamnan); here *Scia* is an adjective, and the form seems to indicate that Adamnan took the nom. to be *Sci*.

Gen.—cum plebe Sceth, AU 668; cum plebe Scith, Ann. Tig. 668; cath for feraibh Sciadh, 'a battle (won) over the men of Skye,' AU and Ann. Loch Cé 1208. (The O.Ir. gen. of *Scētis* would be *Sciatho*, if *Scētis* is an *i*-stem.)

Dat.—imbairecc i Scii, 'a fight in Skye,' AU 701. (The doubling of *i* denotes length.) ó Sci, 'from Skye,' *P.S.*, p. 128; *Anec.* i. p. 6; do Scí, 'to Skye,' *ib.*

In late M.Ir. we have 'clar skeith,' *i.e. clar Sgì*, 'the surface of Skye,' rhyming with 'neith,' *i.e. nì* (Dean of Lismore). Later Ruaidhri MacMhuirich has *clàr Sgìthe*, rhyming with *dìreadh*, *dìobhail*. At the present day the nom. is regularly *an t-Eilean Sgiathanach* (*Sgitheanach*); *an cuan Sgì*, 'the (narrow) sea of Skye,' is applied to a part of the Minch; compare the Norse *Skíð sund* in Hák. Sag. (*Celt. Zeit.*, i. 452). Mary Macpherson, the Skye poetess, uses *Eilean Sgiathanach* once, elsewhere poetically *an t-Eilean Sgiathach* and *Eilean nan Sgiath*, 'winged isle.'

Adjectival forms are: 'ort ocht scuru Scithach,' 'he

destroyed eight encampments of the men of Skye';[1] modern *Sgiathanach*, used now in Skye itself and in other places in the west; *Sgitheanach*, the form which I have always heard in the east, and which is found in the west also. The quantity of the first syllable indicates shortening 'in hiatus,' *i.e.* *Sgitheanach* is formed as from *Sgī*, the *ī* being shortened before the following vowel, and *th* inserted to divide the syllables; compare *cnò*, a nut, pl. *cnothan*. *Sgiathanach* seems due to folk etymology from *sgiath*, a wing.

Dumna, as we have seen, is first mentioned by Pliny, and must have come down from Pytheas. It is the feminine form of E.Celt. *dubnos*, *dumnos*, deep, and means 'the deep-sea Isle.' Macbain identified it with the Long Island, *i.e.* the Outer Hebrides, from the Butt of Lewis to Barra Head, and indeed no other island or island group suits the position ascribed to it by Ptolemy. It has not, however, been observed that the name actually occurs in the older Irish literature. We note in the first place that *Dumna*, which is an *a*-stem (fem.), would become regularly in O.Ir. *Domon* or *Doman* (*i.e.* *domn* with the indeterminate or 'svarabhakti' vowel between *m* and *n*); the genitive would be *Domnae*, later *Domna*. Now according to the copy of the *Leabhar Gabhála* contained in the Book of Leinster (written about 1150), one section of the people of Nemed, who determined to leave Ireland after the trouble with the Fomoire, went to Domon and to Erdomon in the north of Alba.[2] Omitting Erdomon for the moment, we at once

---

[1] Kuno Meyer, *Über die älteste irische Dichtung*, p. 40, v. 21.

[2] 'Luid Matach 7 hErglan 7 Iartach, . i . tri maic Beóain, co Domon 7 co hErdomon i tuascirt Alban'; 'Matach and Erglan and Iartach, the three sons of Beóan, went to Domon and to Erdomon in the north of Alba,' LL 6 b. The corresponding passage in the Book of Ballymote, written about 1390, is 'Luid Matdhach 7 Erglan 7 Iartacht . i . tri maic Beóain maic Sdairn co Dobhur 7 co hIrrdhobur a tuascirt Alban' (27 a 31). In a subsequent passage it is stated of the Tuatha De Danann, 'tangadar a Gregaib 7 gabsat crich 7 fearann a tuascirt Alban . i . ag Dobur 7 ag Ordhobhur'; 'they came from Greece and seized territory and land in the north of Alba, at Dobur and at Ordobur,' BB 32 a 10. Still later, in the 17th century writings of Keating and of Michael O'Clery, the terms become

recognise in Domon the Dumna of Pliny and of Ptolemy. In the next place, a thirteenth century poem addressed to Ragnall, son of Gofraigh, king of Mann and of the Isles, including the Outer Hebrides, speaks of Ragnall as having broken the gate of *Magh Domhna*, 'the Plain of Domon.'[1] The poet's exact meaning is not very clear, as is often the case in this style of panegyric, but at any rate we have here the authentic genitive case of *Domon*, the ancient Dumna. *Magh* recalls the *machraichean* or sea-plains of the Hebrides.

As to Erdomon, the prefix *er*, often written *ir*, is the O.Ir. *ar*, E.Celt *are*, meaning 'on, near,' sometimes as Kuno Meyer has pointed out, 'east of.' A very clear instance is *Are-clūta*, 'On-Clyde, Clydeside, Strathclyde,' which appears in O.Ir. as *Erchlúad* (*do Bretnaib hErclúaide*, of the Britons of Strathclyde).[2] *Er-domon* thus means 'near Domon, next to Domon,' or possibly 'east of Domon,' and I take it to be the old designation of the Inner Hebrides, the group of islands from the north point of Skye southwards to the Mull of Islay. Elsewhere the Book of Leinster records an expedition which took place in A.D. 568, to *Iardomon*, the particular objective of which was 'Sóil ⁊ Ili,' now *Saoil* and *Ìle*, anglicized as Seil and Islay.[3] We have noted that 'Erdomon' may be written 'Irdomon'; if a scribe had this before him, he would readily miscopy it as Iardomon.

The Western Isles are reputed to have been the haunt of the race known as the Fomorians,[4] who are represented

---

Dobhar and Iardhobhar. The corruption of the old version is progressive: *ir* and *or* are both forms assumed by *ar*, E.Celt. *are*; *iar* is exactly opposite in meaning, 'behind, back of: west.' Skene adopted the late forms Dobhar, Iardobhar. (*Celtic Scotland*, i. pp. 173, 174.)

[1] Printed in *Celtic Scotland*, iii. p. 410. Ragnall is styled in the same poem 'rí an Domnan,' which seems to mean 'king of Domnan.'

[2] *Leabhar Breac*, 238 a 3 and 13; Hogan, *Onom.*; and Kuno Meyer, *Zur keltischen Wortkunde*, 190.

[3] Fecht i nIardomon, . i . i Sóil ⁊ i nIli, la Colman mBec mac nD(iarmato) ⁊ la Conall mac Comgaill—LL 24 b. AU has at 567(568) 'Fecht in Iardomhain,' 'expedition against Iardoman.'

[4] The Fomorians (*Fomóraigh, Fomóirc*) were of the race of Ham, according to LU 2 a 31—2 b 1. They were huge and ugly: in his combat

in the old accounts as pirates, ravaging the coasts of Ireland and laying the people under tribute. One of their kings was *Indech mac Dé Domnand,* 'Indech, son of the goddess Domnu,' described as 'a man possessed of arts and accomplishments,' which were no doubt of a magical nature.[1] We are also told that Danann was slain by Dé Domnann of the Fomorians.[2] The connection of the Fomorians with the Isles shows that here we have to do with a goddess of the deep sea, who was the tutelary divinity of the Isles and the divine ancestress of their ancient kings. This Domnu is distinct from the goddess of the Irish *fir Domnann* (p. 26).

However the Thule of Pytheas is to be identified, the Thule of Tacitus is Shetland, and so also with Ptolemy. Müller is of opinion that he indicates in particular the largest of the group, called Mainland. The Irish geographer Dicuil, who wrote about A.D. 825, says that 'Thile is the most remote island of the Ocean, lying north-west of Britain, and having its name from the sun (*a sole*), because the sun makes his summer solstice there.' He had heard minute details about it from certain clerics, who had lived therein from the first of February to the first of August. Dicuil's Thile appears to be Iceland, to which, as is known otherwise, Irish clerics were in the way of voyaging, seeking 'a desert in the sea.' 'Other little isles also there are,' says Dicuil, 'wherein lived for the space of a hundred years hermits, voyaging thither from our Scotia (Ireland), but now they are empty of anchorites by reason of Norse robbers (*causa latronum Nortmannorum*).'[3] In M.Ir.

with Ferdiad, Cuchulainn waxes 'as huge as a *fomóir* or a seaman (*fer mara*); their appearance is described in LU 89 b 32. They are 'loingsig na fairgge,' 'ship men of the sea,' LL 6 a 39. The vowel of the second syllable is sometimes short, but more often long, e.g. *fomóra* rimes with *comóla*, LL 7 a 32, and the quantity is reflected in G. *famhair*, a giant, with open *a* in *-air*. The word is therefore not from *muir*, sea.

[1] 'Fer con dánaib 7 con eladnaib ésen, LL 9 b 10; BB 33 a 17.' 'Indech rí na Fomórach' slew Ogma mac Eladan in the first battle of Mag Tured, LL 9 b 3; 11 a 13; in the latter passage he is styled 'Innech mac Dó Domnand' by Flann Manistrech (*d.* 1056).

[2] LL 11 a 26; BB 35 a 43.

[3] Dicuil, *De Mensura Orbis Terrae*, 7; 8, 11, 14.

Thule is *Inis Tile.* Its soil had magic properties : if a man stood thereon, his feet clave so fast to it that he could not move from the spot. Three druid kings reigned therein.[1] Monach Mór (the great Magician), son of Balbuadh of Inis Tile, appears in a genealogy of MacLeod.[2]

---

[1] *Trans. of the Ossianic Soc.*, iii. 118 ; Plummer, *Latin Lives of the Irish Saints*, i. clviii. ; Adv. Lib. MS. xxxiv., *Bruighean Caorthuinn* ; Adv. Lib. MS., lii. 8 b ; 10 a.

[2] Reeves, *Adamnan's Life of Columba*, p. 437.

CHAPTER II

EARLY NAMES (*continued*)

On the western side of Albion, which he calls the north side, beyond which is 'the Ocean called Dvē-caledonios,' Ptolemy places the estuary of Clōta, and the mouths of Longos, Itis, and Nabaros. Since Nabaros is certainly the Naver in the north of Sutherland, the position of the intermediate rivers is apt to be hazy.

Clōta, mentioned earlier by Tacitus, is the Clyde, well known both in Irish and in Welsh. The O.Ir. nominative was *Cluad*, gen. *Cluaide*, but Adamnan gives a still earlier form in *Petra Clōithe*, a part translation of *Ail Cluaide*, Rock of Clyde, Dumbarton Rock. The modern Sc.G. is *Cluaidh* or *Abhainn Chluaidh*. In O.Welsh Clōta becomes *Clūt*, later *Clūd*, whence the English form Clyde. Gildas the Sage, according to one *Life* of him, was born in Areclūta, 'a region which has taken its name from a certain river, which is called Clūt.' Areclūta is Strathclyde, in O.Ir. *Erchlúad*, as has been noted already. The forms in Gaelic and Welsh show that Clōta represents *Cloutā*; compare *Boud-icca*, 'Victorious Lady,' and O.Ir. *búad*, victory; W. *būdd*, profit, gain. The root is *clou*, wash, whence Lat. *cluo*, I purify; *clo-āca*, a sewer; and the Italian river *Clu-entus*. Like many other river names, Clōta is really the name of the river goddess, meaning 'the washer, the strongly flowing one,' or such. A similar idea is found in the name of its affluent the Cart, connected with Ir. *cartaim*, I cleanse. Clyde is sometimes compared with the Welsh river Clwyd, but the resemblance is deceptive, for Clwyd implies an E.Celt. *Cleitā*, *Clētā*, from a different root altogether.

The position of Longos is difficult. Macbain equated it with Loch Long, G. *Loch Long*, 'Loch of Ships,' which was called by the Norsemen *Skipa-fjörðr*, 'Ship-firth.' The northern branch of Loch Alsh, in Ross-shire, is also *Loch Long*, 'Loch of Ships,' and the river at its head is *Abhainn Luinge*, 'Ship-river.' This would be an exact parallel, if we assume *long*, ship, to be a native Celtic word, and not, as seems more likely, a loan from Lat. (*navis*) *longa*, a warship. Skene, on the other hand, made it 'the river Add (at Crinan), known to the Highlanders as the *Avon Fhada* or long river.'[1] Here Skene made certain assumptions: (1) that Longos is Latin; Ptolemy wrote in Greek. The name, however, might have reached him in Latin, and he might have thought it a native word. (2) He assumes that the language of the district in Ptolemy's time was Gaelic; in view of the *p* in Epidii, the local tribe, this is unlikely. (3) The name of the river Add cannot be separated from Dun Add, the famous fort upon it. Now this name appears twice in the Irish Annals: nom. Dun At, AU 736; gen. Duin Att, AU 683. 'Long' is in O. and M.Ir. *fota*; it is unlikely that the Irish writer meant *fota* when he wrote *At*, *Att*. It is true that the MacVurich historian wrote 'eidir Abhuinn Fhada ⁊ Alta na Sionnach a mbráigh Chinn-tíre,' 'between Abhainn Fhada and the Foxes' Burn in the upper part of Kintyre,'[2] but this seems to be mere folk-etymology. At the present day, as I am informed by competent authorities, the river is locally *Abhainn Ad*, with no final *a*, and without the article; the fort is *Dùn Ad*, with a tendency to long *a*; it is pronounced as if it were spelled *Athd*. It is therefore more than doubtful whether Skene's third assumption is justified, and we may safely leave the river Add out of account. We really cannot be sure of the position of Longos. If it is genuine and Celtic, we may compare Gaulish Longo-staletes, the name of a tribe. It is worth noting that Ptolemy gives Longones as the name of the Gaulish tribe Lingones, connected with O.Ir. *ling-im*, I leap.

Eitios (var. Ituos; Lat. Itis) is amended by Müller to

---

[1] *Celtic Scotland*, i. p. 68. By 'Highlanders' Skene means the Gaelic-speaking people of Scotland.

[2] *Rel. Celt.*, ii. p. 162.

Itios. It is the genitive case, presumably of a nominative Eitis or Itis, which Stokes analyzed as i-ti-s, making it an *i*-stem, 'from the root *i*, whence Lat. *i-re*, Gr. *εἶμι*, to go.' This would connect it with O.Ir. *ethaim*, I go, and its various compounds.[1] The gen. ending may perhaps be compared with that in the Ogham *Turan-ias*, *Anavlamatt-ias*, which are regarded as *i*-stems, probably feminine. If Stokes' explanation is right, we may compare for meaning the river *Cingidh*, in Glengarry, from the base seen in O.Ir. *cing-im*, I stride. As to position, it is usually identified with Loch Etive, and Macbain went so far as to say that the names agree, even though the intervocalic *t* of Eitis ought to appear as *th*, whereas Loch Etive is now in Gaelic, *Loch Éite* (not *Éitigh*, as he has it). As a matter of fact, *Loch Éite* (Loch Etive) is Gaelic, and interesting. It represents M.Ir. *Loch Éitchi*[2] (for *Éitche*), and *Éitche* is gen. sing. of *Éitig*, a feminine proper name (declined like O.Ir. *sétig*, gen. *séitche*, a mate, a wife), meaning 'foul one,' 'horrid one.' The lady who had this ugly name was really the goddess of the loch and river, and if we ask why she was so called, we have only to know the stormy and dangerous nature of the loch, and in particular to look at the formidable sea-cataract at its entrance, known as *a' Chonghail*, the Connel. She is still well known as *Éiteag*, a diminutive form, 'the little horrid one.' In literature, *éitig(h)* is coupled with *salach*, foul, of which it is nearly a synonym, and it is not by accident that Eiteag's haunt is traditionally placed in *Gleann Salach*, Foul Glen, beyond Ardchattan. The opposite of *éitigh* is *álainn*, lovely, and the opposite of Loch Etive[3] is *Loch Àlainn*, L. Aline in Morvern. It is therefore incorrect to say that Etive represents Eitis, Itis, etymologically, though it may represent it as regards position. But in the absence of the corresponding modern forms the identification of Ptolemy's rivers of the west coast must be largely a matter of guesswork.

---

[1] See Pedersen, *Vergl. Gram.*, ii. p. 514.

[2] Glenmasan MS., *Celt. Rev.*, i. p. 104 : co Loch n-Eitci ; 110 Glend Eitci.

[3] The English form 'Etive' represents the nom. *Eitigh*, just as *Ceann-ruighe* becomes 'Kin-rive' ; *gh* into *v* is common.

Next to Itis comes Volas Bay, with variants Valsos, Volsas. According to Ptolemy's computation, this bay would be just about the position of Loch Broom ; Macbain, however, preferred to equate it with Loch Alsh, which is much further south. His reason for this was that he believed the name *Loch Aillse*, Loch Alsh, to represent the old form Volsas, but the equation seems doubtful. The combination *-ls-* is so rare in early Celtic that Volas is more likely to be the true reading. Several early Celtic names more or less resembling Volas, and one identical with it, may be found in Holder, but none like Volsas. Volas would not yield *Aillse* in Gaelic, for the final *s* would disappear. While Volas is in all likelihood Celtic, its meaning is doubtful. *Aillse* is a difficult word, but it does not stand alone, for there is another *Loch Aillse* near the source of the river Oykel, on the border of Sutherland and Ross. It may be an abstract noun from *allas*, sweat, meaning 'sweatiness,' with reference to foam and scum ; compare the adjective *aillseach*, sweaty, *e.g.* 'ech odhor aillsech,' 'a dun sweaty horse.'[1] The Kirkton of Loch Alsh is 'an Clachán Aillseach.'

The Nabaros, placed by Ptolemy on the west coast, corresponds in name with the Naver, *Nabhar*, well to the east on the north of Sutherland ; it rises in *Loch Nabhair*, flows through *Srath Nabhair*, and enters the sea at *Inbhir Nabhair* (pron. *I'r N.*). Its ending *-ar-os* appears in *Tam-ar-os*, the Tamar ; *Sam-ar-a*, the Somme ; *Lab-ar-a*, the Laber—there are four Labers in Bavaria—and, in Alsace, the Leber. The base is *nabh*, found in Gr. *νέφ-ος*, *νεφέλη* ; Lat. *nūb-es*, a cloud ; Skt. *nábhas*, wet cloud. The reference is probably to fogs rising from the river.

On the south side (*i.e.* the east side) there are the mouth of the Ila, ὄχθη ὑψηλή, in Latin Ripa Alta, 'High Bank,' estuary of Varar, mouth of Loxa, estuary of Tuesis, mouth of Dēva, estuary of Tava, mouth of Tina, and estuary of Boderia.

Ila is now *Ilidh*, Helmsdale river ; *Bun Ilidh*, 'Ilie-foot' is Helmsdale town ; the strath is *Srath Ilidh*, re-named by

---

1 *Life of Moling*, § 68 (Stokes).

the Norse 'Helmsdale'; the initial *i* of *Ilidh* is short. A Helmsdale man is *Ileach*. The Gaelic form does not represent *Ila*; it might represent *Ilia*, but it may be on the analogy of the numerous stream-names in *-idh*. Ila is an obscure name; whether it should be compared with *Ìle*, Islay, is doubtful; in the latter, *i* is long, as it is also in *Làirig-ìlidh*, the pass at the head of Glenogle, Lochearn Head. A place on the Cromarty Firth, east of Dingwall, is called *Aird-ilidh*. Srath Isla in Forfarshire is now in Gaelic *Srath Ìl'*.

Instead of Varar, perhaps, as Müller suggests, the right reading may be Vararis (gen.); it is represented now by *Farar*, formerly the name of the Beauly river, whose lower part, beginning at Struie, was Strath Farrar. 'The Barronie of Bewlie and all besouth the watter of Forire' is mentioned in 1639 as a division of Inverness-shire.[1] 'Farrar' applies now only to the part of the river above Struie, whose valley is called 'Glen Strath-farrar,' *i.e.* the glen of Strathfarrar. In *Farar* for *Varar* the change of *v* to *f* is according to Gaelic phonetics, and is to be compared with the change of *Votadini* to *Fotudán*, *Verturiones* and *Fortriu*, and with *Foirthe*, Forth, as against W. *Gwerid*. The name is probably Celtic, for *var-* is not uncommon in E.Celt. names, and the formation resembles that of the Gaulish river Arar noted for its slowness and supposed to be connected with W. *araf*, slow. But its meaning is obscure.[2] It is probable that by 'the estuary of Varar' Ptolemy meant the part of the Moray Firth that is inside Chanonry Point.

Half a degree north of the estuary of Varar, and in the same longitude, Ptolemy places the mouth of the river Loxa. If the estuary was reckoned to end at the narrow between the points of Chanonry and Ardersier, the position

---

[1] MacGill, *Old Ross-shire and Scotland*, ii. p. 10. Is 'Forne,' which is given as an old name of the Beauly river, a misreading of 'Forire,' and merely a ghost name?

[2] Macbain suggested comparison with Lat. *varus*, *varius*, and thought that it might mean 'winding, bending' river, which is quite appropriate as regards the last part of its course at present. Nicholson in his *Keltic Researches* proposed *Vo-arar*, 'gentle stream.'

of Loxa corresponds to that of Findhorn, and this appears to be the river actually meant. Macbain was inclined to agree with Skene and Stokes that Loxa was the Lossie, but as he himself points out, the Lossie is much too far to the east to correspond to the position assigned to Loxa, and Ptolemy's information as regards the east coast of Scotland was specially good. The name Findhorn, which means literally 'white Ireland,' was given at a much later period, displacing the older name (p. 230). I offer no opinion as to the meaning of Loxa.

Tuesis may well represent the Spey as to position, but not in name. It is of the same formation as the river names Tam-es-is, Thames; At-es-is, Adige; and may be from the root *tu*, 'swell,' as in Lat. *tumeo*, G. *tulach*.

Kailios, with variants Kelios, Kailniou, Kelniou, is in the genitive case, apparently an *i*-stem, nom. *Kailis*, *Kelis*. If this is the correct reading it is difficult to connect with O.Ir. *cóil*, G. *caol*, W. *cul*, for these require an early *coil-*. The position assigned to this river agrees with that of the Deveron in Banffshire. The name Deveron, like Findhorn, belongs to a period much later than Ptolemy's time, for it literally means 'black Ireland.'[1]

Dēvā, the Aberdeenshire Dee, is the feminine of *deivos*, Gaulish *dēvos*, a god, and was, of course, the name of the river goddess. In Gaelic it is *Uisge Dé*, where *Dé* is the same in form as the Gaelic genitive of Dia, God, O.Ir. *Dea*, *Día*, God, a god; *dee*, 'a pagan divinity.' Ptolemy mentions two other British rivers of the same name, now the Dee of Galloway and the Dee of North Wales. There were also two rivers called Dēvā in Spain, now Deba and Deva.[2] In Ireland there was the river *Dea*, entering the sea at *Inber Dea* or *Inber Dee* (at Arklow), now the Avonmore.[3]

---

[1] Macbain, taking the reading *Kelniou*, *Kailniou*, equated with Cullen, formerly Inverculan (*Reg. Mor.*); but the stream at Cullen is a mere brook, mention of which by Ptolemy is inconceivable.

[2] Holder.

[3] In Irish tradition named after Dea mac Dedad: 'Glas mac Dedad, cuius frater Dea, a quo aband Dea ⁊ inber Dea i crich Cualann'; LL 159 a, near foot. Dedad was the eponymous ancestor of Clann Dedad of Munster.

Again in Scotland we have the important reference by Adamnan to 'the stream which in Latin may be called Nigra Dea'[1] (the black goddess): we note that the river *is* the goddess. Elsewhere the native form of the name is given, when mention is made of 'stagnum Loch-diae,'[2] 'the lake of the black goddess,' identified, no doubt correctly, with *Loch Lòchaidh*, whence issues the river *Lòchaidh*, near Fort William. Another *Lòchaidh*, also, and more correctly, called *Lòchá*, flows through *Gleann Lòchaidh* (*Lòchá*) in Perthshire into Loch Tay, at Killin, styled sometimes *Lòchá Albannach*, 'of Alba' east of Drum Alban, to distinguish it from yet another river *Lòchaidh* which rises near Tyndrum and joins the Orchy, whence it is sometimes called *Lòchaidh* (*Lòchá*) *Urchaidh*. The lochlet at its source, a little to the west of the Perthshire border, is probably the 'stagnum Loogdae,' or Loch Lochy, near which was fought the battle called 'bellum monith carno' (AU 729). With these probably go the *Lòchaidh*, Lochy, of Banffshire, the Black or Lochty Burn, with Inverlochty at its mouth, near Elgin, and the Lochty in Fife and in Perthshire. But for the help given by Adamnan, it would have been impossible to explain with certainty the origin of this remarkable series of 'Black-goddess' streams. There are doubtless still more of the many streams that end in -*(a)idh* which should be similarly explained; for instance *Dubhaidh* or *Duibhidh*, the Divie, near Edinkillie, a tributary of Findhorn, may be for *dubh-dhea*; it is sometimes *Dubhag*, the little black one—a suggestive variant, in view of the many stream names that have this diminutive form. The Galloway river Dee is called in its upper part 'the Black Water of Dee,' but whether this is a trace of a black goddess, or a more modern epithet, is not clear.

Ptolemy's Tava, the Taus of Tacitus, is now *Tatha*, the Tay. In early Gaelic it is declined as follows: nom. *Tōe*; gen. *Tōi*; dat. *Tōi*; acc. *Tāi*, *Tōi*, all dissyllabic.[3] The *Amra Coluim Cille* (Praise of Columba) makes mention

[1] 'Fluvius qui Latine dici potest Nigra Dea' (*Vit. S. Col.*, ii. 38.)

[2] Capitulationes to Book I, p. 110 of Skene's edition.

[3] LU 8 b 28; 14 b; Rawl., B 502, 100 b.

of *Tuatha Tōi*, 'the tribes of Tay,' with the explanation 'na tuatha batar im Thāi,[1] .i. ainm srotha i nAlbain'; 'the tribes that were around Tay, which is the name of a river in Alba.' In another note on the *Amra* Columba is said to have vanquished the lips of the rude (*borb*) folk who were with the high king of Tōe. A Latin tract of the twelfth century has Tae (nom.).[2]

The change of early *Tōe* to *Tatha* in Scottish Gaelic is regular: the long vowel 'in hiatus' is shortened and the syllables are separated by *th* silent, which is not organic but a mere device of spelling. In addition the *ō* becomes *a*, as often, and final *e*, unstressed, becomes an indeterminate sound represented by *a*. Had *Tōe* been treated according to the procedure in Irish Gaelic, the modern form would be *Tua*, a monosyllable. The old form *Tōe*, however, could not have arisen from *Tava*; it presupposes a form *Taviā*. The next stages are as follows: in *Taviā* the *av* became a diphthong *au*, while *-ia* became *-e*, giving *Taue*; then the *au* became *ō*, giving *Tōe*. *Tōe* is to be compared with O.Ir. *tō*, silent; *tōe*, silence, stillness. As a river name it was doubtless primarily the name of a goddess, 'the Silent One.' The corresponding Welsh term is *taw*, silent, silence; and the Welsh name for our Tay was *Tawy*.[3] Later on I shall have occasion to mention an early Irish saint called *Tua*, 'the silent,' whose name became with us *Tatha* and in one district seems to appear in the Irish form *Tua*.

Ptolemy's next stream is Tina, placed between Tava and Boderia (Tay and Forth), thus suggesting the Eden at St. Andrews. That there is any philological connection between the names is unlikely: Tina may be the Haddington Tyne misplaced, or even the Newcastle Tyne which is not otherwise mentioned by Ptolemy (Bede's Tina, Tinus). There are two rivers on the Continent called Tinea, but the meaning is uncertain; one may perhaps compare O.Ir. *tinaim*, 'evanesco,' disappear (*cf.* Turret, p. 446).

Forth is in Tacitus Bodotria, in Ptolemy Boderia with a variant Bogderia; it is uncertain what river the Ravenna

---

[1] Rawl., B 502, 100 b 11 im Thōi.

[2] Skene, *P.S.*, p. 97.

[3] *cf.* *Celt. Rev.*, iv. p. 270.

geographer's Bdora is meant for. For Bodotria, Zeuss conjectured *Bodortia*, and compared the Irish participle passive *buadartha*, 'turbulentus,' turbid. This leaves out of account the form Boderia, and also postulates a long *ō* (to yield *ua* in Irish), while in Boderia the *o* is short. There is, however, a more feasible explanation, as it seems to me. In Ireland the term *bodhar*, O.Ir. *bodar*, deaf, occurs freely in names of places, always with reference to absence of sound, stillness. *Bodar usce* means 'stagnant or sluggish water.'[1] Dr. Joyce, who, however, does not quite understand the usage, gives a number of instances, *e.g.* Glenbower, 'Noiseless Glen'; Drohid-bower, 'Noiseless Bridge,' from a still pool beneath it.[2] Similarly in Welsh *byddr leddf* denotes a letter or syllable that becomes silent. From a comparison of the two forms Bodotria and Boderia, it may be suggested that Bodotria is to be corrected to Bodortia, as Zeuss proposed, corresponding to Ir. *bodartha*, deafened, 'the Deafened One,' while Boderia (? Bodoria) means 'the Deaf One, the Silent One,' and stands in relation to Ir. *buidhre*, deafness, as *Tóe* does to Ir. *tóe*, silence (*cf.* p. 435).

The modern 'Forth' is not etymologically descended from these ancient forms. A twelfth-century notice states that the river bore three names, Froch in Gaelic (*Scottice*), Werid in Welsh (*Brittanice*), Scottewattre in English (*Romane*).[3] Here Werid is for Gwerid. In the Welsh Laws reference is made to an expedition of reprisals to the North by the men of Gwynedd (North Wales), in course of which they came to the banks of Gweryd and disputed who should take the lead through the river Gweryd. Here Gweryd is Forth, and the spot where the dispute took place would be at the Fords of Frew. The Pictish Chronicle states that Kenneth, son of Mael Coluim (*ob.* 995) fortified the banks

---

[1] 'a sháith do bodar usci na Sabraindi dó,' 'he had his fill of the dead water of the Sabrann' (Lee) (*Vis. of MacConglinne*, 19, 11); 'deoch do bodar usci na cuirre,' 'a drink of the dead water of the pool' (*ib.*, 55, 6); 'an guth bodhar mar teillean,' 'the dull tone, like a bee' (O'Curry, *M. and C.*, iii. 357).

[2] *Irish Names of Places*, ii. p. 48 (ed. 1883).

[3] Skene, *P.S.*, p. 136.

of the Fords of 'Forthin,' *i.e.* the Fords of Frew.[1] In a Latin poem of the twelfth century the name is 'Forth'; [2] a prose account of the same period has 'magnum flumen Forthi,' the great river of Forth.[3] It is an extremely curious fact that the name is absolutely unknown in modern Gaelic, either in common speech or in literature. In Old Gaelic, however, an important notice of it occurs which has hitherto escaped attention. This is in a poem, composed before 1100, on Aedán mac Gabráin, who was king of Dál Riata in Alba in Columba's time.[4] The narrative goes back to Aedán's birth, at a time when his father, Gabrán, was engaged in a foray beyond his own bounds, apparently 'at Foirthe.' Further on in the poem, Aedán is styled 'Prince of Foirthe,' a title which is not amiss in the mouth of a bard in view of Aedán's campaigns in that region, and which also recalls the statement in the *Amra* as to the presence of Columba, Aedán's friend and counsellor, in the Tay district. *Foirthe* is an *-ia* stem, representing an old *Vo-rit-ia*, from the root seen in Ir. *rith*, act of running, *rethim*, I run, meaning 'the Slow-running One.' We may compare the

---

[1] *Ib.*, p. 10. [2] *Ib.*, p. 118. [3] *Ib.*, p. 153.

[4] The poem deals with the story of the birth of Brandub, son of Echu, King of Leinster, and of Aedán, son of Gabrán. Echu and his queen were visiting Gabrán, and Brandub and Aedán were born in the same night. According to the tale, which seems to have no historical foundation, both queens bore twins, Echu's wife having two sons and Gabrán's wife two daughters. The latter exchanged one of her daughters for one of Echu's sons without the knowledge of the fathers (see *Celt. Zeit.*, ii. p. 134). The relevant quatrains of the poem are:—

Dochuidsium ar sluaged sel Gabrán ria ngarraid Góidel
ar chrechaib nir ches in fer ruc leis Echaid na noiged.
Dar éis na fer ic Foirthe cia caingen bad chaemsoirthe?
na mna cen chosait chaidhche a n-osait i n-oen-aidche.'

'Gabrán, before whom the Gael shouts, went for a space on a hosting—the man did not stint forays; with him he took Echu, of many guests. Behind the men at Foirthe—what matter could be of happier issue?—the ladies, without once complaining, were delivered in one night.' (Rawl. B 502, 86 a 47.)

Ro ba co cenn naei mís mór ari Forthe na fledól
aróen is Brandub cen brón cen formandur immedón.

'To the end of nine full months the festive Prince of Foirthe (Aedan) was with Brandub without sorrow, without parting.' (*Ib.*, 86 b 46.)

O.Ir. gloss which explains 'perigrina per marmora' by 'trisna foirthiu ailitherdi,' translated 'through the foreign fords.[1]

The Welsh Gwerid may be referred to the same form, which would become *guorid* but for the fact that *o* becomes *e* under the influence of *i* in the next syllable.

On the south Ptolemy mentions three rivers, Abravannus, Dēva, Novios. Dēva, the Dee, has been already discussed. In part of its course it is called the Black Dee, suggesting comparison with Adamnan's Lochdae, Nigra Dea.

Novios is agreed to be, by position, the Nith. It is the same as in E.Celtic *Novio-magos*, 'new plain'; O.Ir. *naue*, *nóe*, *núae*, *nuie*; Welsh *newydd*, new, fresh; cognate with Lat. *novus*. The reference is most likely to the freshness and verdure of the riverside. In modern Sc.Gaelic *novios* becomes *nodha* (as opposed to Mod.Ir. *nua*, *nuadh*), and in Argyll we have *Abhainn Nodha*, the river Noe, flowing through *Gleann Nodha*, Glen Noe, a verdant glen on the north-west skirts of Cruachan.[2] *Noid*, Nude, in Badenoch, is probably of the same origin, with *-id* suffix, representing an early *novant-i-*. Whether the name Novios is represented by the modern Nith, in 1124 Stra-nit, is another question. Sir John Rhys held 'it is from some stage of this last (Welsh *newydd*, new) that we get *Nith*; but it could only happen through the medium of the men who spoke Goidelic.'[3] Macbain says, 'the word Nith is a Brittonic rendering of the old name.' On the other hand, Kuno Meyer compared it with the river *Níth*[4] in Louth, now the Dee, and others have pronounced it impossible that Nith should represent Novios.[5] The truth is that it is difficult to say with certainty what precisely is impossible in the case of a name which we have at the third or fourth remove—a name transmitted from Old British through Welsh into Gaelic and thence into English. We may compare Nithbren, the

---

1 *Thes. Pal.*, i. p. 488; compare 'co forthiu Máil,' LL 52 a 30.

2 For the phonetics, compare *Tōe*, *Tatha*.

3 *Celtic Britain*, p. 199 (ed. 1904).

4 *Voyage of Bran*, i. p. 51.

5 *E.g.* McClure, *British Place-Names*, p. 120.

old form of Newburn, a parish in Fife. Here the second part may well be Welsh *pren*, a tree ; in Gaelic we find *úr-chrann*, a green tree (lit. a new tree), and *núa-chrann* [1] would be quite possible in older Gaelic in the same sense ; the Welsh *newydd* may have been similarly used, in which case Nithbren might mean 'Green-tree.' For the contraction we might compare *Ros-neimhidh*, spelled already in 1225 Rosneth (in *Reg. of Paisley*, Neueth, Neyt, Rusnith), now Rosneath. It would be rash, therefore, to deny that Nith may represent Novios.

Abravannus seems to agree in position with the river Annan, though Skene and Bradley make it Luce Bay. The explanation, given in Müller's edition of Ptolemy, that it is the old equivalent of Welsh *aber-afon*, river-mouth, may be safely dismissed. If the name is correctly written, it might be compared with the continental Celtic river names Obr-incas, Tri-obris, of doubtful meaning. In any case, Ptolemy's name, even if correct, cannot yield Annan. The Ravenna geographer, however, has Anava. As usual with his names, it is hard to say which river he means, but Anava would become in Welsh *Anau*, and in Gaelic *Anu*, with gen. *Anann*. Among the ancient Irish, Anu was the mother of the gods, and the goddess of prosperity ; [2] from her are named *dá chích Anann*, the two Paps of Anu, hills on the eastern border of Kerry. We may compare also Welsh *anaw*, wealth, riches, largess. Annan in fact is the Gaelic genitive case of a British Anau, as Manann is the genitive of a British Manau. Old forms of Annan are : Estrahanent 1124 (W. *ystrad*, strath), Stratanant 1152, Vallis Anandi 1297. The Gaelic poet Alexander MacDonald (*fl.* 1745) has 'eadar Cataibh agus Anuinn,' 'between Sutherland and Annan.'[3] Holder gives a number of examples of Anava as a river name on the Continent.

The only hill specifically mentioned is Mons Graupius in Tacitus' *Life of Agricola*. As the Caledonian leader's

[1] Compare the epithets *nói-cruthach*, fresh of form ; *nói-cride*, of fresh heart—*Fled Bricrend*.

[2] *Cóir Anmann* (*Irische Texte*, iii., pt. 2, 1) ; Cormac's *Glossary*.

[3] P. 102 (1st edition).

name was certainly Calgacus, not Galgacus as given in Tacitus, it may be conjectured that the true name of the hill is Mons Craupius, and as *Craupius* would yield *Crup* in O.Welsh (compare W. *crwb*, a hump, haunch), Mons Craupius is formally identical with *Dorsum Crup*, a place mentioned in the Pictish Chronicle (p. 10) and supposed by Skene to be now Duncrub near Perth. If Skene's identification is correct, there may be some difficulty in further identifying the site of the famous battle with Duncrub: our best authority on the Agricolan occupation is of opinion that the battle was fought north of Tay, while Duncrub is south of it.[1] The name 'hill of the hump,' however, is one that might occur in more than one place.[2]

We have now dealt with the information derived from the earlier writers up to and including Ptolemy. Later writers during the period of the Roman occupation mention the names of three new tribes, the Maeatai, the Verturiones, and the Picti. In 197 Virius Lupus became governor of Britain, and was at once obliged to buy peace at a large sum from the Maeatae, 'because,' says Xiphilinus in his abridgement of Dio Cassius, 'the Caledonians did not abide by their promises, and were preparing to assist the Maeatae.' Writing of Septimius Severus' doings in the North in A.D. 208, Xiphilinus informs us that 'the two most important tribes of the Britons (in the North) are the Caledonians and the Maeatae; the names of all the tribes have been practically absorbed in these. The Maeatae dwell close to the wall which divides the island into two parts and the Caledonians next to them. Each of the two inhabit rugged hills with swamps between, possessing neither walled places nor

---

[1] Dr. George Macdonald, *The Agricolan Occupation of Great Britain* (Journal of Roman Studies, 1919).

[2] A writer in 1723 remarks that at the west end of a moss near Duncrub House 'there is an artificial knoll commonly said to have been a buriall place of the Picts; which has been verified a few years hence in finding thro' digging stone coffins, joints of men's bodies, rings, and pieces of old money.' He adds as a tradition that about half a mile south of this there was a garrison of the Picts on the top of a hill of sugar-loaf shape. Macfarlane, *Geog. Coll.*, i. p. 120.

towns nor cultivated lands, but living by pastoral pursuits and by hunting and on certain kinds of hard-shelled fruits (? nuts). They eat no fish, though their waters teem with all kinds of them. They live in tents, naked and shoeless; they have their women in common, and rear all their offspring (*i.e.* they did not, as the Romans did, practise exposure of undesired infants). Their government is democratic, and they take the utmost delight in forays for plunder. They fight from chariots, and have small swift horses. Their infantry are extremely swift of foot and enduring. Their weapons are a shield and short spear with a knob (apple) of brass on the end of the butt . . . they have also daggers. They can endure hunger and thirst and every kind of hardship. They plunge into marshes, and last out many days with only their heads above water, and in the woods they live on bark and roots, and above all they prepare a certain food such that, if they eat only the bulk of a bean of it, they neither hunger nor thirst.'[1]

The next mention of the Maeatae is in the *Life of Columba*, where Adamnan, in a section headed 'De Bello Miathorum,' tells how Columba prayed for Aedán's victory in a battle which he was then fighting, and in which he lost 303 men. In the next section Adamnan records that two of Aedán's sons, Arthur and Eochaid Find, were slain 'in the above-mentioned battle with the Miathi.' In Tighernach's *Annals* this battle is called *Cath Chīrchind*, 'the battle of Circenn,' that is to say, the district of Angus and the Mearns, and he says that four of Aedán's sons were slain, Bran, Domongart, Eochaid Find, and Arthur.

With regard to the early notice, the difficulty is to decide which wall the Maeatae were close to, that of Hadrian from Solway to Tyne, or that of Antoninus Pius between Forth and Clyde. It is to be noted that Dio writes as if there were only one wall, and the evidence goes to show that the northern wall was no longer held in the time of Severus.[2] If so, the wall which Dio had in view must have been the

[1] Dio, 75, 6, 4.

[2] Dr. George MacDonald, *The Roman Wall*, pp. 381, 400.

southern one, and the Maeatae must have lived between the two walls, while the Caledonians were to the north of the wall between Forth and Clyde. To find them north of Forth in the sixth century is not, after all, very surprising when we consider the great commotions that took place in the fourth century and subsequently. Ireland has many such instances of the shifting of tribes,[1] and in Britain we have the historic case of the migration of the sons of Cunedda from the district between the walls to North Wales about the beginning of the fifth century. Causes that can only be guessed at may well have made the Maeatae settle in the country of their allies, the Caledonians, about the same time or earlier.

That the names Maeatae and Miathi are the same, only at different stages, is certain, for *Maiat-* would regularly become *Miath* in O.Ir., and in Welsh *Maead* or *Mayad*.[2] Adamnan's form is therefore taken direct from Old British into Gaelic. The theoretical Welsh form *Mayad* is interesting in view of Fordun's *Maythi*.[3] Maiatai is to be divided Mai-at-ai, with which we may compare E. Celtic *Gais-atai*, 'Spearmen,' from Gaulish *gaison*, a spear, borrowed into Latin as *gaesum* ; the Gaisatai were not a tribe, but a body of mercenary soldiers ; *Gal-atai*, 'Warriors,' from *gal*, valour, prowess ; *Nantu-atai* (*-ates*), 'Valley-dwellers,' Welsh, *nant*, a valley ; and the Irish tribe *Magnatai* mentioned by Ptolemy. The form of the name is thus thoroughly Celtic, and the parallels given afford good examples of the sort of names formed with this termination : *mai-* (? *maio-*) might denote a distinctive weapon, or a quality on which they prided themselves, or a term denoting locality of some sort. But as the meaning of the base is not known, we can only compare with the Gaulish personal name Maio-rix.

---

[1] One of the most famous is the migration of the Deisi from Meath to Leinster, Munster, and Wales in the third century (*Y Cymmrodor*, xiv. (1901)). Other instances are Corco Ché (*Celt. Zeit.*, viii. 307 *seq.*); Eoghanachta (*ib.*, p. 312 *seq.*); Ciannachta, Gailenga, Delbna, Laigsi, Fothairt (Rawl., B 502, 140 b 30).

[2] J. Morris-Jones, *A Welsh Grammar*, pp. 98 (2), 100 (vi.); Pedersen.

[3] Arturius et Eochodius Find . . . bello Maythorum trucidantur, iii. 31.

Rhys found the name in Dun-myat, an outpost of the Ochil Hills, 3½ miles north-east of Stirling, 'Fort of the Maeatae'; the anglicized form seems to reflect the Welsh rather than the Gaelic form. There is also Myothill, near Denny in Stirlingshire, a conspicuous isolated height. Rhys's further connection of the Maeatae with May Water and May Island is not convincing.

In the year A.D. 297, the orator Eumenius, in course of a panegyric on the Emperor Constantius Augustus, refers to the Britons as being already accustomed to Picti and Hiberni (Irish) as enemies. This is the earliest historical mention of Irish invasions of Britain; it is also the first mention of the Picts. We hear of them next in A.D. 310, when a panegyrist of Constantius Augustus says that the Emperor did not deign to acquire the woods and marshes of the Caledonians, the Picts, and others. Writing of the year A.D. 360, Ammianus states that savage tribes of Scotti and Picti, having broken the truce, were ravaging the parts of Roman Britain in the neighbourhood of the walls (*limites*). Some years later (A.D. 365) Pecti, Saxones, Scotti, and Atacotti harassed the Britons continually. The Picts, he says, were divided into two tribes (*gentes*), the Dicalydones and the Verturiones. The poet Claudian, in lauding the deeds of Theodosius in Britain about A.D. 364 says, 'The Orcades ran with Saxon blood; Thyle (Shetland) waxed hot with blood of Picts.' Further classical references to the Picts may be passed over for our present purpose. The oldest native authority is Gildas, styled 'the Sage' (*sapiens*), who died in or about A.D. 570, and had no love for the Picts. He describes them as coming from the North, from across the sea; savage men they were, with more hair on their villainous faces than decent clothing on their bodies. They left Britain and then returned, swarming from their curachs (*de curucis*), in which they had crossed 'the vale of Tethys' (the sea). A second time they left, and then for the first time settled down in the furthest part of the island, now and again making forays for booty and causing devastation.

Bede, who died in A.D. 735, says that the Picts, 'as is

reported,' came from Scythia in a few ships and landed on the north coast of Ireland, where the Scots would not receive them, but advised them to seek settlement in the neighbouring isle. So they made for Britain and settled in the northern parts. The Picts, he adds, came into Britain after the Britons, but before the Scots.

Nennius (A.D. 800) says the Picts came and occupied the islands called Orcades, and afterwards from these islands wasted many districts, and occupied the north of Britain, and held the third part of Britain up to the time of writing. They arrived in Orkney, says Nennius, 800 years and not less after the time when Eli was judge in Israel (*i.e.* according to modern chronology, about B.C. 300).

The Irish accounts, apart from the Irish version of Nennius, are more concerned with the ultimate origin of the Picts than with the position of their first settlements in Britain. They came from Scythia, were called Agathyrsi, and were sprung from Gelon, son of Hercules. After various adventures they landed in Leinster, helped the King of Leinster by their skill in magic to win a battle, had to leave Leinster, and so sailed northward, settling, according to one account, 'in Tiree beyond Islay,' whence they took possession of Alba 'from the bounds of the Cats to Foirchiu,' This has been explained as 'from Caithness to Forth.' *Foirchiu*, however, cannot be Forth, which we have seen to be *Foirthe* (fem.) in Gaelic. I would suggest as an emendation *Foirthiu*, and translate 'from Caithness to the Fords,' with probable reference to the Fords of Frew on Forth (see pp. 52, 85). Here *Foirthiu* is acc. pl., as in *co forthiu Máil*, 'to the fords of Mál' (LL 52 a 30), from a u-stem. It rimes with *toirthiu*, acc. pl. of *torad*, fruit, produce.[1]

The chief interest of this tale is that it explains Bede's statement that the Picts 'as reported' came from Scythia. Bede's information was from Irish sources, as might be expected. The references to Agathyrsi and to Gelonus show that the whole story was based on Vergil's

[1] See Skene, *P.S.*, pp. 30 *seqq.*, 43 (*i tir iath seach Ile* means 'in Tiree beyond Islay'); also below p. 85.

mention of 'picti Agathyrsi' and 'picti Geloni,'[1] combined with the account given by Herodotus of the descent of Agathyrsus and Gelonus from Hercules.[2] The myth appears to have started from a note in Servius' commentary on Vergil,[3] and it is quite characteristic of the way in which the learned Irish dealt with matters of which they had no real knowledge. The traditions handed down by Gildas and Nennius appear to be of independent origin and to be much nearer the truth of the matter. They receive remarkable confirmation from the evidence afforded by the distribution of the 'Pictish towers' or brochs, which are now proved to have been occupied at least as early as the time of the Roman occupation.

Of the brochs known to exist, Orkney and Shetland possess 145, Caithness has 150, and Sutherland 67—a total of 362. On the mainland south of Sutherland the distribution is approximately 10 in Ross, 6 in Inverness-shire, 2 in Forfar, 1 in Stirling (Loch Lomond side), Midlothian, Selkirk, and Berwickshire respectively—a total of 22. In the Isles there are approximately 28 in Lewis, 10 in Harris, 5 at least in North Uist, 30 in Skye, 1 in Raasay, 5 in Tiree, 1 in Lismore—in all 80. In addition there are 3 on the coast of Wigtownshire. These facts of distribution prove conclusively that the original seat of the broch-builders was in the far North, and that their influence proceeded southwards. Their main strength, during the period of broch construction at least, was in Orkney, Shetland, and Caithness, including the eastern part of Sutherland—that part, in fact, which, as we have seen, was occupied by the Cat tribe. The brochs on the western mainland are few—3 in Sutherland, 5 in Ross, 3 or 4 in Inverness (all in Glenelg), and the three outlying brochs in Wigtownshire.

The broch men evidently found the Isles much more suited to their purposes than the western mainland. That they practised agriculture appears both from the

---

[1] Vergil, *Aen.*, iv. 146; *Georg.*, ii. 115.

[2] *Herodotus*, iv. 10.

[3] Compare Giraldus Cambrensis in Skene, *P.S.*, p. 163. I have not seen the passage referred to by Giraldus.

position of many of the inland brochs and from the results of excavation. But it is equally clear that they used the sea boldly and freely, as any one who has considered the position of the island brochs will admit readily. Among the things said to have been learned from the Picts was *barc dibergi*, 'a pirate ship,'[1] and we may indeed suspect that many, if not all, of the island brochs were pirate holds. This explains the position of the Wigtown brochs, ideally placed for raiding Ireland and Britain. It is not unlikely that it was the depredations of the Picts of the Isles which gave rise to the legends of the Fomorians, whose base of operations was a tower named *Dún Balair*, 'Balar's fort,' in what is now Tory Island off the north-west coast of Donegal. The extent to which piracy was carried on in the West could be illustrated from the frequent references in early records. From the time of Labraid Loingsech (c. B.C. 200) onwards we hear of expeditions against the 'Orcs,' or men of Orkney.[2]

The Christian settlements on the west were often attacked, as at a later period they were by the pagan Norsemen. About the middle of the sixth century Comgall's monastery in Tiree was raided by Pictish robbers.[3] Pirates from the

---

[1] Poem on the Picts, Skene, *P.S.*, p. 42. For *barc* we should probably read *bairc*, 'ships.'

[2] A very old poem, ascribed to Find mac Rossa Ruaid, a royal poet of the first century, in recounting the main exploits of Labraid Loingsech says 'lamair innsi hili Orcc,' 'he ventured on the numerous isles of Orcs'; (Rawl., B 502, 115 b 13 (ed. Kuno Meyer, *Über die älteste irische Dichtung*)). The same poem (115 b 5) says, 'ort ocht turu Tiri Iath,' 'he destroyed eight towers in Tir Iath' (Tiree)—probably brochs. Of Lughaidh mac Con it is said, 'a longus co hiath Alpan. Forbrisfi cethri mórcatha for tuatha Orca,' 'he shall voyage to the land of Alba. He shall break four great battles on the tribes of Orcs' (*Celt. Zeit.*, iii. p. 461). (Lughaidh was King of Ireland c. A.D. 250). In A.D. 580 Aedán mac Gabráin made an expedition against Orcs (*fecht Orc*) (AU). Compare also the pirates of the ancient tale of *Bruigen Da Derga* (*Rev. Celt.*, xxii., xxiii.). Two expeditions by Pictish kings are recorded against the Orcs—in 682 when 'the Orkneys were utterly destroyed by Brude' (Tighern.), and in 709.

[3] 'Gentiles latrunculi multi de Pictonibus irruerunt in illam villam, ut raperent omnia que ibi erant, siue homines, siue pecora.' The 'villa' was 'in regione Heth.' (Plummer, *Lat. Lives*, ii. p. 11.)

Orkneys (*de insulis Orcadibus*) harried Gildas' retreat on an islet in the Severn estuary.[1] Tory Island, off the north coast of Donegal, was devastated by 'a marine fleet' in 612 (AU) or 617 (Tighern.). Donnan and his fifty-two companions were murdered in Eigg in 617 :[2] Columba, who knew the island Picts, had warned him not to settle there.[3] It would appear that the Picts of Orkney and of the Northern Isles remained pagan after the other Picts had become Christian. An instructive sidelight is thrown on the seafaring habits of the Picts by Tighernach, when, under A.D. 729, he records the wreck of thrice fifty ships of the 'Picardaich' off a cape called Ross Cuissini, which may have been Troup Head, inland from which are Cushnie and Little Cushnie.

As further illustrating the sphere of the operations of the 'Orcs,' I may add that an old poem states that 'the sea of the Orcs and the cold sea of the Britons' meet at Coire Bhreacáin,[4] by which is meant, not the strait between Jura and Scarba, but a whirlpool between Rathlin and the coast of Antrim. Dr. Alexander Carmichael records that the sea north-east of the Long Island (Butt of Lewis to Barra Head) was known to the old people as 'Cuan nan Orc,' the sea of the Orcs. This agrees with the statement in the Irish version of Nennius : 'the Britons originally filled the whole island with their peoples from the English Channel to the Sea of Orcs,'[5] *i.e.* the Pentland Firth, but the reference just given brings the Sea of Orcs south to the mouth of the Firth of Clyde.

In connection with the occupation of the Hebrides by the Picts we have the ancient tradition of the migration of the sons of Umor as given in the Book of Ballymote. When the Tuatha Dé Danann came to Ireland, they found the Fir Bolg, 'Bag-men,' in possession, and fought against

---

[1] Mommsen, *Chronica Minora*, p. 109 (Life of Gildas).

[2] AU (which says 'with 150 martyrs') ; Tighern.

[3] *Félire* of Oengus, 17th April, note.

[4] BB 398 b 2.

[5] Rolinsat Breatain in n-insi uile ar tus dia clannaib, o muir n-Icht co muir n-Orc. (*Ir. Nenn.*, p. 30, Ir. Arch. Soc.)

them the battle of Magh Tuireadh. 'Now the Fir Bolg fell in that battle all save a few, and these went out from Ireland fleeing from the Tuatha Dé Danann, and they settled in Aru (Arran) and in Íle (Islay) and in Rachru (Rathlin) and in Britain and in other isles besides, so that it was they who brought the Fomorians to the second battle of Magh Tuireadh. And they were in those islands till the time of the provincial kings,[1] when the Cruithnigh drove them out, and they came to Cairbre Niafear (of Leinster), and he gave them land. But they could not remain with him on account of the weight of the tribute he imposed on them. So they fled from Cairbre to the protection of Ailill and Meadhbh (of Connacht), and they gave them land. That is 'the Migration of the Sons of Umor' (*Immeirci mac nUmoir*). Oengus, son of Umor, was king over them in the east (*i.e.* in the Isles), and from them are named these lands (in Ireland), namely Loch Cime from Cime, son of Umor, and Rind Tamain in Meadhraighe from Taman, son of Umor. Dún Oengusa in Aran is from Oengus; Carn Conaill in Aidhne from Conall; Magh Adhar from Adhar; Magh nAssal in Munster from Assal, son of Umor. Meand, son of Umor, was their poet. And they were thus in strongholds and in isles of the sea round Ireland till Cuchulainn destroyed them.' Then follows a poem on the same subject by Mac Liag, who died in 1016, contained also in the Book of Leinster.[2] Elsewhere the name Umor is spelled Ugmor.[3]

Here, then, Irish tradition recognizes two early strata of population in the West, first the Fir Bolg, who are elsewhere described as a black-haired race, and second the Cruithnigh or Britons. Later the Gaelic conquest and settlement added a third stratum. The only clear trace of the Fir Bolg in Scotland is 'Dun fir Volg'—for 'Dùn Fear Bholg'—in St. Kilda, mentioned by Martin,[4] and meaning, 'the Fort of the Fir Bolg.' Another is less certain. One of the chiefs of the Fir Bolg, named Balar, figures largely in the tradition;

---

[1] *i.e.* the beginning of the Christian era.

[2] BB 30 a 22; LL 152 a.

[3] Bodleian *Dinsenchas*, no. 14 (Stokes).

[4] *Western Isles*, p. 281.

he is styled *Balar bailcbéimnech*, 'Balar of the mighty blows,' and was specially noted for his evil eye. His name is preserved in Carn Bhalair, somewhere in the very north of Ireland.[1] At Ledaig, in Lorne, there is *Dùn Bhalaire*, a site on the high rock near the vitrified fort, but on the opposite side of the public road, without trace of fortification, so far as I could make out. If this name is to be connected with the Fir Bolg warrior, as it may be, it implies a different early form, *Balarios* instead of *Balaros*;[2] it might also represent an early *Balarion*; compare Bolerion (Ptolemy), Belerion (Diodorus), '(?) place of Boleros,' the early name of Land's End in Cornwall or of some place near it.

From all this we may reasonably infer that the Picts really did settle at first in the Northern Isles, and held a position there very similar to that held afterwards by the Norsemen. Thence, like the Norsemen, they gradually extended their power on the mainland and throughout the Isles of the West, and to a less extent along the west coast. The island Picts were seamen and pirates; the Picts of the mainland were to a considerable extent agriculturists. Once they had become lords of the mainland as far south as Inverness, which they probably did at an early time, they would come in contact with the ruling tribe of the Caledonians, and in the fourth century, when the power of the Caledonians had declined, the Picts assumed the leading place among all the tribes north of the Wall of Antoninus, who had been little touched by Roman influence and doubtless considered themselves, as they did in the time of Calgācus, the noblest of the Britons. In the middle of the sixth century, in Columba's time, their capital was still in the North, and they were ruled by

---

[1] 'O Chiarraigh go Carn Bhalair,' 'from Kerry to Balar's Cairn,' occurs in the *Book of the Dean of Lismore*, p. 108 of M'Lauchlan's edition. Dún Balair was in Tory Island.

[2] The popular explanation, which is noted by Pennant, is 'Dùn Bhaile an Rìgh,' 'Fort of the king's stead,' with reference to the vitrified fort of Dùn mac Snitheachan, traditionally supposed to have been a royal residence. But the Gaelic pronunciation is *Dùn Bhalaire*.

a king whose word was law in the Orkneys.[1] While the hegemony of the tribes was held by the Caledonians, the tribes were styled collectively Caledonians and their country was known as Caledonia; when the hegemony passed to the Picts, the tribes formerly called Caledonians were called collectively Picts, and their country—from the far north to the Forth—came to be called in Latin, Pictavia. At a still later date, and for exactly similar reasons, came the further change to Scots and Scotia.

While the coming of the Scots introduced a new element into the population, there is no evidence that the advent of the Picts to power meant anything more than a change of rulers or of overlords. The Caledonians and the others remained racially what they had been. Nor is there any evidence that the Picts of the North differed in race or in customs from the Caledonians, or that the Caledonians differed from the tribes to the south of the walls as they were before the Roman conquest. On the latter point the narrative of Tacitus in the *Life of Agricola* is decisive. Among both Caledonians and Picts, as elsewhere in Britain and Ireland, the ruling race were Celts, tall, fair, or ruddy-haired men, blue-eyed and large of limb, who formed a military aristocracy. Boudicca, queen of the Iceni, as described by Dio, was a good example of the physical type,[2] and the same characteristics appear in the descriptions of continental Celts by classical writers and in the numerous descriptions of nobles to be found in Irish and Welsh literature. Under them, and subject to tax and tribute, and service of various forms, were the pre-Celtic people, forming doubtless the bulk of the population, and themselves of

---

[1] Adamnan, *Life of Columba*, ii. 43.

[2] 'The one who stirred them up most of all and persuaded them to make war on the Romans, and who claimed the leadership and acted as commander-in-chief of the whole war was Boudicca, a British lady of the royal stock, with spirit greater than a woman's. . . . She was very tall, most grim of countenance, and most piercing of look. Her voice was rough, and she had great plenty of yellow hair which came down past her waist. She wore a great twisted collar of gold, and a flowing vest of rich and varied work, and over it a thick mantle fastened with a brooch. So she went ever attired.' (Dio, 62, 2, 2.)

more than one racial origin.[1] The Celtic rulers were influenced by these in respect of marriage customs and other matters, and among the tribes north of the isthmus of Forth and Clyde this influence persisted long after its most striking effects had disappeared elsewhere in Britain and in Ireland.

We have already seen that the old Gaelic term for a native of Britain was Cruthen. At its widest it included all inhabitants of Britain; later it was restricted to the unromanized Britons north of the Wall of Antoninus Pius, while the partly romanized Britons to the south were known by the newer term *Breatan*, pl. *Breatain*, Welsh *Brython*. Thus all tribes north of the Wall were styled *Cruithne* or *Cruithnigh*, and their country was called *Cruthentuath*,[2] 'Cruthen-territory'; whereas in Welsh the corresponding term *Prydain* denotes Britain as a whole. Further, in consequence of the historical accident that the Picts became the leading tribe in that region, all the tribes north of the Wall were styled collectively Picts: the terms 'Cruthen' and 'Pict' became co-extensive and convertible, though they were by no means so originally. It is to the confusion so caused that we owe the statement that 'the Picts must have been the predominant race in Britain in the fourth century B.C.'[3] whereas in point of fact they are not even heard of till about A.D. 300. To the same confusion between 'Cruthen' and 'Pict' is due the almost religious scrupulousness with which the Cruithnigh of Ireland are nowadays styled Picts, contrary to the practice of all the older Irish writers. In Ireland, as in Great Britain, Cruithnigh is the Gaelic equivalent of Pretani: the Irish Cruithnigh were no more Picts than they were Caledonians.

As to the name *Picti*, the forms in which it has been transmitted have to be kept in view. In Old Norse it is

---

[1] For Ireland, see Professor John MacNeill's *Early Irish Population Groups*, Proc. R.I. Acad., xxix., § C, no. 4 (1911).

[2] Arran is said to lie 'idir Alpain ⁊ Cruithentuaith,' 'between Alba and Cruithen-land' (*Acall. na Senorach*, l. 332). Though this is rather a late text, the description must be old. Alba is applied to the Gaelic settlements in Kintyre.

[3] A. Macbain, Skene's *Highlanders of Scotland*, p. 383.

Pettr, in Old English *Peohta*, in Old Scots it is *Pecht*. These all suggest an original *Pect-*, and in Ammianus the term occurs once as Pecti. There are further the Welsh *Peithwyr*, meaning 'Pict-men,' and the personal name *Peithan*. 'Peith' comes from 'Pect,' like 'Gueith' from 'Vectis' (Wight), and *peithyn*, a slab or slate, from Latin *pecten*, a comb. An original 'Pictos' would yield 'Pith' in Welsh, like *brith*, speckled, from *briktos*. It would thus seem that, while the form 'Picti' is certainly Latin, it is based on a genuine native form, and we may compare the Welsh place-name *Peithnant*, of unknown meaning. There is also the Gaulish Pictones, the name of a tribe on the Bay of Biscay, south of the Loire, near the Veneti, whose name appears also as Pectones.

The name Scotti may have been, as Professor John MacNeill thinks, of general application, not designative of a special group or tribe. If they have left any trace in Ireland, it is in the name *Scotraige*, which occurs in the list of vassal septs (*aitheach aicmeda*) in the Book of Ballymote,[1] and there only, so far as known to me. Their geographical position is not stated. Scottus, like Britto, is probably the contraction of a longer form, but no certain explanation of its meaning has been given.[2]

The Verturiones were situated in the midlands of Scotland, north of the wall between Forth and Clyde. They and the southern division of the Caledonians formed the two great divisions of the Picts to the south of the Grampians. As Rhys saw long ago, their name survived in *Fortrenn*, the genitive case of a hypothetical nominative *Fortriu* (dat. *Fortrinn*), which is a gaelicized formation from *Verturion-*. Here, as so often, the Gaelic form came straight from the Old British, before the latter had passed into the stage corresponding to Old Welsh. The Welsh form would be presumably *Gwerthyr*, in the same way as *Mercurius* becomes *Merchyr*, and this is preserved by Symeon of Durham and other authorities who record that in 934 King Aethelstan wasted Scotland with a land army as far as Dunfoeder

[1] BB 255 a 5 (fcs.).

[2] For various suggestions, see Macbain's *Etym. Dict.*, 'National Names.'

(Dunottar) and Werter-morum, otherwise Werter-more.[1] The Gaelic form *Fortrenn* is not quite regular, for an early *e* before *u* usually becomes *i* in stressed syllables, as, for instance, E.Celt. *eburos*, O.Ir. *ibar*, Sc.G. *iubhar*, yew : the irregularity is probably due to the analogy of the numerous Gaelic compounds in *for-*, especially *for-trén*, 'very strong.'

Rhys further connected *Verturio* with Welsh *gwerthyr*, a fortress, comparing *Verterae*, now Brough-under-Stanmore in Westmoreland, where 'Brough' (*i.e. burh*, a fortress) is probably a translation of Verterae.

The basin of the Tay contains many ancient circular fortified dwellings built of dry stone, and resembling the northern brochs in thickness of wall and manner of entrance, but of a style of masonry inferior to that of the brochs. These circular forts or 'castles,' as they are called locally, are not confined to the basin of the Tay ; they are found on the north side of the Forth, from Dunblane westwards through the Vale of Menteith, where they are called 'Keirs,' and round about Dalmally in Argyll.[2] One of them which stands on a commanding bluff near the church of Fortingal, at the mouth of Glen Lyon, is now called *an Dùn Geal*, 'the White Fort,' and is traditionally the residence of Fionn mac Cumhaill. Fortingal is in modern Gaelic *Fartairchill*, in old records Forterkil, etc., evidently meaning 'Forter-church.' In Fife there are Forthar and Kirkforthar, with 'barrows' called 'Lowrie's Knowe' and 'Pandler's Knowe' at, or near, the former.[3] Probably 'Kirkforthar' is not to be taken as exactly parallel to 'Forterkil' ; in the latter *fortar* is adjectival, while in the former it is *kirk* which is adjectival. There is also Forter or Forther in Glenisla, Forfarshire, the site of an ancient castle. With these may be compared Ferter in Barr parish, Ayrshire, all probably Gaelic forms of a term corresponding to Welsh *gwerthyr*, a fortress.

What inferences are we entitled to draw from consideration

---

[1] Compare *Werid* for *Gwerid*, the river Forth. Wertermore, i.e. Werter Moor, corresponds to *Magh Fortrenn*, 'the plain of Fortriu.'

[2] *Proc. Soc. Ant. Scot.*, vol. xi. (Fourth Series, 1912-13), and vol. i. (Fifth Series, 1914-15).

[3] *Old Stat. Acc.*, i. p. 381.

of the material supplied by the writers of the Roman period? They may be stated shortly:—

(1) The tribal names of Scotland are Celtic, and of the same type as those of South Britain; some names—Cornavii and Dumnonii—are common to both. The name Smertae is akin to several Gaulish names.

(2) Most of the names of places and rivers are plainly Celtic, but some are probably pre-Celtic, *e.g.* Ebuda, Thule, (?) Malaios, (?) Ila. Some more names that are probably of this class will be met later. Some names are Latin—Orrea, Trimontion, Victoria; Pteroton Stratopedon is Greek. These indicate permanent occupation by the Romans.

(3) Some of the Celtic names go back, in all likelihood, to the fourth century B.C., the time of Pytheas' voyage, including Orcas and Dumna in the far North. Ebuda and Thule date from the same period. We infer, therefore, that while Celtic was certainly spoken all over Scotland in the first century A.D., strong Celtic influence was already felt in the far North in much earlier times.[1]

(4) Such evidence as we have goes to show that the Celtic of Scotland at this period was of the P-type, like Old British and Gaulish.

Two tribal names contain *p*—Epidii in Kintyre, and Picti in the North. That the Caledonians were of the P-group is indicated by the personal name 'Lossio Veda nepos Vepogeni, Caledo,' 'Lossio Veda, grandson (or descendant) of Vepogenos, a Caledonian.' The two other names of Caledonians which appear on record are Calgăcus and Argentocoxus (Swordsman and Silver-foot), both Celtic.

The consonant group *-nt-*, which appears in Gaulish and in Old British, is preserved in Welsh at the end of words. In Gaelic it was reduced to *t* (*d*) during the Early Celtic period: the reduction was already established before the period of Ogham inscriptions. In the material before us, *-nt-* appears in the tribal names Decantae (Easter Ross), Novantae (Galloway), in the

[1] If it is suggested that Pytheas' sailors, being Gaulish, made up the names themselves, the answer is that (1) the names have persisted; (2) some of the names are apparently non-Celtic.

place-name Carbantorigon (Galloway), and in the personal name Argentocoxos, a Caledonian. Here, however, it has to be kept in view that while *-nt-* is known to have been reduced in Early Gaelic, it cannot be definitely said to have been so reduced in the first century. Ptolemy's Irish names are not decisive on the point, though his river Argita, in Ulster apparently, may be from *argent-*, silver, a not uncommon element in ancient Irish names.

This evidence agrees with the probabilities of the situation. It is unlikely that the Caledonians spoke a language different from their neighbours on the south, nor does Tacitus give any hint of such a thing, though he was interested in the question and had excellent opportunities of knowing about it and referring to it.

(5) Gaelic was introduced into Scotland before the Old British language had definitely passed into the stage of Welsh. This is, of course, matter of history, and there is besides the significant fact that whenever the names recorded in this early period, with which we are dealing, have been preserved in Gaelic, the phonetics show that they were taken straight from Old British, which had not yet become Welsh. This holds for the whole of Scotland, both the southern parts, in which Welsh is historically known to have preceded Gaelic, and for the north and west. The southern names usually show double forms, Welsh and Gaelic, both direct from Old British as :—

OLD BRITISH	WELSH	GAELIC
Re-rigonion	*Penrhyn Rhionydd*	(Loch) *Ryan* : **Ríog-haine ?*
Clōta	*Clūt*	*Cluad* (mod. *Cluaidh*)
Arecluta	*Arclut*	*Ercluad*
(*Alauna*)	*Alclut*	*Ail Cluade*
Votadinoi	*Gododdin*	*Fothudán*
Lemannonios Kolpos	[*Argoed Llwyfain*]	*Leamhain*, *Leamh-nacht*
Calidones	*Celyddon*	*Dùn Chailleann*, *Sídh Chailleann*
Tav(i)a	*Tawy*	*Tóe* (mod. *Tatha*)
Maiatai		*Miathi* (Latin)
Verturiones	(*G*)*werter-mora*	*Fortriu*, *Magh Fortrenn*

To these may be added :—

**Voritia*	*Gwerid*	*Foirthe*

On the West :—

Epidion Akron	(*Pentir*)	*Ard Echde*
Dumna		*Domon*
Malaios		*Muile*
Scitis		*Scí*(*th*)
Ebuda		*Ibdach* (fr. *Ebudā-cos*)

In the North :—

Smertae		*Smeart* (gen. pl.)
Orc-ades		*Innsi Orc* (mod. *Arcaibh*)
Varar		*Farar*
Nabaros		*Nabhar*
Deva		*Dé* (gen. sing.)

There is also the ancient group, *Pretanos*: *Priten, Pryden*: *Cruthen*.

It will not be surprising, therefore, if in the course of our investigation we meet other names in Gaelic dress which are of Old British origin.

CHAPTER III

NAMES IN ADAMNAN'S 'LIFE OF COLUMBA'

DURING the four centuries following on the Roman evacuation of Britain events took place which had great and lasting influence on Scotland. Chief of these were the consolidation of the Scottish or Gaelic kingdom of *Dál Riata* in Argyll about A.D. 500, the advent of the Celtic Church, the prolonged struggle of the native rulers in the south against the Angles of Northumbria, and the coming of the Norsemen at the end of the eighth century. All these events had important effects, political, social, and linguistic, and their results are reflected in the names of places also. Something will be said later that bears more or less on the first three. In the meantime it may be convenient to complete the list of names that appear on record before Norse influence began to be felt, and for this purpose it is necessary to consider the earliest extant document known to have been written in Scotland, namely Adamnan's *Life of Columba*.

Adamnan, ninth Abbot of Hí, was born in 624 and died in A.D. 704. The oldest manuscript of the *Life* now extant was written by Dorbbene, who was elected Abbot of Hí in A.D. 713, and died in the same year. An account of it will be found in Dr. Reeves' edition of the *Life*. This manuscript, of priceless value to Scotland, lies in the Stadtbibliothek or Municipal Library of Schaffhausen in Switzerland, and no effort should be spared to recover it for the country to which it relates and in which it was written.

Adamnan writes in Latin, and the names he mentions are usually latinized in termination, but not uncommonly they are given in native form, and to Gaelic names he not infrequently adds 'Scottice,' *i.e.* 'in Gaelic.' In his time the language was in what is called the archaic stage of Old

Irish, and the forms he gives are in accordance with the very oldest literary forms. Adamnan, for instance, retains E.Celt. *ē* (=*ei*), which later broke into *ia*; also *ō* (=*ou*, etc.), which later became *ua*. He retains *i* in stressed position before *a* or *o* in the following syllable; later *i* became *e* in this position.

Besides the name of the country, which is always 'Britannia,' corresponding to *Alba* in the old sense, he has 3 or 4 tribes, 5 streams, 5 lochs, 3 bays, 15 islands, and 15 other places all in Scotland. In addition he has a number of interesting personal names.

In connection with Britannia, Adamnan has occasion four times to mention 'Britanniae Dorsum,' which is his translation of *Druim Alban*, 'the ridge of Alba'; he defines it as the mountains which divide the Picts from the Scots,[1] for by his time the Scots had all the West. In Britain he distinguishes four peoples—the *Britones* (sing. *Brito*), the Britons, or Welsh of Strath Clyde; the *Scoti Britanniae*, Scots of Britain, *i.e.* the Dalriadic Scots or Gael, as distinguished from the Scots or Gael of Scotia (Ireland); the *Picti*,[2] east of Dorsum Britanniae, and the *Saxones*, by which term he means the Angles of the south-east of Scotland.

As tribes of the Scots he mentions 'genus Gabrani' (*cenél nGabráin*), and 'genus Loerni' (*cenél Loairn*), the tribe of Gabrán and the tribe of Lorne, which will come in their place later. In Skye, Columba met and baptized a man called Artbranan, described as 'primarius Geonae cohortis,' leader or captain of the *Geona* band. 'Cohors' is a military term, and might be used to translate the Gaelic *cath*, 'a battalion': there were traditionally seven battalions of the Fiana of Fionn mac Cumhaill (*seacht catha na Féine*). It is not improbable that the 'Geona cohors' was a body of warriors corresponding to the Irish Fian,[3] and that the term 'primarius' corresponds to *rígféinnid*, 'Fian-chief' (lit. 'king fian-warrior'). There were, of course, 'Fiana' in Britain as well as in Ireland,

[1] *Vita S. Columbae*, ii. 47.

[2] Adamnan reserves 'Picti' strictly for the Cruithne of Alba.

[3] See Kuno Meyer's *Fianaigecht* (Todd Lecture xvi.).

i.e. bands of warriors under one leader, whose relations to the territorial rulers depended largely on their respective strength.[1] Adamnan writes as if the 'Geona cohors' needed no explanation; to us, 'Geona' conveys no meaning, though it may have been a tribal or a territorial term.[2]

The Miathi have been already mentioned (p. 58).

The five streams named by Adamnan are Dobur Artbranani, Aba, Sale, Lochdiae (gen.), Nesa. The first, 'Artbranan's Water' was in Skye, at the spot where Artbranan received baptism. The only place-name in Skye that involves *dobur*, so far as known to me, is Tot-arder in Bracadale, where 'arder' is for *ard-dobhar*, which means, as it stands, 'high-water.' The place is on the sea coast, and so far agrees with Adamnan's description, but the data are insufficient for identification. This is the only instance of *dobur* known to me in the islands; on the mainland, as we shall see, it is common.

In 'Stagnum fluminis Abae,' the lake of the river Aba, we have a latinized form of O.Ir. *ab* (also *aub*, *oub*), a river, gen. *aba*; the river here is the Awe, *Abhainn Abha* (pronounced A'a), from Loch Awe, *Loch Obha*. What is the reason for the difference in vowel I cannot say, but it appears in other instances of *abh*. At the east end of Loch Awe is *Dùn Abhach*, Awe Fort, and a loch on the north-west side of Loch Awe is *Loch Abhaich*, Loch of Abhach, anglicized 'Avich'), *i.e.* Stream-place. There is another *Loch Abha* (Loch Awe) in Sutherland.

Along with the river Sale it will be convenient to take the isle which Adamnan calls Airthrago.

The monastery of Hí needing repairs, oaks had to be taken from the mouth of the river Sale. As the sailors

[1] In the 'Instructions of Cormac' (*Tecosca Cormaic*), among the things that are best for the good of a tribe is 'warrior bands without overbearing' (*fianna cen diúmmus*).

[2] Rhys equated 'cohors' with Ir. *dál*, a division or part, 'frequently used in forming ethnic names.' He further equates it with μοῖρα and suggests that 'Geona cohors' may be a defective spelling of *Genona*, and that the term is equivalent to the Γενουνία μοῖρα of Pausanias (*Celtic Britain*, 1904, p. 91). The 'Genunian Division' appears to have been in the north or midlands of England.

rowed out on their return journey, a west wind sprang up, and they made for the shelter of the nearest island, called in Gaelic (*Scottice*) Airthrago. At the pleading of the monks a most favourable south-east wind blew, which enabled them at once to put to sea. Skene identified the isle of Airthrago with Kerrera, near Oban. Sale he identified with a brook, nameless on maps, which flows into the Sound of Seil. But Adamnan could not with propriety refer to this brook as a river, nor could he refer to its mouth (*ostium*) as a point presumably well known. Further, this Seil is in Gaelic *Saoil*, the Isle of Seil, and appears in the Irish records as *Sóil*, which Adamnan would have written Soil. Adamnan's Sale is the O.Ir. form of *Salia*, a river name which appears as a tributary of the Moselle and elsewhere. Its modern form in Gaelic is *Seile*, which applies to two rivers in Scotland, the Shiel of Moidart and the Shiel at the head of Loch Duich. Adamnan mentions the river Sale elsewhere as *piscosus*, full of fish ; his companions, who were keen fishers (*strenui piscatores*) caught salmon there without nets. We have seen that it was noted for oaks. This is beyond doubt the Shiel of Moidart, of which Alexander Macdonald says :—

> 'Is lìonach slatach cuibhleach breacach
> Seile ghlas nan samhnan.'

'Good for net and rod, for reel and for fish, is green-hued Shiel of the big trout.' Here *breacach* is exactly *piscosus*. In another poem, entitled *an Airc*, 'the Ark,' the instruction is given :—

> 'Dèan àirc de dharach Locha Seile.'
> ('Make an ark of the oak of Loch Shiel.')

The Isle Airthrago, in whose lee the party sheltered, is now Shona, from which the south-east wind brought them along nicely to the point of Ardnamurchan, and so by the west of Mull to Iona. Being no seaman, I have consulted an expert on the point, and he assures me that the wind indicated is quite favourable for the voyage. The name of the isle, which is expressly said to be Gaelic, is the genitive case of *airthráig*, 'foreshore': 'insula quae Scottice vocitatur Airthrago' means literally, 'the isle which is called in Gaelic the isle of the foreshore.'[1] Here, then, we have the pre-Norse name of Shona.

The river in Lochaber 'which may be called in Latin *Nigra Dea*,' black goddess, is undoubtedly that mentioned in the headings of chapters of Book I. as 'stagnum Lochdiae,' which, as has been pointed out already, is the modern *Loch Lòchaidh*, with its river *Lòchaidh*.

The Ness occurs four times, twice in the accusative *fluvium Nesam*, twice in the genitive *fluminis Nisae*. The English form Ness represents the old nominative which is now lost in Gaelic; we have now only the genitive form, in the terms *Inbhir Nis*, Inverness, *Loch Nis*, Loch Ness, *Abhainn Nis*, the Ness. In a poem preserved by the Dean of Lismore, the genitive is *Nise*, correctly.[2] Adamnan's nominative would have been *Nesa*, fem. latinized from *Nes*, fem., which stands for an early *Nesta*. This gives a genitive *Nise*, in the same way as Lat. *cella* gives nom. *cell*, gen. *cille*. The river Nestos, in Thrace, has been compared, but the resemblance may be accidental. It may be noted that Adamnan mentions a man called Nesan who lived in Lochaber, whose name stands for *Nesagnos*, 'Ness-sprung.' There is here no necessary connection with the river Ness as such: Nessán is a well-known Irish name, and we find 'Nissi and Neslug, sons of Nessan,' among the Irish saints.'[3]

Of the five lochs named by Adamnan, three are named after the rivers that issue from them—the lake of the river Aba (Loch Awe), of the Lochdia (Lochy), and of the Nesa

---

[1] In modern Gaelic the name would be *Earraigh* (for *Airthraigh*), and it is the sort of name we should expect to find elsewhere. I know of no instance, however, except perhaps *Eilean Earraid*, off the coast of Mull, known to all from *Kidnapped*, which Dean Munro calls Erray.

[2] 'Gi glwnnyn is me in nynvir nissa,' *i.e.* 'go gcluininn is mé a nInbhir Nise'; *Rel. Celt.*, i. 91. As this is the first line of a Séadna couplet, the last word must be a dissyllable.

[3] Rawl., B 502, 90 b 50; compare Mac Nissae, *Thes. Pal.*, ii. p. 284; Mac Nise, *ib.* p 365; Ness . . . o Chill Nessi, L.Br., 22 b 7; LL 353 c.; Mac Nise, L.Br. 22 b 32; Mac Nisse, *ib.* 22 b 24. Conchobar's mother's name, Ness, has gen. Nessa, a *u*-stem; the river name appears to be an *a*-stem.

(Loch Ness). The others are Stagnum Aporicum or Stagnum Aporum and Stagnum Crog reth.

Skene's proposed identification of Crog reth with Loch Creran in Appin cannot stand in view of the Gaelic *Gleann Craibhrionn*, now *Gleann Creurán*,[1] and the spellings Logh-creveren, Bra-glen-crevirne in Macfarlane.[2] *Crōg* stands for early Celtic *crouc-*; G. *cruach*, a hill, O.W. *cruc*, W. *crug*; compare Adamnan's *Petra Clōithe*, later *Ail Cluade*, W. *Alclut*; Rodercus filius *Tōthail*, later *Tuathal* (nom.), O.W. *Tudgual*, W. *Tudwal*. *Reth* may be compared with O.Ir. *raith*, fern, bracken, later *raithneach*; W. *rhedyn*, fern; O.W. *redinauc*, ferny; O.Corn. *reden*; Gaulish *ratis*. Thus 'Stagnum Crog reth' would mean 'the loch of bracken hill,' and the absence of declension indicates that the term is not Gaelic.[3] The hill in question can only be *Cruach Raithneach*, probably the range of hills marked on the maps as *A' Chruach*, some distance to the west of Loch Rannoch and forming part of the boundary of the district of Rannoch. Early Christian influence on Loch Rannoch side is proved by the names *an Annaid*, the Annet, near the east end, and *Cill Chonnáin*, Killichonan, near the west end, both on the north side of the loch. *Connán* is a diminutive of *Conn*, and is not to be confused with *Conán*. His *díseart* (*desertum*, whence Dysart) was at Dalmally, and was the burial-place of the MacGregors.

Stagnum Aporum means 'the loch of the *apors*,' whence it is also called 'the *aporic* loch' (*stagnum Aporicum*). Macbain referred *apor* to Ir. *abar*, Sc.G. *eabar*, a puddle; but it is much more satisfactory to take it as meaning 'confluence,' as in Apor-crossan, the ancient name of Applecross. A Lochaber man is *Abrach*; Applecross was *a' Chomraich Abrach*. The confluences are those of the Nevis and the Lochy, both of which enter Loch Linnhe close together. Thus Loch Abar is really the old name for what is now

---

[1] Adv. Lib. MS., lx. (Mackinnon's *Catalogue*, p. 64).

[2] *Geog. Coll.*, ii. pp. 154, 516, 596, 536.

[3] Compare 'insulam quae Scottice vocitatur Airthrago,' 'the island which in Gaelic is called (the isle) of Airthráig'; *Life of St. Columba*, ii. 46. Here Adamnan gives the Gaelic term in the genitive.

called in Gaelic *an Linne Dhubh*, 'the Black Pool.' The English 'Loch Linnhe,' it may be noted, is a 'ghost name.' The part of the sea-loch outside the Corran of Ardgour is in Gaelic *an Linne Sheileach*, and *seileach* was explained to me locally as 'brackish,' applied to salt water which has a large admixture of fresh water. This may be correct, and if so it is to be taken in connection with the rivers called *Seile*, which we have been discussing.

The bays mentioned are Muirbolc Mār, Muirbolc Paradisi, and Aithchambas in Ardnamurchan.

Muirbolc is from *muir*, sea, and *bolc*, now *bolg*, *balg*, 'a bag,' and means 'sea-bag,' applied primarily to a rounded sea-inlet. In Ireland a small bay is often called *murbholg*.[1] Muirbolc Paradisi was a port to which seamen from Ireland put in. On a certain day Columba and his companions were travelling in Ardnamurchan, when they met and talked with sailors just arrived from Ireland at Muirbolc Paradisi. The place was therefore in Ardnamurchan or very near it. Skene made it *Port nam Murlach* in Lismore, which is not satisfactory either geographically or phonetically. *Murbholg* becomes in Scottish Gaelic *mur'lag*, not *murlach*, which latter means a dogfish; *Port nam Murlach* is rightly explained locally as 'Port of the Dogfish.' It is important to consider what Irish word Adamnan translates by 'Paradisus,' the natural meaning of which is 'garden.' The garden of Eden is in early Irish *pardus*, or more specifically *pardus Adaim*, but the word in ordinary use for a garden was *lubgort* (*lub-gort*, herb-garden), and an old writer remarks 'fitting it was that Jesus should be laid hold of in a garden, for it was in a garden (*lubgort*), *i.e.* in paradise (*i pardus*) that Adam was seized.'[2] As *pardus* was not likely to be used in a place-name, it is most probable that the word Adamnan is translating is *lubgort*. Now of the many bays around the Ardnamurchan coast, the most likely to be called a *muirbolc* is Kentra Bay on the north side, not far from Acharacle, and on the south side of this fine

[1] Joyce, *Irish Names of Places*, ii. p. 255.
[2] Atkinson, *Passions and Homilies*, l. 3020.

bay there is a township called *Gortan-eorna*, 'the barley-field,' in which *gortan* is the diminutive of *gort*, 'a field, a garden.' This is quite near the river Shiel, which is known to have been visited by Columba. The identification, however, remains conjectural.

In modern Gaelic *muirbolc* has become *murbhlag, mur'lag*; the indeterminate parasitic vowel heard between *l* and *g* of *balg* made the word practically three syllables—*murbhalag*, which became *murbhlag* by syncope. Its diminutive is *murbhlagan, mur'lagan*. Like some other sea terms, these have worked their way inland, and have come to be applied to bays or bights on fresh-water lochs or even rivers. Only two instances of *murbhlag* are known to me. On Loch Earn in Perthshire there is *Ard-mhur'laig*, 'promontory of the sea-bag,' in English Ardvorlich. The promontory is small and so, by consequence, is the bay. Behind it is *Beinn Mhur'laig*, 'peak of the sea-bag'; in Blaeu's MS. maps they are Ardvouirlig and Binvouirlyg.'[1] Both names are repeated on the western side of Loch Lomond—Ardvorlich, Ben Vorlich. The former is Ardiŭŭrlik in 1543,[2] Ardvurlig in a MS. list of names in Arrochar, c. 1800: there is no doubt that the names are the same as those on Loch Earn. There is a small bag-like bay at the western Ardvorlich.

Murbhalgan is more common; it appears now in English form as Murlagan. The only instance of its occurrence on the sea coast is Murlaggan on the eastern side of Loch Long. At the south end of Loch Ness, 'half a myl from Glendo Beg is Mourvalgan upon the said burn of Do';[3] it is Murvalgan in 1770;[4] now Murlagan. Another instance is on the right bank of Spean above Roy Bridge; Murvalgane 1476 (RMS), Moirvalagane 1574, *ib.*, Murlagoun 1552, *ib.* Murlagan on Loch Arkaig is Murlagen 1552 (RMS), Moirvalagane

[1] The Gaelic form given above was given me some years ago by all the older people of the district; among the younger people *Beinn Mhurlaich* was common.

[2] Fraser's *Chartulary of Colquhoun*, p. 391, where the name is printed 'Ardmurlik,' but the MS. reads as above.

[3] Macfarlane, *Geog. Coll.*, ii. p. 556.

[4] *Forfeited Estates Papers* (Scot. Hist. Soc.).

1574, *ib.* In Rannoch there was Murlaggan 1597, *ib.*, Murelagane 1619, *ib.*, Murrullagan in Macfarlane,[1] a spelling which reflects the Gaelic pronunciation very well; 'Mount Alexander or Murlagan *c.* 1751.' The name is now Dun Alastair, just opposite a bag-like bend of the Tummel. In Glen Lochy, near Killin, there is Murloganemore, 1528, *ib.*; in Macfarlane it is Murrulagan (ii. p. 536), now Murlagan, on a bend of the river Lochy. In Balquhidder there is Murlagan on Loch Voil; Murlagane, 1587. In the Trossachs district a writer in Macfarlane says, 'half a myl from Murlagan is Achrai, and 3 myl therfra Keandrochart upon the southeast end of Ardkeanknoken Loch' (ii. p. 567); these other places are now Achray, in Gaelic *Ath-chrathaidh*, 'ford of shaking,' Brig of Turk, and the site of the Trossach-Hotel, in Gaelic *Ard Cheannchnocain*, 'height of head-hillock.' This Murlagan seems to be now obsolete; it was probably the name of the bay at the north-west end of Loch Achray, as the loch is now called.

Muirbolc Mār means 'the big sea-bag'; it was on the island Hinba. Skene, following Reeves, identified Hinba with *na h-Eileacha Naomha* (which he calls wrongly 'Eileann na Naoimh'), off the south-east coast of Mull, and most writers have accepted his identification. There are, however, reasons which are decisive against it. In the first place, there is nothing on this islet which could by any stretch of courtesy be called a large bay or inlet. In the next place, *na h-Eileacha Naomha* was beyond reasonable doubt the site of Brendan Moccu Alti's monastery of Ailech, founded probably some time before Columba had come to Iona. *Ailech*, from *ail*, a rock, means 'a rocky place,' which accurately describes this islet [2] and also the other islet *Garbh-eileach*, 'rough, rocky place,' which gives its name to the little group (Garvellach). Next to *na h-Eileacha Naomha*, 'the Holy Rocks,' is *Cùil Bhrianainn*,

[1] *Geog. Coll.*, ii. p. 597.

[2] Compare the passage in which Brendan, while in Ailech, is said to have been praying near the sea on a lofty rock (cum solus oraret prope mare in rupe eminenti). Plummer, *Lat. Lives*, i. p. 143. (In this text Ailech is given as 'Auerech.')

'Brendan's Retreat' (*secessus Brendani*). There is thus more than a probability that the islet was known as Ailech in Columba's own time, and it is most improbable that Adamnan would call it by another name. Lastly, Columba founded a monastery on Hinba not very long after his coming to Iona, over which he placed his uncle Ernan as superior. Skene thinks that this foundation may have been a restoration of Brendan's monastery of Ailech, but while the site of Ailech is one that might be supposed to have attracted a man of Brendan's type, it is not the kind of place that Columba would favour: it is, in fact, fitted for a penitential station rather than for a self-supporting community such as Columba's monasteries were. That there was a religious establishment on *na h-Eileacha Naomha* is certain from the remains of *clocháin* (stone huts) and the very interesting old cemetery and buildings; the name, too, confirms it.

For Hinba, in my opinion, we must look among the isles which now have Norse names, and whose old names are therefore lost. It is not likely to have been a small island. It was distinguished by a large 'muirbolc.' It was not very far from Iona, for Columba often visited it. It appears to have lain in the track of vessels coming from Ireland, at least that seems to be the inference from the fact that Comgell, Cainnech, Brendan, and Cormac, coming from Ireland to visit Columba, found him in Hinba. The choice seems to lie between Colonsay and Jura. At the south end of Colonsay is Oransay, separated by a channel fordable at low water, and with extensive remains of a medieval priory. It is on Oransay that Columba is traditionally said to have landed first with the intention of settling there. But finding on ascending the little hill that he could see Ireland thence, he went on to Hí, and so the hill still bears the name of *Cùl ri Éirinn*, 'Back to Ireland.'[1] The channel between Colonsay and Oransay widens into a broad bay, with a bag-like horn on the north side about half a mile wide at the mouth. The island of Jura is almost bisected by the deep inlet on

[1] Blaeu has 'Karn culri Allabyn' and 'Karn culri Erin' at the head of Loch Laffan, now Loch Scriodain, in Mull.

its western side, known as Loch Tarbert. At some distance from its mouth the loch contracts, and then widens, thus forming an ideal *bolc* or bag. On its shores are well-known and spacious caves, among them *an Uaimh Mhór*, 'the big cave,' where it was usual for gentlemen who were hunting to stay. Another bears a name very relevant to our purpose, *Uaimh mhuinntir Idhe*, 'the cave of the folk of Hí.' This at once recalls the circumstance, related by Adamnan, that a certain monk, called Virgnous, lived an eremetical life for twelve years on Muirbolc Mār. The parish church, on the eastern side of the island, is called by Dean Monro 'Killernadail,' which agrees with the present pronunciation in Gaelic as heard by me, viz. *Cill Earnadail*. A later spelling is Kilaridil (Blaeu), and *Cill Earradail* is also heard in Gaelic now. The O.S.M. has 'Killernandale.' The name certainly looks as if it meant Ernan's Church, with the addition of Norse *dalr*; we may compare Navidale in Sutherland, in Gaelic *Nei'adail*, 1563 Nevindell, which is for *neimhidh*, 'sanctuary,' with *dalr*, also Kilquhockadale in Galloway. Norse-Gaelic hybrids have to be treated with caution, but some of this type are quite intelligible and authentic: the Norse newcomers sometimes adopted the old name, adding their own generic term. Local tradition has it that Earnadail or Earradail lived in Islay, and left instructions that his body should be conveyed to Jura, and borne onwards until a patch of mist should be seen, and buried in the spot where the mist appeared. His body was landed at *Leac Earradail*, and buried in the parish graveyard at Killernandale. With this tradition may be compared Adamnan's account of Ernan's death: falling ill in Hinba, Ernan was taken at his own request to Hí to Columba, but before coming into Columba's presence, he died between the harbour and the monastery. Here again it cannot be claimed that there is sufficient proof for absolute identification, yet the claims of Jura seem strong.[1]

[1] Mr. Duncan Shaw, a native of Jura, informs me that the island is traditionally known as *an t-Eilean Bán*, 'the blessed isle,' 'the holy isle'—a well-known secondary meaning of *bán*, as also of *fionn*.

As to the name Hinba itself, it is readily explained as a latinized form of O.Ir. *inbe*, 'an incision'; it is once called 'Hinbina Insula.' This again suggests the deeply indented isle of Jura; for the meaning we may compare Eigg.

The island Colosus is mentioned twice. A robber named Ercus (*Ercc*) came from Colosus to Mull by night, to steal seals off a little isle near the coast of Mull. This does not settle the position of Colosus, but the other passage is more helpful. Columba being in Ardnamurchan met there a persecutor of Columbanus, one of Columba's friends, and as the robber was making off with his booty in a boat, Columba, entering the clear green sea water to his knees, appealed to Christ with lifted hands. He then told his companions that the boat would not reach land, and that the crew would be drowned. 'After the lapse of a few moments, even while the day was perfectly calm, behold a cloud arose from the sea, and caused a great hurricane, which overtook the plunderer with his spoil, between the Malean and Colosus islands (inter Maleam et Colosum insulas), and overwhelmed him in the midst of the sea.'[1] A glance at the map will show that the islands between which the robber was wrecked were Mull and Coll,[2] not Mull and the larger Colonsay, as Reeves and Skene made it. Colonsay is in modern Gaelic *Colbhasa*; MacVurich spells it *Colbhannsaigh*; fourteenth century *Coluynsay*; Dean Monro *Colvansay*; whence it is certain that the name is Norse *Kolbeins-ey*, Kolbein's Isle, a name which recurs in Landnámabók. Colosus, then, is Coll, now in Gaelic *Cola*, with *o* open, as in *ola*, oil. Here Adamnan, unlike his usual custom, gives the actual name of the island, not an adjective derived from it; similarly he writes, 'de insula Coloso' and 'in insula Coloso.' If the name is Celtic, the *s* in the body of the word requires explanation, for Celtic intervocalic *s* had disappeared long before Adamnan's time, and when *s* final or medial occurs in O.Ir., it represents an older

---

[1] Skene's translation, p. 52.

[2] So, too, Stokes and Strachan: 'identified by Reeves with Colonsay, but seems rather Coll.' *Thes. Pal.*, ii. p. 276 *n*.

group, *e.g.* *cos* for *cox-a*, foot; *seas* for *sist-o*, stand. But Colosus may be a pre-Celtic name.

Egea insula, 'the Egean Isle,' is Eigg, in Gaelic *Eilean Eige.* It occurs often in the older Irish literature in connection with names of saints, but only in the genitive, so far as I have noted: that it was feminine, is proved from *Donnán Ega huare*, 'Donnan of cold Eigg,' in the *Féilire* of Oengus. Examples are: Combustio Donnáin Ega, the burning of Donnan of Eigg (AU 617); Oan princeps Ego (moritur), Oan, superior of Eigg dies (AU 725); Cummene nepos Becce religiosus Ego, Cummene, descendant (or grandson) of Becc, the devotee of Eigg, died (AU 752). Other clerics connected with Eigg were Berchan Aego (LL 358 f 5); Congalach ó ard Aego (Mart. of Don.; Fel. Gor.); Conán Ego (Fel. Gor.); Enán insi Ego (*ib.*). An article on the martyrdom of Donnan (LL 371 b) begins: 'Ega nomen fontis,' 'Eigg is the name of a fountain'; this curious bit of information is doubtless to be explained by reference to a gloss in the Book of Armagh, which uses *enga* to explain 'aqua supra petram, *i.e.* fons,' 'water above a rock, that is, a fountain.' It shows that the writer recognized the word to have *gg*, which in Greek corresponds to Latin *ng*; hence *Egga* explained as *enga.* This note is mentioned merely as a curiosity; the name is Ir. *eag*, fem., gen. *eaga*, *eige*, 'a notch'; Sc.Gaelic *eag*, fem.; gen. *eig*(*e*); *Eilean Eige* means 'the Isle of the Notch,' with reference most probably to the marked depression that runs across the middle of the island from Kildonan to Bay of Laig. There is another *Eilean Eige*, much notched, off the coast of Arisaig.

Ethica Terra, 'the land of Eth,' also called Ethica insula, 'the isle of Eth,' is Tiree; the sea between Tiree and Hí is called Ethicum pelagus. In other Latin Lives of the Saints it is called Heth, Heth regio, terra Heth. In Irish literature the name occurs, so far as I have noted, only twice, once in the Book of Ballymote (205 a 11), where the Cruithne are said to have gone from Ireland 'to Tīr-iath beyond Islay' (*i tīr iath seach Īle*), and again in Rawl. B. 502, 115 a 5, where Labraid Loingsech is stated to have razed eight towers in Tiree (*ort ocht turu Tīri iath*), both already referred to. The poem in Rawl. is very old. The fact that Adamnan's *Eth-* becomes in O.Ir. *iath* proves that the *e* of *Eth-* is the long E.Celtic *e* (for *ei*), which is retained in the very oldest specimens of O.Irish and in some Irish names in the Latin Lives of the Saints, but which by A.D. 800 had been broken to *ia* when it was followed by a broad vowel in the next syllable. Thus *ēth* became *iath*, as *Cēran* (the saint's name) became *Ciaran*. This, as Kuno Meyer has pointed out, is fatal to Dr. Reeves' derivation from O.Ir. *ith*, gen. *etho*, corn, attractive as it is in view of Tiree's proverbial richness in barley. In the twelfth century Reginald of Durham has *Tirieth*. Subsequent forms are *Tiryad*, 1343; *Tereyd*, 1354; Tyriage, 1390; Tyree (Fordun, *c.* 1385); Tyriage, 1494; Tiereig, 1496. A learned Gaelic poet of the sixteenth century writes *Tír igedh* (*i.e. Tír-igheadh*). In Gaelic now with the people of Tiree it is *Tireadh*; outside the island it is *Tìr-idhe* or *Tìr-ithe*; but one also hears *Tìr-idheadh*, with a distinct stress on *Tir*, which is very like the form used by the sixteenth-century poet. I have also heard *Tìr-iodh*, or *Tìr-eadh*. The Norse form was *Tyrvist* (compare *Ivist*, the Norse form of Uist), but *-vist* can have no phonetic relation to O.Ir. *iath*.[1] The modern *Tiristeach*, a Tiree man, must come from this Norse form. The modern *Tìr-iodh* may be compared with *Magh-iodh* or *Mag-eadh*, near Crieff, anglicized as Monzie; in *Tìr-idhe*, *Tìr-ithe*, the second part is the same in sound as *Idhe*, *Ithe*,[2] the genitive of *Ì*, Iona. The second part of the old Gaelic *Tír-iath* is the same in form as *iath*, a district, region. But the various forms, ancient and modern, cannot be reconciled with each other, and their diversity indicates that the second part is not Gaelic, possibly not even Celtic.

Ilea insula corresponds to our modern *Eilean Ìleach*, *Ilea*

---

[1] O.Norse *vist* means (1) an abode, domicile; (2) food, provisions, viands. *Tyr-vist* seems to point to a folk etymology of *Tir-ithe* as from O. and M.Ir. *ithe*, act of eating, translated by *vist*.

[2] In O.Gaelic *Ie*; the *dh* or *th* of *Idhe* is not organic, but merely a phonetic device for separating the syllables.

being an adjective formed from *Ile*. The name, anglicized as Islay, occurs often in Irish records and literature, usually as *Ile*, sometimes *Ila* ;[1] the modern Gaelic form is *Ìle*, which would represent an old *Iliā*. If the name is Celtic, it might be compared with Gaulish *Ilio-mārus*, a man's name, in which *Ilio-* may be some part of the body, like Latin *īlium*, *īlia*, the flanks of a man or animal : *Ilio-māros* would thus mean 'big-flanked' or perhaps 'big-buttocked.' The Welsh verb *ilio*, 'ferment,' may be compared : the root notion is 'swell.' The peculiar shape of the island lends itself to such an origin for its name ; compare the Irish *Áru*, the isle of Aran, from *áru*, gen. *árann*, 'a kidney.' Ptolemy's river Ila, with initial short vowel (as is proved by the modern Gaelic form) is probably of different origin ; the only names known to me that resemble *Ìle* are *Làirig Ìlidh*, the name of the pass on the northern side of Glen Ogle, Loch Earn, where *Ìlidh* is probably the name of the stream, and *Srath Ìl'*, Strath Isla in Forfarshire.

The island of Hí (Iona) is always in Adamnan *Ioua insula*, whence by a misreading of *u* as *n* has come the popular form Iona. It is likely that the error gained currency, if it did not originate, from the remark of Adamnan on Columba's name, which he says is in Hebrew Iona (Jonah), a dove. The name is difficult and has been discussed more than once. There are two sets of old forms, which had better be taken separately. Adamnan's *Ioua insula* may be taken with practical certainty as formed in his usual style, like *Egea insula*, *Scia insula*, *etc.* : *Ioua* is an adjective formed from the name of the island. This is the form used by Cummine Ailbe, *c.* A.D. 660.[2] With this form go (1) *Eu* of Leabhar na h-Uidhri ;[3] (2) the adjective

---

[1] *E.g.* 'is fuar in gáeth tar Ile,' 'cold blows the wind across Ile' ; rimes with *Cind-tíre* (*Tig. Ann.*, A.D. 624) ; 'Firbolg umorro rogabsad Manaind 7 rogabsat alaile indsi archeana . i . Ara 7 Ila 7 Recca,' 'the Fir-bolg took possession of Man and of other islands besides, Arran and Islay and Rathlin.'

[2] *Vita S. Columbae.*

[3] 'In tan conucaib a chill hi tosuch . i . Eu,' 'when he raised his church first, *i.e.* Eu.' LU 11 b 38.

Eoa in Ann. Ulst.;[1] (3) *Eo* (nom. two syllables) of Walafridus Strabo, *c.* A.D. 831;[2] also found once in Tigernach; (4) *Euea insula* in the *Life of St. Cadroe.*[3] For the phonetics of these we may compare O.Ir. *beo, beu,* 'living,' from E.Celt. *bivos*; here E.Celt. *i* becomes *e* before *o* of the next syllable (like *viros* : *fer*, 'man'), while *v* disappears between vowels. Both these changes took place subsequent to the Ogham period; Adamnan, as often, uses the old vowel: the change from his *Ioua* to *Eoa* indicates that we have to do with an earlier *Ivo-*, which later became *Eo, Eu* as *bivo-* became *beo, beu.* Now *Ivo-* is an element which occurs in a number of Celtic terms: in Oghams[4] it appears as *Iva-* in *Ivacattos* (gen.), M.Ir. *Eochado*; *Iva-geni* (gen.), Adamnan's *Iogen-anus, Eugen* in AU A.D. 667, later *Eogan.* In Gaulish it appears in *Ivo-magus, Ivo-rix.* It is in fact the early form of O.Ir. *eo*, a yew-tree. Adamnan's adjective *Ioua*, however, seems to go back, not exactly to *Ivo-*, but to a derivative *Ivova*, which might mean 'Yew-place,' with which we may compare the Gaulish *Ivavos*, the local god who was the genius of the healing wells of Evaux in France. It was natural to connect the longest-lived of trees with health and long life. 'Patriarch of long-lasting woods is the yew, sacred to feasts as is well known.'[5] Irish literature has a good deal to say about a personage named *Fer hÍ mac Eogabail.* He is described as foster son of Manannan mac Lir, the Celtic sea-god, and druid of the Tuatha Dé Danann, who were ancient gods of the island Celts. Manannan sent him to fetch a certain lady from Ireland.[6] Fer hÍ took her

---

[1] 'Pascha commutatur in Eoa civitate.' AU 716.

[2] 'Insula Pictorum quaedam monstratur in oris/fluctivago suspensa salo, cognominis Eo.' 'On the coasts of the Picts is pointed out an isle poised in the rolling salt sea, whose name is Eo.'

[3] Skene, *P.S.*, p. 108.

[4] See Professor John MacNeill, *Notes on Irish Ogham Inscriptions.*

[5] S. H. O'Grady's translation, 'Sinnser feda fois/ibar na fled fis.'—*Silv. Gad.*, i. p. 245.

[6] Ainm in techtaire cathaig/ba Fer Fí mac Eogabail
dalta do mac Lir na lend/drúi de Tuathaib Dé Donand.

"The name of the warlike envoy was Fer Fi, son of Eogabal, foster son of Manannan of Mantles, a druid of the Tuatha Dé Danann." LL 152 b 21; also BB 395 b 40; Stokes, *Bodleian Dinnsenchas*, p. 43.

from Tara to the mouth of the river Bann, in Ulster, where she was drowned while he was looking for a boat in which to carry her across the sea. This tale, so far as it goes, connects Fer hÍ with the West. Another tale connects him with Munster, where, through his gift of music, he brought about a quarrel among the ruling family which had serious results.[1] Here he is represented as closely connected with a yew-tree, which indeed he had formed himself through his magic. The point of interest is that Fer hÍ means 'Man of Yew,' or better, 'Son of Yew'; his father's name, *Eo-gabal*, means 'Yew-fork.'[2] He appears to have been a tree, or rather a yew, divinity. The inference to be drawn from the whole data, is, in my opinion, that the 'Iouan island,' otherwise *Eo* or *Eu*, means 'the Yew-isle,' and that it may well have been the seat of a yew cultus, of which we may possibly have a trace in the legend of Fer hÍ, the foster son of Manannan, whose home was in the Western Isles. In this connection it is relevant to note the tradition of the Irish *Life* that Columba found druids before him in Hí, and expelled them.[3]

Alongside of the forms we have been considering there are nom. *Í*, *Hí*, *Ia*; gen. *Ie*, *Iae*; dat. *Í*, *hÍ* (once with the article:[4] *a breo dond hI*, 'thou flame from Hí'; LL 194 b 34). It is now nom. *Í*; gen. *Idhe*—*dh* being used merely to separate the syllables; a man of Iona is *Idheach*; the sound between Mull and Iona is *Caol Idhe*. These forms point to an old *Ivia*, a shortened or reduced form of *Ivova*; compare *Britannia*, *Brittia*. This would yield in O.Ir. nom. *Ie*, later *Ia*; gen. *Ie*, *Iae*; dat. *Í*; the nom. *Í*, which is the form in regular use, is the dative used as nominative. That

---

[1] The story, which is too long to tell here, and does not concern our subject, is told in LL 27 a 36; *Silv. Gad.*, i. p. 310.

[2] *I* is genitive case of *eo* (**ivi*), as *bí* is gen. of *beo*. The datives are *iu*, *eu*, *biu*. 'Mairg damsa de, mairgg di Chliu/dia frith Fer Fith inna eu.' 'Woe to me therefor and woe to Cliu, that Fer Fith was found in his yew.' LL 291 b 15; 292 a 22.

[3] Skene, *Celtic Scotland*, ii. p. 491 (where they are called 'bishops,' who were not in truth bishops); O'Donnel's *Life of Columba*, p. 201 (O'Kelleher and Schoepperle, 1918).

[4] Compare 'secht noemepscoip na hIi'; *L. Breac* 24 a.

these forms are contemporary with the series *Ioua* (adj.), *Eo*, *Eu*, appears from Bede's *Hii*, *Hy*, and from the still older *ad Segienum Hiiensem abbatem* of Cummian's epistle, A.D. 634.

In the old literature of Ireland there is an interesting reference to Hí in a period long before Columba's time. Ailech, near Derry, the royal seat of the northern Hy Néill, was called of old Ailech Néit, and later Ailech Frigrenn, 'Ailech of Frigriu.' According to the Dinnsenchas, it received this name because it was constructed by Frigriu, the artificer of Fubthaire, king of Alba, as a place of keeping for Fubthaire's daughter, who had eloped with him to Ireland. The point of interest is that in the Book of Leinster Frigriu is described as 'Frigriu, the artificer of Cruthmag of Cé, in the time of Fubthaire from Hí.'[1] The lady became afterwards the wife of Eochaidh Doimlen, son of Cairbre of the Liffey, and their sons were the famous three Conlas or Collas.

Of the remaining islands, Malea insula (Mull), Scia insula (Skye), and Orcades insulae (Orkney), have been mentioned already.[2] The position of the others is uncertain; the names have evidently been replaced by names of Norse or comparatively recent Gaelic origin.

The island 'which in Latin may be called the Long Island

---

[1] 'Frigriu cerd Cruthmaige Cé/i ró Fubthaire ó hI.' LL 164 a 46. The lore of Ailech is given in LL 181 a and 164 a; BB 399 b. One version in the latter is as follows: 'Frigriu, son of Ruide, went from the isle of Britain to Ireland. He was artificer to Fubthaire, and he took Ailech, Fubthaire's daughter, with him in elopement to Ireland. So Fubthaire went in pursuit of his daughter to Ailech, and Frigriu made a house for her of red yew, and that house was adorned with gold and with silver and with bronze and with gems, so that that house was equally bright by day and by night, and the damsel was taken there for safe keeping. And it is said that she was foster daughter of the artificer, and not his wife, and that she was wife to Eochaid Doimlen and mother of the three Collas. And Fiacha Sraibtine was king of Ireland then.' In the Laud Genealogies (*Celt. Zeit.*, viii. p. 319, 28) the artificer is called Crinden (genitive). In BB 399 b 23 he is called 'Frigriu, son of Ruba Ruad, son of Didul, of the Fomorians of Fir Falga'; (F. mac Rubai Ruaid maic Diduil do Fomuirib Fer Falga). The 'Fir Falga' were somewhere in the Western Isles.

[2] pp. 28, 38, 39.

(*Longa*),' must be a translation of *an Innis Fhada*, and can have nothing to do with the Isle of Luing (Ship-isle), or with the Norse Lingay (Heather-isle). By ' the Long Island ' we mean at the present day the Outer Hebrides, from the Butt of Lewis to Barra Head, but it is not likely that this was what Adamnan meant.

Sainea insula was somewhere off the coast of Lorne. It is equated by Skene with Shuna, east of Luing, but Shuna is a Norse name, and apart from that fatal objection the phonetics are impossible. The brethren of Hí were coming from Ireland, and owing to an unfavourable wind were sheltering in the lee of this island. On the 9th June, St. Columba's Day, they set out for Hí at daybreak, and with a fair wind arrived after the third hour, or 9 A.M. The Sainean isle was not Seil or Luing; it may have been an island somewhere near these, possessing a good anchorage, or it may have been Colonsay.

Elena insula contained a monastery, and was therefore probably a fairly large island. Reeves, followed by Skene, took it to be from Gaelic *eilean*, an island, E.Ir. *ailén*, which is a Norse loan-word, not yet borrowed in Adamnan's time. We may perhaps compare *dóirad Eilinn*, ' the enslavement of Elenn,' under A.D. 678 in the Annals of Ulster.

Ommon, where the presbyter Findchan's offending right hand was buried, is conjectured by Skene to have been Sanda, off the south coast of Kintyre. The pre-Norse name of Sanda is preserved in Gaelic still as *Ābhainn* (in Arran *Ēibhinn*, *Eubhainn*), Fordun's *Awerynе*, Dean Monro's *Avoyn*, but this could not represent Ommon, for *mm* would not be aspirated.

Cainnech left his pastoral staff in Hí by mistake. He remembered it on his way to Ireland, as he was nearing Oidecha insula, and going ashore found it on the turf of a little district called Aithche (gen.).

Aithche is probably genitive sing. of the noun from which is formed the latinized adjective *Oidecha*. In O.Ir. *aithche* is genitive of *adaig*, ' night ' ; *Oidecha*, however, is doubtless to be compared, or equated, with *Odeich*, mentioned in *Senchus fer nAlban* (BB 148) as a division of Islay containing twenty houses (*tech*). It might thus be appropriately described as a little district (*terrula*), and the island in question was probably off it and named after it. Identification must be uncertain, but *Oidecha insula* may have been the isle now called by the Norse name Texa, off the south-east coast of Islay.

Kintyre is in Adamnan Caput Regionis, a literal translation of *Ceann Tíre*. It is one of the names most commonly mentioned in Irish records. That there was another form, Sáil-tíre, 'Land's Heel,' appears from the Norse forms *Saltiri*, *Salltiri*, less correctly *Satiri*.[1] Columba, when visiting Kintyre, spoke with the captain and crew of a ship newly arrived from Gallia (France); the most likely place for the incident would be *Ceann Loch Cille Chiaráin*, 'Head of St. Ciaran's Loch,' now Campbeltown.

Three places in Iona are mentioned. Munitio Magna, a hillock (*monticulus*), where Columba sat, represents *Dún Mór*, 'the Great Fort,' a name now obsolete. 'Colliculus Angelorum, Scottice vero *Cnoc Aingel*,' 'the Angels' Hillock, in Gaelic *Cnoc Aingeal*,' where angels visited the saint, is also, according to Pennant, called *Sìthean Mór*, 'the big fairy knoll.' In Cuul Eilni the double *u* indicates length; we may compare *Eilni* (dat. *Eilniu*) and *Mag nEilni* about Coleraine in Ireland. The meaning is 'Nook (or Back) of Eilne,' whatever that means.

Some places are mentioned as sites of monasteries. There were two in Tiree, one on Campus Lunge, otherwise Campus Navis, 'Ship's Plain' (*Magh Luinge*), founded by Columba and ruled by Baithen; the other on Artchāin, 'the Fair Cape (or Height)' founded by Columba's contemporary, Findchan. As to the former *Brigit Maigi Luinge* is one of the fifteen saints named Brigit in Rawl. B 502, 94 d 45, and there was a chapel named Kilbride on Cornaig Mór, 'a quarter of a mile south of the corn mill, and about the same distance east from Loch Bhasapol.'[2] *Port na Luinge* is on the south side of the island, near Bailemartin township.

---

[1] W. A. Craigie, *Celt. Zeit.*, i. p. 452.

[2] Erskine Beveridge, *Coll and Tiree*, p. 146.

Cella Diuni, the name of a church near Loch Awe, named after the brother of Cailtan, its superior, has unfortunately disappeared from among the numerous *Cill* names of Argyll: I have not met any Irish personal name which this would represent. The monastery of Kailli au Inde was founded by one of Columba's followers called Fintén, whose name survives in *Cill Fhionntáin*, 'Killundine,' in Morvern. 'Monasterii fundator quod dicitur Kailli au inde' seems to mean 'founder of the monastery which is called (the monastery) of Caill aui Fhinde,' *i.e.* 'of the wood of the grandson (descendant) of Findia.'

The position of Delcros, where the peasant reaped in the beginning of August barley that had been sown in mid June, is uncertain. O'Donnell puts it near Derry in Ireland, but from the heading of the chapter in Adamnan it must have been in Scotland. In modern Gaelic it would be *Dealgros*, 'Prickle-point,' either from its shape or from prickly shrubs growing on it. There was a place in Kintyre whose name is given in the genitive as *Delgon*, but this is probably too far from Iona to suit the story. The old name of the point of land at the junction of Earn and Tay was *Rindalgros*, 'Point of Dealgros,' now Rhynd. The name *Dealganros*, 'Prickle-point,' occurs at least thrice: (1) at Comrie in Strathearn; (2) near Blair Athol; (3) near Inverness, where it is anglicised Dalcross.

Ardnamurchan appears thrice: nom. Artda Muirchol; gen. Aithchambas Art Muirchol; dat. in Artdaib Muirchol, where *artda* is pl. of *ardd*, a height, a promontory, already seen in *Art-cháin*. The name means 'Heights of Muirchol' or 'Points of Muirchol.' The first part of *muirchol* is *muir*, sea; Reeves took *col* to be *coll*, hazel, and rendered 'Height of the two sea-hazels' (*Art-dá-muirchol*). The dative form, however, shows that *da* is not *dá*, two, and it is unlikely that so accurate a writer as Adamnan would have written *col* when he meant *coll*—the two words were sounded differently. The same applies to *coll*, destruction, skaith, and to *coll*, a head. There is, however, *col*, sin, wickedness, and *muirchol*, 'sea-wickedness,' might refer to acts of piracy or wrecking; thus *Artda Muirchol* may mean 'Capes of

Sea-sins.'[1] It is worth noting that a king named Talorg, son of Muircholach, appears in the Pictish Chronicle.[2] The modern form is *Ard-na-murchan*, spelled by MacFirbis *Ard na Murchon*; this, as it stands, means 'Point of the Sea-hounds,' i.e sea otters. *Muirchú*, sea-hound, is a well known Irish name—one of the many compounds of *cú*. There is also *muirchat*, sea-cat.[3]

The little bay in Ardnamurchan from which the robber[4] set out on his last voyage was called in Gaelic (*Scotice*) Aithchambas.[5] Here the first part is *áith*, sharp; the second is the O.Ir. form of the modern *camas*, a bay, a derivative of *cam*, bent; the meaning is 'sharp bay,' i.e. narrowing to a point at the inner end.

Coire Salcain was the name of a place where Columba once stayed, and it has been identified with the corrie behind Salachan in Morvern, opposite Aros, which may be correct. *Salcain* (gen. of *Salc(h)an*) is from *sailech*, willow, meaning 'Willow-copse'; compare *Salchaigh* in Glen Dibidale Forest, Ross; *Sauchie*-burn. This is the earliest instance of our Scottish use of *coire*, a cauldron, to denote a more or less circular hollow in the hills, with only one outlet.

The sea gulf *Coire Bhreacáin* is in Adamnan Charybdis Brecani. Already in Fordun's time the term was applied to the tidal whirlpool between Jura and Scarba, but, as has been already noticed, at an earlier period it meant the whirlpool between Rathlin and Antrim, now called *Sloc na Mara*. The tale of the Breccán who was drowned there with all his company of fifty ships is told in the Book of Ballymote; according to one version, he was the son of Partholon; another makes him son of Maine son of Niall of the Nine

---

[1] E. W. B. Nicholson translates '*Artda-muir-Chol*, Col-sea-heights,' with reference to the Isle of Coll. This is impossible, for (1) *muir* would require to be genitive (*mora*); (2) the stress would have to be on *Chol*, which is unstressed; (3) Coll is in Adamnan *Colosus*. *Keltic Researches*, p. 33.

[2] Skene, *P.S.*, p. 7.

[3] *Lives of the Saints*, Stokes.

[4] His name was Ioan, son of Conall, son of Domnall, and he was of the royal race of Gabran, son of Domongart and father of Aedan.

[5] So Stokes and Strachan, *Thes. Pal.*, ii. p. 278.

Hostages, and says that when Columba was sailing by, his rib arose out of the whirlpool to greet his kinsman.[1]

On one of his journeys beyond Drum Alban, Columba baptized an aged Pict called Emchatus and his son Virolecus, at a place near Loch Ness called Airchartdan. This is now *Urchardan*,[2] Urquhart, whence Glen Urquhart off Loch Ness. Here *air* is the preposition meaning 'on, near'; in Ir. and Welsh *ar*, now very often *ur* with us. The second part cannot be explained from Gaelic, and has been rightly equated with Welsh *cardden*, a copse, the whole thus meaning 'On-wood,' 'Wood-side'; compare the common Welsh *Argoed*, from *ar*, on, and *coed*, wood; Gaelic *Urchoill*, 'Orchil,' from *ar*, and *coille*, wood. We have already noted the formation in the Old British *Are-cluta*, O.Ir. *Ercluad*, O.Welsh *Arclut*, Clyde-side. The element *cardden*, as we shall see, is common in Scotland, especially on the east.

Here I may mention the remaining larger islands on the west which appear to have names that are pre-Norse, but are not named by Adamnan. The Isle of Rum is in Gaelic *Rum*, gen. *Ruim*; the adjective is *Rumach*, as in *an Cuilionn Rumach*, 'the Coolin of Rum,' as distinguished from *an Cuilionn Sgitheanach*, 'the Coolin of Skye'; *na h-Earadh Rumach*, 'Harris of Rùm,' as distinguished from *na h-Earadh Ìleach*, 'Harris of Islay,' and *na h-Earadh*, Harris south of Lewis. The genitive is in Irish literature *Ruimm*; Beccan Ruimm, 'Beccan of Rum' died in 677 (AU); Tigernach (676) has 'Beccan Ruimean quievit in insola Britania'; in a note to Gorman's *Félire* it is Ruiminn. Nennius says that the isle of Tanet (Thanet) was called in British 'Ruoihm,' but the readings vary too much for certainty. Rum may be pre-Celtic.[3]

Bute is in Gaelic *Bód*, gen. *Bóid*; Rothesay is *Baile*

---

[1] BB, 398 b; Stokes, *Edinburgh Dinnshenchas*, p. 60 (No. 58); O'Donnell, p. 379; Cormac's *Glossary*.

[2] Skene, following Reeves, says '*Glen-arochdan* is the local pronunciation of the name' (*Life of Columba*, p. 291), but this is not so. He has, as usual, been followed by others.

[3] Stokes says 'identical with Gr. ῥύμβος, ῥόμβος'; *Linguistic Value of the Irish Annals*. The isle is lozenge-shaped.

Bhóid, 'town of Bute'; a Buteman is *Bódach*. It does not appear in Irish literature, but the Norse form is Bót (Hák. Sag.). The suggestion has been made that *Bód* is connected with Ebudae, the Hebrides, but this is quite impossible. A possible explanation is O.Ir. *bót*, fire, if we suppose, as is likely, that the original name was *Inis Bóit*, 'isle of fire,' with reference to signal fires or bale fires. An Irishman of old was styled 'Fergus Bót tar Bregaibh,' 'Fergus Fire over Bregia,' from his burning of that district of Meath. The personal name Buti (a saint who died in or about 521) is explained thus: '*Búti* .i. *beo*, no *Búti* .i. *tene*, ut dicitur "bót fo Bregu"; unde dicitur *bútelach* .i. hignis magnus'; 'Buti comes from *beo*, living, or from *bót*, fire, as in the saying, fire throughout Bregia; whence comes *bútelach*, a big fire.'[1] Elsewhere *bútelach* is said to mean '(a place) where a big fire is made.' The term seems to connect with the English form 'Bute.'

Arran appears often in Irish literature; the short poem in the *Acallamh* beginning 'Arann na n-aighedh n-imdha,' 'Arran of the many stags,' is one of the finest pieces of nature poetry in Gaelic. The name is declined as follows:—

Nom.—Ara, *P.S.*, p. 23; Arand, Arann, Acall., ll. 340, 351.
Gen.—Arann: loingsech Ile ocus Arann, *P.S.*, p. 99; a Aedáin Arann, Rawl., B 502, 86 b 32; sealg Arann, Acall., i. p. 331.
Dat.—Araind: ic Araind uair, BB 52 c 34.

The modern Gaelic is *Arainn*, the old dative form; the adjective is *Arannach*. 'An Coileach Arannach,' 'the Cock of Arran,' is a great rock or boulder at the north end of the island, which gives its name to a small district known as 'the Cock.' A west coast triad is 'a' Chearc Leódhasach, an Coileach Arannach, agus an Eireag Mhanannach,' 'the Hen of Lewis (Chicken Head), the Cock of Arran, and the Pullet of Man (the Calf)'; to see these three in one day was

[1] L.Br. 19 a; Cóir Anmann 262; *Féil. Oeng.*, p. 256 (H. Bradshaw Soc.) O'Davoren's Glossary, *Archiv f. Celt. Lexik.*, ii.

reckoned good sailing—too good to be true. The initial vowel is short: *Arainn* rhymes with *Manainn*, and *Arann* with *abann*.[1] It is therefore a different word from the Irish *Áru*, gen. *Árann*, dat. *Árainn*, the Aran Isles. For the same reason it cannot be equated with Ptolemy's *Adros* (Ἄδρου ἔρημος), for by all analogy Early Celtic *adr-* would yield *ār-*, not *ar-*. The meaning is unknown so far, and the name may be pre-Celtic, but there are several Welsh names which appear similar: Afon Aran in Radnorshire; Aran Mawddy and Aran Benllyn, adjacent hills south-west of Bala; Arenig Fach and Arenig Fawr north-west of Bala. According to Irish tradition Arran was the home of Manannan, the sea god, and another name for it was *Emain Ablach*, 'Emain of Apples.'[2] This is, I suppose, equivalent to making Arran the same as Avalon, the Happy Otherworld.[3]

St. Kilda is in Gaelic *Hiort*, also often *Hirt*, the latter being the genitive used as nominative. A Norse saga has *Hirtir* to denote certain isles beside the Hebrides.[4] About 1370 it is Hert, Hyrte (RMS); Fordun has 'insula Hirth, omnium insularum fortissima,' 'the strongest of all the isles'; Dean Monro has Hirta, 'aboundant in corne and gressing, namely for sheipe,' etc. *Hiortach* or *Hirteach* is 'belonging to Hirt,' 'a native of Hirt.'[5] The name is identical with O.Ir. *hirt*, *irt*, explained by Cormac as *bás*, death, and it has been suggested as 'likely that the ancient Celts fancied this sunset isle to be the gate to their earthly paradise, the Land-under-the-waves, over the brink of the western sea.'[6] It is still more likely that the name was given with reference to the manifold hardships and dangers

[1] A. MacDonald, p. 100 (1st ed.); BB 52 c 34; Acall., l. 351 (Stokes).

[2] YBL 178 a; BB 258 a 39; see also the poem to Raghnall, son of Gofraigh, printed by Skene (*Celtic Scotland*, iii. p. 410), where 'Eamhoin Abhlach' may mean the Isle of Man.

[3] Rhys, *Arthurian Legend*, p. 334, etc.

[4] Professor W. A. Craigie, *Celt. Zeit.*, i. p. 451 (*Bisk. S.*).

[5] *Hiort* is used in Lewis and on the mainland; in Uist they say *Hirt*; for 'in Hirt' they say *an t-Irt*, and a native of Hirt is with them, sometimes at least, *Tirteach*.

[6] A. Macbain, *Place-Names*, p. 177.

connected with landing and living on this remote spot, which in the Hebrides is regarded as a penitentiary rather than a gate to paradise.

As to 'St. Kilda,' Martin says it 'is taken from one Kilder who lived there, and from him the large Well *Toubir-Kilda* has its name.' A Dutch map of 1666 styles the island 'S. Kilda,' while a little to the south-west of Loch Roag in Lewis it has 'S. Kilder,' denoting perhaps the islet of Ceallasaidh.[1] There was no saint called Kilda; the name evidently arose from confusion with the name of the well at the landing place which is still called *Tobar Childa.* This again is Norse *kelda*, Danish and Swedish *kilde*, a well, with G. *tobar*, a well, added in unconscious duplication of meaning. Probably the Dutch fishermen, who were active on the west in the seventeenth century, were in the way of taking in water from this well. The well must have had more than a local fame, for it occurred in a rime known in Easter Ross in my boyhood.[2] In view of the fact that the ancient name of the island is still the only name in use all over the Gaelic-speaking area, it seems right that it, and not a Dutchman's blunder, should be the name on maps.

Hirt occurs also in *an Duibh-hirteach*, 'the black deadly one,' the name of the lonely rock, now bearing a lighthouse, almost due west of the north end of Colonsay. It is regarded much as Hirt is regarded by the people of the Outer Hebrides.[3]

---

[1] The map or chart is one of a collection by Pieter Goos, published at Amsterdam 1663/66, a copy of which is in the Library of the Royal Scottish Geographical Society, Edinburgh.

[2] The rime was 'Tobar Childa challda, allt Chamshroin a lobhair,' and I think we understood that to be the full name, but we had no idea where the well was. Lately the Rev. Dr. Malcolm Maclennan informed me that he heard it from an old man in St. Kilda as 'Tobar ghildeir chaldair, allt chamar nan ladhar.'

[3] Professor Mackinnon, who was a native of Colonsay, wrote an account in Gaelic of a trip to the Du'irteach (*sic*) which appeared in *The Gael,* iii. p. 197. He writes of it as the great bogey (*bòcan*) that was used to keep children in their place. 'If you did not take your food when you were told, it would be left on the Du'irteach. If you asked for anything that they did not wish to give you, it was on the Du'irteach. If you did not do smartly whatever you were told to do, you would be put on the

Other names that probably contain *hirt* are Craighirst in Dumbartonshire and Ironhirst on Ironhirst Moss near Lochar Moss in Dumfriesshire. The latter seems to be for *earrann hirt*, 'portion of death,' 'deadly piece (of land),' with reference to the exceedingly dangerous character of this great bog.

Two isles mentioned in Irish literature cannot now be identified. One is *Inis meic Uchen* in the tale of Cano, son of Gartnán;[1] it may have been off the coast of Skye, or in an inland loch. The other is *Inis Áne*, where the men of Tyrone defeated the men of Alba with the loss of thrice fifty curachs and their crews.[2]

Du'irteach.' He declines it as a feminine noun, nom. *an Du'irteach*, gen. *na Du'-irtich*, dat. *an Du'irtich*. In Uist one says to a teasing person, 'Nach robh thu an t-Hirt,' 'I wish you were in Hirt'; a common threat to a child is 'cuiridh mi Hirt thu,' 'cuiridh mi Hirt air muin mairt thu,' 'I 'll send you to Hirt on a cow's back.' Similar expressions are used in the other Outer Isles.

[1] *Anecdota*, vol. i. pp. 1, 2. It appears to have been a crannog.

[2] LL 182 a 42; 183 a 10.

CHAPTER IV

TERRITORIAL DIVISIONS

In very early times the name of the tribe or population-group served as the name of the territory which they occupied. In Gaul, when Caesar wishes to mention the country of the Sequani, he says 'in Sequanis,' 'among the Sequani,' 'in Sequanos,' 'into the land of the Sequani,' and so forth. Similarly when Ptolemy deals with the British Isles he gives no territorial names but only names of tribes, or population-groups. Tacitus, it is true, mentions Caledonia, the land of the Caledonians, but this is purely a Latin formation. Some of these ancient tribal names survive, not as territorial names, but as descriptive terms attached to a generic term: Dunmyat, 'the fort of the Maeatai'; Dun Chailleann, 'Fort of the Caledonians'; Sìdh Chailleann, 'Hill (or Seat) of the Caledonians'; Carn Smeart, 'Cairn of the Smertae.' Another tribal name, not mentioned by Ptolemy but which may be inferred to have existed in his time and long before it, exists as a territorial name in the dative plural, namely Arcaibh, Orkney, for the older 'i n-Orcaibh,' 'among the Orcs.' Three other names of this type survive as territorial names, *Gallaibh*, Caithness, for 'i nGallaibh,' 'among the foreigners'; *Cataibh*, Sutherland, for *i Cataib*, 'among the Cats'; Galloway for 'i nGall Gaidhealaibh,' 'among the foreign Gael.'

The history of tribal and sept names in Scotland must be always obscure: we have no material for a monograph such as Professor Eoin MacNeill's on *Early Irish Population Groups.* But whatever may have been their development, they have not influenced the existing territorial names to any great extent. The great majority of these are descriptive; some are from personal names. These remarks are not meant to apply to administrative divisions such as parishes and counties, but to the old unofficial districts, such as Badenoch, Rannoch, Angus, Ferindonald, and so on. Parishes are, as a rule, named after the parish church or from its original site, as Kilmuir, 'Mary's church'; Logie, 'at-hollow.' Counties sometimes preserve an old territorial name, as Sutherland, Ross, Moray, Fife, Argyll. These are 'counties' (*comitatus*) in the strict sense of the term, inasmuch as they represent territories which were once under the authority of an earl (*comes*). The others take their names from the seat of the sheriff (*vice-comes*), and are, therefore, strictly speaking, 'sheriffdoms, shires.'

The province of Lothian extended of old to the Tweed, which according to Symeon of Durham divided Northumbria and Loida. The name now denotes the three counties of Haddington, Edinburgh or Midlothian, and Linlithgow, known as 'the three Lothians.' There are some indications of it even beyond Tweed, namely Lothiangill south-west of Carlisle, and Catlowdy east of Canonbie in Cumberland. Mount Lothian, south-east of Penicuik, is Mountlouthen, Mountlouthyen, Mundlouen, Muntlaudewen, Muntlouen, Muntloudyan in Reg. of Neubotle and Chart. of Holyrood. Lothian Burn is a brook from the Pentlands. Some of the old spellings of Lothian are: in Lodoneo, 1098 (Lawrie); de Lodoneio, 1117 (Lawrie); Loeneis, 1158 (Bain's Cal.); Loenes, 1249 *ib.*; Leudonia a. 1164 (*Fragment of Life of St. Kentigern*); Laodonia (Aelred); Loðene (O.Eng. Chron.); Lowthyan (Wyntoun). The oldest tradition as to the origin of the name is given in the *Fragment of the Life of St. Kentigern*, which says that Kentigern's grandfather on his mother's side was Leudonus, a man half pagan, from whom the province which he ruled was called Leudonia. A Welsh MS. of about A.D. 1300 describes him as 'Lleidun llydaw o dinas etwin yn y gogled,' 'Lleidun of Llydaw from Edinburgh in the north.' *Llydaw* is the Welsh form of *Litavia*, Armorica or Brittany. Another MS. of about 1400 has 'Lleudun luydauc o dinas Eidin yn y gogled.' Here *luydauc* is possibly to be compared with W. *llwydd*, success, in which case it would mean 'successful'; it may, however, be a corruption of the older form.[1] Another variant of the epithet is *lueddog, lueddauc*, 'having a host'; compare the M.Ir. personal name *Sluagadach*, which appears in the form *Slogadadh* as the name of a leader of the Bishop of St. Andrews' host, *c*. 1128 (Lawrie).[2] It is notable that Lleidun or Lleudun of Dinas Eidin does not appear in the old Welsh genealogies of 'the Men of the North,' though his contemporaries appear there, and that no Welsh authority mentions who his father was, though the man himself is made a contemporary of Urien of Rheged. Geoffrey of Monmouth, who calls him Lôt of Lodonesia, makes him a grandson of Sichelm, king of Norway; in Fordun he is Loth or Loyth, 'of the family of the leader Fulgentius.'[3] From all this one would infer that Leudonus, if he was a real historical personage, was an incomer from foreign parts, probably Gaul.

It is unfortunate that we have no real tradition of the name Lothian in Gaelic. We know it indeed as Loudy (with English *l*) from the harvesters who used to travel to Lothian even from Sutherland in the earlier part of last century; also the surname Lothian, which was not uncommon in north-west Perthshire, is in Gaelic Loudin.[4] But these are merely adaptations of the vernacular Scots. In Welsh literature it does not occur, so far as known to me, but from the statements of two Welsh scholars it seems to be known in Wales. Mr. Egerton Phillimore wrote, 'Lleudun Luyddog, the Leudonus of the older *Life of St. Kentigern*, whence the territorial name Lleudduniawn, the Welsh form which has got gaelicized and shortened into

---

[1] *Luydauc* also suggests comparison with the note on Confer, the ancestor of the royal house of Strathclyde: 'Confer ipso est vero o litauc dimor medon venditus est,' which seems to mean 'Confer however is from Litauc; he was sold (? he came !) from the Mid-sea,' *i.e.* the Mediterranean Sea (*Y Cymmrodor*, ix.). 'Dimor medon' seems rather Irish than Welsh.

[2] The variants of the whole passage are tabulated in *Archiv. f. Celt. Lexik.*, ii. p. 191; see also *ib.*, p. 168.

[3] Fulgentius or Fulgenius was, according to Geoffrey, a contemporary of Severus, *i.e.* early third century. He is quite mythical.

[4] A part of the old burial-ground of *Cladh Chunna*, near the foot of Glen Lyon, is occupied by Lothians.

Lothian.'[1] Sir E. Anwyl says: 'Leudun Luydawc . . . This name, Lleuddin, is the same as that of the Lleuddin from which is derived the Welsh name Lleuddiniawn (the district of Lleuddin) for the Lothians. It is singularly like a Welsh derivative of Laudīnus.'[2] Here the ending *-awn* is the Latin *-ānus*, as in Christiānus, used in Welsh to form names of districts from names of men, as *Cereticiaun*, now *Ceredigion*, Cardigan, from Ceretic. *Lleuddun*, however, could not come from *Laudīnus*, though it might come from *Laudinus* with *i* short, as *Mauricius* gives *Meurig* and *Meurug*. Holder gives as Gaulish personal names Laudio, gen. Laudionis; Laudo, gen. Laudonis; Laudonius, Laudonia; Laudōn- would give Lleuddun, as Lat. *scōpa* gives W. *ysgub*, G. *sguab*, a besom.

The form Loenes, Loeneis, represents Lodonesia, through Norman-French; hence the poetical Lyonesse?

The additions, from a Welsh source, to Nennius record that Cunedag, an ancestor of Mailcun of Gwynedd, came from the North, from Manau of the Guotodin. This is agreed to have been a district about the head of the Firth of Forth, and it is styled 'of the Guotodin,' *i.e.* 'of the Votadini,' to distinguish it from the other Manau, the Isle of Man. In early Gaelic Manau would become Manu, Mana, with genitive Manann, and it is this genitive that survives in Slamannan, in 1275 Slefmanyn (Theiner), 'hill,' or rather 'moor of Manu,' where Gaelic *sliabh* is probably a translation of Welsh *mynydd*. It is also found in the name Clackmannan, 'stone of Manu'; the stone is conspicuous in the centre of Clackmannan town, and as the spellings from the twelfth century onwards agree in Clac-, Clack-, the first part is probably Welsh *clog*, stone, rather than Gaelic *cloch*, *clach*. Cremannan in Balfron parish, Stirlingshire, 'John de Cromenoc' 1296 (Bain's Cal.), Cromennane 1303 (Ch. Inch.), has been explained as *crìoch Manann*, 'boundary of Manu,' but apart from the difficulty of supposing Manu to have extended so far to the west, the *ch* of *crìoch* would hardly have disappeared so early, and the old forms point to *Crò-meannán*, 'kids' fold.'

[1] *Y Cymmrodor*, xi. p. 51. [2] *Celt. Rev.*, iv. p. 140.

Dalmeny, Dumanie, *c.* 1180, in Linlithgowshire, is Dunmanyn in 1250 and twice in 1296 (Ragman Roll); Dummanyn, Dummany, RMS, App., 2; Dunmanie *alias* Dalmanie, 1662, RMS; there is no *l* in the present vernacular pronunciation. The first part is *dún*, fort, probably gaelicized from Welsh *din*; the second part cannot be from Manu in view of the old forms and the present pronunciation; it may be *meini*, pl. of W. *maen*, stone, for 'stone-fort' would be quite a possible name in a district where the forts are usually of earth, as they are in the Lothians. Manau has been thought to appear in the old Welsh poetry as Mynaw, but the identity is doubtful. Sir J. Morris-Jones has suggested with probability that 'the unintelligible *o berth Maw ac Eidin* of the Book of Taliesin should be *o barth Manaw ac Eidin*,' 'from the region of Manau and Eidin.'[1] In Irish we have 'strages Pictorum in campo Manann,' 'slaughter of the Picts in the plain of Manu,' AU 711, and the place is defined by the Anglo-Saxon Chronicle as between Haefe and Caere, understood by Skene as 'between Avon and Carron.' Whether 'bellum Manand fri Aedán,' 'the battle of Manu against Aedán,' in AU 582, 583, was fought in this district or in the Isle of Man is not quite certain. There was a Dún Manann in Co. Cork,[2] and Mín Manann, 'the smooth plain of Manu' was in Connacht.[3] There is also Manaw in Anglesey, and Glin Mannou appears in the Book of Landaf. As to the name itself, the Isle of Man is in Caesar Mona, in Pliny Monapia, in Orosius Mevania (for Menavia or Manavia), and it is this last (Manavia) which gives Welsh Manau; the base is probably the same as that of *Man-ap-ii*, the name of a tribe in the south-east of Ireland, apparently an offshoot of the *Men-ap-ii* who lived on the lower Rhine, but there is no certainty as to its meaning. Another name for the Isle of Man was *Eubonia*: Nennius has 'Eubonia, id est, Manau.' With this we may compare *Emonia* given by Fordun for Inchcolm in the Firth of Forth: 'monasterium Sancti Columbe in insula Eumonia' (RMS 1440); Emonia,

[1] *Taliesin*, p. 80.
[2] Hogan, *Onom.*
[3] O'Donovan, *Hy Fiachrach*, p. 241 note.

1480, *ib.*; 'monasterium Insule S. Columbe,' 1550, *ib.*; also 'reversio Uloth de Eumania,' 'the return of the men of Ulster from Eumania,' AU 578.

Immediately west of Lothian was a small district called in Latin Calatria or Calateria; its position is roughly fixed by the statement of Ailred that William the Conqueror 'penetrated Lothian Calatria and Scotia as far as Abernethy.' In 1136 Dufoter de Calateria witnessed a charter of David I. Skene identified Calatria with a certain place called Calathros, 'hard promontory,' in the Irish Annals. In 678 (AU) Domnall Brecc, king of Dál Riata, was defeated at Calathros. In 736 (AU) was fought the battle of Cnoc Coirpri (Cairbre's Hill) in Calathros at Etarlindu, between Dál Riata and Fortrenn, when Talorgan, son of Fergus, defeated the army of Muiredach, king of Dál Riata. 'Cnoc Coirpri,' says Skene, 'is now Carriber, where the Avon separates Lothian from Calatria'; this, however, is impossible, for Carriber is a compound term, with the stress on the second part, pronounced locally 'Car-rubber.' In the next place 'Etarlindu' means 'between pools' (*etar lindu*), and there is no spot to which this name would apply found in the neighbourhood. Lastly, it was in this very year that Angus, Talorgan's brother, wasted Dál Riata, seized Dún Add, and burned Creic: it is therefore most unlikely that the Dalriadic king would be in a position to invade Lothian. The site of Calathros must be sought elsewhere, and the name Etarlindu may afford a clue to its position. There is just one place in Scotland now which bears this name, and that is *Eadarlinn*, Ederline, at the south-west end of Loch Awe, so called because it is between Loch Awe and a small loch. Ederline is only about six miles north of Dún Add, in the region where Angus was operating. The promontory which separates the two lochs is now *an t-Sròn Mhór*, 'the big nose.' This may have been the scene of the defeat of the Dalriadic kings.[1]

Skene further identified Calatria with Callendar, near

[1] Near Ederline is the place called *Gocam-go*; it was prophesied of Alasdair mac Colla that he would be fortunate until he planted his standard on the knoll of Gocam-go: the spot was of ill-omen. (See *an Gàidheal*, ii. p. 370.)

Falkirk,[1] early spellings of which are 'de Striuelinschire et de Kalenter,' *c.* 1150 (Lawrie); 'in Carso de Kalentyr,' 1267 (Reg. Neub.); Calantair, 1296 (Ragman Roll); Kalentar, 1362 (RMS), and so on. Calatria is a latinized form of a native name, and the most likely name is the stream-name which appears in G. as *Caladar*, and in W. as *Calettwr*, for an earlier *Caleto-dubron*, 'hard-water.' This is usually in English Calder, also Cawdor, but this is not the only form. Callater in Braemar is Callendar, 1652, 1635 (RMS), but Glencullady 1601 (Ret.), Glencallader in Sir James Balfour's Collections. Callander near Crieff, between the Kelty (*Cailtidh*) and Barwick burns, is Kalentare 1504 (RMS), but in T. Pont's maps Kalladyrs (there were Callander-more and -beg); I have failed to get the Gaelic pronunciation, but there can be little doubt that one of the burns was a 'hard-water'; *Cailtidh*, Kelty, is from *caleto-*, 'hard,' ultimately. Callendar on Spey, nearly opposite Ballindalloch, is 'Calatar super aquam de Spey,' 1541 (RM), and earlier Calledure, *ib.*; in Blaeu Kalladyr. Callander on Teith is not in point, for it is a name transferred from Callander near Falkirk.[2] We may compare, however, Loch Spallander in Ayrshire; Lochspaladar (RMS), 1466; Loch Spalladurr in Blaeu; Lochspallender, 1601 (Ret.). This is evidence for the development of an intrusive *n* in English, but it can hardly be considered evidence for the development at so early a date as 1150, which is synchronous with the spelling Calatria. I think, therefore, that the equation of Calatria and Callander must be taken as not proven. At the same time Calatria was certainly somewhere in this neighbourhood, and it is quite possible that it represents the district east of the Almond contained by the parishes of West Calder, Mid Calder, and East Calder, the last of which is now included in Kirknewton parish. The village of West Calder is on the Calder Burn; the Linhouse Water,

[1] *Celt. Scot.*, i. pp. 247, 291.

[2] The Livingstone proprietor of both places had his lands incorporated into the one barony of Callander, 1549 (RMS); the change was helped by the similarity of Callander to Calindrade, apparently the old name of Callander on Teith.

in its upper part the Crosshouse Burn, was probably once called Calder.

Old tradition has it that Alba—by which is meant here Scotland north of the Forth and east of Druim Alban—was divided into seven provinces by the seven sons of Cruithne, the ancestor of the Cruithnigh or Picts. This is put shortly in the well-known quatrain :—

> ‘ Mōrsheiser do Cruithne clainn
> raindset Albain i secht raind :
> Cait, Cē, Cīrig, cētach clann,
> Fīb, Fidach, Fotla, Fortrenn.
> Ocus is e ainm gach fir díb fil for a fearand.’ [1]

Here Cait, Cīrig, Fortrenn, are all in the genitive case, and the same forms are used in the prose text which amplifies the statement of the quatrain. We may infer that the other names also are meant to be genitives, depending on *rann*, a division, understood. We may infer too that the verse is older than the prose, and that the writer of the latter followed the forms of the verse without troubling to understand them. The translation then would be : ‘ Seven of Cruithne's children divided Alba into seven divisions : the portion of Cat, of Cē, of Cīrech, children with hundreds of possessions ; the portion of Fīb, of Fidaid (?), of Fotla, and of Fortriu. And it is the name of each man of them that is on his land.’ The quatrain is ascribed to Colum Cille, and though we need not take that ascription seriously, it is certainly old. It applies to Cruithentuath or Pictland as it was after the Gaelic conquest of the West and before the union of the Picts and Scots.

Along with this may be taken the account in the twelfth century tract entitled ‘ De Situ Albanie,’ which states that Albania ‘ was of old divided by seven brothers into seven parts. The chief (*principalis*) part is Enegus with Moerne, so called from Enegus, the first-born of the brothers. The second is Adtheodle and Gouerin. The third part is Sradeern with Meneted. The fourth part is Fif with Fothreue. The fifth part is Marr with Buchen. The sixth

---

[1] BB 203 a 36 ; *Irish Nennius*, p. lxvi. ; Skene, *P.S.*, p. 25.

is Muref and Ros. The seventh is Cathanesia on this side and on the further side of the hill (*ultra montem*), for the hill called Mound divides Cathanesia in two. So each of these parts was then called and was in fact a "region," because each of them had in it a sub-region. Thence it is that those seven brothers aforesaid were held as seven kings, having under them seven petty kings (*reguli*). These seven brothers divided the kingdom of Albania into seven kingdoms and each of them ruled in his own kingdom in his own time.'

The writer proceeds to give the bounds of the kingdoms as related to him by Andrew, bishop of Caithness (d. 1184), and in so doing he varies his account by making the seventh kingdom Arregaithil, Argyll, instead of Cathanesia.[1] Apart from this he is evidently giving the old tradition—of which he seems to be unaware—in another form, suited to later conditions. The correspondence between the two accounts will be best seen by placing them side by side, though by so doing we anticipate a little.

Province of	Cīrech	Angus and the Mearns.
,,	Fotla	Athol and Gowrie.
,,	Fortriu	Strathearn and Menteith.
,,	Fib	Fife with Fothreue.
,,	Cē	Marr and Buchan.
,,	Fidaid (?)	Moray and (Easter) Ross.
,,	Cat	Caithness and (S.E.) Sutherland.

We may now take the districts separately. *Cīrech* means in Gaelic, 'crested,' from *cìr*, a comb, a crest; Crus mac Cīrig, 'Crus, son of Cīrech,' was the chief warrior of the Cruithnigh.[2] In the Pictish Chronicle the name is given as *Circinn* and *Circin*, which is the genitive of *Cīrcenn*, 'Crest-headed' (*P.S.*, p. 4), and this corresponds to some of the other old forms: Cath Chirchind, 'battle of Circhenn' (Tighern., 596); cath Maigi Circin, 'battle of the plain of Circen' (YBL fcs., 192 b 30); Magh Circinn i nAlbain (Mac Firbis—Hogan). Alongside of these we have a form *Gergenn*: Eoganacht maigi Dergind (read *Gergind*) i nAlbae, 'the

[1] Skene, *P.S.*, p. 136. There are other variations. [2] *Ib.*, p. 41.

Eoganacht of the plain of Gergenn in Scotland' (Rawl., B 502, 148); Eoganacht maigi Gergind i nAlpae (LL 319 c); Eoganacht maige Gerrghind a nAlbain (BB 172 b 4); Eoghanacht mhuighe Geirrghinn (Keating, ii. 386); Cairbre Cruthneachan a Muigh-gearrain, 'Cairbre C. from Maghgearrain' (*Celt. Scot.*, iii. 475); defunctus est Palladius in Campo Girgin in loco qui dicitur Fordun (Colgan). The fact that Fordun was in 'the plain of Girgen' shows that Girgen was the name of the Mearns or rather that the Mearns was in Girgen.

We have thus three forms of the name—Círech, Círchenn, Ger(r)genn, of which the first two go together. The last form, Gerrgenn, suggests comparison with the Irish name Gerrchenn, 'Short-head.' In the *Táin Bó Cualnge* a man of this name appears as father of the warrior Muinremur, 'Thick-neck,' who is styled Muinremur mac Gerrchinn, with variants—all in the genitive—*Gerginn*, *Gercinn*, *Gerrcinn*, *Eirrginn*, *Errcinn*; and nominative *Gergend*. There is also Gerrchenn mac Illadain, with variants Gerchenn, Gerrgen, Cerrcen, and, in the genitive, Gerrce, Errge.[1] Here 'Cerrcen' seems to be owing to confusion with the name *Cerrchenn*, 'Wry-head' (Tighern., A.D. 662). This comparison leaves little doubt that the Irish writers who used the form *Mag Gergind*, etc., understood it as 'Gerrchenn's Plain.' With regard to the other forms, the first *i* of *Cīrig* is long, and if as I have assumed, *Cīrcinn* is the genitive of *Cīrchenn*, 'Crest-headed,' with its first *i* also long, it is difficult, if not impossible, to correlate it with *Gerginn*.

This ancient province between Tay and Dee was, according to Irish tradition, the home of a branch of the Eoganacht, that family of Munster origin from whom, according to our oldest authority, was descended Oengus, king of Alba, that is, as I suppose, Angus, son of Fergus, who died in 761. Skene remarks that Angus 'appears to have been the founder of a new family';[2] he certainly broke the succession

---

[1] Thurneysen, *Die irische Helden- und Königsage*. Compare *serrgind* for *serrchind*, Acall., l. 6924; *Táilgenn* for *Táilchenn*, 'adzehead.'

[2] *Celt. Scot.*, i. p. 288.

of Pictish kings, and brought the family to which he belonged into a position that it had not occupied before. It is significant that the twelfth century writer gives the premier place among all the provinces of Albania to 'Enegus,' and it may well be that it owed that position to the fact that the district of Angus, now Forfarshire, was specially connected with Angus, son of Fergus. Another possibility, which deserves consideration, is that the name may derive its origin from the tribe of Oengus of Dál Riata, a theory which becomes more attractive when taken along with the facts relating to the adjacent province of Gowrie.[1] The upland part of Forfarshire is in Gaelic *Bràigh Aonghuis*, 'Angus' upland,' and the lower ground, forming part of Strathmore, is *Machair Aonghuis*, 'Angus' plain.' The earliest mention of Angus as a province is in A.D. 938, when the Pictish Chronicle records the death of Dubucān, mormaer of Angus. Dubucan is formed from *dub*, black, in the same way as *Flanducán* is from *flann*, red; the modern anglicized form is Dugan.

The other division of Cīrchenn or Gerrgenn is the Mearns or county of Kincardine. The name appears first in the Pictish Chronicle, which states that 'viri na Moerne, 'the men of the Mearns,' slew Malcolaim, son of Domnall, 'in Fodresach, id est, in Claideom,' 'at Fetteresso, that is, in the Swordland.'[2] This was in A.D. 954. A MS. of about 1100 says that Pleidias, *i.e.* Palladius, served God at Fordun in the Mearns (*isin Māirne*).[3] MacVurich gives *an Mhaoirne* as nominative, which, allowing for normal change, is the same as the dative *Māirne*, and as the genitive *Moerne*. The name is from M.Ir. *maer*, now *maor*, W. *maer* (from Lat. *maior*), a steward, an officer, with the abstract suffix seen in M.Ir. *bardine*, 'the bardic art,' 'a bardic composition'; W. *barddoni*, bardism; *apdaine*, the office of an abbot; then, abbey lands (Appin). The term does not occur in Irish or Scottish Gaelic; in the former 'stewardship' is *maoracht*, in the latter *maorsachd* or *maorsainneachd*; but Welsh has *maeroni*, and it is from this that *Maoirne* appears

[1] See p. 112. [2] Skene, *P.S.*, p. 10. [3] LU 4 a 39.

to have been borrowed.[1] *An Mhaoirne* thus means 'the Stewartry,' and the implication is that the district so called was ruled or administered by an official who held delegated authority. The origin of the name of the parish of Mearns in Renfrewshire is the same. The first mention of it is in the Register of Paisley, where Roland of Mernes witnesses a charter of about 1177 ; in 1179 (Reg. Glasg.) it is Meornes. The adjacent parish in Ayrshire is Stewarton. With these may be compared the Stewartry of Kirkcudbrightshire, so called from the fact that Archibald, Earl of Douglas, in or about 1372 received in perpetual fee all the Crown lands between Nith and Cree, and appointed a steward to collect his revenues and administer justice there.[2]

The ancient province of Fotla, which is one of the names of Ireland, appears later as Athol with Gowrie. Athol means 'new Ireland,' and will be referred to later. The name of Gowrie survives in the Carse and Braes of Gowrie and in Blairgowrie, or more properly Blair in Gowrie, as distinguished from Blair in Athol. The province or earldom extended from the head of Strathardle to the Firth of Tay, and included the 'manors' of Scone, Cubre (Coupar Angus), Forgrund (Longforgan), and Stretherdel (Strathardle).[3]

In the Book of Taliesin there is a poem to Gwallawc, son of Lleenawg, of the race of Coel and an ally of Urien in his wars against the sons of Ida. The poem, which deals with Gwallawc's battles, appears to be connected entirely with the North, and at the end foretells the defeat of the *Peithwyr* or Picts. 'Ravens shall wander in Prydein (Pictland), in Eidin, in Gafran, in the quarter of Brecheiniawc.' On this passage Sir E. Anwyl remarks, 'the quarter of Brecheiniawc (is) probably a Brycheiniog of the North,'[4] *i.e.* not the Welsh Brycheiniog, which is now Brecon or Brecknock. The name of this latter province is derived from its founder

---

[1] The old Scottish term for the office of a *maor*, however, was 'marnichty.' In *The Duke of Argyll* v. *Campbell of Dunstaffnage*, one of the Lords of Session held that the office of marnichty was equivalent to baron baillieship (*Court of Session Cases*, 1912, p. 496).

[2] Sir H. Maxwell, *History of Dumfries and Galloway*, p. 117.

[3] RMS, i.; Charter of Malcolm IV., confirmed by Robert I.

[4] *Celt. Rev.*, iv. p. 264.

Brachan or Brychan, a prince of Irish origin, on his father's side at least. According to the Welsh genealogies he had eleven sons and twenty-five daughters, one of whom, Luan or Leian, was the wife of Gabrán, king of Dál Riata in Scotland (d. 558), and mother of Aedán.[1] It is, however, generally agreed now that there must have been more than one noble of this name in early times whose history has been merged in that of Brachan of Brecheiniawc in Wales, and that there was a Brecheiniawc in the North is supported not only by this poem to Gwallawc but also by the name Brechin in Forfarshire, gen. *Brecini* (for *Brechini*) in the Book of Deer, and in the Pictish Chronicle, *Brechne.* This is probably a shortened Gaelic form on the analogy of the numerous Irish names in *-ne*, such as Brefne, Conmaicne, Cuircne; it may be compared with *Maoirne*, the Mearns. The Gafran mentioned in the poem was evidently a district, and the only district anywhere which can be meant is Gowrie, spelled Gouerin in Skene's *P.S.*, p. 136; Goverine, 1306 (Bain's Cal.), etc. Gabrán, king of Dál Riata, appears in the Welsh genealogies as Gafran; in Skene's *P.S.* as Gouren, Goueran; in Wyntoun as Gowrane. This again leads to the question whether the district may not be named after the man, as often happened, *e.g.* Kyle, Cowal, Lorne, Angus. We know no details of Gabrán's life before about 537, when he began to reign; if he lived to the age of seventy, he would have been then about fifty-two, for he died about 559. His son Aedán was closely connected with the British before he became king of Dál Riata in 574 and afterwards; he seems to have been brought up among them. If Gabrán's wife was the daughter of a Brachan of Forfarshire, he himself may have ruled the district adjacent to Forfarshire on the west, namely Gafran or Gowrie, before he went to Dál Riata. At that time (537) his son Aedán,

[1] Keating says that Gabrán's wife was named Ingheanach (*bean Gabhráin, Ingheanach a hainm*), and others have said the same thing after him. This, however, is an error caused by careless reading of the old tale about the birth of Aedán and of Brandub in Rawl. B 502, 81 b 21, in which it is said, 'maccach dano ben Echach, ⁊ ingenach ben Gabráin,' 'now Echu's wife bore (only) sons, while Gabrán's and bore (only) daughters,' p. 53, n. 4.

born in 533, would have been about four. With regard to the latter, Skene points out that for five years before he became king of Dál Riata (574) he had reigned somewhere else—south of the Firths of Forth and Clyde, he thinks. In 596, however, we find Aedán fighting a battle in Círchenn, and we wonder what it was that took him there. He must have had some claim: was he claiming what he considered to be his own patrimony? Was this the district over which he had once ruled, and which his father Gabrán had ruled before him? The hypothesis suggested serves to connect and explain the facts.

The province of Fortriu becomes later the two provinces or 'region' and 'sub-region' of Strathearn and Menteith. The latter is now in Gaelic *Tèadhaich*, without the article; the upland part was given me—rather doubtfully—as *Mon Tèadhaich*, 'the high ground or hilly part of Tèadhaich' (in Perthshire *monadh* is always *mon*, and now means specially 'hill pasture'). A Menteith man is *Tèadhach*, *e.g.* in *Dail an Tèadhaich*, 'the Menteith man's holm,' at the foot of Làirig an Locháin in Glen Lyon, where the leader of some Menteith raiders was killed by an arrow shot from the old tower (*caisteal*), the site of which is still traceable. Old forms are Meneted (*P.S.*, p. 136), Manethet 1264; Meneteth 1286; on seals Mentet, Meneteth, Menetet (Bain's Cal.). The final *d*(*t*) represents Welsh *dd*, Gaelic *dh*, and *-ted*, *-tet* is evidently the name of the river Teith. In Gaelic this would be presumably *Tèadh*, but the name is now completely lost; at least my experience was that of four Gaelic-speaking men born near Callander, two of whom were over 80 and had an excellent knowledge of the place-names, none knew the Gaelic form of either Teith or Forth. *Tèadhaich* means 'region of Teith,' a Gaelic formation in *-ach* from the British *ted*; Menteith is probably shortened from *minit-ted*, 'Teith Moor,' but the meaning of Teith is obscure to me.

The old province of Fife became later Fife and Fothreve. Fife, *Fīb*; gen. *Fībe* in 'clann Conaill Cirr . i . Fir ībe (for Fhībe),' 'the descendants of Conall Cearr (son of Eochaidh Buidhe, son of Aedán, son of Gabrán), that is, the men of

Fife.'[1] The dative is *Fīb* in the Book of Deer and in 'Cu-sīdhe a quo Clann Con-sīthe a Bhīb (for bhFīb),' 'Cu-sīthe from whom are Clann Con-sīthe in Fife.'[2] In modern Gaelic it is *Fìobh*; the Welsh or British is not recorded. The meaning of *Fīb* is quite obscure ; it may be a personal name, and it is not impossible that we have it in *Brude uip* (*P. S.*, p. 5), one of the thirty kings of the Picts who are said to have been called Brude, also in the name of a subsequent king, *Vipoig*. This again suggests comparison with *Vepogenos* of the inscription recording the gift of 'Lossio Veda nepos Vepogeni, Caledo.' The difficulty, however, is that it is doubtful whether O. British *p* could become *b* (*bh*) when taken over into Gaelic.

The district of Fothreve formed a deanery, containing roughly the parishes of what is now Kinross and West Fife. Early spellings are Fotriffe, 1070-1093 ; Fotherif, Fothrif, 1128 (Lawrie); Fothryffe, 1363 (RMS), Fothrik, 1450 (RMS) ; Blaeu has Forthridge Muirs north of Inverkeithing ; in the records it is usually coupled with Fife. The name is a Gaelic form of O. British *Vo-treb-*, 'sub-settlement,' in modern Welsh *godref*, 'small town, lodgement' ; *Godre Fynydd*, 'small town of the hill,' is at Aberllefeny in Merioneth ; *Godre Dewi*, 'St. David's Godref,' is in Caermarthen, south of Newcastle Emlyn. Fothreve seems to have stood in the same relation to Fife as the Mearns did to Angus.

The position of the two ancient provinces of Cē and Fidach is uncertain ; one of them extended from Dee to Spey, including Marr and Buchan and what is now Banffshire ; the other corresponded to Moray and Easter Ross, the large district from Spey to the Dornoch Firth, but it does not seem possible to determine which of these was Cē and which was Fidach. Cē might be supposed to survive in the name Keith in Banffshire, which is well known in Gaelic as *Cé* and *Baile Ché* ; the difficulty is, however, that in view of the old spellings of Keith—Ket, Keth, 1203 (Reg. Mor), etc.—the Gaelic spelling should be *Céith* rather than *Cé*,

[1] BB 140; Skene, *P.S.*, p. 315 note 7; A. O. Anderson, *Early Sources*, i. p. cliv.

[2] Skene, *Celt. Scot.*, iii. p. 481.

the sound of *Cé* and of *Céith* in modern Gaelic being very similar. A version of the ancient Irish tale of Frigriu, who eloped to Ireland with the daughter of Fubthaire of Hí, styles him 'the artificer of Cruthmag Cé,' and this may mean 'the Cruithnean plain of Cé,' but throws no light on its position.[1] There is a Loch Cé in Roscommon, twice referred to by Adamnan as 'Stagnum Cei.'

If *Fidach* is in the genitive case, as seems fairly certain, it is treated as a palatal stem. The Gaelic *fid*, now *fiodh*, and the Welsh *guid*, *gwydd*, are favourite elements in personal names. An early king of the Picts was named *Guidid*, 'woodsman,' which may be the British original of this same name, for the Gaelic speakers knowing the meaning of *Guidid*, would turn it into Fidaid, inflecting on the analogy of the common Gaelic name, *Eochaid*, gen. *Echach*. There is a Glen Fiddich in Banffshire, given to me in Gaelic by a good authority as *Gleann Fithich*, which cannot be 'Raven's Glen,' for *th* of *fitheach*, Ir. *fiach*, is not radical.

Marr is in M.Ir. *Mar*, gen. *Mair* (AU 1014); also gen. *Marr* (Book of Deer). If the form *Marr* is correct, as it seems to be, judging from the modern Gaelic, it would represent an early *Mars-*, whence Stokes compared it with the Italian tribe *Marsi* and with *Marsigni*, the name of a tribe in Bohemia, which means 'sprung from Marsos.' It is worth noting that Mar was a British personal name, borne by a son of Ceneu, son of Coel.

Moray is now in Gaelic *Moireabh*; a Moray man is *Moireach* for *Moireabhach*, as *bantrach*, a widow, is for *ban-treabhach*, 'a female householder.' The Irish records have gen. Murebe, 1032 (AU); Muireb, 1085, *ib.*; Moreb, 1130, *ib.*; Moriab, 1116, *ib.*; the Pictish Chronicle has "in Moreb"; Wyntoun has Murrave; the Norse form is Mærhafi, sometimes Morhæfi; in Latin Moravia. *Moireabh* is a compound of *mor*, the compositional form of *muir*, the sea, and *treb*, a settlement, and stands for an early Celtic *mori-*

[1] Kuno Meyer in his *Contributions* took *Cé* here to mean 'this,' *i.e.* 'Cruthmag here in Ireland,' as, of course, it might. But Frigriu is expressly said to have been from Britain (*inis Bretan*) (BB 399 b 40); or of the Fomorians of Fir Falga (in the Western Isles) (BB 399 b 23).

treb-, 'sea-settlement,' 'seaboard settlement,' becoming in Gaelic *Moirthreabh* ; it is the *th*, now quite silent, which is represented by *h* of the Norse form. We may compare the Gaulish *Morini*, 'sea-board folk,' the name of a tribe who lived on the shore of the English Channel. The province of Moray at one time extended to the western sea, for we find the western part of Inverness-shire styled 'Ergadia quae ad Moraviam pertinet' '(the part of) Argyll which pertains to Moray.'

By Ross is meant that part of the present county which is east of the Wyvis range, ending in Tarbat Ness. In old Gaelic and still in Irish *ross* means both a promontory and a wood, both follow naturally from the original meaning of 'something forthstanding,' for it comes from *(p)ro-sto-*. In Welsh it means a moor, heath, plain. The meanings 'promontory' and 'moor' are both applicable, for the district as a whole is a promontory and a very large one, while at no very distant date it was a moor and it still contains a number of names such as Muir of Ord, Delny Muir, Tullich Muir, Muir of Fodderty, and so on, where 'muir' translates Gaelic *blàr*. Ross occurs fairly often with us in the sense of 'promontory,' but it usually applies to smallish promontories ; the largest elsewhere is *an Ros Muileach*, 'the Ross of Mull,' which, it may be noted, has the article, while the county name has not. There is, however, nothing decisive in favour of either meaning.

On the sunny side of the Cromarty Firth, between the river Averon on the east and *Allt na Làthaid* near Dingwall on the west, is *Fearann Domhnaill*, Ferindonald, 'Donald's land,' the home land or *dùthchas* of the Clan Munro ; it is co-extensive with the parishes of Alness and Kiltearn. According to a tradition which may be substantially correct the ancestors of the Munros came from Ireland, from the foot of the river Roe in Derry, whence the name *Bun-rotha*, giving *Mun-rotha* by eclipsis of *b* after the preposition *in*, in the same way as *i mBun-locha* gives Munlochy. The river is *Roa* (two syllables) in O'Donnell's *Life of Colum Cille*, and would be *Rotha* with us. A Munro is always *Rothach*, 'a Ro-man,' in Gaelic. There is another Ferindonald in Skye.

The province of Cat, which included south-east Sutherland—the 'Southland' of the Norsemen of Caithness—as well as what is now Caithness, has been mentioned. The 'hill called Mound' which divided the old province in two is now the Ord of Caithness, in Gaelic *an t-Ord Gallach.* The modern 'Mound' is an embankment which carries the public road over the tidal water of the river Fleet, constructed in 1816 and pierced with four arches and sluices for the passage of the river and sea water. The district on the north side of the Dornoch Firth and Kyle of Sutherland, corresponding roughly to the parish of Creich, was formerly called Ferincoskry. A writer in Macfarlane's Collections states: 'there is a part of Sutherland within the parish of Creigh called Chilis or Ferrin-Coskarie, which is eighteen miles in length, lying upon the north side of the river of Port ne Couter (Port a' Choltair, 'Coulter Ferry,' now the Meikle Ferry) and Oikell, where there are hills of marble.'[1] This district often occurs on record in various spellings, and though the Gaelic form is now forgotten the name represents *Fearann Coscraigh,* 'Coscrach's land.' The personal name *Coscrach* or *Coscarach,* is from *coscar,* victory, and means 'conqueror'; it was once fairly common in Ireland, but does not seem to have been much used in Scotland. Drostan mac Cosgreg, 'Drostan, son of Coscrach,' according to the Book of Deer, came from Hí with Colum Cille to Aberdour in Buchan.

The Reay Country in the north of Sutherland is now *Dùthaich Mhic Aoidh,* 'Mackay's Country,' but the older name for it is *Meadhrath,* spelled *Meghrath* by MacVurich, now applied to the parish of Reay. This is the form which I have heard generally, but there is another form, *Mìodhrath,* also in use now, and this is the form used by the Reay poet, Rob Donn, who makes it rime with *gnìomh sin.* In his poems it is spelled *Mìoghradh,* gen. *Mìoghraidh.* The second part is *ràth,* a circular fort; a hill south of Reay village is called *Beinn Ràtha.* The first part is most probably O.Ir. *mid-,* mid, middle, as in the compound *meadh-bhlath,* lukewarm, pronounced often *mìodh-bhlath*;

[1] *Geographical Collections,* iii. p. 101.

the meaning is thus 'mid-fort,' probably with reference to its position midway between the waters of Halladale and Forss. The formation is like that of Medio-lānon, probably 'mid-plain'; Medio-nemeton, 'mid-temple, mid-shrine.' About two miles east of Reay village is *Dùnrath*, Dounreay, which may mean 'fort-rath,' with reference to a broch near it; this is probably more likely than to postulate an E.Celt. *Dūno-rāton* or *Dūno-rātis*.'[1] The *a* of the (unstressed) second syllable of *Meadhrath*, *Mìodhrath*, *Dùnrath*, is open, indicating that it was originally long.

We may now mention such of the other principal divisions as have not been dealt with.

Bàideanach, Badenoch, is from *bàithte*, drowned, submerged, with suffixes *-n-ach*, 'the drowned land,' 'submerged land,' so called because it was subject to inundation from Spey, as it still is occasionally, notwithstanding that Spey has been banked; compare *an Fhéith Bhàithte*, Febait, 'the drowned bog,' in Ross-shire. *Bàideanach*, like *Raithneach*, is not declined, nor does it take the article.

Garioch, in Gaelic *Gairbheach*, is a large district in Aberdeenshire; it is not declined, and does not take the article, though the fact that in English it is commonly called 'the Garioch' rather indicates that it took the article when Gaelic was general in the county. *Cath Gairbheach* is 'the battle of Harlaw.' An old writer says that 'the Garioch is so called as being a rough ground, for so the word in Irish signifies.' The formation is uncommon, for the term seems to come from the abstract noun *gairbhe*, roughness, with the suffix *-ach*, meaning 'place of roughness.'

Formartin, the district of Aberdeenshire between Ythan and Don, is *ferann Martain*, 'Martin's land,' in a note on Oengus' *Félire* at Nov. 3 (see p. 318). In 1286 Reginald de Chen was *firmarius* of the thanage of Fermartyn. The name is to be compared with *Talamh Martuinn*, 'Martin's land,' in Uist, Strathmartin in Perthshire, Ferindonald in Ross, and Ferincoskry in Sutherland, but there is nothing to show who the eponymous Martin was.

[1] *Celt. Rev.*, ii. p. 238.

Buchan, the district north of Formartin, is *Bucain* (gen.) in the note just mentioned, and *Buchan*, *Búchan* (gen.) in the Legend of Deer; *Buchain* (dat.) in the Pictish Chronicle; one of the Norse mythical sagas mentions *Búkan-síða*, Buchan-side; the *u* appears to be long. There are also Buchan with Buchan Burn and Hill on Loch Trool in Kirkcudbright, on record as 'foresta de Buchane' 1526 (RMS); Buchan on Carlingwark Loch, Kirkcudbright; the barony of Buchquhane in Strathore, Fife, 1530 (RMS), now apparently obsolete; with which may go *Buchantaigh*, Buchanty in Glen Almond, on record as Buchondy 1428 (RMS), and Buchany in Kilmadock parish, Buchny 1511 (RMS). *Beinn a' Bhuchanaich* is behind Daviot Station, Inverness. These are difficult names and may be not all of the same origin. The district names may be connected with W. *buwch*, a cow.

Lennox denotes primarily the district about Loch Lomond, but it was used in a much wider sense. The name of the river from Loch Lomond is *Leamhain*, gen. *Leamhna*, Leven, meaning 'Elm-water.' Its plain, the Vale of Leven, was *Magh Leamhna*, and Loch Lomond was of old *Loch Leamhna*, all connected with Ptolemy's Lemannonios Kolpos. There is a river of the same name in Kerry, the Laune, and there was a district called *Leamhain*, gen. *Leamhna* (M.Ir. *Lemain*, *Lemna*) in the north of Ireland; *inis Leamhna* is an islet in Belfast Lough. With us also *Leamhain* is used as a district name, as in *Mormhaer Leamhna*, 'lord of Lennox,' in the poem by Muiredach Albanach.[1] A Lennox man is *Leamhnach* now and in the Irish records; *Mormhaer Leamhnach* means 'Lord of the Lennox men'; *Lemnaig Alban*, 'the Lennox men of Alba,' were ultimately of the same origin as the Eoghanacht of Munster.[2] This explains 'the earl of Liuenath,' 1237 (Bain's Cal.); 'earl of Levenath,' 1244, *ib.*; meaning 'earl of the Lennox men,' 'Leuenax,' 1199 (Reg. Pasl.), 'Levenax,' 1259 (Bain's Cal.) is the English plural of *Leamhnach*. The Lennox men are still *Leamhnaich* in Gaelic. MacVurich's term for Lennox is *Leamhnacht*, formed like *tòiseachd*, etc., and the Dean of

[1] Skene, *Celt. Scot.*, iii. p. 455.

[2] BB 41 b 41.

Lismore wrote it phonetically *Lawenacht* (*Rel. Celt.*, ii. 91) in the early part of the sixteenth century, but I have not found any really old instance of this form.

'Stormont,' says an old writer, 'is that part of Perthshire which lyes betwixt the rivers Tay, Yla, and Erich (Ericht), and hath upon the north and north-east Athol and Strath Erich.'[1] This corresponds to the present bounds of the district. Another and older authority states that 'Stormonth is devyded be the river of Tay in two parts, viz. West and East Stormonths.' West Stormonth included Logy Almond, Little Dunkeld, Auchtergaven, Luncarty, and Kinclavin. East Stormont included Caputh, Lethendy, Cluny, Lundie, and Bendochy.[2] From the record notices of West Stormont this latter account appears to be correct for the earlier period. In the text of the law called 'Claremathan' it is given in Latin as Stratyeymund and in Scots as 'the Starmunth' (*Acts of Parl. Scot.*, i. 373). Other early spellings are Starmonde, 1376 (RMS), 'via qua itur versus *le* Starmonth, 1374/7 (RMS); Wyntoun has 'In the Stermond at Gasklune' (Bk. IX., i. 1561). The name is a compound of G. *stair, stoir*, 'stepping-stones over a bog or in a river,' sometimes, too, a rude bridge, and *monadh*, a moor, and means 'stepping-stones moor'; it is perhaps impossible now to determine what particular moor gave rise to the name.

Rannoch is in Gaelic *Raithneach*; it is not used with the article nor is it declined; the Dean of Lismore has *a crith rannyt*, for *a crích Raithneach*, 'from the bounds of Rannoch.' The meaning is 'bracken region'; compare the Old Welsh *Tref-redinauc*, explained as *villa filicis*, 'fern ville.'[3] Near Loch Tummel, at the lower end of the Rannoch basin, is *Both-reithnich*, Borenich, 'bracken booth,' showing inflection, and vowel as in Welsh. I have already suggested that Rannoch may be of British origin.

Argyll is in M.Ir. *Airer Gáidel*, later *Oirer Gháidheal*, spelled in the Dean of Lismore's book *orreir zeil*. The

[1] Macfarlane's *Geog. Coll.*, iii. p. 221.

[2] *Ib.*, ii. p. 571.

[3] *Cambro-British Saints*, p. 50.

meaning is given correctly as *margo Scottorum*, 'coastland of the Scots,' *i.e.* of the Gael. 'Harald, bishop of Ayr-gaythyl' is on record in 1228 (RM). In a poem of the sixteenth century it is *Eir-ghaodheal,*[1] now *Earra-Ghàidheal.* As the Dalriadic kingdom was sometimes called Alba, a natural variant is *Oirer Alban*, which occurs in one of the lays of Deirdre. In its widest extent Argyll meant all the western coast from the Mull of Kintyre to Loch Broom ; the section north of the point of Ardnamurchan was called the *Oirer a Tuath*, and that to the south of it *Oirer a Deas*, the northern and the southern coastland respectively. In a restricted sense Argyll denotes the district between Lorne and Cowal, bounded on the north by the lower part of Loch Awe, Loch Avich, and Loch Melfort, on the east and south-east by upper Loch Fyne, and on the south by Loch Gilp and the Crinan Canal—Neil Munro's 'real Argyll.'

The four chief tribes of Dál Riata were the tribe of Gabrán, son of Domongart ; the tribe of Comgall, son of Domongart ; the tribe of Loarn Mór, son of Erc ; and the tribe of Oengus, son of Erc. No trace of the first and last mentioned remains in the place-names of the West ; I have suggested that Gabrán's name is to be found in Gowrie ; the others have given name to the districts of Cowal and Lorne. The personal name Loarn, older Loern, pronounced as two syllables, stands for an older *Lovernos*, 'fox' ; a chief of the Gaulish Arverni bore the name of Lovernios, 'son of the fox,' and an inscription in Caernarvonshire reads 'Fili Loverni Anatemori.' The word is now obsolete in Welsh, but the Book of Llandaf has 'villa crucov leuirn,' or 'leuguirn,' 'the vill of the hillocks of the foxes' (p. 262) ; Cornish *lowern* ; Breton *louarn*. In modern Gaelic Lorne is *Latharn* or *Latharna*, the *th* serving merely to separate the syllables. A Lorne man is *Latharnach*, and the town of Oban is *an t-Òban Latharnach*, 'the little bay of Lorne.' In Aelred's account of the Battle of the Standard he records that the third division consisted of the men of Lothian (Laodonenses), the Islesmen, and the men of Lorne

[1] Adv. Lib. MS., XXXVI, 79 b.

(*Lavernani*).[1] Larne, in Antrim, is of different origin, being traditionally derived from Lathair, an early Irish prince, who received this district as his portion ; the early spelling is *Latharnu*.[2] The parish of Latheron or Latheron-wheel, in Caithness, is in Gaelic *Latharn a' Phuill*, 'Latheron of the Pool,' or 'of the Hole' ; in 1287 (Theiner) it is Lagheryn and Laterne. Here the *th* (*gh*, *t*) is almost certainly radical (not, as in *Latharn*, Lorne, a mere phonetic device) ; the name is probably from *lath*, W. *llaid*, mire, whence come Ir. *lathach*, Sc.G. *làthach*, mire, puddle. There is a place called Lorne on the north side of Loch Earn ; Donylawrne, 1542 (RMS), Donylawren, 1564, *ib.*, later Donylairne, is now *Dùnaidh*, Dounie, near Ardgay, Ross-shire.

The tribe of Loarn was of old in three divisions—*Cenél Fergusa*, 'Fergus' Sept,' *Cenél Cathbath*, 'Cathbad's Sept' ; and *Cenél n-Eachach*, 'Eochaidh's Sept,' all named after sons of Loarn (P.S., p. 313). 'Jugulatio generis Cathboth,' 'slaughter of the sept of Cathbad,' is recorded in 701 (AU).

Cowal is in Gaelic *Comhghall*. An interesting trace of the old fourfold tribal division appears in the description of the church of Kilfinnan in Cowal as 'ecclesia Sancti Finani que est in Kethromecongal' 1253 (Reg. Pasl.), 'the church of St. Finan which is in the quarter of Cowal.' According to the writer of the New Stat. Account (1843) this parish was at that time 'well known in the district as Kerry or Ceathramh,' *i.e.* 'the fourth part.'

The part of Lorne now known as Morvern was of old called Kinelvadon,[3] *i.e. Cineal Bhaodain*, 'Baodan's tribe,' from Báetán (Bóetán, later Baodán), grandson of Loarn Mór. The name Baodán occurs in *Cill Bhaodáin*, 'Baodán's church,' in Ardgour ; in *Baile Bhaodáin*, 'Baodán's Stead,' in Ardchattan ; and in *Suidhe Bhaodáin*, 'Baodán's Seat,' in Gleann Salach, near Ardchattan, a boulder which was broken up many years ago. In these instances *Baodán* is doubtless the name of a saint, not to be identified with

[1] Skene, in the preface to Fordun (*Historians of Scotland*, v. ii. p. liii.), makes them Lennox men, wrongly.

[2] Joyce, *Irish Names*, I.

[3] *Orig. Paroch.*, ii. pt. I., pp. 188, 189, 190 ; Skene, *Celt. Scot.*, i. p. 264.

Báithen, the successor of Columba, whose name would now be *Baoithein*. The writer of the New Stat. Account of Ardchattan unfortunately confused Baedan with Medan, an error which has been faithfully copied ever since : Modan stands for *M'Aodhán, i.e. Mo Aodhán* (earlier *M'Aedán*), 'my Aedán,' as in Kilmodan, Glendaruel, which is in Gaelic *Cill Mhaodháin*.[1]

Morvern itself is badly spelled Morven on maps; in Gaelic it is *A' Mhorbhairn*, gen. *na Morbhairne*; a poem of the sixteenth century has *san Mhorbhairn*, 'in Morvern'; in the next century the nominative is written *an Morbhairni*.[2] On record it is Morvarne, 1476 (RMS), etc. As *o* in the first syllable is short and in stressed position, the first part cannot be *mór*, big; it is the compositional form of *muir*, the sea, as in *mormhoich*, Ir. *murbhach*, 'sea-plain,' Welsh *morfa*, from *muir* and *magh*, a plain. The second part is *bearn*, also *bearna*, a gap, and Morvern means 'sea-gap,' with reference to the great indentation of Loch Sunart which makes Morvern a peninsula and nearly an island. *A' Mhorbhairn* in fact is the pre-Norse name of Loch Sunart. With it we may compare *Loch Shubhairne*, Loch Hourn, spelled by the Dean of Lismore *sowyrnni* for *Subhairne*. At the head of Loch Hourn is *Coire Shubh*, 'Corry of fruits,' or 'of berries.' Its water, *Allt Coire Shubh*, enters the loch, which lies here in a tremendous *bearna* or gap. *Subhairne*, therefore, is for *subh-bhearna*, 'berry gap.' The phonetics are the same in both cases, and they are quite regular, *e.g. samh*, summer, compounded with *seasg*, dry, yeld, results in Irish *samhaisg*, a heifer. The variation in the nominative between *morbhairn* and *morbhairne* corresponds to the variation in *bearn, bearna*. The number of districts on the west coast which are named after sea inlets is rather notable.

Cenél Sétna, 'Sétna's Sept,' in Islay, was named after Sétna, son of Fergus Beag son of Erc; the sept was known

---

[1] The record forms of Baile Bhaodáin and Cill Bhaodáin always have *b*—Balliebodan, Kilbedan (Ardchattan); Kilbodane, Kilbedane (Ardgour).

[2] Adv. Lib. MS., XLII, 23 a; LII 34 a.

later as *Cenél Concridhe*, so named after Sétna's grandson (BB 148 b 47; P.S., p. 310).

Between Loch Creran and Loch Leven is the pleasant district of Appin, G. *Apuinn*, gen. *Apunn*, *na h-Apunn*, from M.Ir. *apdaine*, abbacy, hence concretely 'abbey land.' It probably belonged to Moluag's community of Lismore. There were many districts of this name,[1] but the two best known in the North were this one and Appin of Dull, near Aberfeldy, and they were distinguished in Gaelic as *Apuinn mhic Iain Stiubhairt*,[2] 'John Stewart's son's Appin, *Apuinn nan Stiubhartach*, 'the Stewarts' Appin,' and *Apuinn nam Mèinnearach*, 'the Menzies' Appin,' respectively.

In the northern part of Appin is the little district of Duror, styled in Gaelic *Dùror na h-Apunn*, 'Duror of Appin.' It is named after its river, called in Macfarlane 'Avon Durgur.' Old spellings are Durdoman (for Durdowar), Durwoin, Durgune, Durgwyn, etc. (RMS, i., and Robertson's Index), in all of which *n* should be *r*. The name is from *dūr*, hard, and *dobhar*, water, stream, giving *dùrdhobhar*, 'hard water,' 'rocky water,' a formation like the Welsh *Caleddfrut*, 'hard stream,' and like our Calder.

With Duror, as regards origin, goes *Mórar*, in the west of Inverness-shire. This name is on record as Mordhowar, Moreovyr, Morowore, and is very plainly *Mórdhobhar*, 'big water'; it is the name of the stream from Loch Morar, which though short is of considerable volume—it seems, in fact, to be the biggest *dobhar* in Scotland. Both these district names, originally names of streams, are doubtless old British names taken over into Gaelic.

Lochaber has been already noticed. Moidart, Arasaig, and Knoydart are all Norse, and their pre-Norse names are unknown; they are primarily names of sea-inlets ending in *fjörðr*, firth, and *vík*, bay. Further north, Loch Alsh, Loch Carron, and Loch Broom are also primarily names of sea-inlets which have become names of districts. Applecross, however, is different; it is in Gaelic *a' Chomraich*,

[1] Skene, *Celt. Scot.*, ii. p. 393.

[2] From John Stewart of Lorne, who flourished in the middle of the 14th century, known as *Iain Dubh nan Lann*, 'Black John of the sword-blades.'

gen. *na Comraich*, 'the sanctuary,' 'girth,' of Maelrubha who founded the monastery of Applecross in A.D. 673, and who died in 722, and is buried there. The girth is said to have had a radius of six miles, and it appears to have been marked by stone crosses.[1] The other great Comraich of the North was that of St. Duthac (*Dubhthach*) of Tain, and it was likewise marked by stone crosses, the position of which is known.

The north-west part of Loch Broom parish is called *Cóigeach*, gen. *na Cóigich*, 'the place of fifths,' a collective term. There is tradition of an old division between three brothers—not five—who used to meet at a great stone called *Clach na Comhdhalach*, 'the trysting stone.'

Part of Assynt, in Sutherland, is called *Ailbhinn*, Elphin, 'rock-peak.' There is another Elphin in Co. Roscommon, but it is stressed on the second syllable and written in O.Ir. *ail find*, 'white rock.'

[1] The stump of one of the crosses is still to be seen; the cross itself, as I was told on the spot, was smashed by a mason about fifty-five years ago, when the school was being built.

CHAPTER V

GENERAL SURVEY OF LOTHIAN

BEDE, who died in A.D. 735, says that Britain 'studies and confesses one and the same knowledge of the highest truth in the tongues of five nations, namely the Angles, the Britons, the Scots, the Picts, and the Latins.' Elsewhere he says that Oswald received under his sway all the nations and provinces of Britain, which is divided into four languages, those of the Britons, Picts, Scots, and Angles. Adamnan, who died in A.D. 704, makes it clear that whatever language the Picts spoke, it was different from Gaelic, for he mentions that Columba used an interpreter on two occasions, once in Skye, once in some place on the mainland. In Skye he was addressing a man of good position, in the other instance a 'plebeian.'

As I have already pointed out, the evidence at our disposal goes to show that at the time of the Roman occupation the language current all over Britain was Celtic of the P-group, that is to say, Old British, represented now by Welsh. Within this there may have been, and very probably were, dialectic differences, as there were in Gaul. The Roman occupation had the inevitable effect of separating and alienating the unconquered tribes from the Britons of the Roman province; in the fourth century the former were the bitter enemies and scourge of the provincials. Their relations were somewhat similar to those between the Lowland Scots and the Gaelic clans at a later period, with the difference that the Lowland Scots had the power of the Crown to help them; the provincials, once the Roman protection was withdrawn, were almost defenceless. This separation, which began when the language was still in the Old British stage, with full inflections, continued while the language was passing into its modern form, a period involving very great linguistic changes. That the change should proceed at the same rate or on precisely the same lines among the two sections of the population would be quite too much to expect; especially when it is remembered that the language of the southern Britons was considerably influenced by Latin. The cleavage between north and south is reflected in the difference that emerges in the national names: the old *Pretani* or *Britanni* are now the tribes of the north, *Prydein*; the romanized, or partly romanized, tribes are now *Brython* (*Britton-es*). It is therefore not difficult to understand how the language of the Picts comes to be reckoned as differing from that of the Britons.

The Britons of the region that is defined, roughly speaking, as between the two Walls—from the Forth and Clyde isthmus to the Tyne—were, of course, Brythons; they were known as *Gwyr y Gogledd*, 'the Men of the North,' and much of the earliest Welsh poetry is concerned with them and their doings. The genealogies of their ruling families who lived in the fifth and sixth centuries are extant, and have been printed more than once.[1] There are two great branches. The first is that of the kings who ruled from *Alclut* (Dumbarton), and of these the earliest who is known to history is Ceretic Guletic, the Coroticus to whom St. Patrick addressed his epistle. In Columba's time this line was represented by Riderch Hael (Rodercus Largus of Adamnan), 'Riderch the Liberal,' who was fifth in line from Coroticus; it continued as the line of kings of Strathclyde. The other ruling family was that of Coel Hen, whose period seems to be about A.D. 400, and whose name is preserved in the division of Ayrshire called Kyle. The family seem to have been originally rulers of Ayrshire and Galloway; in the sixth century the princes of Lothian, from Forth to Tweed and as far south as Carlisle, belong to this line. About the end of the Roman occupation, the region between the Walls was ruled by a prince named Cunedag (later Cunedda), son of Aetern, son of Patern

[1] *Y Cymmrodor*, ix.; *Archiv f. Celt. Lexik.*, i.; Skene, *P.S.*, ii. p. 454.

Pesrut (of the red mantle), whose Roman connection is obvious. His mother is said to have been a daughter of Coel; in the 'Deathsong of Rhun,' Etern's sons and the Coeling (descendants of Coel) are styled kinsmen.[1]

At this period, soon after A.D. 400, the pressure from the Picts and Scots became so heavy that, according to Gildas, they seized all the northern part of the land up to the Wall, displacing the natives.[2] Gildas is, of course, referring to Roman Britain, and the Wall is the southern Wall, not that between Forth and Clyde. Similarly the Irish Nennius says: 'immediately (after the Roman evacuation) the power of the Cruithnigh and Gaels rose over the middle of Britain, and they drove the Britons out up to the river which is called Tin (Tyne).'[3]

These statements illustrate the other in the additions to Nennius that 146 years before the reign of Mailcun, his ancestor Cunedag, with his eight sons, came from the North, from Manau of the Guotodin, and drove out the Scots from Guenedot with very great slaughter. Mailcun, later Maelgwn, was the powerful king of Gwynedd (*Guenedot*), or North Wales, who died in A.D. 547, and Manau of the Guotodin was the district round the head of the Firth of Forth, whose name remains in Slamannan and Clackmannan. The inference is that from about A.D. 400 there was a gap in the succession of rulers between Forth and Tyne, during which the district was under the Picts and Scots. How long this state of matters lasted we have no evidence; very little is known for certain of the history of the district during the fifth century. It would appear, however, that towards the end of this century the Britons began to recover, and it is to this period, the end of the fifth century and the beginning of the sixth, that the activities and exploits of Ambrosius and Arthur belong, the former in the south, the

---

[1] Sir J. Morris-Jones, *Taliesin*, p. 210.

[2] Omnem aquilonalem extremamque terrae partem pro indigenis muro tenus capessunt' (*Excidium*, 19). This was the third *vastatio*; in consequence of it the Britons appealed to Aegitius (Aetius), consul for the third time. Aetius' third consulship was in A.D. 446.

[3] *Irish Nennius*, p. 72.

latter, as I believe, in the north. The chronicle attached to Nennius gives A.D. 537 as the date of the battle of Camlan and the death of Arthur. By the middle of the sixth century we find native rulers firmly established from Forth to Tyne, and according to the genealogies, they are of the line of Coel, to which, as we have seen, Cunedda belonged on his mother's side. In A.D. 547 the Anglic prince Ida, son of Eobba, founded the kingdom of Bernicia, and thenceforward there was a severe and continuous struggle between the native rulers and the foreign aggressors. In this contest two of the British princes stand out conspicuous, Urbgen (Urien) of Rheged, and his son Owein, who according to tradition was the father of Kentigern. The position of Rheged has been much discussed, but from the fact that Carlisle was situated in it, Rheged cannot have been in the north about Dumbarton, as Skene thought.[1] Urien was killed treacherously while besieging Medcaut (Lindisfarne), probably about A.D. 585. The date of Owein's death is uncertain; he was not slain by Flamddyn (Deodric), as Skene says, but slew him.[2]

The final effort against the Angles at this period was made not under the leadership of a British chief, but under Aedán mac Gabráin, the king of Dál Riata. Aedán was born in A.D. 533. His mother was a British princess, and he was closely connected with the British in his younger days. Born some four years after the death of Arthur, he must have been well acquainted with the story of his exploits, and it is specially notable that he named his eldest son Arthur—the first Gael, so far as we know, to bear that name. After he became king in 574, his activities were not confined to the West: he fought the Miathi in Circhenn and lost there two sons, one of whom was Arthur; [3] he is styled 'prince of Forth'; he is said to have had a residence at Aberfoyle.[4] Now in A.D. 602-3, at the age of seventy,

[1] See Sir J. Morris-Jones, *Taliesin*, p. 64 *seqq.*

[2] *Celt. Scot.*, i. 158; *Taliesin*, pp. 90, 187.

[3] So Adamnan; Tighernach, under the year 596, states that he lost four sons in that battle.

[4] See p. 225.

Aedán appears as head of a great confederate host against the Angles under their king Aedilfrid. Among his followers was Mael-umae, son of Báetán mac Muirchertaigh,[1] a famous Fian-leader (*rígféinnid*) from Ireland, who slew Eanfrid, brother of the Anglic king. The decisive battle was fought at Degsastan, which Skene has identified with Dawstone in Liddesdale. Aedán's host was routed, and the victory of the Angles made them masters of the east from Forth to Tweed and of Galloway. The result was finally reversed five hundred years later at Carham (1018).

There is no proof of Teutonic settlement in Lothian in the sixth century, though there may have been raids. No Anglic or Saxon grave of the pagan period, nor yet any example of Anglo-Saxon art of the pagan period, has been found in the south of Scotland.[2] The Angles accepted Christianity in A.D. 627, under Edwin, who succeeded the victor of Degsastan ; it was during his reign that settlement took place. The process was marked by the settlement in 681 of Trumwin as bishop at Abercorn.

In this connection some difficulty is presented by the operations of the Ulster king, Báetán mac Cairill, who is recorded to have cleared Manu of foreigners (*ó gallaib*), 'so that its sovereignty belongs to Ulster thenceforth ; and in the second year after his death the Gael abandoned Manu.' [3] Báetán's death is recorded in AU at 581 and also at 587. In 582 and 583 the Annals of Ulster record battles in Manu fought by Aedán mac Gabráin. Reeves was of opinion that Aedán's battles (or battle) took place in Manu of the Gododdin, about the head of the Firth of

---

1 Mael-umae died in 610 (AU). He was for a time with Columba in Iona (see *Celt. Scot.*, ii. p. 494). His genealogy is given in Rawl. B 502, 140 a 38, where he is styled 'Maelhuma in rīgfēinnid' ; also in LL 349 g : 'Maeluma mac Baétain maic Muirchertaigh qui fuit filius Ercae quae fuit filia Loairn' (cf. BB 219 f 38). In Rawl. 140 a 10 he is styled 'Maelhuma heros no garg,' 'Maelhuma the hero or the fierce.' He was counted a saint (LL and BB as above).

2 Compare Professor G. Baldwin Brown, *Proc. Soc. Ant. Scot.*, xlix. p. 335 (1915).

3 *P.S.*, p. 128, 'is leis glanta Manand o gallaib' ; so BB ; LL 330 b and the Laud genealogies (*Celt. Zeit.*, viii. p. 327) omit *o gallaib* ; the latter has *Manu*.

Forth, which agrees with the other notices of him.[1] Skene would assign Báetán's campaign to the same district,[2] not to the Isle of Man, which has the same name as the other district both in Gaelic and in Welsh. Here we may note that the poem on the kings of Ulster, preserved in the Book of Ballymote, makes the conqueror of Manu to have been Cairell, not his son Báetán, and adds that Cairell died of grief at Arran, in the Firth of Clyde: the cause of his grief is not stated.[3] While this indicates some uncertainty as to agent, the mention of Arran may perhaps be taken as evidence that the Manu in question was Manu of the Gododdin: Arran would be in the track of a voyage thence to Ulster. On the other hand Tighernach and the Annals of Ulster have at 577 the entry, 'first adventure of the men of Ulster in Eumania' (or Eufania), and again at 578, 'return of the men of Ulster from Eumania.'[4] Further, the Welsh Annals record at 584, 'a war against Eubonia.' Now Nennius says of the island between Ireland and Britain that it is called 'Eubonia, id est Manau,' *i.e.* the Isle of Man. If these entries refer to the wars of Báetán as they may well do, notwithstanding the difference in dates, they would seem to prove beyond question that his operations were in Man, not in West Lothian. I say 'seem' because there is still the curious fact that Fordun and others apply the term 'Emonia' to the Isle of Inchcolm in the Firth of Forth, and there is just the possibility that this name is ancient and related to the adjacent district of Manu. We have, however, no other evidence that this Manu was ever called Eubonia, Eumonia, or Emonia, or indeed that the application of this term to Inchcolm is really ancient.

---

1 *Vita S. Columbae*, p. 371. 2 *Celt. Scot.*, i. p. 160.

3 BB 52 c 33: 'A cuic fo . V . Choirill calma/do clainn Muredaig mind sluaigh/marb in fer do mhugaigh Manaind/do chumaidh ac Araind uair.' 'Five times five years the reign of valiant Cairell, of Muredach's children, diadem of a host; dead is he who vanquished Manu, of sorrow at cold Arran.' The initial short *a* of *Araind* (rhyming with *Manaind*) proves that Arran in the Firth of Clyde is meant; the initial *a* of the Irish Aran is long.

4 'Primum periculum Uloth in Eufania' (577 AU); 'reuersio Uloth de Eumania' (578 AU). Tighernach has Eumania in both.

The Book of Leinster (190 a) makes mention of a tale, now lost, entitled '*Sluagad Fiachna maic Báitáin co Dún ṅGuaire i Saxanaib*,' 'The Hosting of Fiachna son of Báitán to Dún Guaire in Saxon-land.' This Fiachna was a son of Báetán mac Cairill, king of Ulster; he was a famous warrior, and was killed in A.D. 626. *Dún Guaire* is the Irish form of the British *Din Guayroi*, the native name of Bebbanburch, now Bamborough, the capital of Bernicia. It is doubtless to this expedition that we are to refer the notice in the Annals of Ulster (A.D. 623), '*expugnatio Rātho Guali la Fiachna mac Báetáin*,' 'the storming of Ráth Guali by Fiachna son of Báetán.'[1] I mention this as not only a matter of interest in itself, but as also showing that, if the Britons were quelled by the result of Degsastan, the Gael were not.

The number of Old Welsh place-names that still survive between Forth and Tweed indicates that the native Britons long persisted under the rule of the Angles. A rough provisional list from the Ordnance Survey Map of one inch to the mile gives the following numbers for the various counties: Linlithgow, 20; Edinburgh, 52; Haddington, 32; Berwick, 42; Roxburgh, 52; Selkirk, 22; Peebles, 43.

As for the Britons of Strathclyde, their line of British kings lasted till the early part of the eleventh century. The last of them, Owein the Bald, helped Malcolm, son of Kenneth, at Carham, and seems to have died soon afterwards. He was succeeded by Duncan, Malcolm's grandson, and when Duncan became king of Scotland, Strathclyde practically ceased to form a separate kingdom, though its people continued to be styled officially *Walenses* (Welshmen) till about the end of the twelfth century. We may probably assume that Welsh was still spoken in Strathclyde when the *Leges inter Brettos et Scottos* were drawn up in the reign of David I. (1124-1153). In 1305 Edward I. of England ordained 'that the customs of the Scots and the Brets be

[1] Compare Kuno Meyer, *Fianaigecht*, xiii. (Todd Lect., xvi.). Bebbanburch was so called after Bebba, wife of Aedilfrid. In 627, four years after the burning of Bamborough, King Edwin's royal residence was 'Ad-gefrin,' now Yevering, on the river Glen, near Wooler.

henceforth prohibited and disused.'[1] The 'ancient laws of Galloway' are referred to in an Act of Robert I.[2]

We may now consider shortly the periods at which Gaelic was introduced into the south of Scotland, and to what extent it is represented in the place-names. The historical facts as regards Lothian—using the term in the wider sense to cover the district between Forth and the English march—are given by Skene, and need only be summarized. It is not likely that Gaelic had any hold here before A.D. 600. During the centuries of Anglic supremacy the only part in which it may have extended is the region of Linlithgow and Edinburgh. After Degsastan, the policy of Aedán mac Gabráin slumbered for nearly 250 years, and then it was revived when the kingdoms of the Picts and the Scots were united under Kenneth MacAlpin in A.D. 844. He invaded Lothian six times, burned Dunbar and seized Melrose. His successors, each in turn, followed the same policy. Dunedin fell to Indolb about 960. Finally Malcolm, son of Kenneth, routed the Northumbrians in the great and decisive battle of Carham in A.D. 1018, and fixed the boundary of Scotland practically as it stands now. The persistent vigour with which this policy was carried on for 170 years to ultimate success is one of the most remarkable things in history.

It was during this period, probably from about A.D. 960 onwards, that Gaelic came to be current in Lothian; there is some evidence that it extended beyond the present boundary of Scotland.[3] Few Gaelic names of landed men in Lothian have reached the charter period, but among them is Macbeth of Liberton in the reign of David I. Symeon of Durham, writing of events soon after 1060, mentions a certain man of great authority beyond the river Tyne, called 'Gillomichael,' that is 'the Lad of Michael, by

---

[1] Bain's *Calendar*, ii. p. 458. [2] RMS, i. p. 457.

[3] Gaelic place-names are found in the north of England. For Gaelic personal names see *An English Letter of Gospatric* (*Scott. Hist. Rev.*, vol. i. p. 62). (Thorfynn mac Thore, Melmor for *Mael Muire*, Kunyth for *Cinaeth.*) The date is c. 1067-1092, and the district is Allerdale in Cumberland. A number occur in the Pipe Rolls (Cumberland).

contrariety, for he would have been more justly named Lad of the Devil'—for his want of respect to English clerics. The wife of Radulph, son of Dunegal of Stranit (Nithsdale), was Bethoc, *Bethóc*, now *Beathag*, a well-known woman's name in Gaelic. This lady possessed several manors in conjunction with her husband, and in particular she possessed the property of Rulebethok, now Bedrule, *i.e.* Bethoc's Rule, in Roxburgh. The old forms of Bedrule are: Badrowl, 1275 (Reg. Glas.); Rulebethok, 1280, *ib.*; Bethocrulle, 1306-1329 (RMS), later Bethrowll, Bedroule (*Orig. Paroch.*, i. 347); of these Rulebethok is Gaelic in form, as is also Rule Herevei or Rule Hervey, now Abbotrule in Roxburgh. These two names are contemporary: Radulf, Bethoc's husband, flourished about 1140; Rule Herevei appears *c.* 1165. Gillemunesden *c.* 1200 (Reg. Glas.), on Lyne Water in Peebles, means 'the den (narrow valley) of St. Munn's servant,' *Munn* being a contracted form of *Mo-Fhindu*, St. Fintén (p. 307). Some part of Vlfkelystun, now Oxton in Lauderdale, belonged to Gilfalyn, *i.e. Gille Faolain*, 'St. Fillan's servant' (Reg. Glas.).

The site of the Nunnery of North Berwick was called Gillecalmestun, a. 1228 (Chart. Monial. de North Berw.). Congalton in Dirleton parish, Haddington, is from the ancient Gaelic name Congal. Similarly there are Gilchriston in Haddingtonshire, and Gilmerton near Edinburgh, from *Gille Criosd*, 'Christ's servant,' and *Gille Moire*, 'Mary's servant.' Names beginning with *gille* are characteristic of the period after A.D. 1000, when *gille* begins to displace the older *maol*, 'shaveling, servant.' An important list of names of men living around Peebles about A.D. 1200 contains Gille-Micheil twice, Gille-Crist twice, Gille-Moire twice, Cristein twice, Padruig twice, Gille-Caluim the smith of Peebles, Padinus (*Paidín*), Bridoc (*Brídeóc*), alongside of Welsh names—Queschutbrit (St. Cuthbert's servant), Cospatricius (St. Patrick's servant), Cosmungho, and Cosouold. With these go English names such as Adam son of Edolf, Randulf of Meggett, Mihhyn Brunberd, Mihhyn son of Edred.

A working list of Gaelic place-names taken from the Ordnance Survey Map of one inch to the mile, supplemented by some record names, contains the following numbers for the various counties, including Peebles: Linlithgow, 66; Edinburgh, 89; Haddington, 46; Berwick, 29; Roxburgh, 16; Selkirk, 21; Peebles, 99. A number of these are of some importance in one way or another. Eddleston, near Peebles, appears on record first as Penteiacob, a purely Welsh name, meaning 'Headland of James's house.' In the twelfth century it is Gillemorestun, the 'toun' or 'baile' of Gille Moire, St. Mary's Lad. Before 1189 it was granted to Edulf, son of Utred, and was thenceforward known as Edulfstun, Eddleston. This illustrates the change from Welshman to Gael, and from Gael to Saxon. In Eddleston parish there appears on record a burn called Aldenhisslauer (Reg. Glas.). This I conjectured to represent the Gaelic *Allt an Eas Labhair*, Burn of the loud Waterfall, and on inquiry I found that the burn can be readily identified—it formed part of a march—and that it has a noted waterfall now called 'Cowie's Linn.' Near Penicuik is a well from which the old people sometimes desire to drink when ill: it is called 'the Tippert Well,' from Gaelic *tiobart*, a well. Close to the Castellan of Dunbar is a knoll called 'Knockenhair,' *i.e. Cnoc na h-Aire*, 'the Watch-hill,' suggesting Gaelic occupation of the old fort of Dunbar. There is another place of the same name near Carco, Sanquhar, in Dumfriesshire. Near Lauder, on the way to Blyth, there was an old cairn, the stones of which had been scattered long ago—they are now collected into a cairn once more. Before I had seen the place, it was described to me as very stony: its name is Clacharie, which is Gaelic for a stony place. At Abbey St. Bathan's, in Berwickshire, there is an old ford on the Whitadder; the river bank at this ford is called Shannabank. There can be little doubt that Shanna is G. *Sean-áth*, old ford; if so, it was 'old' at the time of the Gaelic occupation. The Gaelic term for a common, common pasture, is *coitchionn*, spelled phonetically in the Book of the Dean of Lismore (*c.* 1515) *colchin, catchin, cotkhinn*. In the thirteenth century, Fergus, son of Gilbert, earl of Strathearn, declares that 'the land which is called Cotken in Kathermothel was in the time of all my ancestors free and common pasture to all the men residing around the said pasture, so that none might build a house in the pasture, or plough any part of it, or do anything that would interfere with the use of the pasture.'[1] 'Kathermothel' means 'Fort of Muthil' (*cathair Mhaothail*), and the common was 'the moor of Over and Nether Ardoche, called Cathkyne-muir,' in Perthshire,[2] spelled also Catkin, Cathkin. In the parish of Borthwick, Midlothian, is Catcune, on the right bank of Gore Water, in 1296 'Thomas de Catkune,' etc.;[3] in Blaeu Cathkin; in Retours Catcun, Cotcun, Cattun, Caltun, Caltoun. Here, then, we have a glimpse of the Gaelic people settled round their *coitchionn* or common pasture at some time, probably not very long before the time of Thomas de Catkone. There appear also 'the land of the Lord of Borthwick called Catkunyslandis, under the burgh, territory, and burgage of Lauder.'[4] Lauder still possesses extensive common lands, which may be the lands here referred to, but I am not sure that the reference is not to Catcune in Borthwick. The name might be expected to be fairly common, but the only other instance known to me is Cathkin in Carmunnock, near Glasgow. Other names indicating settlement are those containing *baile*, a stead, *achadh*, a field. Of the former there are five in Linlithgow, five in Midlothian, four in Haddington. Of the latter, one is in Linlithgow, four in Midlothian, one in Haddington. Of all these the most interesting is Balantrodach, now Arnieston, in Temple parish (Baltrodoc, 1306, Reg. Hol.); Blantrodoch (Reg. Calch.), for *Baile nan Trodach*, Stead of the Warriors. It belonged to the Knights Templar, who had a chapel there in the time of David I., and there can be little or no doubt that the name was given with reference to the Knights, who fought for the Holy Sepulchre in the Crusades—a valuable indication of

---

[1] *Chartulary of Lindores*, p. 31.
[2] Retours.
[3] Bain's *Calendar*.
[4] RMS., 1501.

Gaelic activity in Lothian about the middle of the twelfth century.[1]

As the Gaelic element in this district has not been much studied hitherto, it will be useful to give a fairly full selection of the Gaelic names in the different counties.

Bonjedward, 'Jedward-foot,' from *bun*, which is common in the sense of 'river-mouth,' as in *Bun Abha*, Bonawe, 'Awe-foot.' It seems that the Gaelic people took Jedward to be the name of the river, instead of Jed, plainly a case of Gaelic supervening on English. Cappuck, the site of a Roman camp, may be Gaelic; compare *Capaig*, Caputh in Perthshire. Craigover, Maxton, may be *Creag ghobhar*, 'goats' rock,' or *Creag odhar*, 'dun rock'; the name occurs also in Midlothian. Cromrig may contain *crom*, 'bent.' Eildrig, Elrechill, 1511 (RMS), is probably the common Elrig, G. *eileirig*, 'a deer-trap.' Falnask, Falnish 1511 (Orig. Paroch.) is probably Gaelic. Along with it goes on record Tandbanerse, the first part of which is *tòn b(h)àn*, 'white rump,' partly translated by the second part. The term 'Kip,' which occurs often throughout this region (Lothian) is *ceap*, 'a block,' as in Edinkip for *eudann (an) chip*, 'hill-face of the block.' Blaeu puts Tomleuchar Burn at the head of White Esk; it is for *tom luachra*, 'clump of rushes.' Roxburgh

Altrive, Eltryve 1587, etc. (RMS), is probably *alt*, a height, later a burn, and *ruighe* a slope; Kinrive, in Ross-shire, is in G. *Ceannruighe*, 'head of the slope,' *gh* between vowels becoming *v*, as often. Bellenden, so in 1624 (RMS), is to be compared with Bellendean in Roberton parish, Roxburgh, perhaps also with Ballindean, Ballindain 1459 (RMS), in Fife, and Bandean, Ballindane 1468 (RMS), in Perthshire; *baile*, a stead, often becomes *Bel* in unstressed position in English (not in Gaelic), as Belhelvie for *Baile Shealbhaigh*, 'Sealbhach's stead,' in Aberdeenshire, Belmaduthy, for *Baile mac Duibh*, 'MacDuff's stead,' in Ross, Selkirk

[1] Baltroddie, in Perthshire, is for *Baile troide*, 'stead of (the) quarrel (or combat).'

etc.; Ballindane' is probably for *baile an deadhain*, 'the Dean's stead.' Capel Fell is probably 'horse fell' (*capull*). Catcraig, 'wildcat rock,' may be either Gaelic or Welsh, and so may Clockmore, 'big rock,' a hill of 2100 feet on the left side of Megget.[1] Craigdilly may be for *creag tulaigh*, 'rock of the eminence'; *tulach* is also *tilach*. Craiglatch is 'rock of the *latch*,' *i.e.* boggy streamlet or puddle, a term common in Lothian, and used by Scott in *Guy Mannering*; compare *Loch Laid*, 'loch of the puddle,' Abriachan, Inverness: *laid* is the genitive singular of *lad* or *lod*, a puddle. Cramalt Craig may be G. *cromallt*, 'bent precipice,' or W. *crwmalt*, with the same meaning. Dalgleish is *dail g(h)lais*, 'green haugh,' perhaps taken over from W. *dol-las*, with the same meaning; compare *Dail ghil*, Dalziel, '(at) white haugh'; both are in the dative case.

Deuchar occurs also in Peebles and in Tannadyce parish, Forfar; it is probably for *dubh-chàthar*, 'black, broken, mossy ground.' Essenside, on a tributary of Ale, is probably from *easán*, 'a little rapid.' Glengaber, on Yarrow, is 'goats' glen'; compare Glengaber in Kirkconnel parish, Dumfries. Glenkerry, on Tima, is Gaelic; its second part might be explained in various ways. Jeshur Loch occurs in Macfarlane, probably *deisir*, 'of southern aspect.' Muckra appears to be for *mucrach*, 'place of swine'; compare Muckera in Ireland (Joyce). Salenside, on Ale, may contain *sailín*, 'a little heel,' *i.e.* spur of land.

Berwick

Auchencraw is Aldenecraw, 1333 (Bain's Cal.), apparently Gaelic but rather doubtful as to meaning. Aldcambus, in Cockburnspath, now Old Cambus on the Ordnance Survey Map, is Aldcambus, *c.* 1100 (Lawrie); Aldecambus, 1126, *ib.*; Aldcambhouse, 1298 (Ragman Roll); Auld Cammos, 1601 (RMS); Old Cammes (Macfarlane). 'Cambus' is doubtless G. *camas*, old G. *cambas*, a bend in a river, a bay; Aldcambus is an old parish name, and the ruins of St. Helen's Church there are close to a small bay. The traditional explanation of 'ald' as 'old' is probably right, as in Oldhamstocks, Aldehamstoc, 1127 (Lawrie), the name

[1] It is more likely to be Welsh *clog*, 'crag, cliff, precipice.'

of the parish adjacent to Cockburnspath on the north. Blanerne is probably *baile an fhearna*, 'alder stead.' Bogangreen may be for *bog an g(h)riain*, 'gravel bog,' *i.e.* resting on gravel or near gravel. Bondriech is 'foot of hill face' (*drech, dreach*). Boon seems to be simply *bun*, 'bottom, foot.' Cowdenknowes is Coldenknollis, 1559 (Lib. Melr.); Coldunknowes and Coldin- in Blaeu; here 'Cowden' stands for *colltuinn, calltuinn*, hazel, as it usually does in Scots; the name is a hybrid, meaning 'hazel knolls.' Dron Hill is *dronn*, a hump; compare Dron, a hill in Longforgan parish, Perth; also the name Dumfries. Knock, for *cnoc*, small hill, occurs in Duns and in Gordon. The Long Latch in Coldingham is 'the long boggy rivulet.' Longformacus, Langeford Makhous, *c.* 1340 (Johnston), is 'Maccus' *longphort*,' *i.e.* encampment or hut, dwelling. Longskelly Rocks, off the coast, contains *sgeilig*, a reef, as in *Sgeilig Mhícheil*, 'Michael's reef,' off the coast of Kerry; *long* may be English or it may be G. *long*, ship. Poldrait was the name of a croft at Lauder 'between the Kirkmyre and the land called Gibsonisland,' 1501 (RMS); compare 'the land in Hadingtoun called Sanct Androisland in Poildraught' (Ret.); the first part is *poll*, a pool or hollow; the second part is probably *drochaid*, a bridge, causeway, as in Frendraught, Ferendracht in Reg. Arbr., 'bridge land,' Aberdeenshire. Powskein, on a tributary of Cor Water, Tweedhead, is for *poll sgine*, 'knife pool'; compare *Inber Scéne*, 'estuary of the knife,' the old name of the mouth of the Kenmare River in Ireland, from its resemblance to a knife slash; also Loch Skene, Dunskine. Ross Point, Ayton, is *ros*, a cape, promontory.

This county is full of Gaelic names. Blairie (Wood) is from *blàr*, 'spotted,' probably through a derivative *blàrdha*; compare Muie-blairie, 'dappled plain,' in Ross-shire. Breach Law, off Manor Water, may be from E.Ir. *bréch*, a wolf, as in *bréachmhagh*, 'wolf-plain,' common in Ireland as Breaghva, etc. (Joyce). Cowden Burn is 'hazel burn' (*calltuinn*). At Crosscryne there is a crossing over a hill, but I have no evidence for a cross there; the first part I therefore take to be G. *crosg*, a crossing over a ridge; but Peebles the second part is uncertain. The hill of Crosscryne was one of the limits of the territory in Scotland ceded to Edward III. after the battle of Dunbar. As Wyntoun has it :—

> 'At Karlinlippis and at Corscryne
> Thare thai made the marches syne.'

The other place is now Carlops, 'the carlin's loups (leaps),' perhaps a translation of *Léum na Caillich*, 'the hag's leap.' Dundreich, 'fort of the hill-face,' compare Bondreich. Fingland is for *Finn glend*, later *Fionnghleann*, 'white glen, 'fair glen'; the term occurs four distinct times in the county, once also Finglen, which is the same. Garvald and Garrelfoot, 'source of the Garvald,' *c.* 1210 (Orig. Paroch.), is for *garbh-allt*, 'rough burn'; the primary meaning of *allt* is 'steep, precipice,' and it is rather notable to find it used in the modern sense of 'burn' so early as 1210. Glack, on Manor Water, is *glac*, a dell, common in the north; in Macfarlane it has the article 'the Glack,' as it has in Gaelic; Glack 1390/1406 (Orig. Paroch.). 'Glen' appears over thirty times; some of the instances may be Welsh. *Ord*, a hammer, a hammer-shaped hill (*i.e.* rounded like a throwing hammer) appears in Kirkurd, Ecclesia de Orda, 1170/81 (Reg. Glas.); Ecclesia Orda, 1186, *ib.*; Urde, 1306/29 (RMS). The original would be *Cell-uird*, which might very well mean 'church of the order,' *i.e.* of church service, as is probably the meaning of *Cell Uird* in Ireland,[1] but here we have one of the Kirkurd hills called 'Hammer Hill,' also Loch-urd and Lady-urd; compare Kinnord, Aberdeenshire; Ord of Caithness, etc. Logan is 'little hollow,' now *lagán* in Gaelic. Quilt is for *cuilt*, 'nook,' as in *a' Chuilt Raithnich*, in Perthshire, 'the bracken nook.'

ıddington Achingall is for *Achadh nan Gall*, 'the strangers' field.' Balgone, Balnegon 1337 (Bain's Cal.), Balgon in Blaeu, is for *Baile na gCon*, 'hounds' stead,' with eclipsis of *c* after *n* of the article genitive plural (*na ncon*). Balgrenagh 'in the tenement of Nodreff,' 1336 (Bain's Cal.), is for

[1] Hogan, *Onomasticon*.

Baile greanach, 'gravelly stead'; compare Greanagh, 'gravelly stream,' near Adare in Limerick; also *Greanaich*, Grennich, 'gravelly place,' in Strath Tummel. Balnebucth, Balnebuth, Balnebouch, Bellyboucht (Reg. Neub.), somewhere near Esk, appears to mean 'stead of the poor' (*bocht*), compare Dirie-bucht near Inverness. Balnegrog, 1336 (Bain's Cal.), appears to be for *Baile na gCnoc*, 'stead of the hillocks'; if so, this is an early instance of *cnoc* being pronounced *croc*, as it is often in modern Gaelic. Barbauchlaw is Balbaghloch, 1336 (Bain's Cal.), for *Baile Bachlach*, identical in form with *Baile Bachallach*, near Churchtown, Cork. *Bachlach* comes from *bachall*, a crozier (Lat. *baculum*), and Joyce explains the Irish name as probably 'church land, belonging to a bishop.' This may be the explanation here; an alternative is that the land was held in respect of the custody of a pastoral staff.[1] There is another place of the same name in Linlithgowshire. Broxburn and Broxmouth are on record 'inter presmunetburne et broc,' 'between Pressmennan burn and Brock,' 1153/65 (Chart. Melr.); Broccesmuthe, 1094 (Lawrie); Brooksmyth in Blaeu; here, however, we have to do, not with G. *broc*, badger, but with A.S. *bróc*, now 'brook.'

Clackerdean, Bolton, may contain *clochar*, 'stony land'; the latter part is doubtful. Cockenzie is Gaelic doubtless, perhaps for *Cùil C(h)oinnigh*, 'Kenneth's nook'; I have not seen early forms. Cowie Burn is 'hazelly burn,' from *coll*, hazel, adj. *colldha*. Deuchrie is probably for *Dubh-chàthraigh*, 'place of black broken mossy ground,' with syncope of the middle syllable (unstressed) as usual; compare Deuchar, and *Cnoc Dubhcharaigh* in Ross-shire. Doon Hill is 'fort hill'; Drem is for *druim*, a ridge. Dunbar is 'summit fort,' probably taken over from British *din-bar* with the same meaning. Dunskine, however, is probably purely Gaelic for *Dun-sgine*, 'knife fort.'

Garvald is 'rough burn.' Glenbirnie is probably for *Gleann Braonaigh*, 'glen of the oozy place,' with the usual

---

[1] *Bachlach* also means 'a shepherd, a rustic'; but this is not likely to be the meaning here.

metathesis; compare Birnie in Moray, Brenach, 1291 (Bain's Cal.), and Culbirnie. Glenterf, 1458 (RMS), is for *Gleann Tairbh*, 'bull's glen.' Gullane is probably for *gualainn*, 'at-shoulder'; 'ecclesia de Golyn' (Reg. Dunf.). Kilduff appears in the *Life of St. Kentigern* as 'a hill which is called Kepduf,' *i.e. ceap dubh*, 'black block'; one infers that the other places called 'Kip' are also Gaelic. There is also Kippie Law, a hill near Dalkeith. The Latch, near Gifford, is a boggy rivulet near the fort on the Witches' Knowe. Leckmoram Ness begins with *leac*, a slabstone, a flat rock; compare Legbernard, supposed to be now Leadburn in the parish of Penicuik (Lawrie, p. 384). That Legbernard was in this parish is proved by 'the lands of Hallhouse and Leckbernard within the parochin of Pennycook,' 1653 (Ret.); as Hallhouse is now Halls, south of Penicuik, Legbernard was probably near it. Phantassie is probably *fàn taise*, 'slope of softness,' *i.e.* wetness; there is also Phantassie in Fife, near West Wemyss.

Portmore is from *port*, a place, a dwelling, a hold, meaning 'big fort,' with reference to the finely preserved fort called Northshield Rings. In Powsail Water, Drummelzier, the first part is *poll*, a slow stream, in Scots *pow*, as in the Pow of Inchaffray, etc. The second part is probably *sail*, willow; the name may be compared with Powlsaill on Tummel (Ret.), and for sense with Sauchieburn. According to tradition it is beside Powsail, at the foot of a thorn-tree, a little below the churchyard, that Merlin is buried. An old prophecy was—

'When Tweed and Pausayl meet at Merlin's grave,
Scotland and England shall one monarch have.'

It was believed to be fulfilled by a strange and sudden rising of the waters on the day when James VI. ascended the English throne.[1]

Tarbet, on the islet of Fidra, is for *tairbeart*, an isthmus, a portage; more correctly Tarbert.

[1] *Orig. Paroch.*, i. p. 205.

Auchencorth, Auchincorth 1604 (RMS), Auchincorthe Midlothian 1608 (Ret.); Auchincreoch, 1653, *ib.*; Auchincroich, 1675, *ib.*; Achincorc Blaeu; probably for *Achadh na Coirthe*, 'field of the standing stone'; the spelling of 1675 looks like *Achadh na Croiche*, 'gallows-field.' Auchendinny, Aghendini, 1335; Aughendeny, 1336; Aghendeni, 1337 (Bain's Cal.), is from O.Ir. *dind*, gen. *denna*, a height, a fortress, or from *dindgna*, gen. *dindgnai*, of the same meaning; compare Auchindinnie, Gartly, Aberdeen; Baldinnie, Ceres, Fife; Blairindinny, Kennethmont, Aberdeen, Denny, etc. Auchinoon may be for *Achadh nan Uan*, 'lambs' field'; compare Balnoon, Inverkeithing, Banff; Strathnoon, in Strathdearn, is different, being in Gaelic *Srath-nìn*. Balerno, Balhernoch 1280 (Bain's Cal.); Balernauch, 1283 (Acts of Parl.); Balernaghe, 1296 (Bain's Cal.); Ballernache, 1375 (RMS), for *Baile Airneach*, 'sloe-tree-stead'; compare Ballernach, Nevar, Forfar; Balernock near Gareloch-head; Airneachan in Co. Cavan (Joyce).

Balgreen near West Calder, also in Ecclesmachan, Linlithgow, may be either *Baile Griain*, 'gravel-stead,' or, which is less likely, *Baile Gréine*, 'sunny stead.' Balleny is probably for *Baile Léanaidhe*, 'stead of the damp meadows'; compare Lenzie, Kirkintilloch, Moylena in Ireland. Braid, Braid Hills, is for *bràghaid*, dative of *bràighe*, 'upper part.' Camilty Burn is for *Camalltaidh*, 'crooked little burn,' which describes it. Cammo is Cambo, Cambok in 1296 (Bain's Cal.), and is to be compared with Cambow, Cambou, 1374 (RMS), Cambok, 1506, *ib.*, in Fife, and with Cambou, 1324 (RMS), Cammak, 1512, *ib.*, in Kincardine; all are from G. *cam*, O.Ir. *camb*, bent, and they may be the same as *an Camach*, 'the bent place,' on Tummel, opposite Bonskeid. Carberry, Crefbarrin *c.* 1143, 1150 (Reg. Dunf.); Alexander Crabarri, 1311 (Bain's Cal.); Carbery (Chart. Hol.), seems to be a compound of *craobh*, branch, tree, and *barrán*, a top-fence or hedge, palisade, meaning 'branch fence,' or the like. Colzium, at the head of Camilty Burn, is for *cuingleum*, 'defile leap,' a rather common term for a narrow gorge in a stream. Corstorphine,

Crostorfin, *c.* 1130, 1142 (Chart. Hol.). Dunbar, with poetic licence, stresses it on the last syllable—

> 'He has tane Roull of Aberdene
> And gentill Roull of Corstorphine.'

The meaning is 'Torfin's crossing,' rather than 'Torfin's cross,' with reference to a crossing over the hill. Torfin is a personal name; Macbeth gave Bolgyne, which belonged to the son of Torfyn, to God and to St. Serf of Loch Leven (Lawrie).

Dalry, now part of the city, may mean 'king's meadow,' like *Dail an Rìgh*, Dalry, near Tyndrum. But *Dail Fhraoigh*, 'heather dale,' is possible (*fraoch*, old gen. *fraoigh*). Craigenterrie, of which I have no old spelling, may be for *Creag an Tairbh*, 'bull's rock'; compare Craigmarry. Craigentinnie is most probably for *Creag an t-Sionnaigh*, 'the fox's rock'; the fox survived near Edinburgh till fairly recent times; compare Ardentinny on Loch Long; Ardatinny, 'the height of the fox' (Joyce). Craiglockhart, Craiglikerth, Craiglokart, Craiglokert (RMS I); here the second part is probably one of the many forms assumed by *longphort*, an encampment, etc.; compare Barrlockart in Wigtownshire. Currie is the name of a parish and village and also of a place in Borthwick parish; the former is Curri, 1336 (Bain's Cal.). These are dative of *currach*, a wet plain; compare Currochs near Crieff; also Curroks on record in Lesmahagow, now Corehouse, whence also Cora Linn on Clyde at Corehouse. In Ireland the forms Curra, Curragh, Curry are common (Joyce). Drum occurs by itself and in Drummore, 'big ridge'; Drumsheugh, Drumselch 1507; 'the common muir of Edinburgh, once called the forest of Drumselch,' for *Druim-seileach*, 'willow-ridge'; Drumdryan, for *Druim-draighinn*, 'blackthorn-ridge.' Dunsappie, east of Arthur's Seat, is to be compared with Torsappie, formerly Torsoppie, Thorsopyn, 1282 (Acts of Parl.); the second part is genitive of *sopach*, 'place of wisps' or tufts of grass; the lines of the fort on Dunsappie are traceable; when the name was given by the Gaelic speakers, the interior was full of rank grass, as it is now;

compare Coolsuppeen for *Cúil soipín*, 'nook of the little wisp,' in Ireland (Joyce), and Dalsupin (RMS I) in Ayrshire. Galla Ford, near Harperrig Reservoir, is for *geal-àth*, 'bright ford,' not the same as Gala Water. Garvald, in Heriot, is 'rough burn.' Glencorse, Glencrosk 1336 (Bain's Cal.), etc., is 'glen of the crossings' (*crosg*); there are three different old crossings, all of them now rights-of-way. Glenwhinnie, Stow, may perhaps be compared with Dalwhinnie, *Dail-chuinnidh*, 'champions' dale,' from E.Ir. *cuingid, cuinnid*, a champion; compare *Gleann Cheatharnaich*, Duthil, 'warrior's glen.' Lennie, Corstorphine, is Lanyn, 1336 (Bain's Cal.), probably as in Balleny. Leny, near Callander, is different, being in Gaelic *Lànaigh*. Loganlee, Glencorse, may be for *lagán liath*, 'grey hollow.' Longford, West Calder, is Lomphard in Blaeu, for *longphort*, encampment, etc.; compare Lumphart Hill in Aberdeenshire. Malleny is probably for *magh léanaidhe*, 'plain of the damp meadows,' like Moylena in Ireland. Muldron is 'height of the hump' or humps. Tipperlinn means 'well of the pool.' Torr, a rounded hill, appears in Torbane, 'white torr'; Torduff, 'black torr'; Torphin, 'white torr'; Torsonce, of doubtful meaning, and others. Windy Gowl, on the south side of Arthur's Seat and elsewhere, is doubtless a part translation of *gobhal na gaoithe*, 'windy fork,' which well describes the gap at Arthur's Seat.

Auchenhard is for *Achadh na h-Airde*, 'field of the height.' Linlithgow. Balbardie, Balbardi 1336 (Bain's Cal.), Balbairdy in Blaeu, is probably for *Baile a' Bhàird* or *Baile nam Bàrd*, 'bard's (or bards') stead.' Another possibility is *Baile Bàrda*, 'stead of the guard, watch, garrison'; for the former compare Pithoggarty near Tain, Balhaggarty in Aberdeenshire, where the *-y* is hard to explain, and where the second part would normally be '-haggart' for *-shagart* (gen. pl.); compare also Balbairdie in Kinghorn parish, Fife. Ballencrieff, Balincref 1296 (Bain's Cal.), Balnecref 1335, *ib.*, is for *Baile na Craoibhe*, 'stead of the tree'; compare for meaning Trapren, Traprain. There is another Ballencrieff in Haddington, in Blaeu Bancreef, Banckreif. Balvormie is obscure to me. Bangour, Bengouer, 1336 (Bain's

Cal.), is for *beann g(h)obhar*, 'goats' peak.' Benhar is probably for *beann charra*, 'peak of the rock-ledge'; compare *Drochaid Charra*, Carr Bridge. Beugh is for *beitheach*, 'birch-wood.' Binning, ecclesia de Bynynn (Reg. Dunf.), is for *binnean*, a little peak, very common in place-names. Binns, Bynnes 1336 (Bain's Cal.), is for *beinn*, dative of *beann*, with English plural. Binny is dative of *binneach*, 'peaked place.' Bonhard, Balnehard 1296 (Bain's Cal.), is 'stead of the height'; compare *Achadh na h-Airde.*

Breich is to be compared with Balnabreich, in Kincardineshire, and with Balanbreich, an ancient castle in Flisk parish, Fife, on a steep bank overhanging the Firth of Tay, also with 'the barony of Ballinbreich' in Ross-shire. As this last is *Baile na Bruaich*, 'bankstead,' in Gaelic, it is probable that 'breich' in the other names stands for *bruaich* as genitive or dative. Broxburn, the basin of which used to be known as Strathbrock, may be 'badgers' burn,' in view of the combination Strathbrock, which would naturally mean 'badgers' strath'; if so, it is different from Broxburn in Haddington.

Cockleroy, in Blaeu Coclereuf, is pronounced Cockelroy; the second part is probably *ruadh*, red; the first part is doubtful.[1] Craigengall is for *Creag nan Gall*, 'rock of the foreigners,' *i.e.* English. Craigmailing is probably for *Creag Mhaoilinn*, 'rock of the bare round hillock,' a term which is found in *Maoilinn*, anglicized as Moulin, near Pitlochry. Craigmarry is Craggenemarf, a possession of which part was granted by King David I. to Holyrood at its foundation; Craggenemarfe, 1391 (RMS); Craigmarvie (Ret.). It is for *Creag nam Marbh*, 'dead men's rock'; compare Blairnamarrow, 'dead men's moor,' in Kirkmichael parish, Banff; 'the cairn of stones called cairnne na marrow, *alias* Deidmanniscairne' in Logie Easter, Ross-shire, 1610 (RMS). 'Craigmarvie' corresponds rather to *Creag mharbhaidh*, 'rock of slaughter,' and may be a later form. Drum, 'ridge,' appears in Drumbeg, 'little ridge'; Drumbowie, 'yellow ridge' (*buidhe*); Drumcross, 'ridge of the crossing (or cross)'; Drumforth, Drumelzie,

[1] Perhaps G. *cachaileith*, a gate.

and Drumbrydon are doubtful in the absence of old forms; Drumshorland is in Blaeu Druymshorling, also doubtful; Drumtassie is probably for *druim taise*, 'ridge of wetness'; compare Phantassie. Dundas, Dundass 1425 (RMS), may be for *Dùn deas*, 'south fort'; compare *Ràth-thuaith*, Rahoy, 'north rath,' in Argyll. Duntarvie is for *Dun-tarbhaidh*, based on *tarbh*, a bull; there were several places in Ireland called *Tarbhda, Tarbga*, genitive *Tarbgi*, meaning practically 'bull place,' which in modern Sc.G. would be *tarbhaidh*; compare Tarvie, near Garve, in Ross-shire, and Glen Tarvie, Strathdearn.

Echline, 'Radulphus de Echelyn' (Chart. Melr.); Echlinge, 1449 (RMS), is the dative of *eachlann*, 'a horse enclosure, paddock'; compare Aghlin in Leitrim (Joyce). Gavie-side, on Breich Burn, is to be compared with Pow-gavie, Inchture parish, Perth, and Pur-gavie, Lintrathen; it seems to be genitive of *gàbhadh*, danger; compare *Tigh-chunnairt*,[1] 'house of peril,' on the left bank of Lyon a little below Bridge of Lyon, close to a pool called *Linne Lonaidh*, in which many people have been drowned. Glen-puitty, near Dalmeny, is from *puiteach*, gen. *puitigh*, 'potty, place of pots,' *i.e.* holes or hollows; a number of names in Ireland and Scotland come from this base. Irongath is Arnegayth, 1337 (Bain's Cal.); 'Iron' is seen in Irongray, Ironlosh, Iron-macannie, Ironhirst, etc., and usually represents *earrann*, 'a portion,' though sometimes it appears to be a contraction of *ard na*, 'height of the,' followed by a fem. noun. The second part is *gaoth*, 'marsh,' fem., gen. *gaoithe*, and the name is for *Earrann (na) gaoithe*, 'portion of the marsh.' Kinglass is for *ceann glas*, 'green head.' Logie is for *Logaigh*, later *Lagaigh*, dative of *logach*, 'place in the hollow.' Muckraw means 'swine place.' Ryal represents *riaghail*, a rule; compare *an Riaghail*, a glen off Glen Lyon, spelled Regill, 1502 (RMS); Regal Burn at the head of Glengavel Water, Avondale, Lanarkshire, in 1478 Regalegill (RMS); le Rigale 1478 (Ant. A. & B.), now Raggal in Boyndie parish, Banff—Thomas Ruddiman, the Latin grammarian, was born appropriately on the farm of Raggal;

[1] The MacGregor pipers lived here.

Carse-regale in Wigtownshire, 1546 (RMS), was apparently near Kilquhokadaill. In Ireland there was *Ros Riagla*, 'promontorium sive collis regulae' (Onom.), 'cape or hill of the rule.' The special meaning of *riaghail* in names of places is not clear to me.

Tannach is for *Tamhnach*, a green or fertile field, especially in waste or heathery ground; it is common in our place-names from Caithness southwards.

On these names generally it may be remarked that they are of a simple, straightforward type and fairly modern in form, being as a rule what may be called 'phrase names.' It may be noted further that the Gaelic *inbhear*, 'inver,' occurs in Innerleithen, Inveresk, Inverleith, denoting formerly what is now Leith, and in the obsolete Inverwieedule (RPSA), 'the inver of Wedale,' *i.e.* the junction of Gala and Tweed. 'Inver' is usually followed by the name of the stream, but in the last instance it is followed by the English term Wedale, just as in Gairloch we have Inverasdale for Inveraspidale, 'the inver of aspen-dale,' a Norse term. It is to be inferred, therefore, that here Gaelic settlement followed English settlement. Incidentally also we may infer that the name 'Gala' was not yet established for what is now the Gala Water; the latter, as the old forms Galche, Galge, Galue (Lib. Melr.), Gallow (New Stat. Acc.), clearly show, is simply 'gallows water.' In Innerleithen, Inverleith, and possibly in Inveresk, the second part—the stream name—is British, not Gaelic.

In the next place we may consider what traces of the Celtic Churches, British and Irish, are to be found between the Firth of Forth and the Border. The most definite of these traces are the place-names that involve the names of saints, and the question that arises is what inferences are we entitled to draw from such names? The question is difficult, but some points are fairly clear. Under the Celtic Church in Scotland missionary activity emanated from monasteries ruled by abbots, who were completely independent except in so far as a daughter monastery might be subject to the parent monastery. All foundations were made with leave of the king or local lord or both, and that leave, to be effective, was accompanied by a grant of land, made to God and to the abbot, and, in the case of a new foundation, to the cleric who was to be head of the new church; *e.g.* Iona was granted to God and to Columba; Deer was granted to God and to Columba and to Drostan. After Columba's death such grants would be made to God and to Columba and to the abbot of the time and to the head of the new church. The resulting commemorations were in reality a recognition of ownership. Iona came to be styled *Hi Choluim Chille*, 'Columba's Iona,' in the sense that it belonged to Columba. A foundation from Iona made in Columba's time might come to be named after him or after the cleric who presided over it. A foundation made in Adamnan's time might still be named after Columba as representing Iona, or it might bear Adamnan's name or the name of the cleric who was left in charge of it. On Loch Tay there is an ancient site called *Cill Mo-Charmaig*, 'my Cormac's church,' on a small promontory called now *Ard-Eodhnáig*, formerly *Ard Eodhnáin*, 'Adamnan's cape': this, as I understand it, may be taken to mean that the church was founded from Iona in Adamnan's time, and that the land was gifted to Adamnan and to Cormac, the cleric in charge. It may be added that these church names appear to have been popular rather than official.

Another factor that may have to be taken into account is migration. The Iceland *Landnámabók* tells how a Norseman named Orlyg, migrating from the Hebrides to Iceland, received from his bishop, who was named Patrick, consecrated earth and other things necessary for founding and equipping a church. Orlyg landed in Iceland, erected a church, and dedicated it to Colum Cille, as directed by the bishop. This took place before A.D. 900. Haldor, son of Illugi the Red, built a church thirty ells long, roofed it with wood, and dedicated it to Colum Cille and to God. Elsewhere also newcomers may have taken with them the noted saints of their former districts.

It will thus be seen that commemorations of Celtic saints form proof of influence by the Celtic Church, but that as a basis for dating the precise period of that influence they have to be treated with caution. The fact that a church bears the name of Columba does not necessarily imply that it was founded by Columba, or even that it was founded in his time.

Under the influence of the Church of Rome dedications in the modern sense of the term began. Thus, as Bede tells, Nechtan, king of the Picts (706-724), dedicated a church to St. Peter. The Book of Deer records a gift of land made 'for the consecration of a church of Christ and of Peter the Apostle, and also to Colum Cille and to Drostan' about A.D. 1132. Another is 'to God and to Drostan and to Colum Cille and to Peter the Apostle.' The commemorations of St. Oswald (Kirkoswald and Kirkcarswell) are obviously dedications in the modern sense, *i.e.* the churches were put under the protection of the saint.

The earliest trace of the Celtic Church in this region is found in the legend of Darerca or Moninne or Sárbile, whose death is recorded in 517 or 519 (AU). This lady is styled 'of Cell Sléibhe Cuilinn' in Co. Armagh, and according to some accounts she came to Britain, accompanied by her maidens, and founded seven churches, including one on Dunpelder, now Traprain Law, in Haddingtonshire, and another on Dunedene, now the Rock of Edinburgh.[1] The only trace left of her mission, or rather of the tradition of it, is the name Castra Puellarum, Castellum Puellarum, 'the Maidens' Castle,' applied at one time to Edinburgh Castle.

Jocelin's *Life of St. Kentigern* states that Kentigern abode eight years in Lothwerverd, where he constructed a cross of sea sand, some real or supposed traces of which apparently existed in Jocelin's own time. The place is now Lochquhariot, and St. Kentigern's Well there is mentioned in 1534 (RMS). There are also St. Mungo's Well in the minister's garden at Penicuik, and St. Mungo's Well at Peebles. St. Kentigern's Bog was in the parish of Cockpen, 1580 (RMS). About 1200 the priest of Eddleston bore the Welsh name of Cosmungho.

Kentigern was traditionally associated with St. Serf,

[1] Skene, *Celt. Scot.*, ii. p. 37.

whose seat was at Culross in Fife. Sydserf, near North Berwick, is Sideserf, 1290 (Bain's Cal.); William de Sideserfe, 1296 (Ragman Roll); here 'side' stands for *suide*, later *suidhe*, a seat: Sydserf is St. Serf's Seat. 'Sanct Serffis Law' or Serflaw was in the barony of Abercorn, 1526 (RMS), in Linlithgowshire.

St. Machan, who is commemorated in Ecclesmachan in Linlithgowshire, is said to have been a disciple of St. Cadoc of Llancarvan; if so, he was contemporary with Kentigern.

During the short activity of the Irish Church from Iona in Northumberland (635-664), the monastery of Melrose was founded by Aedan of Lindisfarne, and the mixed monastery of Coldingham in the time of his successor Fínán. The superiors of these monasteries were Angles, and whatever Irish character they may have had disappeared after 664. Nevertheless, to the influence of Iona must be due the commemorations of Baithene, Columba's successor, in St. Bathans or St. Bothans, the old name of the parish of Yester, 'ecclesia collegiata de Bothanis,' 1448 (RMS), etc., and in Abbey St. Bathans, 'ecclesia sancti Boythani,' 1250. Patrick MacGylboythin of Dumfries signed the Ragman Roll in 1296; his father's name means 'servant, or lad of St. Baithene.'

Kilbucho in Peebles is Kilbevhoc, *c.* 1200 (Chart. Melr.); Kylbeuhoc, *c.* 1200 (Reg. Glas.); Kelbechoc, 1214/49, *ib.*; Kylbocho in Boiamund; Kilbochok, 1376 (Chart. Mort.); Kilbouchow, 1475 (Chart. Hol.). The saint commemorated here is Begha, probably the nun called Begu by Bede, who lived in the time of Aedan of Lindisfarne and of Hilda. She is not mentioned by Irish writers so far as I know, but the termination *-oc* in her name is the affectionate diminutive common in names of Irish saints; her connection was with the Church of Northumberland. Gillebechistoun or Killebeccocestun, *c.* 1200 (Chart. Melr.) in Eddleston parish, Peebles, means 'the *toun* of St. Begha's servant'; 'toun' is doubtless for an earlier *baile*. 'St. Bais wall' (well) at Dunbar, on record in 1603 (RMS) may commemorate 'the very mythical Irish saint Bega, whose name is preserved in St. Bees.'[1]

[1] Plummer's *Bede*, vol. ii. p. 248; i. p. 431.

The cult of St. Bride was strong in the south, especially among the Douglas family, and the altar of St. Bride at Melrose is mentioned about 1368 in connection with one of them (Orig. Paroch., i. 335). Traquair was formerly known as Kirkbryde and St. Bryde's parish (Orig. Paroch. and Ret.).

Inchcolm in the Firth of Forth is 'St. Columba's Isle.' The monastery of Inchcolm was founded by Alexander I. and Walter Bower, Abbot of Inchcolm. In 1123 King Alexander, driven upon the island by a storm, was entertained by a hermit who served St. Columba in a small chapel, and thereafter, in gratitude for his preservation, founded an Augustinian abbey on the isle in honour of St. Columba. St. Columba's Well at Cramond, 'between the lands of the common of Cramond and the sea shore' is mentioned in 1601 (RMS).

Under the church of Dalmeny were an altar and chapel of St. Adamnan (Ret.); this may have been Adamnan of Coldingham.

The form of Dalmahoy, near Edinburgh, indicates that it contains a saint's name, and the name must be *Tua*, gen. *Tuae*. It means 'the silent one,' from an early *Tovios*, and there were four saints of that name,[1] one of whom, also called Ultan of Tech Tuae, is commemorated in the Calendars of Oengus and Gorman at December 22. The earlier form of *Tua* in Irish would be *Tóe*, which, according to the practice of Scottish Gaelic, would become with us *Tatha*. *Tua* is the form in Oengus's *Félire*, composed in the ninth century, and it is probably this, rather than the older Tóe, which appears in Dalmahoy, for *Dail mo Thuae*, 'my Tua's meadow.' In the north, however, the saint's name must have been introduced in the earlier stage, for we have it on Loch Awe side in *Ċill Mo-Thatha*, 'Ḳilmaha'; near Callander in Perthshire is *Loch Mo-Thatháig*, 'Loch Mahaick,' with the affectionate diminutive form *Tathág*; Abergairn Church, in Aberdeenshire, is in Gaelic *Cill mo Thatha*, and *Féill mo Thatha*, 'St. Tua's fair,' used to be

[1] Rawl. B 502, 94 d 5; LL 368 g.

held there.[1] In *Tatha* the *th* is used merely to divide the syllables.

Eccles in Berwickshire means 'church,' and may represent W. *eglwys* rather than G. *eaglais*, as also in Ecclesmachan. G. *ceall*, cell, church, in its dative form *cill* appears in several names, of which Kilbucho has been mentioned. Kinleith, in Currie parish, is 'ecclesia de Kildeleth' (Reg. Dunf.); Kildeleth, 1327 (Chart. Hol.); Keldelethe, 1372 (RMS); 'the glebe of the parish church of Currie *alias* Kildleithe,' 1609 (Ret.); 'Killeith in the parish of Currie,' 1663 (Ret.). As the church of Currie stands on the Water of Leith, is it possible that the old forms stand for 'Kil de Leth,' 'the church on Leith?' The Kell Burn which joins Whitadder, near Priestlaw, is a part translation of *Allt na Cille*, 'the church burn.' There is also Kill Burn, otherwise Church Burn, at Traquair.

The Old Stat. Account mentions the Tower of Penicuik, 'on an eminence above the Esk, and about half-way betwixt the village and the present house of Pennycuick.' The writer adds that the old name was Terregles, for which I find no other authority.[2] 'The Maynes of Pennycook with their tower called Reglis' appears in 1613, 1647 (RMS); in 1675 (Ret.) it is Regills. This appears to be for *reclés*, 'a cell, oratory, close,' and if so it is of importance as showing that here a cleric of the Irish Church lived and wrought. It is the only instance of *reclés* that I know of in our names of places, if it is an instance.[3]

Romanno, in Newlands parish, Peebles, is Rumannoch, 1266 (Ex. Rolls and Acts of Parl. Alexander II.); Rothmaneic and Rumanach (Chart. Hol.). In the middle of the twelfth century a carucate of land in the fief of Rothmaneic,

---

1 For this information I am indebted to Mr. F. C. Diack.

2 The writer is not quite trustworthy on such a point; *e.g.* he makes Mountlothian into Monkslothian.

3 A note on Oengus's *Félire* (Oct. 11), states that St. Cainnech of Achad Bó, the contemporary of Columba, had a *reclés* in Cell Rígmonaid (the church of St. Andrews), *i.e.*, a chapel or oratory named after him (atá reclés dó hi Cill Rígmonaid i nAlbain). This, of course, is not proof that St. Cainnech visited St. Andrews.

with pasture for a thousand sheep, was granted to the canons of the Holy Rood at Edinburgh.[1] The name is for *Ràth Manach*, 'the monks' rath,' with reference to the great rath on the high ground above Romanno. It may have been given after the grant to Holyrood, but there is the possibility of an earlier religious settlement of the Celtic Church. The terraces at Romanno are similar to those on the eastern and south-eastern slopes of Arthur's Seat near Holyrood. They were evidently designed for purposes of cultivation, and both sets may owe their origin to the monks of Holyrood.

[1] See further *Orig. Paroch.*, i. p. 193.

CHAPTER VI

GENERAL SURVEY OF DUMFRIES AND GALLOWAY

Of the history of Galloway and Dumfriesshire—the land of the Novantae and the Selgovae—in the centuries following the Roman evacuation we have but little definite knowledge. That the district remained British may be taken as a matter of course: names such as Ochiltree prove that here too, as in Lothian, the old British language passed into its Welsh stage.[1] It may also be reasonably assumed that before it came under the dominion of the Angles at some time in the seventh century it was ruled by native princes, but who these were is rather a problem.

In the latter part of the sixth century Urien of Rheged and his sons formed the bulwark of the Britons against the pressure of the Angles. After the death of Urien and his son Owein their place in the leadership was filled by Aedán of Dál Riata; it would seem that among the native ruling families there was no one left capable of making effective resistance. Aedán's defeat at Degsastan was followed by the Anglic conquest of Lothian and Galloway, not, however, of Strathclyde. The two former regions seem to hang together. The question arises, where was Urien's province of Rheged? This is a much-discussed problem, as to which the most various views have been held. Skene would

[1] The evidence of personal names is equally important. In 1124, the lord of Stranit (Nithsdale) is *Dunegal*, for O.W. *Dumngual*; in H36 his sons are Radulf and Duuenald, the former English, the latter Gaelic or Gaelicized, later *Domhnall*, 'Donald': *Dumngual* and *Domhnall* both represent E.Celt. *Dumno-valos*, 'world-ruler.' Similarly, compare the forms of St. Connel's name (p. 169); also the name *Gille-gunnin*, of which the first part is Gaelic and the second part is the Welsh form of Gaelic *Finnén* (p. 165). Compare too the term 'Gossock' (p. 178).

place it in the north near Dumbarton; it has also been placed in the south about Dumfriesshire, in Cumberland, in Lancashire, and even in Wales. The latest discussion is by Sir J. Morris-Jones,[1] and he has helped the solution by showing that Rheged contained Carlisle. To this important fact there may now be added another which serves further to define the position of Rheged. The name itself occurs beyond doubt in Wigtownshire, where, at the head of Luce Bay, we have Dunragit, Dunregate 1535 (RMS), meaning 'Fort of Rheged'; the site of the old fort is on a rounded eminence called the Mote of Dunragit. The inference is that Wigtownshire was in Rheged, and if Wigtown, then also the rest of Galloway and Dumfriesshire. References in the old Welsh poetry about Urien bear this out: 'When he returned in the autumn from the country of the men of the Clyde, no cow lowed to her calf'—so clean, that is to say, was the sweep made of the Strathclyde cattle by his raiders; 'a battle when Owein defended the cattle of his country, a battle in the ford of Alclud.'[2] Urien is styled, as Sir J. Morris-Jones points out, 'Urien of the *echwydd*,' 'lord of the *echwydd*,' 'shepherd of the *echwydd*,' a term which he explains as a flow of water, a tidal current, a cataract. It seems fairly certain that in this case the *echwydd* is the Solway, which is noted for the violence of its tides;[3] compare *Trácht Romra* (p. 161).

On the issue of Degsastan depended the fate of two districts or provinces, Urien's province of Rheged and the province of Lothian, which had been in alliance with Urien, and as we have seen, Rheged included Dumfries and Galloway. Had Aedán been victorious the result might well have been the political fusion of Urien's province and even Lothian with Dál Riata at that time. This was the last important event of his life; he died three years thereafter at the age of seventy-four.

---

[1] *Taliesin*, pp. 64 *seq.*, where a summary of the views held is given and the derivation of 'Rheged' discussed.

[2] Skene, *Four Ancient Books of Wales*, p. 363; compare p. 350.

[3] For Sir J. Morris-Jones' own view see *Taliesin*, p. 68.

During most of the seventh century and the whole of the eighth the Angles were the superiors of the whole region. It seems very doubtful, however, whether they made much settlement in Galloway, and their overlordship may not have been very effective at any time, especially toward its latter stages.

Communication between Ireland and Galloway must have been common enough from the earliest times, and there are some ancient references which illustrate this. One of them is as follows: in the time of Concobar mac Nessa—that is to say in the early part of the first century—Néide son of Adna, the chief poet of Ireland, went to Eochu Echbél in Kintyre to finish his studies. After he had spent some considerable time there, he proceeded homewards. His route was first to Kintyre, *i.e.* presumably the Mull, and thence to *Rind Snóc*. Then he went to *Port Ríg*, and thence across the sea to *Rind Roiss*, a cape in Island Magee north of Belfast Lough, thence to Larne, and so on.[1] Rind Snóc was somewhere in the Rinns of Galloway. Wyntoun, quoting from Barbour's *Bruce*, says: 'Fra Wek anent Orknay till Mullyrryssnwk in Gallway,' the old equivalent of 'frae Maiden Kirk till John o' Groat's.'[2] A Latin *Life* of St. Cuthbert states that Cuthbert, along with his mother and others, came from Ireland in a curach of stone and landed at a port in the Rinns which is called Rintsnoc.[3] Skene thought that Rintsnoc was Portpatrick,[4] but this is not so. Néide came from Kintyre to Rind Snóc, and then went on to Port Ríg. Four places in Scotland are called—in their anglified spelling—Portree or Portrie, one in Skye, one at Carradale in Kintyre to which Bruce came from

[1] LL 186 a; *Rev. Celt.*, xxvi. p. 4.

[2] *Orygynale Cronykil*, ii. p. 363 (*Hist. of Scotland*).

[3] Miro modo in lapidea devectus navicula, apud Galweiam in regione illa quae Rennii vocatur, in portu qui Rintsnoc dicitur applicuit. In cuius portus littore curroc sancti Cuthberti lapidea adhuc perdurasse videtur; *Libellus de Nativitate Sancti Cuthberti*, cap. xix. (Surtees Society, vol. viii.). As the stone curach was apparently extant when the *Life* was written, it is possible that it may yet give a clue to the position of the port.

[4] *Celtic Scotland*, ii. p. 203.

Bute, one in Great Cumbrae, and one at Portpatrick.[1] It was to this last that Néide came after leaving Rind Snóc, which was therefore north of *Port Rig*. In fact *Port Rig*, 'king's port,' must have been the old name of Portpatrick; the name was probably connected with Rerigonion. It is to be noted that Wyntoun's—and Barbour's—expression means 'Mull *of* Rinn Snóc'; this suggests as a possible explanation that Rinn Snóc was the name of the double promontory from what is now Corsill Point southwards. In any case we must suppose that Néide landed at some point in the northern end of the promontory. Why he took this apparently roundabout way, instead of going straight from Kintyre to Ireland, we are not told, but it may be conjectured that it was in order to visit the seat of the British prince of the district, perhaps at Rerigonion.

Another old reference is found in the tale of Lugaid mac Con, who was king of Ireland about A.D. 250. Lugaid was banished from Ireland, and like other princes in similar plight he went to Alba, where he was received by the king, who on learning his position promised him help from himself and from the Britons. So all the ships and galleys and barks that were on the coast of the Britons and the Saxons were gathered to *Port Rig* in Alba, and with them a vast flotilla of curachs. So great was the fleet that men said it was as it were one continuous bridge between Ireland and Alba.[2] This *Port Rig* must be the one which is now Portpatrick.

These tales, which are semi-historical, belong to the pre-Christian period. The earliest Christian establishment in Scotland of which we have definite knowledge is that of Ninian at Candida Casa, now Whithorn in Wigtonshire,

---

[1] Portree in Skye is often pronounced in Gaelic *Port-righ*, as if 'king's port,' and the name is supposed to date from a visit of James v. in 1540. The unsophisticated Gaelic pronunciation of Skyemen, however, is *Port-righeadh*, and the second part is clearly from *righ* or *ruigh*, 'fore-arm,' common in our place-names as 'slope,' 'ground sloping up to a hill'; Mid.Ir. *rig*, gen. *riged* (*i.e. righeadh*).

[2] LL 289 b; *Silva Gadelica*, 313 (Gaelic), 352 (English). The anachronism of 'Saxons' in Britain at this period is characteristic of the later versions of tales.

founded probably about A.D. 400, and dedicated to St. Martin of Tours. The historical Ninian, as opposed to the Ninian of Ailred's *Life*, is rather a shadowy personage, and our knowledge of him is derived from Bede, who wrote about three hundred years after Ninian's time. Ninian, says Bede, was a Briton. He was regularly trained at Rome. Further, he preached the gospel to the Picts who lived to the south of the Grampian mountains, who, under his instruction, 'forsook the error of idolatry and received the faith of truth.' Even these statements of Bede are qualified by his cautious 'as they relate,' showing that they were based on tradition which he considered worthy of credit. Early religious sites were not chosen at random; they were often in the neighbourhood of the seat of the local prince or king, and the position of Candida Casa suggests that Ninian was there with the approval of the British ruler of the district, whose consent and protection had been secured. That his work among the southern Picts was not permanent in all cases appears from the references of his younger contemporary Patrick to the apostate Picts who shared with the British of Strathclyde and the pagan Scots of Alba in the spoil of Christian captives from Ireland.[1] But that failure—which may well have been only partial—did not mean the eclipse of Candida Casa. That remained long an important centre of religious life and learning, widely known and much frequented. The Irish style it *Teach Martain*, 'Martin's House';[2] *Rosnat*, which is a diminutive from *ross* and means 'Little Cape'; *Magnum Monasterium*, 'the great monastery'; and *Futerna*, the latinized Gaelic form of the old English *Hwiterne*, 'White House.' Some of the most eminent of early Irish clerics were trained there, the last on record of them is Findbarr of Magbile (Moyville) who died in 579. He came, we are informed, to study under Mugint—a Briton.[3] From about 730 till about 800 there were Anglic

[1] *Epistle to Coroticus*.

[2] For religious establishments styled *teach*, see Hogan, *Onomasticon*.

[3] See Skene, *Celtic Scotland*, ii. p. 46. With Mugint compare Meigant; Rees, *Essay on the Welsh Saints*, p. 269.

bishops at Whithorn, 'and beyond these,' says William of Malmesbury, 'I find no more anywhere; for the bishopric soon failed, since it was the furthest shore of the Angles, and open to the raidings of the Picts and Scots.' The name of Ninian was honoured to the end: between 782 and 804 Alcuin writes 'to the brethren of Saint Ninian of Candida Casa,' and causes a silken shroud (*velum*) to be sent for the body 'of our father Nyniga.'[1] It is clear that long before Hí was founded by Columba, Candida Casa formed a very important link between Ireland and Scotland.

During part of the seventh century intimate relations existed between the kingdom of Northumbria and the monastery of Iona. Osuald, son of Aedilfrid, spent long years of exile there and learned to speak Gaelic; when he came to the throne, it was to Iona he turned for men to instruct his people in the faith. Short as was the time during which the Celtic Church lasted in Northumbria (635-664), great work was accomplished: 'Thenceforward,' says Bede, 'very many began to come to Britain day by day from the country of the Scots, and to preach the word of faith to those provinces of the Angles over which Osuald reigned. . . . Churches therefore were built throughout the land . . . along with their parents the children of the Angles were instructed by the Scots their teachers in the studies and observance of regular discipline.'[2] In 685 the Northumbrian power was broken by the Picts at the battle of Nechtansmere and king Egfrid was slain. His body, according to William of Malmesbury, was buried in Iona, where Adamnan was then abbot. If this is true, as there is no reason to doubt, it is a very remarkable testimony to the veneration in which the mother of Northumbrian Christianity was held, the Synod of Whitby notwithstanding. Egfrid's successor, Aldfrid, 'was then in exile among the islands of the Scots for the study of letters'; part of the time, at least, he spent in Ireland where he was known as Fland Fína mac Ossu (d. 705). He was an accomplished Irish poet, and verses and wise sayings

[1] Lawrie, *Early Scottish Charters*, p. 3.

[2] *Historia Ecclesiastica*, iii. p. 3.

ascribed to him are extant. Adamnan calls him his friend (*regem Aldfridum amicum*),[1] and visited him twice at his court, obtaining on the occasion of his first visit the release of sixty Irish captives carried off from Bregia in 684 by the Angles. On this journey we are informed that he put in at *Trácht Romra*, 'Strand of (the) mighty sea' (*ro-muir*), by which is meant the Solway. This appears to have been the usual route from the west to Northumbria, and it may have been the route followed by Aedán before Degsastan. When Bede concluded his history in 731, he records that, 'the Scots that dwell in Britain are content with their own territories, and plan no snares or deceits against the nation of the Angles.' Of the other nations he reports that the Picts had at that time a treaty of peace with the Angles. The Britons 'for the most part oppose the nation of the Angles with the hatred natural to them'; they refuse to accept the Roman date of Easter; in part they are independent, but in some parts they are reduced to the servitude of the Angles.

The number of dedications to Gaelic saints is sufficient proof of the activity of the Irish Church in this quarter, silent as the Irish records are upon the subject, and something may be learned from considering them briefly.

To *Brigid* there are two dedications in Wigtown, three in Kirkcudbright, and two in Dumfriesshire, in the forms Kirkbride, Kilbride, Kirklebride (for Kirk-kil-bride). Brigid of Kildare died about 526; her name became so popular that many saints bore it after her time—fifteen are on record[2]—so that we have no certainty that these commemorate the famed contemporary of St. Patrick. The dedications to Patrick himself are probably long after his time.

Kirkmahoe, in Dumfriesshire, is Kirkemogho, 1319 (Bain's Cal.), Kirkmahook, five times in the same charter, 1428 (Reg. Glas.); Kirkmocho, 1430 *ib.*; Kirkmacho (RMS,

---

[1] *Life of Columba*, ii. p. 46. Adamnan presented Aldfrid with his book on the Holy Places. He himself was presented with many gifts on his visit to Aldfrid (*Hist. Eccl.*, v. 15).

[2] Rawl. B 502; 94 d 30; LL 368 g.

i., App. 2). In Ireland there is Timahoe in Queen's County, and in Kildare, representing *Tech Mo-Chua* ; of the fifty-nine saints of that name this was Mo-Chua mac Lonáin, who died in 657. Without further data it might be assumed that this is the name commemorated in Dumfriesshire. There is, however, a Kilmahoe in Kintyre of which the Gaelic form is extant, namely *Cill Mo-Chotha*, and this is decisive that the saint's name in this case was Mo-Choe (two syllables) ; in the spelling *Cotha*, the *th* is used to divide the syllables as usual in Scottish Gaelic. Eight saints of that name are on record,[1] the earliest of whom was Mo-Choe of Aendruim on Loch Cuan (Strangford Loch), who died in 497 (AU), and is said to have been the son of Brónach, daughter of the Miliucc who held Patrick in bondage. In BB 214 b 7 he is called Mo-Chai. An abbot of Aendruim who died in 917 (FM) was named Maelcoe. Though it would be rash indeed to assume that the church of Kirk-mahoe was founded by Mochoe of Aendruim, it may have been founded by a member of that monastery. In 1164 (Reg. Glas.) king William granted to the church of St. Kentigern in Glasgow, 'Gillemachoi of Conglud with his children and his whole following' ; the name of this *homo nativus* means 'Mo-Choe's lad.' In Dumfriesshire we find Michael McGilmocha and Achmacath McGilmotha among the chief men of the lineage of Clen Afren in 1296 (Bain's Cal.).

Kirkmadrine, in Sorbie and Stoneykirk parishes, has the stress on *-drine*, which therefore represents the saint's name. No such name appears in the Calendars, but the tract on the mothers of the Irish saints informs us that 'Dina, daughter of the king of the Saxons, was the mother of the ten sons of Bracan, king of Brachineoc, of the Britons, namely, Mogoroc, *i.e.*, Draigne of Sruthair,' etc.[2] Brachan

---

[1] Rawl. B 502, 94 a 27 ; LL 368 2 ; in the latter the alternative spelling Mochuae is given, and Mochue is a variant in Oengus' *Félire*, June 23 ; compare *Tóe*, *Tua*, p. 152.

[2] Di*na* *ingen* rí Saxan m*athair* deich mac mBraca*n* ri Brachineoc do B*ret*naibh . i . Mogoroc ⁊ (read . i .) Draigne Sruthra, Moconoc Cilli Muc*r*aissi ⁊ i *n*Gngelenga (*read* i nGailengaibh) a nDelb*na* Beathra, Cairi*ne* Cilli Cairin,

of Brecheniauc is a prominent, if somewhat shadowy, figure in early Welsh history, and is said to have been the son of an Irish prince named Cormac; he seems to have lived *c.* A.D. 500. *Draigne* becomes *Draighne*, with *gh* silent, in Mid. and Mod. Gaelic, and would naturally be *Drine* in English. Kirkmadroyn, the spelling in Macfarlane (ii. p. 81, etc.) represents the Gaelic pronunciation. Kirkdrine, in Kirkmaiden parish, contains his name without the honorific *mo* or *ma*, my. Whether he was a son of Brachan or not, we may take it that he was a Briton by origin, and we may suppose that he was connected with Whithorn. It is worth noting that the Welsh tradition makes Kynon, a grandson of Brachan, and Run—possibly a son—clerics in 'Manan' or 'Mannia,' by which may be meant either the Isle of Man or the district of Manu in Lothian.

Kirkmaiden parish, in the Rinns, and Kirkmaiden in Glasserton parish, are dedicated to Medana, whose day fell on 5th July, and may therefore be taken as a latinized form of *M'Etáin*, for *Mo-Etáin*, the virgin of Tuam Noa, whose day is the same. Etáin's period appears to have been early, but it is not clear that she was, as Skene thought, the same as Moninne or Darerca, who died in 517 or 519 and is reputed to have founded three churches in Galloway, one of which was 'Chilnecase.'[1] 'John Makelatyn' (1424,

---

Iast i Leamnachaib Cliab, Elloc Cilli M'Ellóc ic Loch Garman, Dirad Edair *Droma*, Dubain ailit*ir* Maes na hImirgi, Cairp*ri* ailitir i Cill Cairp*ri* i Sil Forandan, Paan i Cill Paain i nOsragaibh, Caeman ailitir i Cill Caemain i nGeisilli 7 in aliis locis, 7 mathair Mobeoc Glindi Garg, araba mac sein Brachain maic Brachineoc ut dicunt alii—BB 213 b 39. The corresponding list in LL 372 d has 'Mogoroc Sruthra' with 'i. Dergne' written above; but Mogoroc of Sruthair (? in Co. Dublin) and M. of Dergne in Wicklow were different persons (Rawl. B 502, 93 e 29; LL 368 b; BB 227 f 21). I take BB to mean that Draigne is another name for Mogoróc of Sruthair. In l. 4 above, LL reads 'Iast is lemnachaib Al*ban*,' from this together with 'i Leamnachaib Cliab,' it is clear we should read 'Iast i Lemnachaib Alban,' 'Iast in Lennox of Alba.'

In the Welsh list of Brachan's sons the only name found in the above list is Kynauc sanctus, corresponding to Moconoc (*Arch. f. Celt. Lexik.*, i. p. 524).

[1] *Celtic Scotland*, ii. p. 37.

Bain's Cal.) appears to represent *Mac Gille Etáin*, 'son of Etáin's servant.'

Kirkcowan, Kirkewane 1485, in Wigtown, appears to commemorate a saint *Eoghan*, and as Eogan of Ard-sratha is stated in his *Life* to have been trained in Whithorn,[1] there is a strong probability that it is he who is commemorated here. Eoghan was the son of bishop Erc of Slane, who died in 513, and his period is therefore the first half of the sixth century.

Kilphillan, in Wigtown, is dedicated to one of the saints called *Fáelán*, *Faolán*, 'little wolf,' a reduced form of *Fáelchú*. According to a note in the *Leabhar Breac*, p. 90, Faelán of Rath Érenn in Alba and of Leix in Ireland was son of Oengus, son of Natfraech, and as Oengus was baptized by Patrick and seems to have been alive after 483,[2] it is quite possible that his son is the Faelán who is said to have been trained by Ailbe of Emly, who died in 534 or 542 (AU). Faelán, son of Cáintigernd (Kentigerna), daughter of Cellach Cualann of Leinster, came to Alba about 717, and settled on Loch Alsh in Ross-shire, where he and his mother and his uncle Comgan are all duly commemorated. While there is no proof of the matter, it is probable that the Faelán of Galloway was the same as the saint of Kilallan in Renfrewshire.[3] Gilbert McGillelan or McGillolane (*Mac Gille Fhaolain*) is on record as Chief of Clan Connan in Galloway in the reign of David II. (1329-1370).[4]

Kilblane in Kirkmahoe and Kilblain or Kirkblain in Caerlaverock commemorate the saint who is styled in Gorman's *Félire* 'Bláán buadach Bretan,' 'triumphant Bláán of the Britons,' a highly significant designation. Bláán was bishop of Kingarth in Bute; Dún Blááin, Dunblane, is stated to have been his chief monastery (*cathair*); he is commemorated also in Kintyre (Southend), in Inverary

---

1 Plummer, *Vitae SS. Hiberniae*, cxxvi.

2 Oengus was killed in the battle of Cell Osnad, which took place in the reign of Lughaidh, who reigned from 483 to 512.

3 Compare Kilallan, p. 193.

4 RMS I, App. II.

parish, and at Lochearnhead (*caibeal Bhlathain*). There is a Kilblain near Old Meldrum, Aberdeen. Bláán's tutor (*aitte*, *oide*, Lat. *nutricius*) was Cattán, who is said to have been his uncle and who was a contemporary of Comgall (d. 600) and Cainnech (d. 598).[1] This would make Bláán flourish about the end of the sixth century. He was doubtless an Irish-trained Briton, and his work appears to have been mainly among the Britons of Galloway and Fortriu ; Bute may have been British in his time. Gilcomgal mac Gilblaan was witness, along with Gilendonrut Bretnach, 'the Briton,' and others, to an early grant by Radulf, son of Dunegal in Dumfries.[2]

Kirkgunzeon in Kirkcudbright, Kirkwynnin *a.* 1200 (Johnston), is supposed—and doubtless correctly—to contain the Welsh form of *Finnén*, a diminutive of the name of Findbarr of Moyville, whose death is recorded 'quies Uinniani episcopi' in 579 (AU). As has been noted already, he was for some time in Whithorn. Findbarr was also known as Findia ; a note on Oengus at Sept. 28 on the two saints called Findia says 'alii dicunt comad he dobeth in Futerna isna Rendaib,' 'others say that it was he who was in Futerna (Whithorn) in the Rinns (of Galloway).' Kylliemingan near Kirkgunzeon church, is probably for *Cill m'Fhinnéin*, 'my Finnén's church.' Chapel Finzian in Mochrum is mentioned in Macfarlane (ii. p. 88) ; on Blaeu's map it is Chappell Finnan, now Chapel Finian.

Colum Cille, Columba of Iona, is supposed to be commemorated in Kirkcolm in Wigtown, which has St. Columba's Well ; St. Columba's Chapel and Well are in the parish of Caerlaverock (d. 597).

Donnán of Eigg (d. 617) is commemorated in Kildonan and Chapel Donan in Wigtownshire.

The parish church of Wigtown is styled 'ecclesia S. Macuti (Machuti),' 1451, 1495 (RMS), and 'ecclesia S. Mathuri,'

---

[1] *Celtic Scotland*, ii. p. 133. Stokes in his edition of Gorman's *Félire* says of Bláán, 'ob. 510' ; I do not know his authority for this.

[2] Bain's *Calendar* ; 'Gilendonrut' conveys no sense to me ; 'Dunegal' is the Old Welsh *Dumnagual* corresponding to O.Ir. *Domnall*, G. *Domhnall*, anglicized 'Donald.'

1326 (RMS). The former is doubtless the correct form, as is borne out by the name Killiemacuddican in Kirkcolm parish. This is for *Cill Mo-Chudagán*, St. Mochutu's Church ; the form of the diminutive is that seen in *maccucán*, 'sonnie' ; *Artacán*, *Artagán*, 'little Art' ; *Flanducán*, *Flannagán*, 'little red man,' from *fland*, red, and many other names. The saint is doubtless the famous Mochutu of Rathan and Lismore, who died in 637. *Giolla Mo-Chuda* is the Irish style of MacGillycuddy of the Reeks, but I have not met the name in Scotland. The saints who follow are not dated.

Kirkinner in Wigtownshire is 'ecclesia Sancte Kenere de Carnesmall in Galwedia,' 1326 (RMS), probably a latinized form of *Cainer* ; there were several saints of that name (*v.* p. 275). Her day in the Aberdeen Breviary is Oct. 29.

Killumpha in Wigtownshire is Killumquhy in 1545 and 1594 (RMS) ; the saint appears to be *Imchad*, *Iomchadh*, gen. *Iomchadha*, of Cell Drochat in the Airds of Ulster. John Maklunfaw in Blareboy, barony of Mureith, 1509 (RMS), is probably *mac Gille Iomchadha*, 'son of Imchad's servant,' the modern McLumpha.

Kilquhanatie in Kirkcudbright is Kilquhonide in 1525 (RMS) ; the saint may be *Connaith* or *Connait* (Gorman and Mart. of Donegal), with which may perhaps be compared 'Joannes filius Gilchonedy,' otherwise 'Gilchomedy,' in Lanarkshire (RMS I).[1] There is *Cill Chonaid*, anglicized Killiehonnet, in Brae Lochaber.

Kirkmabrick in Wigtown and Kirkcudbright is in Blaeu and elsewhere (RMS) Kirkmakbrick, which is doubtless the better form. I think that we have here to do with *Aedh mac Bric*, a bishop who died either in 589 or 595 (AU). For the dropping of final *c* of *mac* we may compare Polmadie, of old Polmacde, near Glasgow ; Dunmaglas in Strath Nairn, G. *Dùn mac Glais* ; Belmaduthy in the Black Isle, G. *Baile mac Duibh*, etc. Strictly the name should be 'Kirkmikbrick' (*maic*, *mic*, gen. of *mac*), but 'mak' would easily arise in the unstressed position. The other alternative

[1] Compare Gilquhammite, Gilquhomytty, 1541, 1542 RMS.

would be to take it for 'church of the sons of Breac' (*mac*, gen. pl.), but though this group form of dedication is well established, this particular group does not seem to occur.

Closeburn in Dumfriesshire, a. 1200 Kylosbern (Johnston), Killeosberne 1300 (Bain's Cal.), is supposed to commemorate Osbern, an English saint; there was, however, *Osbran*, 'stag-raven,' anchorite and bishop of Cluain Creamha in Roscommon, who died in 752.

Kirkcarswell in Rerwick parish is Kyrassalda 1365, Kirkassudie 1329/70, Kirkcassail 1537, Kirkcossald 1567, Kirkcaswell 1571, -cossald 1602 (RMS), -castel in Macfarlane (ii. p. 58), -arsell in Blaeu. This is St. Oswald's kirk; see Kirkoswald, p. 188.

Kilquhockadale is Kilquhokadaill 1573 (RMS); it is the same as Kilcock in Ireland, for *Cill Chuaca*, 'Cuaca's church'; she is given by Gorman at Jan. 8. For -dale, which is Norse *dalr*, a dale, compare Killernadale in Jura and Navidale in Sutherland; Kilquhockadale is 'the dale of Cuaca's church.'[1]

Killasser in Stoneykirk is 'Lassair's church,' like Killasser in Mayo for *Cell Laisre* or *Cell Lasrach* (Hogan). Gorman has three female saints of this name, which means 'flame,' and eleven named Lassar. Compare *Cill Lasrach* in Islay.

Kirkcormac in Kirkcudbright commemorates some one of the saints called *Cormac*, the earliest of whom was bishop of Armagh and died in 496. Another famous Cormac was Cormac Ua Liatháin, a contemporary of Columba and abbot of Durrow, styled by Gorman 'Cormac léir,' 'industrious, devout.' It was he who voyaged to the Orkneys, and this dedication may well be to him. This and other churches in Galloway belonged at one period to Iona, for King William the Lyon granted to the Church of Holyrood in Edinburgh the churches or chapels in Galloway which pertained to the right of the abbacy of Hij Columchille, namely the churches of Kirchecormach, of St. Andrew, of Balencros, and of Cheletun (Chart. Hol.). These churches

[1] Compare 'Kyrkecok in the diocese of Glasgow' (*Chart. Hol.*).

are all in Kirkcudbright; Cheletun is Kelton, in which Kirkcormack is now included; Kirkanders is now included in Borgue; Balencros is now Barncrosh in Tongland.

Bean, the saint of Kirkbean in Kirkcudbright, may be St. Bean of Kinkell and of Wester Foulis in Perthshire (*v.* p. 310).

Kirkennan in Buittle and Minigaff is Kirkynnane 1454, Kirkenan 1458, Kirkennane 1490 (RMS), Kirkcunan 1611 (Ret.). The saint is uncertain, but he can hardly be Adamnan, whose name appears in 'Duncan McGillauenan,' 'Duncan son of St. Adamnan's servant,' one of the chief men of Clen Afren in 1296.

Oengus records in his *Félire* at 24th November *Colman Duib Chuilinn*, who is stated in the notes to have been 'in the Rinns, *i.e.* from Dún Reichet and from Belach Conglais in Leinster, and from other places.' [1] One is naturally inclined to place this Colman in the Rinns of Galloway, and to equate Dún Reichet with Dunragit—an equation which is unobjectionable phonetically. There was, however, a district in Roscommon called the Rinns, and there appears to have been a plain in Connacht called Mag Rechet. Thus Dún Reichet, Colman's seat, may have been in the Rinns of Roscommon. He was a younger contemporary of Comgall of Bangor, who died in 602.

Murchereach (? Murchertach), priest of St. Carpre of Dunescor, witnessed a charter of Edgar, son of Dofnald, of the church of Dalgarnoc (Chart. Hol.). Gorman mentions four saints named Cairbre (Cairpre, Coirpre), one of whom was bishop of Mag Bile (Moyville), another was bishop of Cúil Rathin, and a third, Cairbre Cromm, bishop of Clonmacnois.

Ecclefechan, 'ecclesia Sancti Fechani,' is Egilfeichane, 1507 (RMS); Eglisfechane, 1510 *ib.*; Egilphechane, 1542 *ib.*, etc. It is probably to be compared with Llanfechan in Montgomeryshire; the second part, which is supposed to be the name of St. Féchín of Fore, may however be the mutated form of the Welsh adjective *bechan* (fem.), 'little,'

---

[1] 'Colman Duib Chuilinn isna Rennaib . i . O Dhún Reichet 7 ó Belach Conghlais il-Laighnibh et ab aliis locis,' *Fél. Oeng.*, p. 246.

as in Llan-fair-fechan, 'Little St. Mary's Church.' In this case the meaning would be 'little church' (W. *eglwys fechan*).

Ecclefechan is in the parish of Hoddom, where, according to his *Life* by Jocelin, St. Kentigern (d. 612) built churches and placed his see for a time before transferring it to Glasgow. He is commemorated in the adjoining parish of St. Mungo or Abermilk. Crosmungo in Wauchopedale is on record in 1641 (RMS). Though a Welsh triad places Cyndern Garthwys, *i.e.* Kentigern, as chief bishop at Penrionyd (p. 34), no commemorations of him seem to occur in Galloway.

Jocelin explains his name as 'capitalis dominus,' 'head lord,' and later adds 'for *Ken* is in Latin *caput* (head), and *tyern* in the language of Scotland (*Albanice*) is in Latin *dominus*,' thus making it a compound of Gaelic *cenn* and *tigern*. It is, however, British, probably for earlier *Cintu-tigernos*, 'first lord'; compare Welsh *cynben*, prince, for *Cintu-pennos*, 'first head'; the alternative is *Cuno-tigernos*, 'hound-lord'; compare *Cuno-glasos*, 'tawny hound.' Jocelin explains his popular name Munghu, now Mungo, as 'carissimus amicus,' 'very dear friend,' apparently as if from Welsh *mwyn*, kind, dear, and *cu*, dear, amiable; but the formation is not clear to me.

His disciple *Convallus*, O.W. *Conguall*, later *Cinvall*, *Cynwall*, Ir. *Conall*, is commemorated in Kirkconnel in Kirkcudbright and in the various foundations bearing that name in Dumfriesshire; we may compare *Lann Cinvall* in Wales (Lib. Land.). But his chief seat may have been Dercongal or Darcungal (Lib. Melr.), 'Congal's oak-copse,' which in the twelfth century became the site of the abbey of Holywood, founded by John, Lord of Kirkconnel. The 'holy wood' was Congal's oak-copse (*doire*); compare *Preas Ma-Ruibhe*, St. Maelruba's copse, in Contin, near Strathpeffer. 'Thomas and Andrew of Kirkconeval' appear in 1304 (Bain's Cal.). 'Gilcomgal,' son of Gilblaan, means 'servant of Congal'; Adam McGilleconil was one of 'the chief of the lineage of Clen Afren' in 1296 (Bain's Cal.); his father's name means 'Congal's servant.'

Killintringan is for Sanct Ringan's Kirk, taken over into Gaelic from Scots, and the form is obviously late. I take this and most other commemorations of St. Ninian to be dedications of the later type, dating probably from the twelfth century, when the monastic system of the Celtic Church, of which Columba was the head, was changed to the diocesan system, and when it became politic to find a sanction for the change. This sanction was supposed to be found in Ninian's mission.

The ancient term *annaid*, O.Ir. *andóit*, Annat, a patron saint's church, or church which contains the relics of the founder, occurs in Annat Hill, on the farm of Kirkland of Longcastle, Kirkinner; Annatland, near Sweetheart Abbey, New Abbey; Ernanity, spelled by T. Pont 'Ardnannaty,' in Crossmichael, for *earrann* (or *ard*) *na h-annaide*, 'share (or height) of the Annat,' compare the Annaty Burn near Scone.

Clachán, a stone cell, is common, *e.g.* Clauchaneasy for *Clachán Iosa*, 'Jesus' Kirk,' near Loch Ochiltree, the Clachans of Whithorn, of Dundrennan, of Girtoun, of Glenluce, of Kirkcolm or Kirkcum, of Stranraer, and of Invermessan. This is not the same as modern G. *clachan*, stones, which has dull *a* in the second syllable. It is often made in English 'Kirktown,' 'Kirkton,' *e.g. an Clachán Aillseach*, 'Kirkton of Loch Alsh.'

Relic Hill in Kirkmahoe may contain *reilig*, a cemetery.

The results of all this may be summed up shortly. There is some evidence for the presence of the early Welsh or British Church, operating presumably, partly at least, from Whithorn among a Welsh-speaking people. The great majority of the dedications are characteristic of the early Irish Church, which had relations with Whithorn, and though it would be unsafe to infer that all the saints commemorated actually laboured in this region, it is nevertheless probable that some of them did visit it and found churches. As to the period of greatest activity, we may recall the words of Bede, 'thenceforward (*i.e.* after 635) very many began to come to Britain day by day,' etc. This ingress was into the kingdom of Northumbria, of which

Galloway and Dumfriesshire formed part; the route to Northumbria was by the Solway. At that period the language was doubtless Welsh, and the question arises how far the influx of Irish clerics may be expected to have influenced the native speech. As regards the common people, that, as it seems to me, would depend partly on the continuity of the missionary effort, partly on the founding of Gaelic-speaking monastic communities, and partly on the amount of peaceful penetration by laymen. On the two latter points we have no information, and the absence of reference to monasteries suggests that none was founded in early times by the Irish Church. As to the first point, it is likely that the progress of the Irish Church in this region was checked by the decision of Whitby. On the other hand, it is not unlikely that during the seventh century some, perhaps many, of the nobles came to know Gaelic, and that from them it may have spread to some small extent among their people. In this connection it is worth noting how little the names of saints commemorated appear to have been influenced phonetically by Welsh, but this may be accounted for by subsequent gaelicizing, *e.g.* Killiemingan alongside of the Welsh Kirkgunzeon.

A point to be specially noted with regard to the dedications is the close connection that exists between this region and Kintyre: Bláán, Donnán, Faolán, Mochoe, Brigid, Colum Cille, are common to both districts. This is likely to be more than a coincidence.

In 731, as Bede records, the Scots who dwelt in Britain were content with their own territories, and planned no snares or deceits against the Angles. At this time Galloway was an Anglic province, with an Anglic bishop at Whithorn; there is no hint of anything in the nature of a Scottish conquest or general settlement, either from Ireland or from Scottish Dál Riata, though there may have been Gaelic immigrants. Soon, however, great political changes took place. In 736, the year after Bede's death, Angus son of Fergus, king of the Picts, wasted the Scottish kingdom of Dál Riata, took its chief fortresses, and bound with chains two sons of its late king, Selbach. Five years later Angus dealt Dál Riata a 'smiting' (*percussio*) from which she did not recover for a long time, as a result of which the surviving Dalriadic nobles must have lost their position and have been forced either to leave their country or become vassals of the Picts. Disaster also befell Northumbria when in 756 king Edbert's army was almost wholly destroyed on the march homewards from Dumbarton, where he had been operating against the Britons of Strathclyde. After this, says Green, there were fifty years of anarchy in Northumbria.[1] Such a situation would very naturally lead to migration from Dál Riata to Galloway, now masterless and open, and though there is no reliable record of the events, the subsequent history suggests that migration actually took place.

The *Gall Gháidhil*, 'Foreign Gael,' whence the name Galloway, are mentioned for the first time in 852-3, when Aed, king of Ailech—near Derry, the seat of the northern kings—gave battle to their fleet. They are described as 'Scots and foster-children of the Norsemen, and sometimes they are actually called Norsemen.' Further, they were 'men who had renounced their baptism; they had the customs of the Norsemen, and though the real Northmen were bad to the churches, these were far worse.'[2] In 856 the Gall Gháidhil helped Maelsechlainn, king of Ireland, against the Norsemen. In the same year Aed of Ailech, who claimed the kingship of Ireland against Maelsechlainn, defeated the Gall Gháidhil in Tyrone. In 857 the Norsemen defeated Caittil Find, Kettil the White, with his Gall Gháidhil in Munster. In 858 Cerball, king of Ossory, who was on the side of Aed of Ailech, defeated the Gall Gháidhil in Tipperary. Here, then, we have the Gall Gháidhil coming to Ireland in ships and fighting—doubtless as mercenaries—under Caittil Find, a Norseman, on the side of the king of Ireland against his rival, who was helped by Norsemen. Nothing further is heard of them in connection with Ireland, but they appear in the Hebrides. Eigg, the scene of

[1] *A Short History*, p. 41.

[2] *Fragments of Irish Annals*, Skene, *P.S.*, pp. 403, 404.

Donnán's martyrdom, was in the territory of the Gall Gháidhil; so, too, was Bute, where Bláán was bishop in Cend-garad (Kingarth).[1] They were, of course, in Galloway: *Aldasain*, Ailsa Craig, is described in the Book of Leinster as between Gall Gedelu (acc.) and Kintyre.[2] They were also probably in Carrick and Kyle. From this it may be inferred that the Gael and their language were established in Galloway before the coming of the Norsemen, and also probably that these Gael were connected with the other Gall Gháidhil of the Inner Hebrides.[3]

As a territorial term, Gall Gháidhil settled down to mean Galloway, the Hebrides as a whole being known as *Innse Gall*, 'Isles of the Foreigners.' In 1034 Suibne, son of Cinaeth, king of the Gall Gháidhil, died.[4] Rollant mac Uchtraigh, king of the Gall Gháidhil, died in 1199, and his successor, Ailín mac Uchtraigh, in 1234. In these cases, as Skene has pointed out, the term is territorial and denotes Galloway. It may be worth mention as a curiosity that kings of the Gall Gháidhil are made contemporary with Fionn mac Cumhaill in the rather late compilation called *Acallam na Senórach*, 'the Colloquy of the Ancient Men.'[5]

---

1 'Téit iarum Donnán cona muintir in Gallgáidelu ⁊ geibid aitreb ann'; 'then Donnan with his community goes to (the) Gall Gháidhil and takes up his abode there.' *Féil. Oeng.*, Apr 17, note. The context shows that Eigg is meant.

2 'Aldasain . i . carrac etir Gallgedelu ⁊ Cendtíri i n-a camar (for *comar*) immuigh,' 'Aldasain is a rock between Galloway and Kintyre, facing them out (in the sea)'; LL 371 b; now nom. *Allasa*, other cases *Allasan*.

3 Skene's views on the Gall Gháidhil are given in *Celtic Scotland*, i. pp. 311, 312; iii. p. 292. In vol. i. he makes the term apply primarily to the people of Galloway, whence it extended to the Islesmen; in vol. iii. he reverses this view, and seems to hold that the name was applied to the people of Galloway after it was applied to the Islesmen. He does not consider the question how there came to be *Gael* in Galloway.

4 In the thirteenth century Dubhghall mac Suibhne (Dufgallus filius Syfyn) was Lord of Kintyre; *Reg. Pasl.*, pp. 120-22. The name is preserved in *Caisteal Suibhne*, Castle Sween, on Loch Sween in Knapdale. Cf. *Orig. Paroch.*, ii. pt. I. p. 40.

5 Samaisc, Artúr, and Inber, sons of the king of Gall Gháidhil, were drowned by the Smirdris of Loch Lurgan (*Acall.*, l. 4560); Bressal, son of Eirge, king of the Gall Gháidhil, slew Mac Lughdhach, Fionn's grandson (l. 7951)—Stokes' edition.

From the Gaelic *Gall Gháidhil*, primarily the name of the people, was formed in Latin the district name Galwedia (*Chron. of Man*), as Er-gad-ia was formed from *Oirer Gáidheal*, 'coastland of the Gael,' Argyll. The *w* has been thought to indicate the influence of the Welsh form *Gal-wyddel*, and this may be so, but the supposition is hardly necessary: the *o* of 'Galloway' represents an indeterminate vowel developed in Gaelic between the two parts of the compound (Gall*a*-ghaidhil), and after this vowel *gh* would readily become *w*. Other forms are Galwegia, where *g* represents *dh*; Galweithia; Galweia, where *dh* has disappeared. All these are practically the same. Fordun has a different form, Galwallia, and allowing for the disappearance of *ll* in Scots, and the presence of the indeterminate vowel after *Gal-*, this comes very near the modern Gallowa'.[1]

Here we may consider the so-called Picts of Galloway. I have already referred to the brochs of Wigtownshire as evidence that the marauding Picts of the far north found the Rinns to be a suitable base for their raids on the shores of Britain and of Ireland. Of the three brochs of which traces remain, one has been excavated;[2] its site is on Loch Ryan, north of Loch Insh, and near Craig Caffie, for *Creag Chathbhaidh*, 'Cathbad's Rock,' an Irish name.[3] The excavation went to show that there had been no prolonged occupation, and the relics found were few and unimportant. The other sites are at Ardwell Point and near Stair Haven; all were doubtless pirate holds more or less

---

[1] The name Galbraith goes to show, as has been noted (p. 14) that there were 'foreign Britons' as well as 'foreign Gall.'

[2] For account of the excavation, see *Proc. Soc. Ant. Soc.*, xlvi. (1912), p. 189.

[3] Nom. *Cathba, Cathbad* (for an earlier *Cathub*), gen. *Cathboth, Cathbad, Cathbaid*; in *Rel. Celt.* ii. p. 220, *Cathfaidh* (for *Cathbhaidh*), pronounced 'Caffie.' There is in Sutherland *Loch Chathbhaidh* (Loch a' Chaphi 6-inch Ordnance Survey Map, 30), with a very small islet called *an Annaid*. The Galloway name M'Haffie or M'Caffie is 'Cathbad's son.' It may be shortened from *Mac Gille Chathbhaidh*, 'son of St. Cathbad's servant,' for 'John Mackilhaffy,' appears in a list of Wigtown men given by Wodrow, iv. p. 22.

temporary in character and probably of the fourth or fifth century. These brochs could not have been built without consent of the British chiefs, who may be conjectured to have had their share of the spoils. In the ancient Irish tale entitled *Orgain Brudne Ui Dergae*, 'the Sacking of Ua Dergae's Hostel,' Ingcél, son of the king of Britain, is the leader of the band of pirates who storm the hostel and slay king Conaire.[1] This tale relates events that are supposed to have happened early in the first century, and is instructive on the subject of piracy in early times.

We need not suppose that the Picts who raided from these brochs had any permanent influence on Galloway, and we may pass on to consider the question of certain Picts supposed to be styled 'Niduari' by Bede in his *Life of St. Cuthbert*. Here it is best to begin by giving the actual facts that are on record. Of the three ancient *Lives* of St. Cuthbert the oldest, which is anonymous, was written between 698 and 705; the second by Bede, in Latin hexameters, some time later; the third, in prose, is largely based on the first, and was written by Bede before 721. Cuthbert died in 687. The reference in question occurs in all three, in course of a tale about the saint. On a certain occasion he went by sea, with two companions, to the land of the Picts, starting on the day after Christmas. When they had reached their destination a storm arose, which lasted three days, during which they began to suffer from hunger. After prayer by the saint they found three large cutlets of dolphin on the shore. On the fourth day the storm ceased, and they set out for home. The passages relative to the Picts in the respective *Lives* are as follows, in order of priority:—

(*a*) 'At another time he set out with two of the brethren from the same monastery, which is called Mailros, and sailed to the land of the Picts, where they arrived in safety at Mudpieralegis. There they stayed some days.'

---

[1] LU 99 a; O'Curry, *Manners and Customs*, iii. p. 136 *seq.*; *Rev. Celt.*, vols. xxii., xxiii.

(*b*) 'Meantime he is borne by ship to the coasts of the Picts. . . . And now on the fourth day the south winds cease, and with joy they make for the haven of safety across the smooth seas.'

(*c*) 'For on a certain time he set out from his own monastery on some necessary business, and came by sea to the land of the Picts, which is called Niduari, two of the brethren accompanying him.'[1]

Here it is evident that Bede had the older account before him when he wrote. The corrupt *Mudpieralegis* (which might be read *Mudwieralegis*) is obviously some place-name; it is so much longer than Bede's *Niduari* as to demand some explanation, which I am not fit to give, of the palaeographic relation between them. As to the Picts in question, there is not the slightest suggestion anywhere that they are other than the Picts of whom Bede, and the anonymous writer too, speak elsewhere, and who are distinctly said to be north of the Forth. If Bede and the author of the oldest *Life*, both of them careful writers, had been referring to Picts situated in a region so remote from Pictland as Galloway, it is inconceivable, to me at least, that they should not have mentioned the fact. The adverse 'south winds' of Bede's metrical *Life* are additional proof that he was thinking of some place in the north, most likely in Fife, which cannot now be identified. Skene, however, reviving a conjecture made by the editor of 1841, regarded

---

[1] 'Alio quoque tempore, de eodem monasterio quod dicitur Mailros cum duobus fratribus pergens, et navigans ad terram Pictorum, ubi Mudpieralegis prospere pervenit. Manserunt ibi aliquot dies.'—*Vita Anon.*, sec. 15.

'Pictorum interea puppi defertur ad oras.

. . . . . .

Jamque die quarto laeti cessantibus austris
Blanda salutiferum capiunt trans aequora portum.'

Metrical Life, ch. ix.

'Quodam etenim tempore pergens de suo monasterio pro necessitatis causa accidentis ad terram Pictorum, quae Niduari vocatur, navigando pervenit, comitantibus eum duobus e fratribus.'—*Vita S. Cudberti*, c. xi.

The construction of the first sentence from the oldest *Life* is faulty; perhaps it should read 'pergens, navigavit,' etc.

'Niduari' as a tribal name derived from the Nith of Dumfriesshire and meaning 'Picts of the Nith.' In support of this view he has a long note in his first volume,[1] in which he quotes Bede as having written, 'ad terram Pictorum, qui Niduari vocantur,' 'to the land of the Picts who are called Niduari.' Now these are not the words of Bede, and even as an emendation they are discredited by the older *Life*. Curiously enough Skene places the 'Picts of the Nith' in Kirkcudbrightshire; 'the traces of this visit,' he remarks further, 'have been left in the name of Kirkcudbright, or Church of Cuthbert.'[2] From the narratives it is clear that the saint founded no church on that occasion, and there is another Kirkcudbright in Ayrshire, at the spot where Tig joins Stinchar, which used to be known as Kirkcudbright Inner-tig, a place which Cuthbert is not likely ever to have visited.

So far, apart from the broch-builders of the Wigtown coast, we have found no real trace of Picts in Galloway; we come now, however, to something definite. Certain English chroniclers, who wrote in Latin, repeatedly mention the Picts of Galloway, mostly in connection with the period of the battle of the Standard in 1138. Richard of Hexham, writing before 1154, speaks of the Picts who are commonly called Galwegians; he himself regularly calls them Picts, and the character he gives of them is of the worst. Reginald of Durham, in the latter half of the same century, makes the very interesting statement that Kirkcudbright is in the land of the Picts, and that the people there speak the language of the Picts—the language of Galloway at that time being of course, mainly at least, Gaelic. In his *Life of St. Kentigern*, written about 1190, Jocelin mentions 'the land of the Picts now called Galweithia.' Other English chroniclers of the same period, however, use the term 'Pict' in its usual denotation, and style the Galwegians 'Galwenses.' The term used in all charters of David I. is Galwenses, Gawenses, etc. No Scottish or Irish chronicle,

[1] *Celtic Scotland*, i. p. 133.
[2] *Celtic Scotland*, ii. p. 209.

so far as known to me, makes mention of Picts in Galloway. The short and easy way of dealing with the matter would be to refuse altogether to accept the statements of these English outsiders, as indeed the late Dr. Macbain was at one time inclined to do. Latterly he somewhat modified his view, but I think he always regarded the Picts of Galloway as one of what he used to call 'the three frauds of Scoto-Celtic history,' and in that, in the main, he was right. The population of Galloway was never Pictish, if by Picts we mean the real Picts—the early tribes of the far north of Scotland. Nor yet was it ever Pictish in the sense that it at any time came under Pictish rule or hegemony; it never formed part of Pictavia. Nevertheless there was ground for the statement, and Galloway tradition throws an entirely new light upon it.

Dr. Trotter, in the valuable books which he called *Galloway Gossip*, mentions and describes at some length, a certain 'breed' among the Galwegians who were called 'Creenies'; they were reckoned 'foreigners,' and were considered to be descendants of the Irish 'Picts.' Most of them were in the Rinns. Now 'Creenie' is plainly *Cruithnigh*, the plural of *Cruithneach*, and nothing is more likely, one may say certain, than that the 'Creenies' were immigrants from the Cruithnean part of Ulster, facing Galloway. To the English chroniclers the terms 'Cruithnigh' and 'Picti' were synonymous, so they naturally used the latter in Latin. But seeing that these 'Creenies' formed only a part of the population, there remains the question why the Galwegians as a whole were called Picti by some at least of the English writers. The tradition preserved by Dr. Trotter gives the answer, and indicates that it was not by way of compliment. The 'Creenies,' he informs us, were also called 'Gossocks.' In the name Gos-patrick and other names of persons, *gos* represents Welsh *gwas*, a servant—'servant of Patrick'; Gosmungo, 'servant of Mungo,' and so on. In the *Book of Llandaf* a certain cleric bears the name of *Guassauc*, 'servant.' As a common noun *gwasog* means 'a servile person, a person in a servile condition,' and it is this term which has been preserved in

Galloway tradition as another name for the 'Creenies.' It was the name by which they were known among the British population and it indicates their status in the community. Far from owning the soil, the so-called Picts of Galloway were serfs of the Britons or at best rent-paying vassals: when applied to the Galwegians as a whole the name was a term of opprobrium.[1] In Ireland the Cruithnigh were vassals of the Gael.

Though we have no means of knowing at what precise period these Cruithnigh came to Galloway, it is evident that they came when Welsh was still spoken there. Their own language at the time of their migration was certainly Gaelic.

Another Galloway 'breed' mentioned by Dr. Trotter is the Fingauls, described as tall, well made, fair-haired, and blue-eyed, with 'wunnerfu feet for size.' They were commonest in Saterness, Co'en, Borgue, Whithern, and Kirkmaiden, and they were reckoned to be descendants of the Norsemen. Dr. Trotter's authority conjectured that they had come from the Isle of Man. The term is pronounced with stress on the second part, so that it represents not *Fionnghall*, 'fair stranger,' but *Fine Gall*, 'tribe of strangers.' 'Fine Gall' was a well-known district in Ireland, coinciding with the part of Co. Dublin north of the Liffey, so called because it had been settled by the Danes of Dublin. The Fingauls of Galloway may have been immigrants from this quarter, and if so their advent may have been as late as 1014, when the Danes and Norsemen were broken at Clontarff, near Dublin.

While it is difficult to speak positively as to the time and the manner in which Gaelic came into Galloway, it seems possible that the process may have begun as early as the seventh century, while it is probable that settlements from Argyll were made about the middle of the eighth century; there was also immigration from Ireland, probably

[1] In 1259 a Dumfries man slew a miller of the town who had called him 'Galuvet' (a Galwegian) in the sense of 'thief.' 'Pict' would have been even worse. Bain's *Calendar*, i. p. 428; *Acts of Parl. of Scotland*, i. p. 87.

at more than one period. As in all cases where the speech of a district changes, there must have been a long period during which two languages were spoken and when many of the people were bilingual. There must have been a considerable body of Gaelic speakers in the middle of the ninth century, and Gaelic seems to have been firmly established by the time of David I.

In the nomenclature of the region up to the river Nith, Gaelic is markedly predominant. There is a fair, though comparatively small, number of British names, some of them practically unchanged, such as Penpont, 'bridge-head,' Leswalt, 'grass enclosure,' Drumwalt, 'grass ridge,' Trostrie, 'cross-stead,' corresponding to the G. Baltersan, for *baile tarsuinn*, in Holywood. Others are more or less disguised. East of Nith there are still a good many Gaelic names, such as Enzie in Westerkirk, from *eang*, a nook, gusset, primarily something angular or triangular; compare 'the Enzie' in Banffshire. Dalbate in Middleby is for *dail bhàite*, 'drowned dale,' *i.e.* subject to flooding; compare Feabait in Ross-shire. Conrick in Sanquhar is for *comhrag*, M.Ir. *comrac*, a meeting, confluence; compare Conrick in Badenoch. Glencorse in Closeburn is 'glen of the crossing' (*crosg*), like Glencorse in Midlothian. Three Mullach Hill in Hutton is 'hill of three tops,' a part translation. Auchencairn is 'field of the cairn'; and so on. British names east of Nith are not uncommon, as Glen-tenmont, from O.W. *-monid*, hill, and perhaps *tan*, fire, meaning 'fire-hill'; Pennersax in Middlebie, 'head of the Saxons'; Pumpla, from *pump*, five; compare G. *Cóigeach*, 'place of fifths.'

When the *Old Stat. Account* was written there were remains of a 'Druidical Temple' on Graitney (Gretna) Mains, and one of the stones—the largest—was commonly called Lochmaben Stone.[1] The correct form of the name appears in 1398 (Bain's Cal.) when at a meeting held at Clochmabanestane between commissioners for Scotland and for England it was agreed that the men of Galloway, Nithsdale, Annandale, and Crawford Muir should meet the wardens of the

[1] *Old Stat. Acc.*, ix. p. 528.

west March for redress at Clochmabanestane. Thereafter it was the regular trysting place for business connected with the March. The stone is a granite boulder, 18 feet 2 inches in circumference and 7 feet 6 inches high, but only one of its companions remains now.[1] 'Clochmabane' means 'Mabon's stone'; Scots 'stane' translates G. *cloch*, and that again is probably for an earlier W. *clog*, rock, or *maen*, stone. *Mabon* represents E.Celtic *maponos*, a boy or male child; *Apollo Maponos* was the sun-god of the British Celts, and his name appears in inscriptions in the north of England. Two of the personages in the Mabinogion tale of Kulhwch and Olwen are Mabon son of Mellt, a hero, and Mabon son of Modron, whom Rhys equates with the sun-god. The name was borne by others; it occurs twice in the Book of Landaf. Here, however, its association with a stone circle is in favour of a mythological connection; Clockmadron, 'Modron's stone,' in Fife, is stressed on *dron*, and is therefore not from Modron. *Mabon* appears also in Lochmaben, whether with a similar reference is not clear.

In 1304 'the fosse of the Galwegians and the brook (*rivulus*) running thence into Lydel' are named as part of the bounds of Cresope (Bain's Cal.), now Kershope on Kershope Burn, a tributary of Liddel, which forms part of the border between England and Scotland. This 'fosse' may be the ancient march now known as the Catrail,[2] but the part of the Catrail nearest to the Kershope Burn is, according to the one-inch Ordnance Survey map, right at the head of Liddel Water, about ten miles away.

'The Deil's Dike' in Galloway and Dumfriesshire starts at a point on the east side of Loch Ryan, runs discontinuously 'between the desert and the sown' through the uplands, and ends near Dornock. The name is like *Saothair an Daoi*, 'the devil's work,' applied to a trap-dike in

---

[1] *Report on Ancient and Historical Monuments of Scotland*, vii. (Dumfries), p. 92, where a photograph of the stone is given.

[2] 'Catrail' is said to be stressed properly on the second syllable; its meaning is obscure, but it may be compared with Powtrail, the name of a head-stream of Clyde.

Colonsay. Other terms for 'the Deil's Dike' are 'the Picts'' or 'Celtic' Dike: the Devil, the Romans, the Picts, and Michael Scott are often credited with works of which the origin is or was unknown.

The Gaelic names all over are not of a very old type; they are for the most part straightforward phrases, such as Allfornought for *all fornocht*, 'stark-naked rock,' Barsolus, 'bright top,' like Resolis, 'bright slope,' in Ross-shire, Kirminnoch for *ceathramh meadhonach*, 'mid quarter.'

Ericstone near Moffat is Arykstane in Barbour's *Bruce* and is either for *clach na h-éirce*, 'stone of the atonement' (*éiric*), or *clach an eireachta*, 'stone of the assembly'; for the part translation compare Coldstone in Aberdeenshire, earlier Codilstane, 'trysting stone,' from G. *comhdhail*, M.Ir. *comdál*, a tryst.[1] Arnmannoch in Kirkpatrick Irongray is for *earrann nam manach*, 'the monks' portion.' Belgaverie in New Luce is for *baile geamhraidh*, 'winter town,' *i.e.* a good wintering place;[2] there is another Balgaverie near Barr, Ayrshire, and there is *Baile Geamhraidh* on Barr river in Morvern; Balgeuery, 1386 (RMS), also Balgyuery, near Kinghorn in Fife, is the same, and from the places that go with it on record it appears to be now Balwearie.[3] Bentudor in Rerwick is probably for *beinn an t-súdaire*, 'the tanner's (or sutor's) hill.' The Cree river, 'aqua de Creich,' 1326 (RMS), etc., is for (*abhainn na*) *crìche*, '(river of the) boundary.' Carnsmole is Carnysmul, 1371, 1372 (RMS), Carnusmoell *ib.*, for *carnas maol*, 'bare rocky hill.' Craigdasher in Dunscore is 'sun-facing rock' (*deisir*). Dornock is the same as Dornoch in Sutherland and means 'place of handstones,' *i.e.* rounded pebbles,

[1] Near Duntulm in Skye, says Pennant (*Tour of 1772*, p. 304), is '*Chock* (read *Cnoc*), *an eirick*, or, the *hill of pleas*: such eminences are frequently near the houses of all the great men, for on these, with the assistance of their friends, they determined all differences between the people.' Here the word meant is certainly *eireacht*, an assembly: the custom of holding 'parliaments on hills' was common in Ireland, and in Scotland also.

[2] Compare Pennant, *Tour of 1772*, p. 308; *Old Stat. Acc.*, xvi. p. 164, 172 note, 174 note 2.

[3] Yet *Balweri*, 1260 (RPSA); they may have been two places.

but the *ck* may indicate Welsh origin—O.W. *durnauc*, W. *dwrnog*, with the same meaning. Dorniegills on Megget Water may be an English plural of G. *dornaigh ghil*, 'white pebble place'; compare Dornie on Loch Long, etc. Duncow, older Duncoll, is 'fort of hazels' (*coll*). 'The Dungeon' of Buchan is *an daingean*, 'the fastness.' Knocknassy in Kirkcolm is for *cnoc an fhasaidh*, 'hill of the stance' or 'resting-place,' like Teanassie near Beauly. For Laight on Loch Ryan (*Leacht Ailpín*) see p. 198. Loch Eldrig in Penninghame is 'loch of the *eileirg*,' or 'deer trap.' Loch Neldricken is practically the same, only that here we seem to have a diminutive *eileirgín*. The initial *n* may be the article (*an*), or it may be a trace of the old neuter gender of *loch*, as in *Loch nEachach*, Loch Neagh in Ireland. Orchars on the Black Water of Dee is the English plural of M.Ir. *orchar*, Sc.G. *urchair*, a cast, a shot, with reference, perhaps, to some feat, actual or legendary, of casting.[1] Lamford Hill in Carsphairn is 'encampment hill' (*longphort*). Uroch, Balmaghie, is for *iubhrach*, 'yew place.'

Sometimes a prefixed noun occurs, used adjectivally—an old usage—as in Conness, 'dog waterfall' (*con-eas*), in St. Mungo; Countam, 'hound knoll' (*con-tom*); Dunesslin, 'fort of the fall-pool,' in Dunscore; Laggish, 'hollow-haugh,' 'haugh in the hollow' (*lag-innis*), Wigtownshire; Cullendoch, 'holly-vat' (*cuilionn-dabhach*), of old spelled 'Culyn Davach' (RMS, i., App. 2), where *dabhach* is not the old land measure but a vatlike hole or hollow; compare Burn of Vat in Aberdeenshire—a translation of G. *Allt na Dabhaich*, which latter occurs at Ledaig in Lorne. The Old Stat. Account of the parish of Tongland mentions that salmon cruives were locally called 'doachs,' *i.e. dabhachs*.

Names formed with *-ach* suffix (Gaulish, *-ācum*) are rather rare, as Ardoch, 'high place'; Beoch, 'birch-wood'; Breccoes (pl.), 'dappled place' (*breacach*); Capenoch, 'place of tillage plots,' or 'of tree-stumps' (*ceapanach*);

[1] O.Ir. *airchur arathir*, 'projecting part of a plough' (lit. 'front-cast of plough'), glosses *temo*, a plough-beam; Sg. 26 b (*Thes.*, ii. p. 48). In place-names therefore it may refer to a projecting spur of land.

Galdenoch, 'place of coltsfoot' (*gallan*); compare Gallanach, near Oban; for *ll* as *ld* in English, compare Clochfoldich, Grantully, in G. *Cloich Phollaich*, 'the holed stone' (cup marked).

Eclipsis is found, but by no means often, as Pulnagashel for *Poll na gCaiseal*, 'pool of the bulwarks'; Craigenveoch for *Creag na bhFitheach* (or, as in Irish, *na bhFiach*), 'the ravens' rock'; for eclipsis see p. 239.

Traces of Irish settlers are found in Barnultoch for *Barr nan Ultach*, 'height of the Ulstermen,' in the Rinns, and in such names as Irischgait, Erishauch, Irlandtoun, Irishfauld, Erischbank, in Kirkcudbright and Dumfriesshires (RMS and Ret.). The Britons appear in Drumbretton near Kirtlebridge, in Glenbarton in Annandale, and in Drumbreddan in the Rinns; compare 'Gilendonrut Bretnach,' witness in Dumfries (Bain's Cal., ii.); John McBretny, burgess of Quhithern in 1532 (RMS), for *mac an Bhreatnaigh*, 'son of the Briton.'

In so far as one can distinguish between the names of Ireland and of Scotland, the Galloway names seem to go with the latter. To take some instances: from Cape Wrath to Loch Leven, the boundary between Argyll and Inverness-shire, the regular term for an eminence of no great height is *tulach*. South of Loch Leven *tulach* becomes rare; the term in use is *barr*, 'a top.' In the Galloway region *tulach* is very rare, though it does occur, *e.g.* Fintloch for *Fionn-tulach*, 'white height'; the regular term is *barr*. Here Galloway goes with Argyll. *Sliabh*, a mountain, is common in Ireland, very rare in that sense in Scotland; it does not seem to occur in Dumfries or Galloway; *Beinn*, a peak, rare in Ireland but common in Scotland, does occur there. Again the term *eileirg*, 'a deer-trap,' is not uncommon in Galloway as Elrick; it is found in Argyll and in the east from Inverness southwards, especially in Perthshire, but though the term itself appears in O.Irish in the sense of 'ambush,' it does not seem to appear at all in Irish names of places. Lastly, the old land measures of Galloway are those of the west—and certain other parts—of Scotland. I have found no trace of the davach, which is so common in the north, but there are abundant traces of the farthing land, *e.g.* Fardingjames, ' James's farthing land ' ; the half-penny land, *e.g.* Gar-leffin (*leth-pheighinn*) ; the pennyland, *e.g.* Pennyland in Kirkmahoe ; Pindonnan, ' Donnan's pennyland ' ; Pinhannet, ' pennyland of the Annat ' ; the merkland, *e.g.* Merkland in Dunscore ; Shillingland and Two-merkland are in Glencairn ; Poundland in Dunscore. I have not observed the ounceland (*tìr-unga*), which appears as Terung in the Isles. The *ceathramh*, or quarter, is common, usually as Kirrie- ; ' Quarter,' which occurs occasionally, is probably a translation. Another term, indefinite as to extent, is *earrann*, ' a portion,' anglicized as ' Iron ' in Irongray, Ironhirst, etc. It occurs also in Menteith in connection with names of churchmen, *e.g.* Arnvicar, Arnprior. Upper and Nether Cog, on Crawick Water, represent *cóig*, *cóigeamh*, a fifth part, like the Coigs of Strathallan and of Strathdearn.

CHAPTER VII

GENERAL SURVEY OF AYRSHIRE AND STRATHCLYDE

AYRSHIRE formed part of the territory of the Damnonians; later it appears to have become a separate province. Coel, the ancestor of its line of rulers, flourished, if we may judge by the Welsh genealogies, about A.D. 400; his seat was most probably in the central division of the county, called after him Kyle (Cul, 1153, Reg. Glasg.), Cil, 1164, *ib.*, in Irish Cuil). 'There is a tradition,' says the *Old Statistical Account*, 'though it is believed very ill-founded, that Coylton derives its name from a king called Coilus, who was killed in battle in the neighbourhood, and buried at the church of Coylton.'[1] The southern division, Carrick, is Karric, 1153 (Reg. Glas.), from Welsh *carreg*, a rock, borrowed into G. as *carraig*; it appears to have been connected more or less closely with Galloway. Cunningham, the northern division, is Cunegan, 1153 (Reg. Glasg.), Cuninham, 1180, *ib.*, later Conynham, Conyham, Conynghame, Cunyngham; the origin is doubtful, but the expression in Taliesin 'Coel ae kanawon' means 'Coel and his whelps,' not 'Kyle and Cunningham.'[2]

The Irish records yield very little as to Ayrshire, and there is no reference that can be called ancient. The town of Ayr is *Inber-air* in 1490 (AU), which agrees with the modern Gaelic *Inbhir-àir*. A chief of the Macleans married Rioghnach, daughter of Gamel, lord of Carrick, about 1300.[3] The lady's name is Gaelic, though her father's name is English.

---

[1] Vol. i. p. 101, also p. 96; xix. p. 457.

[2] Sir E. Anwyl, *Celt. Rev.*, iv. p. 265.

[3] Rioghnach inghean Gamhail mormair Cairrige; Skene, *Celtic Scotland*, iii. p. 481.

It is of interest to know that one of the thirteen wonders of Britain was a quern which ground constantly, except on Sundays, near Mauchline in Kyle.[1] It was heard working under ground. To witness the tradition there is Auchenbrain, *c.* 1200 (Lib. Melr.) Acchenebron for *Achadh na Brón*, 'field of the quern,' about three miles from Mauchline.

Of the saints commemorated in Ayrshire, Brigid of Kilbride, Donnan of Chapel Donnan, and Findbarr of Kilwinning, have been mentioned in connection with Galloway.

The saint of Colmonell is Colman Elo, abbot of Lann Elo, now Lann Eala, in King's County, who died in 611. There is also Kilcolmanell, *Cill Cholman Eala*, in Kintyre.

Kilkerran is for *Cill Chiaráin*, Ciarán's Church. It is in the parish of Dailly, which appears as Dalmulkerane, 1404 (Chart. of Crosraguel), 'St. Ciarán's servant's dale'; near New Dailly is Dalquharran, but whether the second part here is for Ciarán is doubtful. Of the twenty-two saints called Ciarán the most famous are Ciarán mac an t-Saeir, 'son of the carpenter,' abbot of Cluain mac Nois, who died in 549, and Ciarán of Saigir, bishop and confessor, 'the senior of the saints of Ireland,' who is said to have been before Patrick. The *Life* of the former states that Colum Cille took earth from his grave when he was setting out for Iona, and in passing through the gulf of Coire Bhreacáin cast some of it into the sea when his ship was in danger. There is another *Cill Chiaráin*, Kilkerran, in Kintyre, and probably both are dedicated to Ciarán of Cluain mac Nois.

Kilmarnock, still known in Arran Gaelic as *Cill Mhearnáig*, commemorates *Mo-Ernóc*, contracted into *M'Ernóc*; *Ernóc* is a diminutive of *Ernéne*, itself a diminutive form, translated by Adamnan *Ferreolus*, 'little iron man.' Ernóc might equally well be from Ernán, Ernín, Ernine, all names of saints. Twenty-two saints were called Mo-Ernóc, (Rawl. B 502, 93 h 1), and Reeves held that Mernoc of the two Kilmarnocks (Ayr and Cowal) is the saint whom Adamnan

[1] 'In treas ingnad dēag, brō for bleith do greas im Machlind i Cuil, acht dia domnaig; fo talmain imorro do cluintear'; *Irish Nennius*, p. 119.

mentions as 'Erneneus filius Craseni,' who died in 635 (AU), and whose day is Aug. 18.[1] The Aberdeen Martyrology, as Reeves mentions, gives Mernoc of Kilmernoch at Oct. 25, which suggests that the saint commemorated is Ernáin of Midluachair, whose day in Gorman is Oct. 26.

Kilkenzie, for *Cill Chainnigh*, in Carrick, 1506 (RMS), is a dedication to *Cainnech*. It is Kilmechannache, 'my Cainnech's church' in a charter of Robert I. (RMS). There were four saints of that name, but this is probably the great Cainnech of Achadh Bó, in Queen's County, who died in 599 or 600. He is said to have accompanied Columba on his visit to the court of Brude, king of the Picts, and Adamnan mentions him as an honoured guest of Columba in Iona. His name is preserved in Kilchennich in Tiree, and he had a *reclés* or chapel in St. Andrews.

Killochan, on Girvan Water in Carrick, appears to be Killunquhane, 1505 (RMS), and if the old spelling is correct, the dedication is to *Onchú*, gen. *Onchon*, styled by Oengus 'splendid descendant of the sage' (Feb. 8); he belonged to Connacht, and was specially noted for collecting relics of saints. Hence comes the old surname McClonnachan or McClannochane, for *Mac Gille Onchon*.

Kirkoswald commemorates king Oswald, who was slain in 641 in battle with the pagan Penda. In 1180 and again in 1185 (RMS), a charter by the Earl of Carrick was witnessed by Gilleasald McGilleandrys, 'servant of Oswald, son of the servant of Andrew.' The Welsh form, Coso(s)uold, appears as the name of a man near Peebles about 1200 (Reg. Glasg.).

The Charters of Holyrood record a donation made by Uchtred,[2] son of Fergus (of Galloway) of the church of St. Constantin of Colmonell (Colmanele) 'which is now called Kirkcostintyn,' with the chapel of St. Constantin. This is probably the Constantin whose conversion is recorded in 588 (AU), styled by Oengus 'rí Rathin,' 'king of Rathin,' and by Gorman 'Constantin Britt buanraith,' 'Constantin the Briton of lasting grace.' He is said to have been a

[1] Reeves, *Life of Columba*, p. 26.
[2] Uchtred fl. 1166 (*Bain's Cal.*, i. p. 14).

king of Cornwall who left his kingdom to become a monk under Mochutu of Rathin, where he succeeded Mochutu as abbot. He is also said to have passed over to Scotland, and to have suffered martyrdom in Kintyre, where he is commemorated in Kilchousland. The church of Govan is dedicated to him. In the notes to the *Félire* of Oengus he is called Constantin son of Fergus, king of Alba, but this Fergus died in 820. Another Constantin, son of Aed, was king of Alba from 900 to 942, when he became a cleric (*baculum cepit*), resigning the kingdom to his son Maol Coluim ; he died in 952, and he is probably the Constantin whom the annotator of the *Féilire* has confused with Constantin of Rathin.

Kilbirnie is supposed to be 'Brendan's Church,' and this is supported by St. Brennan's Fair which used to be held there on 28th May. The festival of Brénainn, son of Findlug, famous for his seven years' voyage, was on May 16 (old style). For the formation compare Kilbranne 1494 (RMS), now Kilbrannan in Islay. Culbirnie or Kilbirnie, near Beauly, is in Gaelic *Cùil-bhraonaigh*, '(at) oozy nook,' from *braon*, a drizzle, ooze ; Birnie in Morayshire is 'Brennach' before 1200, which is simply *G. braonach*, a moist place ; the dative-locative is *braonaigh*, which becomes Birnie in Scots by the usual metathesis.

Balmokessaik, 'my-Kessog's stead,' 1541 (RMS), in Carrick, appears to commemorate St. Kessog (Cessoc) of Luss.

Kilphin in Ballantrae parish may be the same as *Cill Fhinn*, Killin ; the name 'Gillebert mac Gillefin' *c.* 1166 (Bain's Cal.) suggests a saint named *Finn*, shortened from *Finnén*.

Kilwhannel in Ballantrae parish is to be compared with Craigmawhannal near Loch Macaterick, and may commemorate St. Connel.

St. Quivox is 'parochia S. Kevoce' 1547, 'Sanct Kevokis' 1591 (RMS). The Aberdeen Breviary gives at Mar. 13 a life of a fictitious female saint Kevoca. As that is the day of Mo Chóemóc (later Mo Chaomhóg) of Liath Mór or Liath Mo Chóemóc in Tipperary, it is clear that he is the saint commemorated, and that the compilers of the Breviary were misled by the ending *-óc*, which they took to be feminine as in modern Gaelic. Mo Chóemóc died in 656 (AU). He was a Connacht man, but spent most of his life in Munster. He appears to have attained a great age: his *Life* brings him into relations with Cainnech of Achadh Bó, who died about 600.[1]

Kilcais is placed by Blaeu north of Ayr and near West Sancher (Sanquhar); it appears often on record and was apparently not far from Prestwick. This may be the church called 'Chilnecase in Galweia' said to have been founded by Monenna.[2]

The old term Annat occurs in Pinhannet, *Peighinn na h-Annaide*, 'pennyland of the Annat,' in Barr, and in Annet at Skelmorlie.

Crossraguel (also in Glassford, Lanark) is probably Riagal's Cross (see *Additional Notes*).

The connection between the saints commemorated in a district and the old personal names of the district is so interesting that I may mention some additional instances. Gillemernoch, 'McErnoc's servant,' brother of Gilleasald, 'Oswald's servant,' witnessed a deed by tho Earl of Carrick about 1185.[3] Malcolm Gilmornaike or Gilmernaykie appears in the reign of Robert I.; he belonged either to Ayrshire or to Wigtownshire.[4] Gillecrist mac Gillewinnin witnessed a charter 'de Colmanele' about 1166[5] in the same document the preceding witness is MacGillegunnin. Bran, son of Macgillegunnin, witnessed a charter by Christian, bishop of Whithorn (Chart. Hol.). Here *-winnin* and *-gunnin* both represent the Welsh form of Finnén of Kilwinning and Kirkgunzeon, and the mixture of Gaelic and Welsh goes to show that Welsh had not yet been quite displaced by Gaelic, or at any rate that Welsh had been

[1] His genealogy is given in Rawl. B 502, 91 f; Oengus Fel., p. 96. A verse in LL 353 (foot) and 357 (left margin) credits him with having lived 414 years. His *Life* in Latin is given in Plummer's *Vitae SS. Hiberniae*, vol. ii. His name is an affectionate form of Cóemgen, and is latinised Pulcherius.

[2] Skene, *Celtic Scotland*, ii. p. 35.

[3] RMS II. [4] RMS I. [5] Chart. of Holyrood.

spoken recently. Another witness about 1180 was Gillecrist Bretnach, 'Gillechrist (servant of Christ) the Briton,' of Kyle or Carrick.[1]

In 750 Edbert of Northumbria added the plain of Kyle to his kingdom; Carrick had doubtless formed part of it before that date—it would have gone with Galloway. Thus in the latter part of the eighth century the political situation in Ayrshire was similar to that in Galloway. Everything goes to show that the introduction of Gaelic and the decline of British followed much the same course in both districts. The dedications are of the same type; the system of land measures is the same—Pin- and Pen-, for *peighinn*, a pennyland, are specially common; the Gaelic names are of the same class, with the same Scottish flavour. Gaelic names are fewer in Cunningham; in Carrick and Kyle they are as plentiful as in Galloway. The British names are few in comparison. The more important streams have retained the old names. Carrick and Kyle are British. Troon is W. *trwyn*, nose, cape; it occurs six times at least in names of capes in the Lleyn peninsula, Carnarvon, *e.g.* *Trwyn y Gwyddel*, 'the Gael's Cape.' Pant, in Stair parish, is Welsh for 'hollow, valley'; Guelt, in Cumnock, is W. *gwellt*, O.W. *guelt*, grass, pasture. A number of names begin with Tra-, Tro-, for W. *tref*, a homestead, vill, and they tend to go in clusters, *e.g.* close to Barbrethan, 'the Briton's height,' in Kirkmichael parish, are Threave (*i.e.* *tref*) and Tranew, while in the neighbourhood are Troquhain and Tradunnock.

We come now to Strath Clyde. References to the Britons of Alba are fairly common in Irish literature, but it is not always possible to be certain which branch is meant. Among these are the tales about Béinne Britt, 'Béinne the Briton,' who is said to have led a host of British at the battle of Mucramha, when Art, son of Conn, was slain in the middle of the third century. He and his son Artúr are often mentioned in connection with Fionn and the Fiana, and he is of some interest to us because the old

---

[1] RMS II.

sennachies and bards claim him as an ancestor of MacCailín, the Gaelic style of the Duke of Argyll.

We are on firmer ground when we come to the battle of Strathclyde (*cath Sratha Cluatha*), recorded among the battles fought by Dathí or Nathí, the last pagan king of Ireland.[1] This was doubtless the expedition which formed the subject of the lost 'chief tale' (*primscél*) entitled 'The Harrying of Strath Clyde (*argain Sratha Cluada*).[2] Dathí's reign began in A.D. 405, and the expedition may well have been undertaken to signalize his accession. Patrick was born, according to Professor Bury's reckoning, in 389, and he himself tells that he was taken captive to Ireland at the age of sixteen. If, therefore, he was born at or near Dumbarton, as Irish tradition states, there is good ground for believing that he was taken captive on the occasion of this same expedition of Dathí. What was implied by such a foray is illustrated by Patrick's words, 'I went into captivity to Ireland with many thousands of persons, according to our deserts, because we departed away from God, and kept not his commandments, and were not obedient to our priests, who used to admonish us for our salvation.'

Some time after Patrick had returned to Ireland as a missionary in 432, the soldiers of Coroticus, the British king of Strathclyde,[3] made an expedition to Ireland, in course of which they slew or captured a number of Patrick's newly baptized converts. When Patrick sent a letter requesting the return of the baptized captives and of some of the booty, his messengers were jeered at, and the captives were shared with Coroticus' allies, 'the Scots and the apostate Picts.'

---

[1] YBL 192 b 25. The prose tract entitled *Baile an Scáil* ascribes this battle to Dathí's son, Ailill Molt, but as this tract omits Dathí altogether from the list of kings there must be an error at this point. (*Celt. Zeit.*, iii. p. 463.)

[2] In 871 the Norsemen of Dublin harried Strathclyde after they had taken Dumbarton, but this is not likely to have formed the subject of an Irish 'chief tale.'

[3] Styled in the *Book of Armagh*, 'Coirthech rex Aloo'; in O. Welsh 'Ceretic Guletic.'

In the latter part of the next century Rhiderch Hael or Hen (' the liberal,' or ' the old '), king of Strathclyde, was among those who helped Urien of Rheged against the Angles, and it is likely that Aedán mac Gabráin had the support of the Strathclyde Britons at Degsastan. Between 682 and 703 the Britons were active in the north of Ireland. In 682 they fought the battle of Ráth Mór Maighe Line (Moylinney) in Antrim. In 697 they joined the men of Ulster in ravaging the plain of Murthemne in Louth. In 702 they slew Irgalach, son or grandson of Conang, king of Bregia, in the isle off Howth which is now called Ireland's Eye.[1] In 703 the Britons were defeated in the battle of Magh Culinn, in Co. Down, when the son of Radhgann, ' the enemy of God's churches,' was killed.[2] This is the last instance on record of a British expedition to Ireland. In 756 Edbert of Northumbria and Angus, king of the Picts, made terms with the Britons at Dumbarton.[3] The rest of their history is given by Skene so far as it is on record, but we may note a Welsh statement relating to the year 890 that ' the men of Strathclyde, those that refused to unite with the English, had to depart from their country, and to go to Gwynedd (North Wales).' There they were given permission to settle in certain parts that had been taken by the English, whom they ultimately drove out.[4]

The traces of the Church in Renfrewshire are similar to those in the south-west generally. Kilmalcolm, Kilmacolme 1539 (RMS), may commemorate Colum Cille or some other Colum ; St. Bride's chapel is in Kilbarchan, St. Oswald in Cathcart. Polloc was dedicated to St. Convallus, who is said by Fordun to be buried in Inchinnan. Inchinnan may contain the name of Finnén, *i.e.* Findbarr of Maghbhile. Kilallan is ' church of Fáelán ' or Fillan, whose chair and well are there. The fact that his fair was held in January goes to identify him with Fáelán of Cluain Moescna in Meath, who is commemorated on Jan. 9, but of whom nothing authentic seems to be known.

---

[1] AU. [2] AU. [3] Symeon of Durham.
[4] A. O. Anderson, *Early Sources*, i. p. 368.

'Kilmaloog under the barony of Renfrew,' 1377 (Reg. Glasg.) commemorates Moluag, probably of Lismore.

The saint of Kilbarchan is Berchán, whose fair was held on the first Tuesday of December, and this identifies him with Berchán of Cluain Sosta, whose day was Dec. 4. In the *Félire* of Oengus he is referred to as 'Fer dá lethe,' 'the man of two sides,' because, as explained in the Martyrology of Donegal, he spent half his life in Alba and the other half in Éire. Gorman and the Martyrology of Tamlacht have a Berchán at Aug. 4, and the latter states that he was of Cluain Sasta, perhaps in error. The Lammas fair at Tain was called *Féill Bearcháin*, 'Sancti Barquhani,' 1612 (RMS), and the autumn market at Aberfoyle bore the same name. Berchán of Eigg is commemorated on April 10.

Paisley is *Paislig* in Gaelic at the present day, and on title-pages of Gaelic books: 'an Baile Phaislig,' 1848; Paislic, 1800; Paislig, 1908.[1] 'The Abbot of Passelek' appears on record in 1296.[2] From this it is clear that the latinized form *Pasletum* has arisen from a misreading of *c* as *t*: the two letters are practically indistinguishable from each other in the early records. *Paislig* is from Lat. *basilica*, a church; M.Ir. *baslec*, a church; a churchyard, cemetery; dat. *baslic*; in Ireland there are Baslick in Roscommon and Baslickane in Kerry. The O. Welsh is *bassalec*, whence 'ecclesia de Basselek' in the Deanery of Newport. The change of initial *b* to *p* is rare in Old Gaelic, but compare Lat. *bestia*, Ir. *peist*, a monster; Lat. *brassica*, cabbage, Ir. *praiseach*, pottage[3]; also Mid. Eng. *bras*, Sc.G. *prais*, brass.[4] Paisley may be of British origin.

Lanarkshire shows comparatively few old dedications. St. Bride is commemorated in Kilbride, and in St. Bride's chapel on Kype in Avendale; her well is in Dunsyre. She was the patron saint of Douglas and of the House of Douglas. The churches of Govan and of Crawfordjohn

---

[1] *Typographia Scoto-Gadelica*, pp. 19, 80, 183.

[2] *Calendar of Close Rolls.*

[3] Pedersen, *Vergl. Grammatik*, i. p. 235.

[4] In modern Gaelic initial *b* is sometimes *p* in dialect, *e.g.* in *bìoball*, bible; *boilcein*, a round, thick-set person.

were dedicated to Constantin. Kentigern is mentioned as the saint of Lanark in 1147-1164 (Chart. Dryburgh).

Cambuslang is dedicated to Cadoc of Llancarvan, a grandson, according to his *Life* and the Welsh genealogies, of Brachan of Brecheniauc, now Brecknock. The *Life* of St. Cadoc [1] states that he crossed over to Ireland and stayed there three years. He visited 'Lismor Muchutu,' where he came to be called Muchutu, after the founder, on account of his holiness and humility. As Mochutu of Rathin and Lismore died in 637, he was later than Cadoc, but the remark as to the name by which Cadoc was known in Irish is of interest. Cadoc also visited St. Andrews in Albania 'which is commonly called Scotia'—the name 'St. Andrews' is, of course, long subsequent to St. Cadoc. On the way back he came to a certain town 'citra montem Bannauc,' *i.e.* to the south of the hill Bannauc, which is said to be in the centre of Albania, and here he was instructed in a vision to remain in this country for seven years. In course of digging the foundations for his monastery he found a collar bone of some hero of old, so huge that a warrior on horseback could ride through it. Cadoc will not eat or drink till he finds from God who this was. That same night he is told in a vision that the ancient giant (*veteranus gigas*) will rise at the first hour of the day. He does rise, a giant huge and terrific, and explains that he is Cau, styled Pritdin,[2] the Cruthen or Briton, otherwise Caur, 'giant,' who had been king of a district in the north, and had come to this place to plunder and to ravage, where he had been killed. He is now in Hell, in evil case. Cadoc reassures him, promising him another chance in this world to atone for his sins by good works, and at once sets him to work as a digger (*fossor*). Cadoc's monastery appears to have been in a province called in the *Life* 'Lintheamine.' His churches north of Forth are Kilmadoc and St. Madoes. O. Welsh *bannauc*, Welsh *bannog*, from *banna*, horn, peak, means 'peaked,' 'mons Bannauc' is 'the peaked hill, or range of hills';

---

[1] Rees, *Cambro-British Saints*.

[2] Kaw o Brydein, 'Kaw from Pictland,' is prominent in the Welsh tale of 'the Hunting of Twrch Trwyth' in the Mabinogion.

compare *Banau Brycheiniog*, 'the Brecknock Beacons,' literally 'the Brecknock Peaks or Horns.' Skene identified 'mons Bannauc' with the Cathkin Hills in the parish of Carmunnock, of old Cormannoc, which would mean on this hypothesis 'peaked circle,' 'peaked close.' Apart from the somewhat unsatisfactory meaning of this it is unlikely that the reference is to a range so insignificant as the Cathkin Hills; the implication of the passage is rather that a well-marked dividing range is meant. 'Mons Bannauc' represents *Minid Bannauc*, and this most probably denoted the hilly region, abounding in peaks, which forms the basin of the river Carron in Stirlingshire, from the northern side of which flows the Bannock Burn.[1]

Corsbasket, 1426 (RMS), now Basket, Blantyre, may be for *cros Pascaid*, the form which would result in Gaelic from O.Welsh *Pascent*, 'Pascent's cross.' A saint of this name is given as a son or grandson of Brychan.

The old name of the church of Carluke was Eglismalescok, 1147 (Lib. Calch.), Eglismaleshoghe, 1319 (Bain's Cal.), Eglismalesoch, 1321 (Reg. Glas.), -malesoke (RMS, i., App. 2.). We may perhaps compare St. Loesuc of Brittany.[2] There is an old site in North Knapdale called Kilmalisaig.

Lesmahagow is Lesmahagu thrice in King David's charter of 1144, and in the charter of the same year by the bishop of Glasgow confirming the grant to Kelso; in 1158 (Lib. Calch.) it is Lesmagu. The personal name *Gille Mohagu* appears several times in early charters of this district: Gilmalagon mac Kelli (in error for Gilmahagou), 1147/60, *ib.*; Gilmagu mac Aldic, 1180/1203, *ib.*; Gilmehaguistoun in Fincurroks, *i.e.* near Corehouse in Lesmahagow parish, 1208/18. These forms establish the vernacular pronunciation of the twelfth century, when Gaelic was doubtless spoken in the parish. On the other hand King David's charter of 1144 gives the saint's name in Latin as Sanctus Machutus,[3] and this is the regular form afterwards. King

[1] Compare p. 293, n. 2. [2] *Rev. Celt.*, xxx. p. 134.

[3] The king grants his firm peace to such as, on account of peril to life or limb, flee for refuge to the cell or reach within the four crosses surrounding it,' ob reverentiam Dei et Sancti Machuti,' *Lib. Calch.*

Robert the Bruce, in 1315, made certain grants 'to God and to the Blessed Virgin Mary and to St. Machutus and to the monks of Lesmachu' for the purpose of providing eight waxen candles, each of one pound of wax, around the tomb of St. Machutus (RMS).

It is clear that Mahagu and Machutus are independent and different names. Machutus was apparently a British saint. There is or was a church of Machutus in the Deanery of Abergavenny in Monmouthshire, called in Welsh, *Lann Mocha*,[1] with which may be compared Lesmachu in the charter referred to. In Brittany he is St. Malo, called in the common speech of the district St. Mahou, and latinized as Maclovius.[2]

In 'Ma-hagow' as now pronounced the *a* of *hagow* is long and sounded like *ai* in English 'maiden'; *gow* is sounded *go*. The name can hardly be other than *Mo-Fhécu*, later *Mo-Fhégu*, the reduced affectionate form of the name of Féchín of Fobhar or Fore. With *o* elided, *Mo-Fhécu* became *M'Écu*, later *M'Égu*, which is seen in the short forms *Lesmagu*, *Gil-magu*. Lesmahagow is for *Lios Mo-Fhégu*, 'my-Féchín's enclosure.' St. Féchín's day was Jan. 20; in Oengus' *Félire* he is *Mo-Écu* (three syllables). He died in 665 or 668 (AU).

Annathill, near Mollinburn, in Newmonkland parish, is the only instance of Annat which I have noted in Lanarkshire.

The British element that remains in the names of Strathclyde is not very large, but it includes a number of the important names, such as Renfrew, Lanark, Glasgow, Partick, Drumpellier. Stream names are largely British, as Clyde, Daer, Nethan, Medwin, Calder, Cander, Elvan. On the 1-inch Ordnance Survey Map about fifty names in Lanarkshire and about twenty in Renfrewshire might be claimed as British. Of Gaelic names, if we take the same basis, there are about one hundred and fifty respectively. Names beginning with *achadh*, a field, are common—

---

[1] *Book of Llandaf*.

[2] Professor Loth deals with Maclovius in *Rev. Celt.*, xxx. p. 141; he considers Machutus to be a derivative from the first part of the compound *Macc-lovo-*.

there are over twenty in Lanarkshire—but *baile* is rare. There are few traces of the system of pennylands, merklands, etc., which is so evident in Galloway and Ayrshire: Glespin, in Crawfordjohn, probably stands for *glas p(h)eighinn*, 'green pennyland.' The number of names in the Glasgow district which begin with *Gart* is notable, and may be due to British influence, though of course *gort*, *gart* of Gaelic and *garth* of Welsh both mean 'field, enclosure.' A curious name in Carnwath is Gowmacmorran; it can hardly be other than the Scots form of *Goll mac Morna*,[1] the famous Fian leader and rival of Fionn, with some term before it which has dropped. This is the only reference to the Fenian tales which I have noted in the south.

The charter of William I. granted about 1205 to the burgh of Ayr, and confirmed by Alexander II. and David II. (1367, RMS), prescribes that 'toll and customs due to the burgh shall be given and received at Mache and Karnebuth (Karnebothe, 1367) and Loudun and Krosnecone (Krosnekone, 1367) and Lachtalpen.' This last point is described elsewhere as in Wigtownshire, adjacent to the lands of Carrick (1319, RMS), and is now represented by the name Laight, on the east side of Loch Ryan, close to the Carrick border; the exact spot is marked by the stone styled on the Ordnance Survey Map 'Taxing Stone' on the county march. The name is for *leacht Alpín*, 'Alpin's grave,' and may well commemorate the burial-place of Alpin, the father of King Kenneth, who is said to have been killed in Galloway about 841.[2]

Krosnecone is now Corsancone on the Nith, in New Cumnock, close to the Lanark march. In the poem 'Does haughty Gaul invasion threat,' Robert Burns has:

> 'The Nith shall run to Corsincon,
> And Criffel sink in Solway,
> Ere we permit a foreign foe
> On British ground to rally!'

---

[1] In Barbour's *Bruce* 'Golmakmorn'; in Sir David Lindsay 'Gowmakmorne.'

[2] A. O. Anderson, *Early Sources*, i. p. 270; Skene, *P.S.*, pp. 149, 172, 288: *Celtic Scotland*, i. p. 292.

In 1398 it was agreed that Englishmen born with Scottish fealty (*i.e.* who were Scottish subjects), should dwell as far from the march as Peebles, Crawford, or Corsincon (Bain's Cal.).

It is for *cros na con*, 'the hound's crossing,' with reference probably to the crossing over the hill of Corsancone, for there is no evidence for a cross having stood at the place. *Cù*, a hound, is here feminine, as it was and is in Irish, though masculine now in Scottish Gaelic; compare Acchenebron for *achadh na brón*, p. 187.

Loudun is the hill of that name on the Lanark border, and is doubtless British; it may be for O.Celt. *Lugudūnon*, 'fort of Lugus,' for the god Lugus was worshipped in Britain (Welsh *Lleu*) and in Ireland (*Lugh*) as well as on the Continent. Mache has been identified with the Maich Burn on the northern border of Ayrshire, on the road from Ardrossan by Dalry to Paisley, while Karnebuth is supposed to be Cairn, on the border of Ayr and Renfrewshires, on the great road from Ayr to Glasgow.[1] It is not necessarily the same as Karnbothe which, as we shall see, is mentioned in connection with Rutherglen.

Cassillis is Castlys, 1363 (RMS), also Caslis, Cassells, an English plural of M.Ir. *casel*, later *caiseal*, a stone wall, a stone fort, from Lat. *castellum*; it is correctly rendered by *castra* in 'de terris de Castris,' 1404 (RMS). Compare Cashel Point on Loch Lomond, the site of a broch; Cashleyis, 1625 (Ret.), in Dumbartonshire; Drumquhassil, *ib.*; Cashley, G. *Caisligh*, probably for *Caislibh*, in Glen Lyon, near which are the sites of several ancient circular forts, locally called *Caistealan nam Fiann*, 'the castles of the Fianna.' Near Cassillis House is Dunree, meaning possibly 'king's fort.' Craignaught, Dunlop, is 'naked rock,' like Allfornought in Galloway. Craigsaigh, near Loudon House, is for *creag saidhe*, 'bitch's cairn'; compare Alltsaigh in Glen Urquhart. Drumranny, near Girvan, is for *druim raithnigh*, 'bracken ridge'; near it is Kilranny. Drumbrochan, Cumnock, is 'porridge ridge,' 'gruel ridge';

Ayr.

[1] *Charters of the Royal Burgh of Ayr*, p. xxv.

compare *Coire Bhrocháin* on Cairngorm, 'porridge corrie,' and *Coille Bhrocháin* on Bonskeid, Pitlochry, 'porridge wood'; here *brochán* is applied to a mixed-up, 'through-other,' sort of place, or to soft sludgy ground. Dalry may mean 'king's dale' or 'heather dale' (*dail fhraoigh*, Ir. genitive of *fraoch*). Finnarts, Ballantrae, is 'white cape,' with English plural; compare *Dubhaird*, Duart, 'black point.' Largs is English plural of *learg*, a slope; Largs, near Heads of Ayr, appears in 'mussa inter Largas at Bethoc,' 'the moss between Largs and Beoch' (birch-wood) in the reign of William I. (Lib. Melr.), along with other places in Maybole parish, such as Culenlungford, 'nook of the encampment (longphort),' Tunregaith for *tòn ri gaoith*, 'rump to wind,' Greenan Castle for *grianán*, sunny knoll. Old spellings of Loch Spallendar show that it is for *loch spealadair*, 'mower's loch' (p. 106). Loch Recar, near Loch Doon, seems to be for *loch an reacaire*; the *reacaire* was the person in the train of a bard who recited the poem composed by the bard himself. Mahago Rig, in Cumnock, seems to be Monyhagane, 1629 (Ret.), Minnihagan, 1658, *ib.*, in which case the spelling should perhaps be *-hague*, *-hagau*; compare Gilmalagon (p. 196), meaning 'St. Mahagu's moss.' Minnishant, near Monk-wood, Maybole, means 'the holy (*sèanta*) moss (*mòine*)' or 'shrubbery' (*muine*); the Holy Loch in Cowal is in G. *an Loch Sèanta*. Sorn, a parish name, is *sorn*, a kiln, from Lat. *furnus*. Stair is *stair*, stepping-stones, sometimes 'rough bridge'; Stairaird, above it, is 'upper Stair'; both are on the river Ayr; so too Starr, at the head of Loch Doon. The term is common in place-names.

Renfrew. Ardoch in Eaglesham is 'high place,' from *ard*, high, with *-ach* suffix, a common name; it occurs in Lesmahagow. Bardrainney is for *barr draighnigh*, 'top of blackthorn thicket'; the nominative *draighneach* is seen in Drynoch, Skye; *Draighnigh*, Drynie, in the Black Isle, is locative. Bardrain in Kilbarchan is *barr draighean*, 'top of black-thorns.' Craiglunscheoch in Kilmalcolm is *creag loingseach*, 'rock of the mariners' or 'exiles.' Clochodrick in Kilbarchan is Clochrodric in twelfth century (Chart. Pasl.); Clochrodryge, 1456 (Reg. Glas.); Clochordruck, 1680 (Ret.) and means 'Roderick's stone.' This may be the stone referred to in the legend which tells how Rodric, king of the Picts, was slain in battle and how a stone pillar was erected to commemorate the event and called Westmering; a great part of the land thereabout was taken in hand by the king who slew Rodric, and called West-mereling land.[1] As this tale is fabulous in any case, it is possible to think of Rodercus Largus or Rhydderch Hael, the king of Strath Clyde, as the source of the name. Dargavel in Erskine is probably for *doire gabhail* (now *gobhail*), 'copse of the fork.' Duchall in Kilmalcolm is *dubh chail*, 'black wet meadow' or 'flat.' Enoch in Eaglesham is *eanach*, a marsh. Greenock is well known in Gaelic as *Grianáig*, dative of *grianág*, a sunny knoll, parallel to the masculine *grianán* of the same meaning. There is another *Grianáig*, Greenock, near Callander in Perthshire, and a third in Muirkirk, Ayrshire. Gourock may be *guireóc*, *guireág*, a pimple, with reference to the rounded hillocks there. Moyne, Neilston, is *mòine*, moss. Selvieland is probably a part translation of *fearann Sealbhaigh*, 'Sealbhach's land,' like Belhelvie in Aberdeenshire.

Auchnotroch in Lesmahagow is for *achadh nan otrach*, 'field of the dung-heaps'; compare Kilnotrie for *coille an otraigh*, 'wood of dung,' *i.e.* where cattle were wont to lie, and Talnotry, for *talamh an otraigh*, 'dung land,' which occurs twice, all in Kirkcudbright. In the latter the reference is probably to a manured infield. Auchtygemmell in Lesmahagow is Auchtigammill, 1533 (RMS); the second part is the English personal name Gamel, which was not uncommon in the southern parts of Scotland; for the first part compare Terraughty or Torachty (Blaeu) in Troqueer parish, where it may be for *ochtamh*, an eighth part. The second part of Auchtyfardle is not clear. Airdrie is probably the same as Airdrie in Ardclach, Nairn, Lanark.

[1] Skene, *P.S.*, p. 155. The 'mering' part may be an attempt to explain the name Mearns in Renfrewshire.

which is in Gaelic *Ardruigh*, 'high reach,' 'high slope.' Balornock near Glasgow is Budlornac, 1172 (Reg. Glas.), Buthlornoc, 1186, *ib.*, it may be pure Welsh, *bod Louernoc*, 'Louernoc's dwelling,' the proper name meaning 'little fox' (O.W. *louern*); if so, *bod* here became *both* when taken over into Gaelic. Cladance in East Kilbride, and Claddens near Cadder are for *cladhán*, a little ditch, with English plural; two or three places of this name are on the Roman Wall, Cleddans east of Duntocher and east of Kirkintilloch, and 'the Cleedins' near Falkirk. Corehouse is the English plural of *currach* or *corrach*, a marshy plain. Fincurrocks in Lesmahagow occurs in 1208/18 (Lib. Calch.), meaning 'white marshes,' from cotton-grass, probably; 'Robert del Corrok' appears in 1259 (Bain's Cal.). 'Johannes de Bennachtyn de la Corrokys' is on record in 1362 (RMS); in 1363 he is 'Johannes Benauchtyne dominus de Corrokes,' *ib.* 'Richard Bannachtyn dominus de Corhouse' appears in 1459, etc. (RMS). 'John Bannatyne of Corhous' appears in 1662, *ib.*; other Bannatynes of Corhous are on record in the interval.[1] The ruins of the old Bannatyne stronghold, Corra Castle, stand right on the brink of the cliff above Corra Linn; here and in Corra Moor close by 'Corra' is for *corrach*; compare Currochs near Crieff, and Currie in Lothian. Cambusnethan is 'Neithon's bight' or 'bend.' At this bend in the river Clyde stood the old church of Cambusnethan, and the name doubtless commemorates a Welsh saint of that name. Cambuslang is Camboslanc, 1296 (Ragman Roll), Cameslong, 1319 (Bain's Cal.), Cambuslange, 1379/81 (RMS), etc., representing *camas long*, 'bight of ships.' There is an exceptionally fine *camas* on Clyde at the village of Cambuslang; the tide comes right up to it, and boats can come to it on a rising tide. At the period when the name was given the conditions probably permitted the light vessels of those days to come so far on a high tide and no further. We may compare *Camas Luinge*, anglicized Camusluinie, 'ship's bight,' on a

[1] The first Bannatyne near Lanark who appears on record is Nicholas de Banaghtyn, who flourished in 1300 (Bain's *Cal.*).

bend of the river Elchaig at the head of Loch Long, 'loch of ships,' in Kintail; here *luinge* is genitive sing. of *long*.[1] Near Conchra on Loch Long is *Camus Longart* (*longphort*), 'bay of the ship-stations.'

Cairntable, 'Kaern de Kaerntabel,' *c.* 1315 (RMS), now the name of the hill, seems to have been originally the name of the cairn on the top;[2] in which case the second part is probably Welsh *tafl*, a cast, from Lat. *tabula*, whence G. *tabhal*, a sling; compare Orchars (p. 183). In Drumclog the second part is probably W. *clog*, a rock, crag, cliff. Drumsargard or Drumsharg[3] is Drumsirgar a. 1200 (Orig. Paroch.), 'William de Drumsyrgarde,' 1296 (Bain's Cal.); 'Mauricius de Dromsagard,' 1312, *ib.*; de Drounsagard, *ib.*, Drumsargarth, 1360 (Ex. Rolls), Drumsergare, 1440 (RMS), -sergart, *c.* 1453, *ib.*, -sargart, 1474, *ib.*, -schargart 1496, *ib.*, -sargat, 1512, *ib.*, -sargat, 1540, *ib.*, -sargat, 1547, *ib.*, -sargatt, 1581, *ib.*, -sargit, 1661, 1663, *ib.* These forms taken together go to indicate that the second part is a compound of *gart*, a field, probably *siar-ghart*, 'west-field'; compare Schiregartane 1451 (RMS), Schargartoun 1598, *ib.*, in Blaeu Shergaden, in Macfarlane Schirgartoun, for *siar-ghartan*, 'west little field,' now apparently represented by Garden near Cardross. Dalziel is for *dail ghil* (dative), 'white mead.' Gartsherrie is for *gart searraigh*, 'colt-field.' Logoch in East Kilbride means 'hollow place' (*logach*), whence the common Logie, for *logaigh* (dative), now *lagaigh*. Mannoch Hill at the head of Nethan is for *Cnoc nam Manach*, 'monks' hill,' close to it is Priest Hill. Mollinburn, Cadder, is a part translation of *allt an mhuilinn*, or such, 'mill-burn.' Polmadie, near Glasgow and on the Renfrew march, is Polmacde in a charter granted between 1179 and 1189 by King William I., specifying the bounds of the privileges of the burgh of Rutherglen; in the ratification

[1] A boat of ten tons can come up to Camas Luinge at very high tides. My information as to Cambuslang was got by a friend from Mr. Geddes of the Humane Society, who had himself gone from Glasgow Green to Cambuslang and back by rowing boat, watching the tide both ways.

[2] There are two cairns (*Ord. Gaz.*).

[3] *New Stat. Acc.*

by James VI. in 1617 (RMS), it is Polmadie.[1] The name may also appear corruptly as Polmalfeith, 1479 (RMS).[2] In compounds with *mac*, the dropping of *c* in anglified forms is common (p. 166). The part following *ma*(*c*) may be compared with the second part of Dundee (p. 220), and the Irish surname O'Dea, Dee, for *O Deaghaidh* ; there is also the old sept name *Maic Ditha*, 'sons of Dith' (pron. Dee).[3] Gorman has Trénóc macDeith, Trénóc son of Deth, at Mar. 22. The meaning is 'pool' or 'hollow of the sons of Daigh,' or some such personal name. Powbrone in Avondale is for *poll brón*, 'quern water'; compare Auchenbrain. Poniel in Lesmahagow is for *poll Néill*, 'Neill's water,' and Ponfeigh in Carmichael is for *poll an fhéidh*, 'the stag's water.' This use of *poll* in the sense of a slow stream, so common in the south-west, is rather Welsh than Gaelic. 'Villa mineschedin' (*read* inine-) appears in 1172 (Reg. Glasg.), in 1174 'villa Inienschedin.' Here the first part *inine, inien*, is for *inghine*, gen. sg. of *inghean*, a daughter, the lightly sounded *gh* being disregarded in the charter phonetic spelling ; the second part, which might be read *schedni*, is a proper name in the genitive, probably the early Irish *Setna*, later *Seadna*. Thus 'villa inine schedin,' (? schedni) is for G. *baile inghine Seadna*,' the vill of Seadna's daughter.' Later in the Register it is 'villa filie Sedin,' 1179, 'villa filie Scadin,' *c.* 1186, where Latin *filie* takes the place of G. *inghine*. About 1186 it is written also Schedinestun, reference to the lady being dropped. In 1226 it is enjoined that 'the provost

---

[1] The bounds in the early charter are : 'de Neithan usque Polmacde et de Garin usque Kelvin et de Loudun usque Prenteineth et de Karnebuth ad Karun.' In 1617 they are : 'a Nethan ad Polmadie et a Garin ad Kelvin et a Lowdoun ad Preinteinethe et a *lie* Carneburgh ad Carroun.' '*Lie* Carneburgh' may be Carnbroe in Old Monkland, Carnbrw, 1489, Carnebrow, 1634 (RMS), with 'the Cairnhill' adjacent ; it is for *carn brogha*, 'cairn of the brugh,' or 'mansion,' perhaps a fairy dwelling ; compare Cairnborrow in Aberdeenshire, Carnbrowyis, 1406 (RMS).

[2] In 1479 King James III., confirming a charter of Robert II., appointed the Earl of Argyll his lieutenant and special commissary 'from Carnedrome to Polgillippe and from Polmalfeith to Loch Long' (RMS).

[3] Woulfe, *Irish Names and Surnames*, 1923 ; *Expulsion of the Dessi*, ed. Kuno Meyer.

and bailies of Rutherglen take toll and custom at the Cross of Schedenestun.' Later with change of *n* to *l* the name becomes Schedilstoune, Scheddilstoun, now Shettleston. Tinto appears in 'Karyn de Tintou,' 'Kaerne de Tintou,' *c.* 1315 (RMS); in Macfarlane it is Tyntoche once, Tynto thrice; in Scots, Tintock, as also in the Retours; it is for *teinteach*, 'place of fire'; the cairn on its top was doubtless a beacon cairn. Compare Tintock north-east of Kirkintilloch, and Carntyne for *carn teineadh*, 'fire cairn,' near Glasgow.

CHAPTER VIII

GENERAL SURVEY OF SCOTLAND NORTH OF FORTH

WE have now surveyed in a general way the country south of the Wall between Forth and Clyde from the linguistic point of view; there remains something to say on the position north of the Wall, in what is called Pictland or in Gaelic *Cruithentuath.* Bede informs us that in his day the Firth of Forth divided the territory of the Angles from that of the Picts, while of old—long before his time—the Britons were divided from the Picts on the west by the Firth of Clyde. As in Bede's time and long before it the chief seat of the Britons of Strathclyde was at Dumbarton, on the north side of the estuary, it is not to be supposed that their territory ceased there; it probably extended round about Loch Lomond—the Lennox district—and into Menteith, while their linguistic influence may have extended well beyond the bounds of their territory.

In the Irish semi-historical literature there is an account of how Cruithneachán mac Lochit maic Cinge (or Inge) went from the sons of Míl, *i.e.* the Gael of Ireland, with the Britons of Fortriu to fight against the Saxons, and won land for themselves, namely Cruithentuath, and stayed among the Britons. This, says a writer in the Book of Ballymote, took place in the time of Erimon, that is to say, not long after the Gael arrived in Ireland. We are further told that the newcomers cleared a swordland (*claid-eamtír*) for themselves among the Britons (*itir Breatnaib*), first in the Plain of Fortriu, thereafter in Magh Ċircin, *i.e.* the Mearns and Angus. Elsewhere the tale informs us that the work of clearance was done by the Gael under compulsion (*ar éigin*), for when the ancestors of the Gael were engaged in their wanderings on the Continent, eighteen warriors of Scythia joined them, and it was prophesied that once the Gael had settled down, they would provide lands for their friends and allies.[1] This tale appears to have elements that are very old: it may refer to a time when the Irish joined the people of Fortriu in making war not upon the Saxons but upon the Romans. Its immediate purpose is to explain the presence of Cruithnigh in the midlands of Scotland, which it does by asserting that they were a colony from Ireland, presumably from Ulster, and unhistorical as this explanation may be, it would appear reasonable enough to an Irish writer who was familiar with the tradition of the expulsion of the men of Ulster, by whom we are to understand the Cruithnigh of Ireland, by Cormac in the third century. Much more remarkable is the part assigned to the Gael of clearing land for the Cruithnigh in Alba under compulsion, actually doing forced service as mercenaries, or at least as subordinates, to the Cruithnigh, who in Ireland were vassals of the Gael. This is a matter which the historian would certainly not have invented. It is an unpleasant fact which he feels to demand explanation, and he does explain it by the ingenious device of making the compulsion arise out of an ancient obligation backed by a prophecy: the thing was, in fact, a debt of honour which the Gael were destined to repay. The historical significance of the tale is its recognition of early settlements of Gael in Fortriu and Magh Círcinn—not independent, but subject to the Picts, and serving with the Picts as mercenaries, both south of the Wall and apparently north of it. In this it agrees with other accounts, as we shall see.

This is the only instance in Irish literature, so far as known to me, where express mention is made of 'Britons' north of the Wall. It is, however, worth noting that the Latin *Life* of St. Serf makes Adamnan assign to the saint 'the land of Fife, and from the hill of the Britons to the

[1] I have here made a synthesis of several accounts which are to be found in Skene, *P.S.*, pp. 45, 319, 328 (BB 43 b 49 fcs.); *Leathar Gabhála*, p. 233 (ed. John MacNeill and R. A. S. MacAllister); BB 19 a 37.

mount which is called Ochil.' [1] The position of the 'hill of the Britons' is uncertain, but it was north of Forth. Some further evidence of their presence is the occurrence of *Breatan* in place-names, such as *Clach nam Breatann*, 'the stone of the Britons,' in Glen Falloch at the head of Loch Lomond; Balbretane, now Balbarton, 'the Britons' stead,' in Fife; Drumbarton, 'the Britons' ridge,' in Aberdeenshire. Along with this, there are the place-names and traditions involving the name of Arthur. The Gael carried the tale of Diarmid to Scotland, locating the scene of his hunting of the Boar and his tragic death in many parts of the North, as for instance at Beinn Laghail (Ben Loyal) in Sutherland, in Kintail of Ross-shire, in Brae Lochaber, and in Perthshire. Similarly the post-Roman Britons took with them wherever they went the tale of Arthur, and as place-names and traditions connected with the Diarmid saga are a sure sign of the presence of the Gael, so Arthurian names and legends are a sign of the presence of the Briton.

The best-known 'Arthurian locality' is Arthur's Seat, Edinburgh. North of the Wall on the West are *Suidhe Artair*, Arthur's Seat, Dumbarton, on the right bank of the Leven; *Beinn Artair* (the Cobbler), at the head of Loch Long; *Aghaidh Artair*, 'Arthur's Face,' a rock on the west side of Glenkinglas, in the same district, with the likeness of a man's profile; Sruth Artair, Struarthour 1573 (RMS) in Glassary, Argyll. In the East there are Arthurstone near Cupar Angus, Arthur's Cairn, Arthouriscairne 1595 (RMS), apparently on the south side of Bennachie, Aberdeenshire; Arthurseat in Aberdeenshire; and *Suidhe Artair*, Suiarthour 1638 (Ret.), now *Suidhe*, in Glenlivet, Banffshire. There is, or rather was, also Arthur's Oven, in 1293 *Furnus Arthuri*, described in 1723 as between the house of Stenhouse (Larbert) and the water of Carron, 'an old building in form of a sugar loaf, built without

[1] 'Habitent terram Fif, et a monte Britannorum usque ad montem qui dicitur Okhél' (Skene, *P.S.*, p. 416). The Register of St. Andrews mentions a place called Munobretun, apparently in Fife.

lime or any other mortar.'[1] It may well be that these are not all connected with the British hero, but most of them probably are so connected, and it is particularly suggestive to find an 'Arthur's Seat' at the head of Glenlivet.

South of the Wall the Old British names have passed into Welsh, though, as we have seen, Gaelic forms of some of them come from Old British direct. There is some evidence that shows very clearly that north of the Wall also a similar process took place. Old British *uxellos*, high, is represented by Gaelic *uasal*, Welsh *uchel*. The latter occurs thrice in the south in Ochiltree (Ouchiltre, 1282, (RPSA)); Ugheltre, 1304, (Bain's Cal.); Uchiltrie, 1406, (RMS), Welsh *Ucheldref*, 'Highstead,' which occurs in Wales now in that form and as *Ucheldre*. North of the Wall we have the Ochil Hills in the counties of Perth and Stirling; an ancient Irish tract records that St. Serf's special seat was at Culross between Mount Ochel and the Firth of Forth.[2] There are also Rossie Ochill, in Forteviot parish, and Catochil, Catoichill 1507 (RMS), near Strathmiglo in Fife, in which the first part may be W. *cat*, a bit, piece, fragment, ? 'high-part.' In the north, the river Oykel between Ross and Sutherland is in G. *Oiceil* (genitive), on record Strath-ochell, 1490 (RMS), Kill-ochell, 1582, *ib.* (for Kyle-, G. *caol*), Strath-okell, 1582, *ib.*, and so on. The Norse form is *Ekkjall*, *Ekkjallsbakki*, 'Oykel-bank,' (Orkn.

---

[1] Macfarlane, *Geog. Coll.*, i. p. 330. It is said to have been 19 feet 6 inches in internal diameter, and 22 feet in height to the round opening at the top, which was 11 feet 6 inches in diameter. The building was circular and rounded towards the top. (Pennant, ii. p. 228.)

[2] Tract on the Mothers of the Irish Saints, BB 212 a-214 b: 'Alina (? Alnia) ingen rig Cruithnech mathair Seirb maic Proic rig Canandan Egipti 7 is e sin in sruth senóir congeibh Cuillenn-ros i nSraith Érenn i nComgellaib etir sliab n-Ochel 7 mur(*sic*) nGiudan'; 'Alma, daughter of the king of the Cruithnigh, was the mother of Serb, son of Proc, king of the Canaanites of Egypt; and he is the sage senior who set up (the religious establishment of) Cuillenn-ros in Strath Earn *in Comgellaib* between Mount Ochel and the sea of Giudan'—BB 214 a 19. The parallel passage from the Book of Lecan is printed by Skene in *Celtic Scotland*, ii. p. 258 n., where S. Serf's mother's name is given as *Alma*, and Culross is said to be *hi Sraith hIrend hi Comgellgaib.*

Saga).[1] The name is not Gaelic, nor is it Norse, but it may go back ultimately to E.Celtic *uxellos*, which had become something like *uckel* by the time at which the Norsemen took it over as *Ekkjall*. If so, *Ekkjallsbakki* represents Ptolemy's *Ὄχθη ὑψηλή*, 'Ripa Alta,' 'high bank.'[2]

Another test is the treatment of E.Celtic *vo*, under, which is in Gaelic *fo*, in Welsh *go*, O.W. *guo*. This is the first part of Gogar in Midlothian, 'ecclesia de Goger' (Reg. Dunf.), Nethergoger, 1335 (Bain's Cal.), Coger, 1336, *ib.*, Gogare, 1392 (RMS); there is another Gogar near Menstrie, Stirling. The second part of both is probably *cor*, as in W. *ban-gor*, 'the upper row of rods; a coping, battlement'; Irish *cor*, a setting, as in *cleth-chor*, 'wattle-setting.' Gogar would thus mean 'a small setting or cast,' with reference to some physical feature such as a small spur or eminence or piece of land; the Gaelic equivalent is *fochar*, as in *Fochar Maigi*, 'small cast (spur) of the plain' (BB 139 a 16).

E.Celt. *ver*, an intensive prefix, is in Welsh *gor*, O.W. *guor*; in Gaelic *for*. It probably appears in Gourdon, Gordoun 1587 (RMS), on the coast of Kincardineshire, with a hill 400 feet high, meaning 'great fort,' like Gordon in Berwickshire.

The first part of Gospartie, Strathmiglo, Gospertie 1507 (RMS), may be W. *gwas*, an abode, dwelling; the second may be a Gaelic extension of W. *perth*, a copse: 'wood-dwelling.' This would correspond to G. *Fas na Coille*, Fasnakyle, for W. *gwas* is G. *fas*.

E.Celt. *-ct-* becomes in Welsh *-th-*, as in *Vectis*, W. *Gueith*, Isle of Wight. The river Nethan in Lanarkshire, twelfth century Neithan, is for an early *Nectona*, 'pure one.' When British names were taken over into Gaelic they were often

---

[1] Compare *Mons Okhél*, p. 208.

[2] Rev. C. M. Robertson, *Celt. Rev.*, i. p. 93. In equating *Ekkjall* with a late form of *uxellos* the difficulties are (1) initial *e*; in view of G. *o*, it seems not unlikely that *e* may be an error; (2) *kk* for *ch*: Norse often makes G. *ch* into *k*, *e.g.* Búkan for Buchan, but there is no example of *ch* becoming *kk*; that Norse *kk* is the correct reading is proved by G. *c* in *Oiceil*, for a single *k* of Norse would become *g* in Gaelic.

gaelicized by the addition of the Gaelic suffix *-ach*, *-ech*, to the name itself or to a shortened form of it. Thus *Abur-nethige* of the Pictish Chronicle, now Abernethy near Perth, has as its second part the genitive of a nominative *Nethech* or *Neitheach* (fem.), which is gaelicized either from *Neithon* directly, or from a British river name from the same root. With it goes Abernethy on Speyside, in Gaelic now *Obar Neithich*, proving that the change of E.Celt. or Old British *-ct-* into *-th-* took place not only in Perthshire but also north of the Grampians. Another example of the same change occurs in the Aberdeenshire river name Ythan, which is the same as the Welsh *Ieithon* of Radnor and Shropshire, for an early *Iectona*, 'talking one,' from the root seen in W. *iaith*, language. For the idea we may compare such stream names as *Briathrachán*, 'wordy one,' *Balbhág*, 'little dumb one.' These names, Nethy and Ythan, cannot be explained from Gaelic, and they show that they were taken over into Gaelic at a period when Old British had passed into the stage corresponding to Early Welsh.

Aberdeen is *Abberdeon* in the Book of Deer, now in G. *Obar Dheathan*, with *dh* silent, as it is regularly after *obar* and *inbhir*—the *th* is used merely to divide the syllables. Here *Deon*, *Deathan*, represents E.Celt. *Dēvonā*, a river goddess name, formed from *dēvos* (*deivos*), a god, in accordance with O.Ir. phonetics, at a period when E.Celt. *ē* (for *ei*) was still preserved, as it is in Adamnan and in O.Ir. of the eighth century. Corresponding to this, we have in early records *Aberden*, *e.g.* *c.* 1180 (Chart. Lind.). Aberdeen means 'the *aber* of the river Don,' for in Gaelic the Don is *Deathan*, rising in *Coire Dheathan*, and flowing through *Srath Deathan*, Strathdon. Corresponding to the modern Don are the early record forms Aberdon, *e.g.* 1172 (Ch. Inch.), 'Simone archidiacono de Aberdoen,' 'Roberto decano de Aberdoen,' 'Matheo de Aberdoen,' c. 1202 (Ch. Lind.). 'Don,' 'Doen,' did not arise from G. *Deon*, *Deathan* : *Deon* and *Doen* represent different linguistic traditions of a common original, the former Gaelic, the latter British. In Welsh the word for god is *duw*, for an older hypothetical *dwyw*, from *deivos* ; Dēvona would be in O.W. *Duion*. In Breton 'god' is *doúe*, M.Br. *doe*. It is some such form that is represented by Don, and also by the river Doon in Ayrshire : both have the same forms in early records—Don, Done, Doyne—and both have as sister stream the Dee (Dēva). Here then we have in effect a doublet, Gaelic Deathan (-deen), British or 'Pictish' Don, parallel to the other doublets G. *Foirthe*, W. *Gwerid*, the Forth, and G. *Tatha*, W. *Tawy*, the Tay.

In Nennius, Loch Lomond is 'stagnum Lumonoy'; the chapter headings have 'De magno lacu Lummonu, qui Anglice vocatur Lochleuen in regione Pictorum'; 'Of the great lake Lummonu, which is called in English Loch Leven, in the region of the Picts.' The name in modern Welsh form would be *Llyn Llumonwy* ; it is from *llumon*, a chimney, a beacon, as in *Pumlumon*, Plynlymon. The tale of Kulhwch and Olwen recounts how 'Kai and Bedwyr sat on a beacon cairn on the summit of Plynlymon.'[1] The ending *-oy*, later *-wy*, is that in *Cornwy*, later *Kernwy*, Cornwall, from *Cornavia*, 'horn-land.' Thus Lomond (*llumon*) is primarily the beacon hill, Ben Lomond ; Lumonoy was the district at its base. The Lomond Hills in Fife are, of course, also 'beacons,' and one has only to look at the peaks of the East and West Lomond to see how well they were suited for that purpose. In Gaelic, Loch Lomond is *Loch Laoiminn*, also *Loch Laomuinn*, and these forms are not from the Welsh *Lumon*, but from an older *loimmon-*, whence both the Welsh and the Gaelic forms ; compare G. *fraoch*, W. *grug*, heather ; the base is that of G. *laom*, blaze. This, then, is another instance of a doublet.

Other survivals of British in the north, such as *aber*, a confluence, *monadh*, hill ground, *carden*, a copse, the stream names Peffer and Calder, etc., will be noticed later.

But north of the Wall, though proof exists that the old British language, or whatever dialect of it was spoken

---

[1] 'Standard,' which occurs in Carrick, Galloway, and Dumfriesshire as a hill name, is probably a translation of W. *lluman*, a standard, confused with *llumon*, a beacon.

there, passed into the stage corresponding to Old Welsh, the vast majority of the names are either pure Gaelic or have been gaelicized. Many of the Gaelic names may have been given at a fairly late period, but not a few are old. As has been pointed out, certain of the northern names, as well as of those in the south, give evidence of having been taken over into Gaelic at a time when the old values of the letters were still distinct enough to admit of their being treated according to Gaelic phonetics. It will thus be desirable to summarize the evidence for traces of early Scottic or Gaelic influence in the north.

Skene's position is that the very first Scottic settlement in this country took place about A.D. 500, and he commits himself further to the statement that 'there is no reason to suppose that prior to [A.D.] 360 a single Scot ever set foot in North Britain.'[1] Now it would be rather extraordinary if it were true that North Britain had remained a *terra clausa* as regards Ireland up to this date, but it is not true. We have the authority of Tacitus that when Agricola was in North Britain, he held discourse with an Irish ruler (*regulus*), who had been forced to leave Ireland on account of civil war and had fled to Britain and thrown himself on Agricola's protection, apparently with some hope of being restored by help of the Romans.[2] It is not likely that this prince came unattended. The occasion may have been the revolt of the *aitheach tuatha*, or vassal tribes, which is traditionally placed just about this period. This was probably not the first Irish prince who had to flee to Britain, and he was certainly by no means the last. Again in A.D. 297 the orator Eumenius states that the Britons were accustomed to Picts and Irish (*Hibernis*) as enemies. These references are by contemporaries of the events.

Irish tradition refers to still earlier times. It claims that Reachtaidh Righdhearg, 'of the red arm,' king of Ireland about B.C. 300, and Labhraidh Loingseach, who came about one hundred years later, were two of the Gaels

---

[1] *Celtic Scotland*, iii. p. 125. [2] *Agricola*, c. 24.

who beat down the lordship of Alba.[1] It has always to be kept in view that by Alba the older writers mean Britain, by no means necessarily North Britain or Scotland. According to Keating, Reachtaidh's exploits were in the north: he imposed a tribute on the Cruithnigh. Tradition is more definite as to Labhraidh Loingseach, who, as we have seen, is recorded to have made expeditions against Tiree, Skye, the people of Orkney, and other places, presumably on the West of Scotland, which cannot now be identified. We may also note the romantic tale of Deirdre and the sons of Uisliu (later Uisneach), who fled to Alba from the wrath of Conchobhar, king of Ulster. 'It was Manannan, son of Agnoe,' according to an old account, 'who settled the sons of Uisneach in Alba. Sixteen years they were in Alba, and they took possession of Alba from Man northwards. And it was they who expelled the three sons of Gnathal, son of Morgann, namely Iatach and Triatan and Manu (Mani, YBL) Lámhgharbh from that territory, for it was their father who had sway over that land, and it was the sons of Uisneach who slew him.'[2] Later versions connect them specially with Cowal and Lorne.

Conn Céadchathach (A.D. 177-212) is credited with having fought a battle in Kintyre.[3]

Later comes the settlement of Cairbre Riata, son of Conaire, son of Mogh Lámha, otherwise—and earlier—known as Eochaidh Riata and Fiacha Riata; later *Riata* becomes *Rioghfhada* (*Rigfota*), 'long-armed.' His father Conaire was son-in-law of Conn Céadchathach (*Conaire cliamain Cuinn*), whom he succeeded as king of Ireland; Conaire was killed in the beginning of the third century; Joyce puts his reign from 212 to 220. After Conaire's death, Cairbre and his two brothers settled in Munster, and thereafter owing to a famine led his people to Ulster, where they settled in the northern part of Antrim, known thenceforward as Dál Riata, 'Riata's share.' A section of them,

[1] The third was Criomhthan Mór, son of Fidach: 'triffiur dochomairt flaith nAlban do Gáidelaib . i . Rechtaid Rigderg et Labraid Loingsech et Crimthand Mór mac Fidaig; Rawl. B 502, 148 b 12.

[2] BB 258 a 58 *sqq.*; YBL 178.

[3] *Celt. Zeit.*, iii. p. 461.

however, still led by Riata, crossed over to Alba and settled on the north side of the Clyde estuary, apparently in Cowal, possibly in Lennox.[1] For the latter part of this account the authority is mainly Bede, who, as is well known, was an extremely careful writer. His authority, again, can hardly have been other than Adamnan, who, as Bede himself informs us, visited Ceolfrid's monastery in his own time.[2] If this was the case, Bede's statement is of great weight, for if any man was in a position to know the facts, that man was Adamnan. At the very least it shows that in the opinion of Bede's informant there had been in the West a settlement of Scots from Ireland long anterior to that by the sons of Erc in the beginning of the sixth century. This settlement also was called Dál Riata; its kings—when it came to have kings in the sixth century—were of the race of Conaire, and the connection between the two Dál Riatas was very close. So close indeed was it, that at one time there was clearly a desire on the part of the Irish Dál Riata to secede to Dál Riata in Alba. Had this happened, the position would have been similar to that of Ulster now, but the problem of the relations between Dál Riata in Ireland and the king of Ireland on the one hand, and Dál Riata in Alba on the other, was settled amicably at the Convention of Druim Ceata in A.D. 575.[3] The settlement in Ulster, it may be remarked,

[1] The position of Cairbre's settlement in Munster is stated thus: 'robí dano tír Cairbri Rigfhoda in tír itáit Ciarraige Luachra 7 Orbraige Droma Imnocht, conad asin dolotar i nAlbain. Gabsat Corco Duibne in tír atat.' 'Now Cairbre Rigfhoda's land was the land wherein are the Ciarraige Luachra and the Orbraige of Druim Imnocht, and it was thence that they went to Alba. The Corco Duibne took possession of the land wherein they are (now)'; BB 140 a 49. North Kerry and Orrery in Cork are the corresponding places now.

[2] Adamnan visited Aldfrid, king of Northumbria, twice, in 686 and 688, and it was probably in 686 that he visited Jarrow; he was then sixty-two and Bede was thirteen. Bede's information was doubtless got later. Adamnan died in 704.

[3] The question at issue has been much misunderstood, but certainly had reference to the political position of Dál Riata in Ireland, not to that of Dál Riata in Alba. The Irish king had threatened to drive the Dál Riata of Ulster and the poets across sea (*tafunn Dáil Riata dar muir 7 tafunn na n-écess*.—Rawl. 502 B, 96 a 15).

was doubtless connected with the operations conducted by Conn and his successors against the Cruithnigh or Dál nAraide of Ulster, for Dál Riata must have been a 'swordland' wrested from them. Cairbre himself is stated to have been slain later by one of the Dál nAraide.[1]

Lughaidh mac Con, who became king of Ireland in A.D. 250, was a contemporary of Cairbre Riata and fought against him when Cairbre was attempting to avenge his father's murder. In consequence he was exiled, and going to Alba in the usual manner of exiles he stayed there for seven years and gained friends and supporters, including Béinne Briot, 'son of the king of Britain.' With these he returned to Ireland, fought and defeated Art, son of Conn, and became king in his stead.[2] Among his chief exploits were a voyage to Alba and four great victories over the men of Orkney.[3] Lughaidh is said to have married a daughter of Béinne Briot and to have had three sons, who were known as the three Fothads, a name which suggests connection with *Fothudáin*, the Gaelic form of *Votadini*. Fothad Canann, one of the three, was a famed leader of Fiana (*rígféinnid*), and is said to have taken possession of lands in Alba. It is interesting to note that Keating derives the house of MacCailín (Argyll) from Fothad Canann, and that the official genealogies have Béinne Briot as one of MacCailin's ancestors. One of the lost 'chief tales' was entitled *Longes Fothaid*, 'the sea expedition (or exile) of Fothad.'[4] The two other Fothads reigned jointly for one year after Cairbre of the Liffey and fell in the battle of Ollarba against the Fiana.

In the earlier part of the fourth century the three brothers called Conla, later Colla, are prominent figures. They were sons of Eochaidh Doimhlein, son of Cairbre of the Liffey, son of Cormac, and their mother was Aileach, daughter

---

[1] LL 38 b, foot.

[2] LL 146 a 27; 288 a.

[3] 'a longus co h-íath nAlpan. Forbrisfi cethri mórcatha for túatha Orca,' *Celt. Zeit.*, iii. 461.

[4] For the Fothads, see LL 190 b; BB 164 b; Rawl. B 502, 155 b; *Fianaigecht* (Todd Lect., xvi.), p. 4 *sqq.*; *Cóir Anmann*, 220.

of Fubthaire of Hí. The eldest, Colla Uais, was king of Ireland for four years (327-331), when he and his brothers with three hundred men had to flee to Alba to escape the vengeance of the son of the previous king, who had been slain by them. They stayed in Alba for three years, doing military service with their kinsmen.[1]

Criomhthann Mór, son of Fidach, is mentioned as one of the three who crushed the sovereignty of Alba. He is mentioned in the important passage in Cormac's *Glossary* which says, 'when great was the power of the Gael in Alba (Britain), they divided Alba between them into districts and each knew the residence of his friend, and not less did the Gael dwell on the east side of the sea than in Scotia (Ireland), and their habitations and royal forts were built there. Whence is named *Dinn Tradui*, *i.e.* the triple-fossed fort of Crimthann the Great, king of Ireland and of Alba to the sea of Icht (the English Channel), and hence also is *Glasimpere* or Glastonbury of the Gael . . . they continued in this power till long after the coming of Patrick (A.D. 432).' But Criomhthann's activities in Britain were probably confined to the south-west of England, and there is nothing to show that he was ever in Scotland.

His successor was Niall Nóigiallach, 'of nine hostages' (379-405), so called because he held a hostage from each of the five provinces of Ireland and four from Britain; elsewhere the four are said to have been from Alba, the Saxons, the Britons, and the Franks.[2] Of him it is said among other things, 'many shall be his deeds on Druim Alban,' *i.e.* the water-shed between the east and the west of Scotland.[3]

Dathí, who succeeded Niall and reigned from 405 till 428, was regarded as a mighty conqueror, and the tradition is that he was killed by lightning at the Alps. His reign, as we have seen, was inaugurated by a great raid on Strathclyde, but he is also credited—and this is notable—with a

---

[1] Rawl. B 502, 141-2; LL 333 a 8. *Celt. Zeit.*, viii. 317 seq.

[2] Rawl. B 502, 81 a 35; 136 b 23.

[3] 'bíet ile a gluinn ar Druim nAlpuind' (*read* nAlpund); *Celt. Zeit.*, iii. 463.

battle in the east of Scotland, the battle of Magh Circin or Gerginn, *i.e.* the Mearns and Angus.[1]

In this connection it is proper to note the statements regarding an early settlement in the eastern midlands, which proceeded from Munster. The two great ruling families of that province were the heads of the Dál nCais and the Eoganacht, descended respectively from Cormac Cas and Eoghan, sons of Oilill Olum, a contemporary of Art, son of Conn. Towards the end of the fourth century, Lughaidh, king of Munster, who belonged to the Eoghanacht branch, took occasion to banish his son Corc, who went to Alba. The romantic tale of his adventures there is contained in the Book of Leinster (287 b 1), but the beginning is missing. On his arrival in Alba he was caught in a snow-storm, was six days without food, and when at the point of death was found by the king's chief poet, Gruibne, whom he had once befriended. Gruibne resuscitated him, and observed an Ogham written on his shield, requesting the king of Alba if Corc came to him by day, to behead him before night, if he came by night, to behead him before day. Gruibne amended the Ogham so as to read that Corc was to receive the king's daughter in marriage, which ultimately he did. He stayed in Alba till three sons were born to him, after which he left with wife and sons and much treasure.[2]

Such is the outline of the romance, which in this form may not be much older than the date of the Book of Leinster itself (*c.* 1150). Gruibne and Corc, however, are twice mentioned in relation to each other in Cormac's *Glossary*, and the second of these passages contains some words of what purports to be Gruibne's welcome (*fáilte*) to Corc, a version of which is given in the Book of Leinster.[3]

---

[1] YBL 192 b 25 *sqq.*

[2] Compare 'Conall Corc and the Corco Luigde'; *Anecdota*, iii. 57. A tale almost identical is told of the Emperor Conrad II. (d. 1039); it is found in *Gesta Romanorum* and in Wyntoun. The Ogham motif is, of course, as old as the tale of Bellerophon.

[3] In *Sanas Cormaic* (no. 688): 'Gorn . i . gai-orn . i . gae orcne . i . aithinde, unde dixit Gruibne oc fáilte fri Corc: immicuirithar gurna gair.'

Cormac mac Culinan, king of Munster and archbishop of Cashel, fell in battle in A.D. 903. The other great work ascribed to him is styled the 'Psalter of Cashel' (*Saltair Caisil*), from which extracts are often made by the later MSS. Among these extracts is an account of the genealogies of the descendants of Eber, including Corc, contained in four of the great MSS., namely, Rawl. B 502, the Book of Leinster, the Book of Ballymote, and the Book of Húi Maine.[1] These all agree in stating that Corc had seven sons, the mother of one of whom was Mongfinn, daughter of Feradach Finn Fechtnach, and this son was ancestor of the Eoganacht of Magh Gerginn. The three first MSS. style Feradach 'king of Cruithentuath' (Pictland), and give the name of Mongfinn's son as Cairbre Cruithnechán, 'Cairbre the little Pict,' or rather 'Pict-sprung'; the fourth makes Feradach 'king of Alba,' and calls the son Maine.[2] Further, Rawl., BB, and Húi Maine add that from the Eoganacht of Magh Gerginn came Oengus, king of Alba. This Oengus was most probably the son of Fergus, who died in 761, and is styled by Tighernach king of Alba' under A.D. 759; there was, however, another king of the same name who died in 832, king of Fortriu.

Corc, son of Lughaidh, is, of course, a perfectly historical character. He became king of Munster, and his grandson Oengus, son of Natfraoch, was baptized by Patrick. That he was actually the founder of the Eoghanacht of Magh Gerginn may be true or it may not. What is certain is that there was a branch of this great Munster family there, and that already in the ninth century it was reckoned to be of old establishment. It may have been these

---

The words here quoted occur in the LL version thus: '(ditnech anaill Chuirc) immaluritar carnd gáir' (LL 287 a 24). 'Ditnech' is corrected to 'ditdech' in the MS. (fcs.). This *retoric* therefore was old when *Sanas Cormaic* was composed; it was probably contained in the Psalter of Cashel.

[1] For a transcript of the relative part of *Húi Maine*, I am indebted to Professor T. F. O'Rahilly.

[2] See p. 221.

Eoghanacht who helped to clear the 'swordland' among the Britons of Magh Gerginn. Their name appears to survive in Balmackewan, *Baile mac Eoghain*, 'stead of the sons of Eoghan,' in the parish of Marykirk, and the name Cairbre is found in Drumforber, Drumquharbir 1539 (RMS), 'Cairbre's ridge,' in the adjoining parish of Laurencekirk. Near it is Conveth, the old name of the whole parish representing Early Irish *coindmed*, modern *coinmheadh*, 'free quartering, billeting'; the district would have been so named because it bore the special burden of quartering the household troops of the lord. Dundee is in Gaelic *Dùn Dèagh*, which seems to mean 'Fort of Daig(h),' for the genitive of *Daig* is *Dega*, in modern spelling *Deagha*. *Daig*, meaning 'fire,' was a rather uncommon Irish name, though *Aed*, 'fire,' was very common. One of Corc's sons was named Daig.

The Irish authorities state that the nobles of Lennox were of the same origin as the Eoghanacht.[1] Muiredach úa Dálaigh, the well-known Irish bard who flourished A.D. 1213 and spent part of his life in Scotland, wrote a poem in honour of Alún, son of Muiredach of Lennox, which has been printed by Skene.[2] In this poem he refers to the coming of Corc to Alba and his marriage to the daughter of Feradach, whom he calls Leamhain. Corc and Leamhain had a son named Maine, from whom were descended the rulers of Lennox. Keating says: 'Maine Leamhna (*i.e.* Maine of Leamhain or Lennox), son of Corc, went from Ireland to Alba, and there occupied territory which is called Magh Leamhna' (the plain of Leamhain or

---

[1] 'Ic hEber condrecait na secht nEoganachta 7 Lemnaig Alban'; 'at Eber meet the seven Eoghanachts and the Lennoxmen in Alba'; LL 318 b 42. 'Eber uero is da claind-sein Dáil Cais . . . 7 Eoganacht Caisil . . . 7 Leamnaigh i nAlbain'; 'as to Eber, of his descendants are the Dáil Cais . . . and the Eoghanacht of Cashel . . . and the Lennoxmen in Alba'; BB 41 b 36.

[2] *Celtic Scotland*, iii. 454, from MacFirbis' *Book of Genealogies*. Another copy, for a transcript of which I am indebted to Professor T. F. O'Rahilly, is in R.I.A., MS. 23 L 17 fo. 15 a. This copy is anonymous, but R.I.A., MS. 23 D 4 (p. 93) gives the first ten quatrains under the name of Muireadhach Albanach, as Professor O'Rahilly informs me.

Leven).[1] The Book of Húi Maine also, as we have seen, makes Maine a son of Corc. That the Gaelic lords of Lennox believed in this descent is indicated by the fact that Alún or Alwyn, the second Earl of Lennox who appears on record, had a son named Corc.[2] Maine, however, was the grandson, not the son, of Corc; his father was Cairbre, and from him were descended two saints, Cummine Fota and Faithlenn.[3] In the poem by Muiredach úa Dálaigh, the Mormaer of Lennox is styled *rí Bealaigh*, 'king of Bealach,' in English now Balloch, at the lower end of Loch Lomond. In early times Balloch was the seat of the lords of Lennox, and it is notable that close beside it is Tullichewan, *Tulach Eoghain*, 'the hill of Eoghan.'

Munster, the homeland of the Eoghanachta, was the great centre and source of the Ogham cult, as Professor John MacNeill has shown in his foundational monograph on the Irish Ogham inscriptions.[4] The total number of known Ogham inscriptions is about 360, of which about 300 are Irish, and of the Irish Oghams five-sixths belong to the counties of Kerry, Cork, and Waterford. Kerry has about 120, Cork about 80, Waterford about 40. In Britain 26 have been found in Wales, 5 in Devon and Cornwall, 1 in Hampshire. Scotland has 1 in Gigha, Fife, Kincardine, Perth, Moray, Sutherland, and Caithness respectively; 3 in Aberdeenshire, 2 in Orkney, 4 in Shetland. 'The distribution of the British Oghams,' says Professor MacNeill, 'clearly corresponds to the region of Gaelic, or, as it was then called, Scottish influence in the period that followed the withdrawal of the Roman legions from Britain.' But it may have begun even earlier. He further

---

[1] *Irish Text Soc.*, ii. p. 386.

[2] *Register of Lennox.*

[3] Rawl. 502 B, 90 g, 91 d; LL 351 c 1, c 24; LB 19 a, 19 b. The genealogies are: (1) *Cummine Fota* mac Fiachna maic Fiachrach maic Duach maic Maine maic Cairpre maic Cuircc maic Lugdach maic Ailella Flainn Big maic Ailella Flainn Móir; (2) *Faithlenn* mac Aeda Damain maic Crimthaind maic Cobthaich maic Duach Iarlaithi maic Maithne (Maine) maic Cairpre maic Cuircc maic Luigdech (Rawl.).

[4] *Proceedings of Royal Irish Academy*, vol. xxvii. sect. C, no. 15 (1909).

gives good reasons for believing that the Ogham cult was reckoned as distinctively pagan and was banned by the Church, and that as a consequence it was arrested by the growing power of Christianity. The distribution of Oghams in Scotland is so sporadic as to indicate clearly that the cult had no centre here, and that it did not endure. The absence of Oghams on the west is due to the fact that the west was settled from the north of Ireland and especially from Ulster, where the Ogham cult never found a footing. The solitary instance in Gigha may be compared with the sporadic instances in Ulster. On the other hand, the presence of Oghams on the east indicates early Gaelic influence from Munster, and confirms in a striking manner the traditions which we have been relating. There remains the question, why there are no Oghams in Lennox, if early settlers from Munster found their way there also. The answer may be that, granting the truth of the tradition, the conditions were different there from those in the north-east: the Christian influence which we know to have existed in the Vale of Leven before the time of Coroticus would have been sufficient to prevent the use of Oghams. Saint Patrick was born about A.D. 365; his father was a deacon and his grandfather was a presbyter.

Here may be noted the distribution of the term *cathair*, gen. *cathrach*, a circular stone fort. In Ireland, says Joyce, there are more than three hundred townlands and towns whose names begin with this term, 'all in Munster and Connaught except three or four in Leinster—none in Ulster.'[1] With us it is very rare, and on the west it does not occur.[2] On the east the furthest north instance which I have met is Corncattrach, for *Coire na Cathrach*, 'Corrie of the Cathair,' adjacent to Shanquhar, for *Sean-chathair*, 'old fort,' Gartly, Aberdeenshire. In Forfarshire there is Stracathro, in 1212 Stracatherach (Johnston), for *Srath-cathrach*: the *cathair* may have been the White Caterthun, or possibly Dunlappie (? for *Dùn-lapaigh* 'Fort of the

[1] *Irish Names of Places*, i. p. 261.

[2] Except in Wester Ross, where it means 'a fairy seat,' like *sìthean*.

bog'), the name of a parish now included in Stracathro. The White and the Brown Caterthun are powerful stone forts of the broch type in the adjacent parish of Menmuir; the first part of the name is doubtless *cathair*, but the second part is obscure to me. It is worth noting that on the south side of the White Caterthun is the Gallows Hillock.

'The ville of Catherlauenoch called Tullibardine' is on early record [1]; the name is for *Cathair Leamhnach*, 'Elm-fort,' and is now represented by Carlownie Hill on the south border of Auchterarder parish, Perthshire. Another instance in Perthshire is Kathermothel, for *Cathair Mhaothail*, 'Muthil Fort,' the old Gaelic name for the important Roman camp of Ardoch.[2] Both of these names came to be used as names of districts.

A seat of the ancient Mormaers of Lennox was called *Cathair*, probably with some distinctive or qualifying term which has not come down to us. Near it was the place of execution, referred to in a charter as *furcas nostras de Cather*, 'our gallows of Cather'; elsewhere as *furcas nostras del Cathyre*, 1370 (RMS), 'of *the* Cathair.' It was in the parish of Kilmaronock, Dumbartonshire, and the name survives in Catter, where, as we are informed by the Old Stat. Account, 'there is a large artificial mound of earth, where in ancient times courts were held; near to which the Duke (or rather Mormaer) of Lennox had a place of residence. There is not now the smallest vestige of the building.'[3] Near Catter House is Drumquhassle, for *Druim (an) Chaisil*, 'Ridge of the Cashel,' or circular stone fort. The *furcae* seem to be commemorated in Crosshill, on Catter Muir. With Catter may be compared Cadder, Kirkintilloch, the site of a Roman fort. 'Foresta de Passelet et Senecathir' appears in 1226 (Theiner).

---

[1] *Charters of Inchaffray*, p. lxxv. (Scot. Hist. Soc.).

[2] *Chart. of Lindores*, 27, etc., 243 (Scot. Hist. Soc.). The district called Kathermothel included Fedal and Cotken, the latter of which appears later (Retours) as the moor of Ardoch.

[3] Compare Macfarlane, *Geog. Coll.*, i. p. 353: 'the house of Easter Catter closs by which ther was once ane old castle belonging to the ancient Earles of Lennox.'

The period we have been considering was one of great activity and enterprise among the Gael of Ireland. That they made settlements in North and South Wales (Venedotia or Gwynedd and Demetia or Dyfed) is well known. The Deisi, who were expelled from Meath in the third century, went to Dyfed; long afterwards they were known to the Irish as the race of Crimthann (*cenél Crimthaind*).[1] The Gael of Gwynedd and Anglesey were driven out by the sons of Cúnedda in the fifth century. The Irish settlements in the south-west of England are referred to in the extract given from Cormac's *Glossary*. In both regions their presence is attested by Ogham inscriptions. During the same period they were familiar with North Britain, and made settlements among the Picts in the midlands of Scotland; it would appear, in fact, that they were rather welcomed, and helped materially to stiffen the native struggle for independence as well as to join in the raids on Roman Britain. The Irish Nennius says that the northern Wall was constructed against the Gael and the Cruithnigh.[2] The tenor of Patrick's Epistle to Coroticus of Strathclyde also makes it clear, as Professor Bury has seen, that the Scots who shared in the booty of Christian captives taken from Ireland were located in Scotland. As in Wales a noble family is recorded to have sprung from the exiled Deisi, so in Scotland there is reason to believe that the mightiest king of the Picts was sprung from a Munster family which had settled in the Mearns about A.D. 400. It would be indeed remarkable if the Irish inroads on Britain in the fourth century were not accompanied by settlement north of the Wall of Antonine.

The coming of the sons of Erc soon after A.D. 500 and the establishment of the Scottish kingdom of Dál Riata were events of first-rate importance; the facts are stated by Skene and need not be repeated. Not less important

---

[1] Kuno Meyer, *Expulsion of the Dessi*; *Y Cymmrodor* xiv. (1901) where their genealogy is given.

[2] P. 64; it was called *Clad na Muicc*, 'the pig's fosse'; also 'doronad clad accu dar in i-insi ri hucht Cruithnech 7 Góedel'; Laud 610, 90 a 1 (quoted in Kuno Meyer's Contribb. s.v. *clad*).

was the introduction of Christianity by Columba into the north and west; according to one authority Columba also visited the valley of the Tay.[1] There are several indications that the Scots of Dál Riata attempted to acquire power in the midlands. Aedán's father, Gabrán, it will be remembered, is said to have made an expedition to Forth, and Aedán himself is styled sub-king or prince of Forth.[2] The Irish life of St. Berach says that Berach came to Aedán's fort, and that Aedán offered up the fort to Berach, even *Eperpuill*, which is Berach's monastery (*cathair*) in Alba. *Eperpuill*, as the Reverend C. Plummer says correctly, is Aberfoyle, on the Forth; the October market that was wont to be held there was called *Féill Barachan* (*Bearchán*), and near Aberfoyle there are what appear to be the remains of a fort.[3] There is the further incident of Aedán's battle in the Mearns recorded by Tighernach and referred to by Adamnan. Aedán's son and successor, Eochaidh Buidhe, is styled 'king of the Picts' in AU, quoting from the ancient Book of Cuanu. The descendants of Eochaidh's son, Conall Cearr, are styled 'the men of Fife' (*fir ibe*, for *Fhibe*).[4] Along with these facts may be taken the remarks on the origin of the name Gowrie (p. 111).

We may here notice the occurrence of one or other of the various names for Ireland which are found in Scotland. In Latin, Ireland is Scotia or Hibernia. Adamnan shows a marked preference for the former, using Hibernia and the adjective Hiberniensis fifteen times, while he uses Scotia, Scoticus, Scotice, sixty times, according to Skene's index. As Scotland became more and more gaelicized, and especially when a Gaelic king came to rule over both Picts and Scots, the term Scotia began to be applied to Pictland, particularly, as Skene points out, to the lowland part east of Drumalban and north of the Firth of Forth. This usage, which resulted in the term altogether ceasing to denote

---

[1] *Amra Coluim Cille* and notes.

[2] See p. 53.

[3] *Irish Lives*, i. p. 35; ii. p. 327.

[4] A. O. Anderson, *Early Sources*, cliv. p. 149; Skene, *P.S.*, p. 315, n. 7. Conall Cearr is also given as Conadh Cearr.

Ireland, is already indicated when Dicuil, about A.D. 825, writes of Ireland as *nostra Scotia*, in contrast with the other Scotia in Britain. Scotia is, of course, entirely a literary term; the terms in actual use among the Gaelic people were Eire or Eriu, Banba, Fodla, and Ealg or Ealga, and of these the first was the most usual. The first three were really names of ancient local goddesses, associated with different parts of Ireland; the last is traditionally said to have been used in the time of the Firbolg, who were pre-Milesian. We might naturally expect to find the literary (Latin) use of Scotia as applied to part of Scotland to be paralleled by a use of the native terms; in other words we might expect to find Eire or its equivalents applied to the Gaelic parts of Pictland. Further, the literary *Scotia* may be expected to have arisen as the reflex of the vernacular term, which would naturally have been in use first. There are reasons to believe that this was actually the case and that at one time Pictland stood a chance of being called Eire, though in the event the old native name of Alba won the day as the equivalent of Scotia.

The tract *De Situ Albanie*, ascribed by Skene to about 1165, ends with the statement that Fergus, son of Erc, was the first of the seed of Conaire to become king of Alba, 'that is, from the mount Brunalban to the sea of Ireland and to Innse Gall. Thereafter kings of Fergus' seed reigned in Brunalban or Brunherę up to Alpin, son of Eochaidh.'[1] Here we have Brunalban instead of the regular Drumalban, with Brunhere as an alternative name. Now whether *brun* is written in error for *drum*, or, as is more likely, represents the Welsh *bryn*, a hill, it is clear that the second part of Brunhere is Eire, uninflected (as it would be if the compound is Welsh), and that we are offered 'Mount of Eire' as an alternative for 'Mount of Alba' or 'Ridge of Alba.' Had this form persisted, we should now be speaking of *Bràghaid Éireann*, 'Breaderin' instead of *Bràghaid Alban*, 'Breadalbane.' The inference

[1] Skene, *P.S.*, p. 137; A. O. Anderson, *Early Sources*, p. cxvii.

is that at one time the district, or part of the district, east of the watershed was called Eire.

St. Fillan (*Faolán*), son of Oengus, son of Natfraoch, son of Corc of Cashel (of the Eoghanacht of Munster), is designated 'of Ráth Érenn in Alba.'[1] This place has been assumed, without any proof, to have been at or near St. Fillan's at the lower end of Loch Earn, but no such name exists there, either now or on record, though the Old Stat. Acc. of Comrie mentions *Dùn Fhaoláin*, 'Fillan's Hill' (or Fort), with St. Fillan's Well on the top of it. *Ràth Éireann* is, as I believe, still extant; it appears on record as 'Raterne in the earldom of Stratherne,' 1488 (RMS); Raterne, 1466 *ib.*, now Rottearns, in the parish of Ardoch. As this is quite outside the valley of the Earn, the meaning must be 'Rath of Eire'—a district. Again in the 'Prophecy of Berchan,' Girig or Grig, who is recorded to have slain and succeeded Aed, son of Kenneth mac Ailpin, fought a battle 'on the fields of Eire' (*ar bhrughaibh Éirenn*).[2] Elsewhere we are told that he died at Dundurn, a fort near the lower end of Loch Earn. This is certainly on the river Earn, but in view of such expressions as *brug Banba*, 'land of Banba (Ireland),' *brugh Bretan*, 'land of the Britons,' etc., the probability is that we have to do with the name of a district. 'Drummondernoch, Drummenerinoch 1595 (RMS) between Comrie and Crieff, is in Gaelic *Drumainn Eireannach*, 'Drummond of Eire.'

We may note further an instance connected with the Irish Church settlement at Glastonbury in Somersetshire. In 971 king Edgar granted certain privileges to Glastonbury and in the record mention is made of a place called 'Bekeria, which is called *parva Hibernia* (little Ireland).' It is described as being *in insulis*, 'among the islands,' that is to say the low insulated lands or Inches near the Abbey.[3]

---

[1] 'Foelan mac Oengusa mic Natfraich, ō Raith Erend i nAlbain 7 o Chill Faelan i Laigis'; note on *Félire* of Oengus, June 20, in *Leb. Breac*, p. 90.

[2] Skene, *P.S.*, p. 88; A. O. Anderson, *Early Sources*, i. p. 367 (where Skene's reading and translation are corrected).

[3] From M'Clure's *British Place-Names*, p. 205.

This is a case of transference, not indeed of the name of Ireland itself, but of the name of a small island called *Bec Ériu*, now Beggery Island, in Wexford harbour. The latter got its name, according to the Irish account, from the bishop Ibar, to whom Patrick said in anger, 'thou shalt not be in Eriu.' 'Eriu,' replied Ibar, 'shall be the name of the place in which I am wont to be,' and he settled in the isle which was so named thenceforward.[1] These names are both connected with religious establishments.

In North Wales, which was once occupied by the Gael, there is a closely parallel instance. There, at the head of Afon Lledr in Carnarvonshire, is *Llyn Iwerddon*, 'Lake of Ireland,' corresponding exactly to our *Loch Éireann*, Loch Earn. Lower down, not far from the Falls of the Conway river, is a hill or place called *Iwerddon*, Ireland. About midway between the two is *Dolwyddelan*, 'Gwyddelan's Meadow,' where *Gwyddelan* is a derivative of Gwyddel, a Gael.

There can thus be no doubt that the name *Éire* was used as a district name in the parts of Britain where the Gael settled; I believe that Strathearn means 'Ireland's Strath,' not 'Strath of the river Earn,' and that Loch Earn (*Loch Éireann*, now *Loch Éir*) means 'Ireland's Loch.' It is not without significance that the purely Gaelic name Strathearn came to displace in part the gaelicized British name Fortriu as the name of the province between Tay and Forth. The south-western part of Fortriu became the Earldom of Menteith, the remainder became the Earldom of Strathearn. Strathearn of old apparently came all the way to the Firth of Forth, for Cuilennros (Culross), 'Holly-point,' is said to be in it.[2] There is a fort called Dunearn in Burntisland parish, Fife.

Marching with Strathearn was the district and Earldom of Athol. In the Book of Deer Atholl is spelled *Athótla* (for *Athfhotla*); in Tigernach's annals it is *Athfhotla* (genitive); the corresponding entry in AU (739) has *Athfoithle*. The Norse form is *Atiötlar*, answering, says Professor Craigie,

---

[1] *Félire of Oengus* (Stokes), p. 119.
[2] See p. 209, n. 2.

to the Scottish *Athwotle*, *Athodel*, etc., from an original *Athfodla*. From these forms it is clear that Atholl represents *ath*, in the sense of Latin *re-* denoting repetition, and *Fótla*, later *Fódla*, *Fódhla*, Ireland; the name means 'New Ireland.' Late forms are *Abhuill* (genitive), in a bardic elegy on Sir Duncan Campbell who died in 1631, and *Afall* (genitive) in the latter part of the same century.[1] The present form is usually *Athall* for all cases, but the Rev. C. M. Robertson tells me that *Blàr an Abhaill* (*Awaill*) is heard for Blair in Athol, and that the Athol men are *na h-Abhaillich* (*Awaillich*).[2]

North of the Grampians Éire occurs thrice. The Register of Moray records 'the church of Eryn, with the chapel of Inuernarren,' *i.e.* of Nairn, and about 1140 King David confirmed to the monks of Urquhart a grant of 'Pethenach near Eren, and the shieldings of Fathenachten.' Pethenach is now, I think, Penick, in Gaelic *a' Pheighinneag*, 'the little pennyland,' near Auldearn church. Fathenachten is probably Fornighty in Ardclach, the Gaelic of which is *Achadh-ghoididh* (? *-ghoide*, from *goid*, theft), apparently a different name for the same place. This Erin was therefore in Nairnshire, and appears to have been the old name for Auldearn parish.[3] In any case it is the name of a district, and *Allt Éireann*, Auldearn, means 'Ireland's Burn.'

We learn that the Abbey of Kinloss possessed a toft in Inverness, Eren, Forres, Elgin, and Aberdeen.[4] Elsewhere separate mention is made of Invereren, the land of the *prepositura* of Invereren, and one toft in Eren, all belonging to Kinloss. Invereren is the lower part of the river Findhorn. It was also the name of the old village of Findhorn, at the mouth of the river, which was swept away by the storm of 1701. *Prepositura* became in Scots 'Grieveship,' and Lachlan Shaw says 'below Mundole, on the

---

[1] *Rel. Celt.*, ii. p. 180.

[2] For *-thfh-* becoming *bh* compare *Srath-thamhuinn* for *Srath-athfhinn*, Strath Aven in Banffshire (*Inverness Gael. Soc. Trans.*, xxiv. p. 165). In *thamhuinn* the *th* is carried over from *srath*; cf. Bohespick, etc.

[3] Lawrie, *Early Scottish Charters*, p. 442.

[4] *Records of the Monastery of Kinloss.*

side of the river (Findhorn), is the Grieship.' The name survives in Greshope. Elsewhere the term is found in connection with places of strength, *e.g.* Cromarty, Forres, Cullen, all of which had castles, and it is more than likely that the *prepositura* of Eren is to be connected with the Castle of Eren referred to in a charter of William I. in 1185 as 'castellum meum de heryn.'[1] The place called Eren in which the toft was situated would have been near this; it was most probably in fact the same as the old village of Invereren. The names Cullerne and Earnhill, near the mouth of Findhorn, meaning 'nook of Eren' and 'hill of Eren,' are further in favour of Eren having the name of a district. The name Findhorn itself is the dative-locative of *Fionn-Éire*, 'white Ireland,' and doubtless refers to the white sands of the estuary. The river is in Gaelic *Éire*, *Uisge Éire*, and its strath is *Strath Éireann*, Strathdearn, 'Ireland's strath.' Near Dulsie Bridge is a large fort called *Dùn Éireann*, 'Ireland's fort,' a name which may indicate that the district extended well inland. Near the head of Strathdearn is a remote little district called 'the Coigs' or 'Fifths,' of which it is said 'tha cóig cóigimh an Éirinn is tha cóig cóigimh an Srath Éireann, ach is fearr aon chóigeamh na h-Éireann na cóig cóigimh Srath Éireann,' 'there are five-fifths in Ireland and five-fifths in Ireland's strath, but better is one-fifth of Ireland than the five-fifths of Ireland's strath.' The saying is old.

Deveron is on record *Douern* and *Duffhern*, meaning apparently 'Black Éire,' as distinguished from 'White Éire,' or Findhorn, but unfortunately the name does not survive in Gaelic. It can scarcely be mere chance that has put *Banbh*, Banff, at its mouth, and, as we have seen, Banba was a name for Ireland. It is also notable that a stream near Deveron on the west is called Boyne, while the patron saint of Boyndie parish, near Banff, was Brendan the Voyager. Tolachherene 1242 (Ant. A. and B.), 'hill of Éire,' appears to have been near Deveron.

Elg or *Ealg* has gen. *Elgga* (LL 45 a 42, 81 b 41) or *Eilgi* (LL 377 b 16); dat. *Eilgg* (LL 49 b 44); Kuno Meyer has

[1] *Charters of Inchaffray.*

compared *Druimm nElgga* in Munster (LL 198 b 4) with *Druimm nAlban.*[1] According to one authority, it means 'pig' (*muc*); others make it 'noble' (*uasal*).[2] With us it occurs as the name of a district in the west of Inverness-shire, commonly known as *Gleann Eilge*, Glenelg. I am assured, however, that the old people considered *Eilg* to be the name of the district, and 'Glenelg' to be 'the glen of Eilg.' This is borne out by the bardic poetry, where we find such expressions as *iath Eilge*, 'the region of Eilg'; *fear finn Eilge*, 'the lord of fair Eilg.'[3] A hill on the border between the parish of Glenelg and that of Glenshiel in Ross is *an Cruachán Eilgeach*, 'the Rick of Eilg.' A Glenelg man, however, is *Eilginneach*, formed probably on the analogy of *deilginneach*, shingles, from *dealg*, a prickle, which seems to be a byform of *deilgneach*, prickly. The form 'Glenelgenie' which appears in Robertson's Index, is curious, if authentic.

Eilginn, Elgin, has been explained by Kuno Meyer as for *Eilgín*, 'Little Ireland,' a diminutive from Elg. This explanation is made the more attractive by the fact that a certain quarter of Elgin, and by no means a recent one, is actually called 'Little Ireland.'[4] The difficulty is that the diminutive in *-ín*, which is common in Irish, is rare with us, and when it does occur—as in *cailin*, a girl—it does not usually double the final consonant. *Eilginn* looks like a locative case, and it may have been formed on the analogy of *Éirinn* and *Albainn*, which are themselves formed on the analogy of *Mumhain(n)*, Munster. Elgin used to be called in Gaelic *Eilginn Muireibh*, 'Elgin of Moray,' to distinguish it from some other place of like name, probably Elg of Glenelg.

Banba is connected with *banb*, now *banbh*, a sucking pig;

---

[1] *Zur kelt. Wortkunde*, no. 42.

[2] 'Ealga . i . Eiriu . i . ealg ainm do mhuic isin tseen-gaidhlig'; 'Ealga, *i.e.* Eriu; in Old Gaelic *ealg* means 'pig': *Coir Anmann*, 243. 'Inis Ealga . i . oilean uasal'; 'Inis Ealga' means 'noble island,' Keat, i. p. 98.

[3] Adv. Lib. MS. LII 43 a; XXXIX 32. The poem is to Macleod.

[4] The Town Clerk of Elgin informs me that the origin of this term is quite unknown.

she was probably a swine goddess. Kuno Meyer did not hesitate to regard both Banff on Deveron and Bamff near Alyth, in Perthshire, as the equivalents of Banba, both meaning Ireland.[1] To these may be added 'Banff with the fulling mill,' 1582, 1587 (RMS), part of the church-lands of Arbuthnot in Kincardineshire, now apparently obsolete. It is true that Banff is *Banb* in the Book of Deer and *Banbh* in modern Gaelic—one syllable. On the other hand, *banbh*, a sucking pig, is not appropriate—one might say it is impossible—as the name of a place or district; the Isle of Muck is not a parallel, for it is *Eilean nam Muc*, Helantmok in Fordun, 'isle of swine.' Besides these there are *Banbhaidh*, Banavie, near Fortwilliam, *Loch Banbhaidh* near Loch Shin, *Allt* and *Gleann Banbhaidh* in Athol, and Banevyn (Reg. Arbr.), now Benvie in Forfarshire. All these might represent *Banbha*, so far as phonetics go; compare Munlochy for *bun-locha*, Dalarossie for *Dail Fhearghusa*. I think, however, that the Athol Banvie is a stream-name like *Mucaidh*, 'pig-burn,' on Loch Tay; the others may be stream names or place-names like Tarvie, 'bull-place,' from *tarbh*, a bull. The Welsh for *banbh* is *banw*, a young pig, and in Montgomeryshire there are confluent streams called *Twrch*, Boar, and *Banw* (on maps Banwy). A Welsh writer says of Twrch, 'many rivers forming deep channels or holes into which they sink into the earth, and are lost for a distance, are so called. A small brook called Banw in the parish of Llan Vigan, meaning 'a little pig,' has been said to be of this family.'[2] In Ireland *banbh* appears commonly as an element of names, but Joyce gives no instance of it used by itself alone or as a stream name. In Wales several streams are called Twrch, none with us.

Yet another name for Ireland is *Fál*, and we have *Dun-fàil*, Dunphail, near Forres, but this may equally well come from *fàl*, a hedge, palisade.

There are some instances of Ireland itself as a place-name,

---

[1] *Zur keltischen Wortkunde*, no. 42.

[2] *Archiv f. Celt. Lexik.*, iii. p. 45.

e.g. in Menmuir parish, Forfarshire, where it is adjacent to Rome: Rowme, Ireland, and Corsbank, 1536 (RMS). This may be a translation, but there is no proof.

To these indications of settlement must be added the influence of the Scoto-Irish Church through its monasteries established in various parts of Pictland. It is clear that whatever may have been the relations of the Church among the Picts to the civil authority, there was a close connection between that Church and the Irish Church, that its personnel was largely Gaelic, and that Picts went to Ireland for training. It is highly significant that in the list of rulers and clerics who signified assent to Adamnan's *Law of the Innocents*, Scotland is represented by Bruide, king of Cruithentuath or Pictland, Curitan, bishop of Rosmarkie, and Ceti, bishop of Hí. Clerics and nobles would have been among the first to learn Gaelic. An interesting glimpse of the transition stage between Pictish and Gaelic is given in the tradition of St. Manirus recorded in the Aberdeen Breviary, where we are told that in consequence of the difficulty caused by diversity of language among the people, Manirus, being excellently skilled in both languages, went to labour at Crathie in Braemar. He was probably a native Pict who had been trained in Ireland.

The difference, so far as it exists, between the place-names of the west and those of the east is due to the historical circumstances. The west was largely and continuously peopled from Ireland and at an early stage became a Gaelic kingdom; for many of the people and for all the ruling class Gaelic was no acquired language, but their native speech. Consequently the change of language was comparatively rapid and general, and the old names were largely displaced. On the east, where settlement was sporadic and the rulers were chiefly Picts, the change was much more gradual. More of the old names were retained or adapted to Gaelic, many were doubtless translated in whole or in part.

In the west again, and also in the north from Caithness to Beauly, strong Norse influence caused a great displacement of old names. In the Long Island, from the Butt of

Lewis to Barra Head, Norse must long have been the predominant language, if not the only one. Here there are indeed many Gaelic names, but all the more important names are Norse and the Gaelic names are of comparatively recent type, being phrase names, *e.g. Allt na Muilne*, 'the burn of the mill.' In the Inner Hebrides names of an older type are found, such as *Sligeachán*, 'shell-place,' *Draighneach*, 'blackthorn-place,' in Skye. Caithness, part of the old Pictish homeland, must have been mainly Norse-speaking for centuries. Sutherland and, in a less degree, Ross are shot through with Norse names.

Bilingualism is no new thing in Scotland. It existed when Celtic began to displace some older tongue before the Christian era; that older tongue had probably displaced one still older. In Galloway, Strath Clyde, and the north Gaelic took the place of British or Pictish. In the far north and the west Gaelic or Pictish was for a time eclipsed by Norse, which in its turn was displaced by Gaelic. The circumstances under which those changes took place can never be known in the way we know how Gaelic is now being displaced by English. Two points, however, are to be noted; the first is that change of language does not necessarily mean change of race: the second is the somewhat remarkable readiness with which languages do change.

Some of the features more or less distinctive of the names of the region north of Forth and east of Drumalban may now be mentioned.

One of the most striking is the frequency of certain terms of British origin, such as *pett*, a piece or portion or share of land, anglicized as *Pit-*; *monadh*, a mountain, hill-ground; *preas*, a copse. To these may be added *gronn*, a mire or marsh, found from Forth to Beauly, and possibly *fother, fetter*, with about the same range.

Among hill terms *barr*, an eminence, is very rare; *tulach, tilach* is very common. *Sliabh* in the sense of 'mountain' probably does not occur. *Carn*, which usually means 'a heap of stones,' 'congeries lapidum,' is often used in the sense of a high, rocky hill; we may compare O. Welsh *carn* explained as *rupes*, a rock, crag, cliff. *Tom* regularly means a rounded hillock or hill, a mound, like Welsh *tomen* ; in Irish *tom* means usually a bush, thicket, as sometimes also with us, as *tom luachrach*, a clump of rushes.

The great native unit of land was the *dabhach* or *dobhach*, fem., a Gaelic term meaning a large tub or vat, the largest of vessels in use ; I have heard it applied to the large tub in a smuggling bothy. The secondary meaning of 'large measure of land' is peculiar to Scotland. But the davach [1] was not so much a measure of land as of 'souming,' *i.e.* it was an amount of land reckoned to support so many head of stock. Burdens on land—*càin*, tax ; *coinmheadh* or 'conveth,' dues of maintenance, and probably the military obligations of *feacht*, 'expedition,' and *sluaghadh*, 'hosting'—were assessed on the basis of the davach.[2] In 1772, in the roll of the Eastern Company of the Strathspey Volunteers, the men were entered according to the davochs of the parish.[3]

The term occurs by itself as Dauch or Doch, usually, however, followed by a descriptive term, as *Dabhach na Creige*, Dochcraig, 'davoch of the rock,' *Dabhach Phùir*, Dochfour, 'davoch of pasture,' and so on. It is rare at the end of names, but there are Phesdo, Fasdawach 1443 (RMS), 'firm davoch,' in Kincardineshire, Fendoch, Findoch 1542 (RMS), 'white davoch,' near Crieff, and Gargawach, 'rough davoch,' in Lochaber. *Leth-dabhach*, a

[1] A full discussion of the davach and its equivalents is given by Skene, *Celtic Scotland*, iii. p. 223. It was reckoned equal to 1 tirung or ounce-land or to 4 ploughgates or to 20 pennylands. A ploughgate or carrucate was 8 oxgangs, so that the davach was equal to 32 oxgangs : but this proportion may not have been invariable, for in MacGill's *Old Ross-shire and Scotland*, ii. p. 31 there is 'that western oxgate or fourth part of the half davoch toun and lands of Newnakill.' Pennant in his *Tour of 1772* (p. 314) says of Loch Broom : 'Land is set here by the *Davoch* or half *Davoch* ; the last consists of ninety-six *Scotch* acres of arable land, such as it is, with a competent quantity of mountain and grazing ground. This maintains sixty cows and their followers ; and is rented for fifty-two pounds a year. To manage this the farmer keeps eight men and eight women servants, and an overseer.'

[2] It appears from the Gaelic entries in the Book of Deer that the king, the mormaer, and the toiseach had, each of them, his own rights in respect of these burdens, *e.g.* each was entitled to dues of maintenance.

[3] *In the Shadow of Cairngorm*, p. 339.

half davoch, is anglicized Lettoch. Lettoch in the Black Isle is Haldach in 1527, Haddoch in 1611, forms which clearly explain Haddo, in 1538 Haddauch, in Aberdeenshire to mean 'half-davoch.' *Trian*, a third part, is rather rare; *Trian-a-phùir*, Trinafour, 'pasture-third,' is in Glen Erichdie, Struan; Blaeu has Trien high up on Black Esk. *Ceathramh*, a fourth part, a quarter, is probably the fourth part of a davoch, or it may be of a half-davoch, in such names as *an Ceathramh Ard*, Kerrowaird, 'the high quarter,' *an Ceathramh Gearr*, Kerrowgair, 'the short quarter,' near Inverness, and Kirriemuir. *Ochdamh*, the eighth part, occurs on the Kyle of Sutherland as Ochtow: 'the wester bovate *vulgo* the Ochtow,' 1589 (RMS). Auchtogorm in Moray is 'the green octave.' Achterblair near Carrbridge is *ochdamh a' bhlàir*, 'the octave of the moor.' In Argyll there are Ochtofad, 'the long octave,' Ochtomore, 'the big octave,' Ochtavullin, 'octave of the mill.'

Other denominations also occur, though as a rule they do not appear on the map. Thus on the south side of Loch Tay there are *Marg na Crannaig*, *Marg Mhór*, *Marg Bheag*, *Marg na h-Àtha*, *Marg an t-Sruthain*, Merkland of the Crannag, Big and Little Merkland, Merkland of the Kiln, Merkland of the Brook. *Baile Mac Neachdain*, Balmacnaughton, on the same side, is known as *an dà fhichead sgillinn*, 'the forty shilling land'; near Ardtalnaig is *an deich sgillinn*, 'the ten shilling land'; Ardeonaig is *an fhichead sgillinn*, 'the twenty shilling land.' In the east end of Fortingal there is *an dà mharg dheug*, 'the twelve merk land.' [1] In Sutherland there is Loch Merkland, and I have heard a field in Altas, Sutherland, called *am plang*, 'the plack.'

Of a number of old personal names which occur, the following may serve as examples. I do not mean to imply that the names are by any means peculiar to this region.

Abhartach, whence the name of the Irish sept *hUi Abartaich* (Rawl. B. 502, 159 *b*), appears in Rossawarty,

[1] For this and much other information about Perthshire I am indebted to Mr. Alexander Campbell of Boreland, Loch Tay.

1508 (RMS), 'Abhartach's cape,' now Roshearty near Fraserburgh. Hence too, *Dùn Àbhartaigh*, Dunaverty, 'Abhartach's Fort,' on the Mull of Kintyre, besieged by Sealbhach in 710.[1] Some *Ailpin* has given name to Rathelpin (RPSA), now Rathelpie, 'Alpin's rath,' in Fife; there is also Cairnelpies (Ret.) in Banffshire.

The old Irish name *Breasal*, 'warrior,' is seen in Donibristle, Donibrysell and Donybrisle in the twelfth century, for *dùnadh Breasail*, 'Breasal's fortress.'[2] From *Bruadar* comes Drumbroider in Stirlingshire, 'Bruadar's ridge'; compare Ballybroder in Ireland (Joyce); Tillybrother in Aberdeenshire may be 'Bruadar's hill.' From *Brandubh*, 'raven-black,' comes *Loch Branduibh*, Loch Brandy, near Milton of Clova, *Sìthean Druim Mhac Bhranduibh*, 'fairy knoll of the ridge of the sons of Brandubh,' near Onich, Argyll, and probably Brandy Burn in Fife. *Gleann Buichead*, Glen Buchat in Aberdeenshire, is 'Buichead's glen'; compare *Dún Buchad* or *Dún Buichead* in Ireland. *Frìth Bhàtair*, Freevater in Ross, is 'Walter's forest,' probably from Walter Leslie, earl of Ross.

O.Ir. *Coirpre*, later *Cairbre*, gives Drumforber, 'Cairpre's ridge,' in Kincardineshire, in 1539 Drumquhariber (RMS). *Cathalán* is probably the second part of Petcathelin, Pethkathilin of early twelfth century (Chart. Lind.), now Pitcaithly, 'Cathalán's share.' From *Cormac*, 'chariot-lad' (*corbmac*) come Balcormock 1428 (RMS), near Abercrombie, and Balcormock 1485, *ib.* near Lundie in Fife, both now Balcormo; another Balcormo in Forfar, so spelled in 1489 (RMS), may be the same.

Fearchar gives Glenferkar, Glenferkaryn (Reg. Arbr.), now Glenfarquhar, 'Ferchar's glen,' in Kincardineshire. From the female name *Finnghuala*, 'white-shoulder' or 'white-shouldered,' comes Finella's Den in Kincardineshire.

---

[1] 'Obsessio Aberte apud Selbacum' (AU). There seems no doubt that Dunaverty is meant, though *dún* is omitted and the natural spelling would be (*Dún*) *Ábartaig*.

[2] O.Welsh *brecl* means 'war' (Zeuss, *Gram. Celt.*, p. 135). The name might thus be claimed as British, but compare Ballybrassil, Ballybrazil, Clonbrassil, etc.—Joyce, vol. iii.

Gartán gives *Srath Ghartáin*, Strathgartney on Loch Katrine, 'Gartán's strath,' also *Làirig Ghartáin*, 'Gartán's pass,' off Glen Etive.

Muirgheal, 'sea-white,' a lady's name, gives Rathmuriel (RPSA), now Murriel in Inch parish, Aberdeen, 'Muriel's rath.' In Glen Nant, near Taynuilt, *Dùn Muirgil*, otherwise *Dùn Meirghil* is probably 'Muriel's fort.'

Mac Gille Eoin, 'son of St. John's lad,' is seen in Balmakgillona, Balmacgillon, *c.* 1200 (Chart. Inch.), now Bellyclone, 'stead of the sons of St. John's lad,' near Inchaffray. The name is now *Mac Gille Eathain*, Maclean. Cupermaccultin, 1150 (Reg. Dunf.), is 'Cupar of the sons of Ultán'; it is Cupermaculty, 1415 (Bamff Chart.), Cultirmacowty, 1525, *ib.*, now Couttie near Coupar-Angus. *Dùn Mac Tuathail*, a fort on the eastern end of Drummond Hill, Aberfeldy, is 'fort of the sons of Tuathal'; Inchtuthil, on Tay, is 'Tuathal's meadow'; Auchtertool in Fife may be 'Tuathal's upland,' for in Ireland Tuathal is anglicized Toole. *Dùn Mac Glais*, Dunmaglass, at the head of Strath Nairn, is 'fort of the sons of Glas,' and *Beinn Mac Duibh*, Ben MacDhui in the Cairngorm range, is 'hill of the sons of Dubh' (black).[1] *Ard Mac Maoin*, on the north side of Loch Katrine, is 'height of Maon's sons' (*maon*, dumb). *Baile Mac Cathain*, Balmacaan in Glen Urquhart, is 'stead of the sons of Cathan.' Tomcrail, opposite Killiecrankie station, is *Tom Mhic Réill*, 'Macneil's knoll'; similarly a great stone on Boreland Farm, Loch Tay, is *Clach Mhic Réill*, 'MacNeil's stone,' and a ferry on Tay below Ballinluig is *Bàta Mhic Réill*, 'MacNeil's Boat,' all with the not uncommon dialectic change of initial *n to r*.

Maol-domhnaich, 'Sunday's lad,' lit. 'Sunday's servant,' occurs in Petmuldonych, 1504 (RMS), 'Muldonych's share,' near Struan; the name survives in the burn name *allt Phit 'al-domhnaich*.[2] *Morgan*, a British name, appears in

---

[1] It may be 'Macduff's peak,' for Macduff (*MacDuibh*), Earl of Fife, held much land in that neighbourhood in early times.

[2] I was informed by the late Mr. John Whyte that in Skye the name *Maol-domhnaich* used to be given to a boy whose maintenance was provided for by the Sunday's collection.

Tillymorgan, Aberdeen; Ramornie, Ramorgany 1512 (RMS), in Fife, seems to be for *Rath Morganaigh*, 'rath of (the) Morganach,' *i.e.* of a man of Clan Morgan of Aberdeenshire.[1]

Nechtán or *Neachdán* gives Dunnichen, of old Dunnechtyn (Reg. Arbr.), in Forfarshire, 'Nechtan's fort,' repeated in Badenoch as Dunachton; the former is considered to have been the scene of the great battle of Dún Nechtain in 685 (AU and Tighernach), where Brude, king of the Picts, routed the Angles. Bunachton in Stratherrick is *Both Neachdain*, 'Nechtan's hut.'

Selbach or *Sealbhach*, 'rich in possessions,' appears in Belhelvie, Aberdeen, 'Selbach's stead'; compare M'Kelvie.

The old name for the men of Ulster was *Ulaid*, gen. *Ulad*, and this is probably found in Rathillet in Fife, spelled Radhulit before 1200, Rathulit 1528 (RMS), for *ráth Ulad*, 'rath of the Ulstermen.' The later term for an Ulsterman is *Ultach*, whence *Dùn nan Ultach*, Downanultich 1539 (RMS), 'the Ulstermen's fort,' in Kintyre; compare Barnultoch in Wigtownshire.

In the west also personal names are fairly common, and as a rule easy to recognize. Glen Finnan in Inverness-shire is *Gleann Fhionghuin*, 'Fingon's glen'; compare *Mac-Fhionghuin*, 'Mackinnon.' Glen Masan in Cowal is 'Massan's glen.'

One other point may be mentioned here, namely 'eclipsis.'

In Irish Gaelic a final *n* is in certain cases regularly carried forward to the next word if that word is closely connected with the preceding word. If the second word begins with a vowel, the final *n* of the preceding word is sounded and written before the vowel; if the second word begins with a consonant, the final *n* disappears before *s*, assimilates to *l*, *m*, *n*, *r*, and combines in various ways with the other consonants. In the case of *c*, *t*, *p* (*tenues*), *nc*, *nt*, *np* are sounded as *g*, *d*, *b*, and are written in modern Irish as *gc*, *dt*,

[1] The Mackays of Sutherland were known as *Clanna Morgainn*, also as *Morganaigh* or *Morganaich*, the plural of *Morganach*.—*Rel. Celt.*, ii. pp. 176, 260.

bp; in the case of *g, d, b* (*mediae*), *ng, nd, nb* are sounded as *ng* (as in English *anger*), *n, m,* and are written now *ng, nd, mb*; lastly, *nf* is sounded as *bh, i.e.* as English *v*, and written now *bhf*. Thus 'in Ireland' is *i nEirinn*; 'of the blades' is *na lann* (for *na nlann*); 'of the dogs' is *na gcon*, pronounced *na gon*; 'of the strings' is *na dtéad*, pronounced *na déad*; 'of the teeth' is *na ndéad*, pronounced *na néad*; 'of the men' is *na bhfear*, pronounced *na bhear*. This process is commonly called 'eclipsis'; the corresponding process in Welsh is called 'the nasal mutation.'

In modern Scottish Gaelic eclipsis survives only in a few isolated instances, but it was once general here also. In the place-names we meet it as the result (1) of the preposition *in* or *an*, 'in'; (2) of the genitive plural of the article, *nan*; (3) of a neuter noun in the nominative singular. As to range, it is not uncommon in the east, rarer in the west, and very rare or non-existent in the Isles. As has been mentioned, it occurs in Lothian and in Galloway.

The record forms of names sometimes show eclipsis where the current forms do not. Gylltalargyn, 1203/24 (RM), is now *Cill Taraghlain*, Kiltarlity, 'Talorgan's church'; initial *g* of the old spelling is due to the influence of the preposition in the phrase *i gCill Talorgain*, which would be in constant use. Similarly we find Gillepedre, 1362 (RMS), for what is now *Cill Pheadair*, Kilpeter in Strathbrora; and Gillecallumkille, 1566 (Orig. Paroch.), for *Cill Chaluim Chille*, 'St. Columba's church,' in the same strath. *Cill Chriosd*, 'Christ's church,' in Mull is Gilzacrest in 1496 (RMS); *Cill Chaomhaidh*, 'Caomhi's church,' near Logierait, Gilliquhamby 1558 (RMS), is anglicized now as Killiehangie. Initial *b* under the influence of the same preposition becomes *m*. Mochastir 1452 (RMS), Mouchester 1524, *ib.*, Moucastell 1579, *ib.*, is now *Both Chaisteil*, Bochastle, 'hut of the castle' or 'chester,' near Callander, with reference to a Roman encampment which is now almost obliterated. *Bonn-sgaod* (also *-sgaoid*), Bonskeid near Pitlochry, is Monskeid, 1511 (RMS), and *Both-reithnich*, Borenich, 'bracken hut,' on Loch Tummel, is Montramyche (*read* Montrainyche),

1508 (RMS). Bunchrew, 'near (the) tree,'[1] Inverness, is Monchrwe 1507, Moncrew 1510, Munchrow 1511 (RMS). The form Airdendgappil, 1351 (Reg. Lenn.), is for *Ard na gCapull*, 'cape of the horses' or 'of the mares,' now Ardincaple on the Gare Loch—an instance of the genitive plural of the article. A good example of the influence of a neuter nominative singular is Mogomar, 1500 (RMS), now *Magh Comair*, 'plain of (the) confluence,' anglicized as Mucomir, in Lochaber at the junction of Spean and Lochy; Mogomar is for *Magh gComair*, a survival from the period when *magh* was neuter.

I shall now give examples of eclipsis in names that are still current in their eclipsed form.

(1) With the preposition *in*: Moness at Aberfeldy is for *i mbun eas*, 'near waterfalls,' lit. 'at waterfalls' foot.' Monessie in Lochaber is for *i mbun easa*, 'near waterfall,' the final *-ie* representing the old genitive sg. ending *-a*. The waterfall is *Eas Chlianáig*, whence the Glaistig took the stones to build Kennedy's house at Lianachan.[2] Munlochy is for *i mbun locha*, 'at loch foot,' or 'near the loch,' *-y* being for the old genitive sg. ending. Benderloch in Lorn, for *beinn eadar dà loch*, 'hill between two lochs,' is in Gaelic *Meadarloch*, where *m*, for *mb*, is all that is left of *beinn* in unstressed position. Muckairn in Lorne, Mocarne 1527 (Orig. Paroch.), is spelled *Bo-càrna* and *Bu-càrna* by the Gaelic poet Alexander MacDonald about 1750,[3] apparently for *Both-càrna*, '(?) hut of flesh'; this form, if correct, would readily become *Mo-càrna* after *in*.

(2) After *nan*, the genitive pl. of the article: Achnagairn, Beauly, is for *achadh na gcarn*, 'field of the cairns.' Achnagullan, Oykel, is for *achadh na gcuilean*, 'field of the whelps'; Amat, close by at the junction of Eunag and

---

[1] For *i mbun chraoibhe*; here the phrase *i mbun* or *a mbun* is used idiomatically as often, in the sense of 'close to,' in the same way as we say 'an cois na fairge,' 'near the sea,' lit. 'at foot of the sea.'

[2] This well-known tale is too long to tell here; see *Carmina Gadelica*, ii. p. 287, another good English version is given by Mr. Donald A. Mackenzie in his *Elves and Heroes*.

[3] *Celtic Review*, iv. p. 303; v. p. 226.

Oykel, used to be called *Àmad na gCuilean* 'Àmat of the whelps,' to distinguish it from *Àmad na h-Eaglaise*, Amat of the church, and *Àmad na' Tuath*, 'Àmat of the laity,' in Strathcarron. *Allt na gCealgach*, on the road from Lairg to Lochinver, is 'the burn of the deceitful men,' in Macfarlane Alt Gellagach and *Aldene-Gealgigh*.[1]

Balnagore in Easter Ross is for *baile na gcorr*, 'stead of the cranes.' Balnaguard near Aberfeldy is *baile na gceard*, 'stead of the artificers.' Bada na Bresoch, Forfarshire, is for *bad na bpreasach*, 'spot of the copses.' Cairnagad in Aberdeenshire is 'wildcats' cairn.' Dalnavert in Badenoch and near Aberfeldy is for *dail na bhfeart*, 'dale of the graves.' Between Pitlochry and Glen Briarachan is *Dail na gCarn*, Dalnagairn, 'dale of the cairns.' *Dail na bhFàd* in Glen Briarachan is 'dale of the sods'; compare Blairnavaid in Stirlingshire. *Féith na gCeann*, 'bog of the heads,' in Kirkmichael parish, Perthshire, is anglicized as Finegand. High up on Ben Lawers is *Lochan na gCat*, 'the wildcats' lochan.' Loinveg in Braemar is in Gaelic *lòn na bhfiodhag*, 'meadow of the bird-cherry trees.'[2]

(3) Of nouns originally neuter the most common in place-names are *allt*, a height, a burn, *ceann*, a head, *comar*, a confluence, *druim*, a ridge, *dun*, a fort, *gleann*, a glen, *inbhear*, a confluence, *loch*, a loch, *magh*, a plain, but very slight traces remain of their influence. Dundurcus, Elgin, might be explained on the theory that the second part is *turcais*, 'boar-place,' formed from *torc* like *tarbhais*, Tarves, from *tarbh*, a bull; *t* becoming *d* after *dun*. But there seems to be no other example of eclipsis after *dun*. Drumgowdrom, Kildrummy, is probably for *druim gcolldroma*,' 'ridge of hazel-ridge,' and Kingoldrum, Kincaldrum, 1505 (RMS), in Forfarshire means 'head of hazel-ridge,' from *coll*, *call*, hazel. *Inbhir nAllt*, Invernauld on the Kyle of Sutherland, means 'confluence of cliffs': the stream which

[1] So called according to tradition because in a dispute about marches, the witnesses on one side swore that they were standing on the soil of a certain estate, having first put some of it inside their brogues.

[2] For this I am indebted to Mr. F. C. Diack.

enters the Kyle here flows through a precipitous gorge ; this old meaning of *allt* is seen, *e.g.*, in *an t-Allt Grànda*, 'the ugly precipice,' the great gorge near Evanton in Ross-shire, and in *an t-Allt Mór*, the precipice near Inverfarigaig. Similarly we have Cumbernauld for *comar* (O.Ir. *combor*) *nallt*, where the meaning may be 'confluence of brooks.'

Monzie near Crieff is Mugedha, 1226/34 (Chart. Inch.), Muyhe, 1226/34 (Reg. Inch.), Moythethe, 1282 (Chart. Inch.), now in Gaelic *Magh-eadh*. The second part is probably for *eadha*, earlier *etho*, the genitive sg. of O.Ir. *ith*, corn, later *iodh*, as in G. *iodh-lann*, a cornyard ; the old name would thus be *mag n-etho*, 'plain of corn,' represented very well by the old spellings, which however rather strangely omit the *n* preserved in the English form. Monzievaird in Strathearn, near Monzie, is Muithard 1200 (Chart. Inch.), Monewarde 1203, *ib.*, Moytheuard 1234, *ib.* ; Moeghauard *c.* 1251 (Skene, *P.S.*), also Morgoauerd (*read* Mog-), in Gaelic now *magh bhàrd*, correctly translated 'bardorum campus,' 'plain of the bards,' in the *Cronicon Elegiacum* (*c.* 1270). As the old form would be strictly *mag mbard* (for *mag nbard*), pronounced *magh mard*, which could not result in 'Monzie-vaird,' it is probable that the latter has been influenced by the neighbouring 'Monzie.'

CHAPTER IX

EARLY CHURCH TERMS

WHEN the three Celtic tribes, the Tectosages, Trocmi, and Tolistobogii, settled in Asia Minor about B.C. 280, in the district called after them Galatia, 'the land of the Gauls,' among their first arrangements was to establish a central council of three hundred to judge cases of bloodshed; other offences were left to the chiefs of districts or tetrarchs and local judges. This great council met at a place called Drunemeton, 'the chief Nemeton' or 'chief sacred place.' In Gaul, by a similar arrangement, according to Caesar, the Druids met at a fixed time of the year in the land of the Carnutes, which was reckoned the centre of all Gaul, and held a session in a consecrated place. Hither came all who had disputes, and these were decided by the judgment of the Druids. The term *nemeton*, 'a sacred place,' was common in Gaul; probably every tribe had one or more such places of judgment and of worship. They were the local habitations of gods; an inscription records that a certain Segomāros constructed a *nemeton* for the goddess Belisama, and when Gaul came under Roman influence we find Augustonemeton, the sacred place or shrine of the deified Augustus. Ver-nemeton, which occurs thrice in Gaul is explained as *fanum ingens*, 'great shrine'; one of the places so called was in Christian times the basilica or church of St. Vincentius. There was a Vernemeton in Britain, a town of the Coritavi, situated near Willoughby on the Wold in Nottinghamshire. The *nemeta* were usually in groves: in an eighth century list of superstitions and pagan rites there is a heading 'de sacris silvarum quae nimidas vocant,' 'concerning shrines in groves which they call nimidae.' In Brittany there was 'silva quae vocatur Nemet,' 'the wood which is called (the wood of) Nemet.' The early Church found it necessary to enact special penalties for making offerings to wells and groves. Among the ancient Celts, then, the Nemeta were holy places, often in groves, used as meeting-places for purposes of judgment; later they remained the objects of superstitious reverence, and became sometimes the site of Christian churches.

Nemeton is the neuter of *nemetos*, sacred, noble, found in the Gaulish personal name Nemetos and in composition with other terms. In O.Irish *nemed* means (1) sacred, noble; (2) sacred place. In the former sense it appears in the expression *in brátha nemed*, 'the dooms of the nobles,' in their capacity of judges. It appears in British personal names, as Nimet, Nemet; Guonemet, 'sub-noble'; Gornivet, 'very noble.' Nemed, according to Irish tradition, was the leader of a colony to Ireland after the death of Partholon and his folk, and the ultimate ancestor of the British people. As a place-name *nemed* seems to be very rare in Ireland; the only instance known to me is distinctly pre-Christian in origin. This is Nemed on Sliabh Fuait, now the Fews Mountains, with regard to which there is an instructive tradition. Fuat, son of Bile, the story tells, in coming to Ireland fell in with an island called Inis Magdena, whose soil possessed the property that no one standing thereon could utter a falsehood. Fuat took with him a sod of this island, and on it dooms were pronounced: if the judgment was false, the sod turned soil upward. This sod, which was called *fót na fírinne*, 'the sod of truth,' was placed on Sliabh Fuait. The spot was brought into relation with Christianity by the further tale as to how Patrick's nag (*gerrán*) lay there and spilled some of the corn which he was carrying: the Christian name for it was *fótán tíre tairngiri*, 'the sod of the Land of Promise.'[1] Here, then, we have an ancient sacred place of judgment, which at a later time was in a manner christianized. In Irish literature *nemed* is not uncommon in the sense of 'holy place, sanctuary, church.' Thus Patrick's *sen nemed*, 'ancient sanctuary,' was at Dunpatrick; a chief is praised because he did not molest church

[1] BB 404 a 31; LL 204 a 16; Fuat is otherwise 'son of Breogan.'

or sanctuary (*ni ra-chráid chill ná nemed*).[1] There is also *fidnemed*, 'a wood sanctuary,' applied sometimes to a pagan shrine, such as that in the Isle of Lemnos.[2] Violation of a *blái-nemed* or sanctuary was punishable by a fine under the Law of Adamnan.

In Scotland the *nemeton* has left its mark very distinctly on our place-names, and its history appears to be the same as in Gaul and in Ireland—an institution originally pagan, taken over by the Church. The Ravenna Geographer mentions Medio-nemeton, 'mid shrine,' which may have been on the line of the Wall between Forth and Clyde,[3] but is not represented among our modern names. Fiacc's hymn to Patrick, composed about 800, begins with the statement, 'Patrick was born at Nemthur,' and a gloss adds that this is the name of a city in North Britain (*i mBretnaib tuaiscirt*), namely 'Ail Cluade,' that is, Dumbarton; another spelling is Nemptor. This O.Irish form is considered to stand for an earlier Nemetodūron, 'stronghold of the Nemet.' Now whether Nemthor was really an old name for Dumbarton or not, there was in the neighbourhood, or at least not far away, a place or district called Neved. Maldoven (*i.e.* Maoldomhnaich), earl of Lennox, granted to his brother Amelec the lands of Neved, Glanfrone (Glenfruin), and other places, and the grant was confirmed by King Alexander in 1225.[4] From another record it appears that the land of Nemhedh lay partly on the eastern side of Loch Long, partly on the eastern side of the Gareloch. The name survives in Rosneath, in Gaelic *Ros-neimhidh*, 'promontory of the Nemet,' on the western side of the Gareloch opposite Row (*i.e. an Rubha*, 'the point'), which latter was doubtless part of the lands of Neved just mentioned. It does not appear how far these lands extended along the east side of the loch in the thirteenth century, but the old parish of Rosneath is considered by the editor of the *Origines Parochiales* to

[1] *Rev. Celt.*, xiii. p. 84. A number of illustrative passages are collected in Petrie's *Round Towers*.

[2] Petrie's *Round Towers*, p. 62.

[3] See Dr. George Macdonald's *Roman Wall in Scotland*, p. 153.

[4] *Reg. of Lennox; Orig. Paroch.*, i. pp. 29, 31.

have extended on that side as far as the neighbourhood of Cardross, that is within less than five miles from Dumbarton.[1] It is thus possible that Nemthor, if it was not Dumbarton, was somewhere within the old parish of Rosneath, and on the Dumbarton side of the water. In any case it is certain that a Nemeton existed in this part of the territory of the Damnonians.

The next instance is Navitie and Navitie Hill near the south-east end of Loch Leven in Fife, spelled Neuechi, Nevathy (RPSA); Nevody, 1477 (RMS). At Navitie Hill is Dunmore 'great fort,' and adjacent to Navitie is Kirkland. The land of Neuechi belonged to the Prior and Convent of St. Andrews. This name is the same as Navity near Cromarty, which will come up later.

In Forfarshire there is the old parish of Nevay, of old called 'Nevyth in Angus' (RPSA). The old church of Nevay is a little more than a mile east of Meigle; besides Kirkton of Nevay there are East and West Nevay and Nevay Park, together extending over a mile in length. Another old parish in Forfarshire is Navar, now Lethnot and Navar, Neuethbarr, 1232 (Reg. Brech.); Nethvar (RMS, App. 2); 'the lordship of Neware,' 1472 (RMS). The first part is doubtless *nemed*; the second part is probably Barr, the short form of the name of St. Findbarr of Cork; compare Newyn Crist in Glen Livet, *c.* 1224 (RM). About a mile from the old church of Navar are the great strongholds called the Caterthuns.

'Nevot in the barony of Alveth (Alva),' 1536, 1573 (RMS) appears to be now 'the Nebit,' a hill north-east of Alva in Clackmannan, but there is not sufficient evidence to claim it as a Nemeton.

Duneaves in Fortingal parish, Perthshire, is Tuneve, 1598 (RMS), Tynnaif, 1598, *ib.*; Tennaiffis, 1602, *ib.*; Tennaffis, 1640 (Ret.); the present Gaelic is *Tigh-neimh'* (for *-neimhidh*); the plural form is due to there being two farms, Duneaves proper and *Tigh-neimh' Ghearr*, 'short (*i.e.* little) Duneaves,' anglicized Tynayere. The name means 'house

---

[1] *Orig. Paroch.*, i., Rosneath Parish and Map.

of the *nemed*'; the farmhouse is opposite the church of Fortingal on a rather remarkable bend of the Lyon, which would form a most suitable place of assembly. A field on the farm is called *Dail mo-Choid*, 'St. Coeddi's dale.' The yew tree at Fortingal church is well known; it was reported by Pennant that its ruins measured fifty-six feet in circumference in 1772. This yew may well have been a sacred tree connected with the Nemeton; another point worthy of mention is that a spot on the farm of Kyltirie, on Loch Tay side, a few miles away, is reputed to be the central point of Scotland.[1]

At Craigrossie in Dunning parish, Perthshire, is Tarrnavie, described in the seventeenth century as 'an artificial knoll, evidently raised and gathered together by men's hands, resembling a ship: whether this has been a work of the Picts or Romans is not well known; however, 'tis rather thought to have been a work of the Romans, it having to this day a Roman name Terrae-navis answering exactly to its form: 'tis commonly called here . . . Terrnavie.'[2] In 1665 (RMS) it is Tarnavie. The first part is probably *tarr*, a paunch, belly, with reference, as often in place-names, to a bulging spur of an eminence; the second part is *neimhidh* doubtless, and the site of the *nemed* was probably on Craigrossie.

The Newe in Strathdon parish, Aberdeenshire, is *le* Newe, 1508, 1513 (RMS), spelled also Nyew; Milne gives the vernacular Scots pronunciation as Nyeow (*y* for *j*, strongly palatal). Gilchrist, earl of Mar, granted to the priory of Monymusk the churches of Loychel, Ruthauen,[3] and Inuernochin (Invernochty or Strathdon). Pope Innocent (1198-1216) agreed to take under his protection all the possessions of Monymusk, including Earl Gilchrist's grant, the churches of St. Andrew of Afford (Alford), St. Marnoc of Loychel, and St. Mary of Nemoth. This does not prove that Nemoth was another name for Invernochty, but as the Newe

---

[1] *Tigh nan Teud*, 'harp-string house,' near Bridge of Garry, has the same reputation.

[2] Macfarlane, *Geog. Coll.*, i. p. 121.

[3] Ruthauen is now Ruthven, in the eastern part of Logie-Coldstone.

is near Invernochty, it is probable that Nemoth is the twelfth century form of Newe. A writer in Macfarlane's *Collections* mentions that on the top of Binnen, which overlooks the Newe, 'is a fountain in the hollow of a rock . . . renouned among the vulgar for marvelous cures. There is said to be a Worm still abiding in it, which if alive, when the patient comes, he or she will live, if dead, they are condemned to die.'[1]

The next instance is in Banffshire, where in the upper part of Glen Livet there is Nevie, once the site of a chapel styled NeuechinChrist, NeuinCrist (RM), 'Christ's sanctuary.' Near it is Gallow Hill; Suie, formerly *Suidhe Artuir*, 'Arthur's Seat,' is some miles further up.

Creag Neimhidh, 'rock of the *nemed*,' is in Glen Urquhart, right above Temple Pier (St. Ninian's).

In Ross-shire there are *Dail Neimhidh*, Dalnavie, *Innis Neimhidh*, Inchnavie, and *Cnoc Neimhidh*, all adjacent to Nonakiln, in Gaelic *Neo'* (for *Neimheadh*) *na Cille*, 'the *nemed* of the church,' where are the ruins of an ancient chapel. When the New Stat. Account was written (1841), there were two glebes in the parish (Rosskeen), one of them at Nonakiln. On Cnoc Neimhidh, above this old church, is a large cairn called *Carn na Croiche*, 'the gallows cairn.' Eastwards is the estate of Newmore, in Gaelic *Neo' Mhór*, 'the great *nemed*,' on record as Nevyn Meikle, corresponding to the Gaulish *Nemetomāros*. It also is in the parish of Rosskeen, and at one time belonged to the church of Tain. In 1275 (Theiner) 'Nevoth and Roskevene' appear as two parishes. It would seem that the whole of this ridge, from the river Averon eastwards to Newmore formed a *nemed* of old. On the opposite side of the Cromarty Firth, high on the ridge of the South Souter above Cromarty, and facing the Moray Firth, is *Neamhaidigh*, Navity, the same name as Navitie in Fife. These names may represent an early *Nemantia*, from the base *nem-* of *nemetón*; compare Gaulish Nemossos, which was another name for Augustonemeton, also Nemausus, now Nîmes. Compare, however,

---

[1] *Geog. Coll.*, ii. p. 22.

Balmadity, Barmuckety, etc., formed from *madadh*, dog, *muc*, swine. Navity was church land and had a chapel. All Cromarty people are familiar with the old belief that the final judgment is destined to take place on the moor of Navity, and Hugh Miller records a very striking instance of its practical effectiveness.[1] These places in Rosskeen and Cromarty may well have been in use among the Decantae.

In the east of Sutherland there is Navidale, G. *Neimhe' dail*, in 1563 Nevindell. Sir Robert Gordon states that there was a sanctuary here, *i.e.* a place that had the right of sanctuary or girth. The formation is Norse, the second part being N. *dalr*, a dale; the Norsemen found the *nemed* there as an important place, and named the dale after it.[2] Off the north coast of Sutherland, east of the Kyle of Tongue, is an islet called on the map 'Neave or Coomb Island.' It has a dedication to St. Columba, and was given me in Gaelic as *Eilean na Neimhe* (? for *E. an Neimhidh*). This may be another instance, and if so it is the farthest north and the only one on an island.

Finhaven was the name of an ancient parish in Forfarshire, now Oathlaw. The hill of Finhaven, with a fort, south-east of the church, appears to be the 'law.' Early forms from RMS are Futhynevynt, 1370; Fothnevyn, Fothenevin, 1374; Futhenevin, 1384. For the first part we may compare Fouthas 1439, Futhes 1440 (RMS), now Fiddes in Kincardineshire, representing most probably *Fiodhais*, 'wood-stance'; Finhaven on this analogy will be for *fid-nemed*, 'wood sanctuary.' As for the final *n*, compare Nevyn, Neuin above; it is very common in charter forms, and can in many instances, as here, be explained only as a trick of spelling; here it has persisted in the anglicized form.

Andóit, now *annáid*, has been already explained as a patron saint's church, or a church that contains the relics of the founder. This is the meaning in Ireland, and it is all we have to go upon. How far it held with regard to Scotland is hard to say: our Annats are numerous, but as a rule

[1] *Scenes and Legends*, chap. xiv.

[2] In *Neimhe'dail*, unaspirated *d* is to be explained on the ground that the first part of the compound ended in *d*, *i.e.* the word was *Nemed-dal*.

they appear to have been places of no particular importance. They are often in places that are now, and must always have been, rather remote and out of the way. It is very rarely indeed that an Annat can be associated with any particular saint, nor have I met any traditions connected with them. But wherever there is an Annat there are traces of an ancient chapel or cemetery, or both ; very often, too, the Annat adjoins a fine well or clear stream, like that sung by Duncan Bàn Macintyre :—

'Fìon uillt na h-Annaid,
Blas meala r'a h-òl air.'

'the wine of the burn of the Annat, its taste was of honey to drink it.' (See p. 436).

In Stirlingshire there is Craigannet, for *Creag Annaide*, 'rock of (the) Annat,' on Carron in the northern part of Kilsyth parish. Near it is Kirk o' Muir.

Longannet Point in Fife is opposite Grangemouth ; 'the rock of Langannand near Culross,' 1587 (RMS) ; the first part may be *lann*, an enclosure, field.

In Kinross, Gouderannet is south of Kinross town, near Loch Leven. I have not met it on record and do not know what the first part is.

In Perthshire, Annaty Burn, for *Allt* (or *Glais*) *na h-Annaide*, is at Scone. 'Annatland with the acres of Tibbermure in the barony of Ruthven,' later of Huntingtower, appears in 1602 (RMS). Annat in Kilspindie parish is in the Glen of Rait, close to the old church and to an ancient fort. In the Braes of Doune, east of Callander, there are Annat and *Allt na h-Annaide*, Annat Burn ; 'Calzecat and Annat in Menteith,' 1508 (RMS) ; the former is *Coille Chat*, 'wildcats' wood.' East of Annat Burn and fairly high up is *Loch Mo-Thathaig*, Loch Mahaick.[1] On the north side of Loch Tay, opposite Ardeonaig, is *Baile na h-Annaide*. The burial-place of this Annat is partly within the present garden, partly in the field just outside it, where the tenant told me that his plough met stone slabs. Another *Baile na h-Annaide* is in Glen Lyon, below Bridge of Balgie ;

[1] See p. 298.

here too there is an ancient burial-place. On the north side of Loch Rannoch, near the east end, there are *an Annaid* and *Allt na h-Annaide.*

Pethannot c. 1195 (Chart. Lind.) was in the south of Kincardineshire, near the sea-shore ; it means 'portion of (the) Annat,' and in the record goes with Pethergus, 'Fergus' portion,' possibly St. Fergus.

In Aberdeenshire there is Andet in Methlick parish.[1] It appears often on record, and had a chapel dedicated to St. Ninian, but this is plainly a secondary dedication. Ennets, Kincardine O'Neil, is probably another instance.

In Morayshire there is Achnahannet in Cromdale parish, 'field of the Annat,' with the site of a little chapel in a field, enclosed by a 'cashel.' Down below near the burn is *Tobar an Domhnaich,* 'Sunday's well.'

In the mainland part of Inverness-shire, Achnahannet is north-east of Duthil Church, with a magnificent spring and what may be the remains of a *clochán.* Groam of Annat, near Beauly, is so called from an Annat which appears to have been in the swampy flat near it. In Glen Urquhart is *Achadh na h-Annaïde.* On the western side of the county there is *an Annaid,* 'the Annat,' at the head of Glen Roy. *Achadh na h-Annaide,* Achnahannet, is on the Fortwilliam side of Spean Bridge. Of it a writer in Macfarlane's *Collections* says : 'There is one little toune where there was a chappell builded of ancient, not two mylls from Kilmanevag . . . . called Achanahannat. . . . And this chappell was a sanctuarie and holie place keipit amongst the Countreymen in the said ancient time.' The writer reports that 'selling and buying wyne, ale, aquavitae, and sundry drinks and merchandice' went on there, and this is confirmed by the adjacent name, *Aonachán,* 'fair place,' 'market place.'[2] There is another Annat on Allt Dotha, west of Corpach, in Kilmallie parish. The name Corpach, 'place of bodies,' may be connected with this Annat.

In the mainland part of Ross and Cromarty, there are *an Annaid* and *Loch na h-Annaide* in Nigg parish. *Achadh na*

---

[1] See p. 319.

[2] *Geog. Coll.,* ii. pp. 170, 524.

h-Annaide, Achnahannet, is in Kincardine parish, on the Kyle of Sutherland and near the ancient church and burial-place of *Cill Mo-Chalmáig*, 'St. Colmóc's Church.' In Strath Conon there are *an Annaid* and *Clach na h-Annaide*, 'stone of the Annat,' near Balnault; also *an Annaid* much further up, opposite Invermany. On the west there are Annat at Torridon and Annat at Kildonan on Loch Broom. In the island of Crowlin, off Applecross, is *Port na h-Annaide*, 'the harbour of the Annat.'

In Sutherland there are *Achadh na h-Annaide* in Durness, and *an Annaid* and *Bàgh na h-Annaide*, Annat Bay, on an islet in Loch Chabhaidh, 'Cathbad's loch,' off Loch Laxford; Cathbad was probably the saint of the Annat.

In Argyll, Annat in Appin appears to have been near the site of the present parish church.[1] The lands of Annat, at Kilchrenan, on Loch Awe, extended to a merkland of old extent, 1594 (RMS). Some miles north-west of Kilchrenan are *Cladh na h-Annaide*, 'the burial-ground of the Annat,' and *Achadh na h-Annaide*, near Taynuilt. Near Diarmaid's Pillar in Glen Lonan, not far from Oban, is *Cladh na h-Annaide*, an old graveyard, described by Angus Smith.[2] On the skirts of Beinn Dobhrain and near the river Conghlais is *an Annaid* with its burn *Allt na h-Annaide*, whose water is celebrated by Duncan Bàn Macintyre. There are traces of an ancient burial-ground and of the site of a chapel.

In the Isles there are *na h-Annaidean*, 'the Annats,' at Shader, Barvas, Lewis. In *na h-Eileanan Sianta*, the Shiant Isles, *i.e.* 'the holy isles,' there is *an Annaid*, and another in the Isle of Killegray in the Sound of Harris, west of which is Pabbay, 'priest isle.' In Skye there are *Clach na h-Annaide*, 'stone of the Annat,' and *Tobar na h-Annaide*, 'well of the Annat,' at Kilbride on Loch Slapin. Another *Annaid* is some miles north of Dunvegan at the junction of the Bay River with a small burn, described by Boswell in connection with Dr. Johnson's Tour. The

---

[1] *Orig. Paroch.*, ii. pt. I. p. 167, gives its bounds.

[2] *Loch Etive and the Sons of Uisnach*, p. 262.

enclosure contained about two acres and apparently had been used as a place of burial. The remains of what was probably a chapel were held by the minister, Mr. MacQueen, to have been a temple of Anaitis.

An offering or sacrifice is in Gaelic *ìobairt*, and the Book of Deer uses this term several times of offerings or gifts of land made to the Church, *e.g.* 'Domnall mac meic Dubbacin robáith na h-ule edbarta do Drostán'; 'Domnall son of Dubbacin's son dedicated (lit. drowned) all the offerings to Drostan.' In the midlands of Scotland it occurs as the place-name Ibert. There are three instances in Perthshire. 'The church lands or glebe called *the Ibert* of the parish church of Monzie' near Crieff are on record in 1642 (RMS), and the name is still extant. In the same district 'the glebe and church lands of the vicarage of Monywaird called *lie* Yburd, between the Water of Turret, the burn *lie* Kelak, and the lands of the lord of Monywaird' appear in 1572 (RMS), later Ibert. The parish of Monzievaird adjoins Monzie on the west. On the left side of Tilt, a little below Bridge of Tilt, Blair Athol, there is a low standing stone called *Clach na h-Iobairt*, 'the stone of the offering.' This perhaps commemorates an offering of land here to the church of Kilmaveonaig, which stands on the same side of Tilt some little way further up.

There is one instance in Dumbartonshire. 'Three and a half acres of land lying near the church of Kilmernok, between Bordland of Kilmernok and Kater, called *lie Iberd*, in the county of Dumbarton' appears in 1591 (RMS). The name is not on the map, but the place was near the left bank of Endrick.

In Stirling there are three. Ibert, near the church of Killearn, is still extant; and so is Ibert near the church of Drymen. 'The church lands of Ybert in the parish of Drymmen' appear in 1600 (RMS), later Ibert. In the parish of Balfron the land of Ibert appears between the lands of Kilfossets and Bent in 1666 (Ret.); the name is obsolete, but probably commemorates a grant of land to the old church of Kilfosset.

The Latin *offerendum*, an offering, an oblation, is taken

over into Gaelic as Early Irish *oifrend*, now *aifreann*, *aifrionn*, the mass, Welsh *offeren*. Inchaffray near Crieff, in Perthshire, the site of an ancient abbey, and still earlier the seat of an establishment of the Celtic Church, is Incheffren *c.* 1190 (Chart. Inch.), Inchaffren, *c.* 1199, *ib.*, and *passim*; the name is in Latin correctly *Insula Missarum*, 'the isle of the masses.' It is a good instance of terminal *n* (here *nn*) being dropped though radical, *i.e.* in English, for the Gaelic form cannot be heard now. It is clear, however, that *oifrend* was used also in the sense of a gift to the Church —in parts of Scotland 'the offeral' still means the collection in church—and it appears in this sense in several place-names over much the same area as *ìobairt*.

In Perthshire the instances are as follows. '*Le Offeris* of Lanark,' 1507 (RMS), lay, it would seem, between Duncraggan and Lanrick at the north-west end of Loch Vennachar. The Old Stat. Account calls the place Offerans, and derives the name from *oir-roinn*, 'the side of the point,' wrongly, of course, for *oirrinn* is a form of *oifreann*, being the old dative used as nominative. At the west end of Loch Achray the meadow at the bridge on the road to Aberfoyle is called *an t-Oirrinn*, and the rock west of it (part of the Trossachs) is *Creag an Oirrinn*: this is probably 'Offroune in Strogartnay,' 1506 (RMS). 'Offeris of Ochtertyre in Menteith,' 1535 (RMS) *alias* Chalmerstoun, 1582, *ib.*, which latter is the modern name, is in the parish of Kincardine. About a mile west of Chalmerston is Offers, near Easter Ross; 'Ros and Offeris in Menteith,' 1535 (RMS). The name of the river Orrin is of the same origin (p. 472).

In Stirlingshire there appears 'the tenandry of Nethir Dischoure called Offeris lying near the church of Kippane,' 1508 (RMS); this was probably the old glebe of Kippen; Dasher is about a mile from the present church. 'Offerandis of Caschlie,' 1545, *ib.*, is represented now by Offrance Moss and other names involving Offerance on the east side of Flanders Moss (*a' Mhòine Fhlanrasach*). On the west edge of Flanders Moss is Offerance, near the Peel of Gartfarran. 'The lands called *lie Offeris* of Schiregartane in the lordship of Menteith,' 1451 (RMS), are described as 'of Lekky,' 1548; Leckie is near the church of Gargunnock. 'Offeris in the barony of Herbertshire,' 1510 (RMS), was near Denny; the name is now obsolete. The Offerances and Offers of Menteith are to be compared with the names Arnclerich, Arnvicar, Arnprior, in the same district, meaning respectively 'portion' (*earrann*) of the churchman, of the vicar, of the prior, and all were doubtless connected with the Priory of Inchmahome in the Lake of Menteith, or with some still older religious establishment.

From early times the Irish saints were wont to retire to some solitary place for contemplation: an early instance is that of Cormac úa Liathain, who attempted more than once to find a desert (*eremum*) in the sea, and actually sailed to Orkney for that purpose.[1] Virgnous or Fergno, another of Columba's company, spent twelve years in a hermitage beside Muirbolc Mār, probably Loch Tarbert in Jura. For such a retreat Adamnan uses the Greek term *erēmos*, whence in English 'eremite, hermit,' but the term which came into common use was the Latin *desertum*, a desert, whence M.Ir. *dísert*, now *dìseart*. Many places in Ireland begin with *dísert*, which is usually followed by the name of a saint. In Scotland there are but few, and the saint's name is not as a rule preserved.

Our best-known instance is Dysart in Fife, traditionally connected with St. Serf, who, according to the Aberdeen Breviary, had his celebrated conversation with the Devil in a cave here.[2] In Forfarshire there is Dysart in the parish of Maryton, formerly a barony and once part of the possessions of Restennot. *An Dìseart* near Pitlochry appears in 1451 (RMS) in connection with Faskel, now Faskally, on Tummel; it is on the estate of Faskally and not far from the mansion-house. The village of Dalmally at the foot of Glen Orchy is, or rather wàs, called *Clachán an Dìseirt*, 'the Kirkton of Dysart,' and the parish of Glen Orchy was sometimes called Dysart.[3] The church contained the

[1] *Life of Columba*, ii. p. 43.

[2] The conversation is reported at length by Wyntoun, Book v.

[3] *Orig. Paroch.*, ii. pt. I. p. 134.

burial-place of the MacGregors. A poem preserved by the Dean of Lismore begins—

'Parrthas toraidh an Díseart.'

'A paradise of fruitfulness is the Hermitage.' The reference is to the precious seed which was committed to it from time to time. The poet styles it *Díseart Chonnáin* 'Connan's hermitage'; also *gardha Chonnáin*, 'Connan's garden.' The Old Stat. Account mentions the well of St. Connan, a quarter of a mile east of the inn of Dalmally, 'memorable for the lightness and salubrity of its water,' and the writer makes some interesting remarks about it.[1] The Aberdeen Breviary says that St. Mund was abbot of Kilmund and Dissert, but the locality of this Dìseart is not known.

The Latin *Roma*, Rome, was borrowed into Old Irish as *ruam*, in the sense of 'cemetery'; it occurs in Oengus' *Félire*, written about A.D. 800: 'is mór Brigit buadach/is cáin a ruam dálach,' 'great is victorious Brigit, fair is her multitudinous cemetery.' This use arose partly from the fact that clerics who visited Rome brought back with them soil from the holy cemeteries which they scattered over the cemetery of their home monastery. Thereafter burial there was burial in the soil of Rome, and the monastery benefited in fame and in burial dues.[2] In the Book of Llandaf we are informed that the isle of Enli, now Bardsey off the Lleyn peninsula, Carnarvon, was called of old 'the Rome of

[1] Vol. viii. p. 351.

[2] An Irish saint is made to say, 'I shall not depart from Rome until I perform thirty fasts, that I may obtain heaven for myself and for every one who shall be in my cemetery.' His companion answers, 'After that the soil of Peter's and of Paul's tombs and the soil of Gregory's grave shall be carried by us in loads to Ireland.' 'And they collected the soil of Peter's tomb and of the tomb of every other apostle and of every great saint that is in Rome, and took it with them to Ireland' (*Life of Colman mac Luachain*; Todd Lecture, xvii. p. 81). St. Lolan had four ass-loads of the soil of St. Peter's cemetery in Rome sent to Scotia for consecrating a cemetery in which his body should be buried. He prayed that whoever should be buried in the said cemetery, or who being ill vowed to be buried there, should receive from God as great indulgences as if he had been buried in St. Peter's cemetery, and finally attain to the kingdom of heaven.—*Aberd. Brev., temp. aest.*, p. cxiii.

Britain,' because of its remoteness and the dangers of the passage, and because of the holiness of the place; 'for there lie the bodies of twenty thousand holy confessors and martyrs.' Similarly Oengus says, 'is Ruam iarthair betha/ Glenn dálach dá Locha,' 'the Rome of the western world is multitudinous Glendalough.' A Scottish Gaelic poem of about A.D. 1600 has ''san ruaimh so sios,' 'in this cemetery below.'

Pennant notes Rome on the east side of Tay above Perth. The farm steading of Rome stood near the left bank of Tay about two miles above Perth bridge, and about a quarter of a mile below the mouth of Almond on the opposite side. The farm was converted into a park in the beginning of last century, and the buildings were demolished. The only place that still bears the name is a net-fishing bothy about a mile and a quarter above Perth bridge.[1] Rome stood in the Abbey lands of Scone, and about one hundred yards due east from the south-east corner of the present palace there was an old burying-ground.[2] This Rome was doubtless in use long before the foundation of the Abbey in 1114.

There appears to have been another Rome at Madderty, on record as 'a piece of land called the Rome,' 1662 (Ret.); 'lie Ron,' 1581 (RMS). It seems to have been close to the present manse.

In the parish of Menmuir, Forfarshire, there is Rome on the Paphrie Burn, on record as Rome, 1517 (RMS), Rowme, Ireland, and Corsbank; 1533, *ib.*, etc. I have not seen the place, and my information is that there are no traces of a cemetery.

Another term for a cemetery is *reilig*, *roileag*, from Latin *reliquiae*, 'remains.' Though still current in common speech, it occurs very rarely in place-names. *Reilig Odhráin*, 'Oran's cemetery,' is in Iona. Ruilick, the Relict 1571 (RMS), is near Beauly. There is also Relic Hill, Relichillis 1624 (RMS), Relicthills 1654, *ib.*, in Kirkmahoe parish, Dumfries.

---

[1] For this information I am indebted to the Earl of Mansfield.

[2] *Ordnance Gazeteer.*

The privilege of sanctuary once enjoyed by certain churches within a definite radius is commemorated by some names. One of the terms for a girth is *comraich*, and the Comraich of Applecross and of Tain have been mentioned. Besides these there are two in North Uist: *Comraich na h-Eaglais*, 'the church girth,' was connected with the church of Kilmuir, dedicated to the Virgin Mary. *Comraich na Trianáid*, 'Trinity girth,' was at *Teampull na Trianáid*, at Carinish, founded by Beathóg, daughter of Somerled of the Isles. The ford near it, between North Uist and Benbecula, is *Faodhail na Comraich*, 'the girth ford,' and a channel therein is *Sruth na Comraich*, 'the girth stream.' Under the Gaelic system the dwellings of bards possessed the right of sanctuary, and in South Uist we have *Comraich nam Bard* at Staoligearry, which was the official residence of the MacMhuirich family, hereditary bards to Clan Ranald.

Another term of the same meaning is *tearmann*, from late Latin *termo*, *termon-is*, a limit. Professor Mackinnon, who was a native of Colonsay, wrote: 'Right in the middle of the strand that separates Colonsay and Oronsay, and covered by the sea for twelve hours of the twenty-four, is *Clach an Tearmainn*, marking the limit to which the sanctuary rights of Oronsay Priory reached. The base of the structure, strongly built with stone and lime, is still entire, but the cross has disappeared.' The local pronunciation of *tearmainn* (gen.) is *tearmaid*, and that this pronunciation was used elsewhere appears from the name Termit in Petty, near Inverness. Termit once belonged to the minister of the parish, and I am informed that the mansion of Holme-Rose now stands on it. Tillytarmont near Rothiemay Station is Tilentermend 1534 (RMS), for *Tulach an Tearmaind*, 'girth hill'; it is over a mile from Rothiemay church.

Between the parish church of Dull and Tirinie in Perthshire is the farm of *Teagarmachd*, Tigirmach 1500 (RMS), Tiggermach 1603, *ib.* This term seems to be based on *teagar*, 'shelter, protection,' with *m* and *acht* suffixes, and probably means 'shelter-place, sanctuary,' with reference to the girth of Dull. Opposite it there used to be a ford on

Tay called *Ath na Teagarmachd*, and a small island near it is *Eilean na Teagarmachd*.

In Iona the place where the dead were landed before burial was called *an Ealadh*, representing O.Ir. *ailad, elad*, a tomb. One of the boundaries of the girth of Lismore was *Clach na h-Ealadh*. The dative case of *ealadh* is *ealaidh*, and this may be the explanation of the name Elie in Fife, called of old 'the Elie,' inland from which is the church of Kilconquhar, 'Conchobar's church.' Eliebank in Peeblesshire and Ellie or Elie near Inchberry on Spey, below Rothes, may perhaps be compared. A later form is *uladh*, and from the idea that graves contained treasure—as sometimes they did—*ulaidh* (dative used as nominative) means 'treasure, hoard' in Scottish Gaelic. *Druim Ulaidh*, Drumullie, is near Boat of Garten, with tradition of a treasure in a neighbouring loch; *Clach na h-Ulaidh*, supposed to mean 'stone of the treasure,' is on *an Linne Dhubh*, the inner part of Loch Linnhe. Craigenholly, presumably for *creag na h-ulaidh*, is near Glenluce Abbey in Wigtownshire.

Suidhe, a seat, is often applied to the place where a saint was supposed to sit in contemplation, and some of these names have the name of the saint still attached to them. Heroes also had seats designated after them, but the only instances which are extant are connected with Fionn and Arthur, and I shall include these among the other 'seats.' Another use is to denote a resting place for funerals. In some cases the *d* (*dh*) of *suidhe*, O.Ir. *suide*, remains or has been restored, as in Sydserf, Suddie, and some of the places called Side may represent *suidhe*.

In Stirling, Seabeg and Seamore, 'little seat' and 'big seat,' are near Bonnybridge, on the line of the Roman Wall, spelled Sabeg, 1450 (RMS), Seybegis, 1506, *ib*. At Seabeg there may have been a Roman fort;[1] I find no proof that *suidhe* was applied to the site of a fort as such, but it may have been, for Suie is the name of a stone fort in Kirkcudbrightshire.[2]

---

[1] Dr. George Macdonald, *The Roman Wall*, p. 219.

[2] Christison, *Forts*, p. 272.

In Perthshire, Suie in Glen Dochart is locally connected with St. Fillan though his name is not attached to it. *Tom an t-Suidhe*, 'the knoll of the seat,' is in Glenshee, and The Seat is the name of a hill south of Carlownie Hill. Seamab is a hill of 1442 feet in Muckhart parish. Severie, east of Callander, is in Gaelic *Suidhe a' Bhritheimh* (pronounced *Suidhe Bhrith'*), 'the judge's seat.' The Seat is a large stone on a knoll in front of the farm-house, and unfortunately it appears to have been injured. The tradition is that the people were judged at this stone, and—when found expedient—hanged at Kilbride, near Dunblane.[1] 'The Judge's Cairn' is some distance to the south-east. This interesting name is one of the very few which commemorate the Judges of old; another is *Eilean a' Bhritheimh*, 'the judge's isle,' off the north coast of Sutherland. There was a possession called Seat, apparently in the lordship of Dunkeld, and 'devoria de Seat,' 'the Dewarship of Seat,' is on record 1641 (RMS). This connection with the Dewar, or custodian of a sacred relic, suggests that Seat is here a translation of *Suidhe*.

Seabeg, 'little seat,' is in Dunnottar parish, Kincardineshire.

In Aberdeenshire, Suie Hill, with Cairn and Wood of the same, are near Kirkton of Clatt; the dedication is to St. Moloch, who is supposed to be Mo-Luag of Lismore. Suie Burn in Tullynessle parish comes from this Suie Hill. 'The lands of Overknockespack (*i.e.* Over Bishophill) . . . called Suyfoord' appear in 1705 (AB, iv. 500). Cividly in Keig parish may have *suidhe* as its first part.

In Banffshire, Suie in Glen Livet is for *Suidhe Artuir* 'Arthur's Seat.'[2] *Tom an t-Suidhe Mhóir*, 'the knoll of the great seat' is in Abernethy parish in Morayshire.

In Inverness-shire, *Suidhe Churadáin*, 'Curitan's seat,' is near the eastern end of Loch Ness, connected with Curitan of Rosemarkie. *Suidhe Ghuirmein*, 'Gorman's seat,' and

---

[1] For this information, and for a photograph of the Seat, I am indebted to Miss Margaret Dewar, of the old family of Dewars of Severie and Annat.

[2] See p. 208.

Suidhe Mheircheird, 'Merchard's seat,' are in Glen Urquhart and Glen Moriston respectively. *Suidhe Chuimein*, 'Cummine's seat,' above *Cill Chuimein*, Fort Augustus, commemorates the abbot of Iona who died in A.D. 669. In Badenoch there is a hill called *an Suidhe*.

In Ross-shire, Suddy in the Black Isle is *Suidhe* simply, without the article. *Suidhe Ma-Ruibhe*, 'Maelrubha's Seat,' is said to be marked by a low stone pillar in a field at Bad a' Mhanaich 'the monk's spot,' at the west end of Loch a' Chroisg (L. Rosque). *Carn an t-Suidhe*, 'cairn of the seat,' is traditionally said to mark one of the spots where the men of Applecross rested, between Kinlochewe and Applecross, when they were carrying home the saint's body.[1]

Suidheachán Fhinn, 'Fionn's seat,' is in Loch Broom parish.

In Sutherland, *Cathair Dhonnáin*, 'Donnan's chair,' is the name of 'two or three huge blocks of stone, in the form of a chair or seat,' in Kildonan parish.[2]

In Argyll, on the mainland, *Suidhe Chreunáin*, 'Creunan's seat,' is at Kilchrenan on Loch Awe. In Glen Salach, near Ardchattan, a large block of stone was called *Suidhe Bhaodáin*, 'Baodan's seat,' but it was sacrilegiously smashed many years ago. Baodan was the saint of Ardchattan.[3] Behind Kilchrenan, the hill on which stands the circular fort is called *Suidhe Cheanathaidh*; there is also *Achadh Cheanathaidh*; I have not met this personal name elsewhere. *An Suidhe* is the name of a hill between Loch Awe and Loch Fyne, on the border of Inverary parish. *Beinn Suidhe*, 'peak of the seat,' is west of Loch Toilbhe (Tulla) and south of Loch Dochart—a hill of 2215 feet. Bacoch's Seat is in Glen Lussa, in North Knapdale; I have not heard

[1] The tradition in Applecross, still extant, is that the saint died at Kinlochewe. The people of Kinlochewe could not carry his body, and four little red-haired men from Applecross had to go and fetch it ('ceithir bodaich bheaga ruadha as a' Chomraich'). I owe this information to Rev. D. Mackinnon, a native of Applecross.

[2] Rev. D. Sage, *Memorabilia Domestica*, p. 73, where the site is described.

[3] See p. 122.

its Gaelic form. *Bealach an t-Suidhe*, 'pass of the seat,' is in Killean parish, Kintyre.

In the Isles, *Aite Suidhe Fhinn*, 'place of Fionn's seat,' is in Portree parish, Skye, overlooking the town on the south-west. *Suidhe Chaluim Chille*, 'St. Columba's seat,' is in *Gleann an t-Suidhe* in Kilmorie parish, Arran; the seat is a cairn. In the same parish are 'two concentric circles, respectively of 12 and 8 stones,' named *Suidhe Choire Fhinn*, 'seat of Fionn's cauldron.'[1] *Suidhe Chatáin*, 'St. Cattan's seat,' is the highest point of Kingarth parish, Bute. *Suidhe Mo-Luáig*, 'Mo-Luag's seat,' is in Lismore. In Mull is *an Suidhe*, Swy 1623 (RMS).

The Lives of the Saints have many instances of a saint bestowing his blessing on a place, and in particular on the spot which was granted him for his church. 'The place where a saint blessed' came to mean 'the place where he had his church,' as for instance 'Abán isé do bhennaigh i cCúil Collaing,' 'Abán, it was he who blessed in Cúil Collaing,' meaning, as Reeves explains, 'who in founding a church gave his blessing to the place.'[2] The place itself was often called 'the blessing,' Mod.G. *beannachadh* or *beannachd*; M.Ir. *bendachtu*, dat. *bendachtain*.

An ancient religious site at Gallon Head in Lewis is known as *am Beannachadh*. *Beannachd Aonghuis*, 'Angus' blessing,' is at the church of Balquhidder, of which Angus is patron. A very primitive chapel on Sùla Sgeir is called *an Tigh Beannaichte*, 'the blessed house.' Bendochy, a parish of East Perthshire, is Bendachtin *c.* 1130 (Johnston); Bendactehin, *c.* 1150 (Reg. Dunf.); Bendaghtyn, 1275 (Theiner); Benachty, 1462, etc. (Reg. Cup.); Bennethy, 1567, etc. (RMS); here the forms in *-in*, *-yn* are dative, the others represent the M.Ir. nominative. Bannety in Strathmiglo is Bannachty, 1595 (RMS). Bannauchtane in Lennox, 1549 (RMS), appears to have been near Edinbarnet, where Blaeu has Bolnachta(n); compare 'Eddenbarnen and Craigbanyoch,' 1591 (RMS). 'Bennochquhy in the *schira* of

[1] *Orig. Paroch.*, ii. pt. I. p. 256.
[2] *Mart. Don.*, p. 282.

Kirkcaldy' appears in 1585 (RMS). In 1635 (Ret.) we have 'the kirklands of Bennachtie' in Kincardineshire, in 1641 spelled Bennethie. In Cunningham there is Bannacht, 1614 (RMS). Lattirbannachy, 1553, Lattirbonachtie, 1588 (RMS), was in Strathearn, meaning apparently 'slope of the blessing.'

Reference to p. 202 will show that the personal name Bannatyne is from a place called Bennachtain (dative). By 'the forest of Bannantyne,' 1664 (RMS), in Aberdeenshire, appears to be meant Bennachie.

In Scotland the custodian of a sacred relic was often called *deòradh*,[1] meaning primarily an exile, outlaw, stranger; it is used to explain Latin *aduena*, a newcomer; *deòradh Dé*, 'God's exile,' meant a pilgrim. The relic of a saint was enclosed in a shrine or covering. Oaths were taken on the relic, and for this purpose it might be conveyed to places at a distance in the care of its custodian, and this practice may well have caused the development of meaning from 'stranger' to 'custodian of a relic.' There is evidence, however, that the relic itself was called *deòradh*: in 1497 (RMS) we find mention of 'the staff of St. Munn called in Gaelic (*Scotice*) Deowray.' With this is to be compared the staff of St. Fillan called 'the Coigerach,' spelled also Quigrich, Quegrich, Quickreith, a name which has caused some curious speculation as to its meaning, but which is simply G. *coigreach*, Ir. *coicrigheach*, one that comes from a neighbouring province or district; a stranger, foreigner. In this latter case the application is plain: the chief function of the custodian of St. Fillan's staff was to follow after stolen goods or cattle wherever they were to be found within the kingdom of Scotland. The staff, whether it accompanied him or not —probably it did—was the effective agent in the business, conferring authority and immunity from harm, and from its travels into foreign parts it was called *an Coigreach*. It was no doubt for a similar reason that St. Munn's staff was called *an Deòradh*. It may be that the term *deòradh* was in all

[1] The term in Ireland was *maer*, *e.g.* 'maer na bachla cochlaige,' 'the keeper of St. Colman Eala's hooded staff' (*Life of Colmán mac Luacháin*, p. 96).

cases originally applied to the relic and that its application to the custodian is an instance of transference; the relic and its keeper went together, and when, for instance, it was said 'thàinig an deòradh,' 'the *deòradh* has come,' confusion between the relic and its bearer would arise readily.

The custodian of a relic held land in respect of his office. *Deòradh a' Choigrich*, the Dewar or keeper of St. Fillan's staff, held the lands of *Iuaich*, Ewich, in Glen Dochart, as appears from a letter of confirmation granted by Alexander Menzies to Donald McSobrell, 'dewar Cogerach,' in 1336.[1] In 1632, etc. (RMS), there are on record 'lie Dewariscroft in lie Suy called Dewar-Vernans-croft, a croft in Auchlyne called Dewar-Nafargs-croft, and a croft in Killin called Dewar-Namais.' Suy is *Suidhe*, 'the seat,' in Glen Dochart, with an ancient burial-ground and remains of a chapel. 'Dewar-Vernan' is for *Deòradh a' Bhearnáin*, 'the Dewar of the little gapped-one,' obviously St. Fillan's bell, for many bells were called Bearnán, *e.g. Bearnán Brighde*, St. Brigit's bell. The old chapel at Auchlyne in Glen Dochart is *Caibeal na Fairg(e)* in Gaelic now;[2] 'Fairg' is probably the genitive sg. of M.Ir. *arc*, or *arg*, gen. *airce*, a shrine, coffer, with *f* prefixed. The Killin croft, 'Dewar-Namais,' appears in 1640 and 1670 (Ret.) as Dewarnamanscroft, Dewarnamaynescroft. It is not known now, but it may be the same as Crettindewar in Acharn near Killin, 1649 (RMS).

A half-merkland called Pordewry in the territory of Inverchapel, in Cowal, was occupied by a 'procurator' who held also the staff of St. Munn, 1497 (RMS). In 1572 Donald Dewar received in feu the forty-penny lands of Garrindewar in Menteith, formerly assigned for the ringing of a bell before the dead persons in the parish of Kilmahug near Callander.[1] This appears to be the same as Carnedewar, near Callander, 1642 (RMS). In 1572 (RMS) we find the kirklands of Strowan (now joined with Monzievaird), with the lands called the Dewarisland, occupied by Thomas Dewar, and the croft called Ballindewar. It is stated that the service

[1] Dr. Joseph Anderson, *Highland Monthly*, vol. ii.

[2] This was given me in 1921 by Mr. Duncan Christie, farmer, Auchlyne, who was then aged ninety-five.

required of the Dewars by the charter on which they held their lands was the ringing of the bell of St. Rowan, which was not the church bell but a fine hand bell.[1] Ballindeoir in Muckairn, Argyll, is close to the site of the old church of Kilvarie; the relic here was called the Arwachyll, *i.e.* *arbhachall*, 'great bachall,' 'great staff,' apparently supposed to be the crozier of St. Maol-Rubha of Applecross. In Kilfinan parish, Argyll, Aiker-in-Deor was a half merk land in Ballimore, 1599 (Ret.). Glenjorrie, for *Gleann (an) Deòraidh*, is east of Glenluce Abbey in Wigtownshire; about two miles southwards is Kilfillan. In Midlothian there are Dewar in Heriot parish and Dewarton in Borthwick parish.

Bachall, a crozier, from Latin *baculum, baculus*, a staff, occurs in several names, denoting either that the lands so called were held in respect of the custody of a saint's staff, or possibly that they were 'crozier-land,' belonging to a monastery or other religious establishment. *Bachall* is now masc., gen. *bachaill*; formerly fem., gen. *bachlae, bachla*. In 1544 (Orig. Paroch.) Archibald Campbell, son of the Earl of Argyll, in honour of the Blessed Virgin and of his patron saint Mo-Luag, mortified to his standard-bearer John Mac Maol-Muire mhic Iomhair and his heirs male, half of the lands of Peynabachalla (pennyland of the bachall) and Peynchallen, extending to half a merkland, in the island of Lismore, with the keeping of the great staff (Bachall Mór) of St. Mo-Luag, as freely as John's father, grandfather, great-grandfather, and other predecessors held of Archibald's predecessors. The land so held is now called Bachall, and the head of the family of Livingstone which holds the land is *Baran na Bachaill*, 'the Baron of the Staff.' About 1128 (Chart. Dunf.) King David confirmed to the church of Dunfermline certain grants made by his father and mother, including Petbachlin, now Pitbachlie for *peit bachla*, 'crozier's share,' near Dunfermline, and Inveresc Minor. The land at Inveresk, on the south side of the Firth of Forth, may be Balbaghloch, 1336, now Barbauchlaw. Another Barbauchlaw (pronounced now -lie) is in Linlithgowshire. Near Methven

[1] Dr. Joseph Anderson, *Highland Monthly*, vol. ii.

in Perthshire are Drumbauchlie and Bachilton, also Culdeesland, all probably belonging to an old religious establishment. Bauchland, for Bachall-land, is near the parish church of St. Martins, Perthshire. In Lochrutton parish, Dumfriesshire, Barnbauchle may be for *barr na bachlae*, 'height of the crozier.' The old name of Wardlaw or Kirkhill parish in Inverness-shire was Dulbachlach, Dulbatelach, Dulbathlach, *c.* 1203 (RM).

One of the officials of the early monasteries was the *fer-léighinn* (*vir legendi*), 'man of learning,' or lector, whose duties were connected with education. In the Book of Deer, Domongart, *fer-léighinn* of Turriff, witnesses a grant to Deer in the eighth year of David I. In 1164 (AU) mention is made of Dubhsídhe, 'black of peace,' *fer-léighinn* of Iona. About 1211 (RPSA) a certain Master Lawrence was Archdeacon and Ferleyn, or master of the schools, of St. Andrews, and as the outcome of a dispute the prior and canons became bound to pay to 'the foresaid Lawrence the *fer-léighinn*) (Ferlanus) and his successors, at the house of the *fer-léighinn* (Ferlanus), for the use of the poor scholars' certain dues. In 1316 'magister Felanus' (for Ferlanus) was 'rector scholarum' of Inverness. In 1157 and again in 1172/99 (Reg. Ep. Aberd.), Pethferlen, Petenderleyn, for early G. *peit ind fhir-léighinn*, 'the lector's share,' appears as part of the possessions of St. Peter's Church, Aberdeen.

Pittentagart in Aberdeenshire is 'the priest's share.' There are also Pithogarty near Tain, and Pittagarty, 1598 (RMS), in Moray, with the same meaning, though the *-y* is difficult to explain. Pettinclcroch, 1489 (RMS), 'the share of the clerics,' was in the earldom of Strathearn. Names like *Dail nan Cléireach* and *Achadh nan Cléireach*, in Ross-shire, are fairly common. *Baile a' Mhanaich*, Balvannich, 'monk's stead,' and *Baile nan Cailleach*, Nuntown, are in Benbecula. Pittenhaglis, 1511 (RMS), in Fife, is for *peit na h-eaglaise* 'the church's share'; compare Pethannot and Pinhannet.

Pitliver in Fife, Petlyver 1450 (RMS), represents *peit libhair*, 'share of the book,' *i.e.* either belonging to the Church or held by the custodian of some special copy of the Gospels. Here *libhair* is the old genitive of *lebar*, *lebhar*, a book. The later form is *leabhair*, often pronounced *leobhair*, and this is probably seen in Pitlour in Kinross-shire, formerly Petenlouir (RPSA), Petinlouer (Reg. Arbr.), but here *peit an lobhair*, 'the leper's share,' is possible, *i.e.* land set apart for the maintenance of lepers.

Three names of considerable interest come from *soisgeal*, gospel. In Stirlingshire there is Pettintoscale 1450 (RMS),[1] Pettyntoskale 1451, *ib.*, Pettyntoskane 1504, 1546, *ib.*, Pantaskin 1617 (Johnston), Pentaskin 1745, *ib.*, now Bantaskin. In Forfar there is Pettintoscall 1410 (Reg. Brech.), Pettintoskell 1472 (RMS), Pettintoskane 1472, *ib.*, Pettintoskell 1533, 1541, *ib.*, Pentoskell 1606, *ib.*, Pinto-scall 1634, *ib.*, Pintescall 1686 (Ret). This name is still extant in Pentaskill Burn, which runs south of Forebank House, within Kinnaird Castle deer park. In Fife there is Pettuscal in the parish of Lathrisk, 1590 (RMS), 'the church lands of the vicarage of Lathrisk called Bantuscall,' 1594, *ib.*, Bantiscall, 1652, *ib.* The lands of the vicarage of Lathrisk may have been near the Mansion House of Lathrisk, now in Kettle parish, but the old name seems gone. These all represent *peit an t-soisgeil*, 'the gospel's share.' The lands may have been so called simply as belonging to the Church, having been gifted as a *screabal soiscéla*, 'gospel penny,' like the places called Ibert and Offerance. On the other hand Joyce refers to the belief that Kilteskill in Galway, *i.e.* *Cill an t-Soisgeil*, was so styled from an ancient copy of the Four Gospels preserved there from primitive ages. This origin, however, while readily intelligible in the case of a church, is less likely in such names as those under discussion. *Linne an t-Soisgeil*, 'the gospel pool,' is in the river Ruel near Ormidale.

O. Ir. *sén*, 'a blessing, a charm,' is from Lat. *signum*, 'a sign,' especially 'the sign of the cross'; hence comes the verb *sénaim*, 'I consecrate, hallow,' with passive participle *sénta*, now *séanta*, 'consecrated, hallowed, charmed,' often in Sc. G. *sianta*. Examples are *a' Bheinn Sheunta*, 'the sacred peak,' in Jura and in Ardnamurchan, *an Loch Seunta*,

[1] See errata at end of vol. ii. of RMS.

Holy Loch in Cowal; *na h-Eileanan Sianta*, the Shiant Isles, 'the hallowed isles'; *an Uaimh Shianta*, 'the hallowed or sacred cave,' in Applecross. Martin gives a most interesting account of Loch Siant Well in Skye, the water of which was believed to cure diseases.

Clachnaharry near Inverness is in Gaelic *Clach na h-Aithrighe*, 'stone of repentance.'[1] At *Cill Chatrìona*, 'St. Catherine's church,' in Colonsay, there is *Clach a' Pheanais*, the 'penance stone,' where the prescribed penance was performed after confession. There must have been many such stones. *Bealach an t-Slèachd*, 'the pass of prostration or genuflexion,' is at Kilmore in Sleat. *Cnoc na Paidir*, 'the hill of the Paternoster,' is at Totescore in Trotternish, where one gets the first sight of Kilbride Church. *Achadh na h-Aifrinne*, 'field of the mass,' is in Strath, Skye.

Petblane, 1643 (RMS), in Daviot parish, Aberdeenshire, seems to be 'St. Blaan's share'; compare Kilblain or Kilblean east of Old Meldrum, in the same district. Saints' names are often found with such terms as *croit*, a croft, *dol*, *dul*, now *dail*, a meadow, more rarely with *baile*, a stead, once only with *dabhach*, a davoch of land.

---

[1] So explained correctly by Mr. James Fraser, minister of Wardlaw, but usually taken to be for *Clach na h-Aire*, 'stone of the watch.' *Wardlaw MS.*, p. 87.

CHAPTER X

Bondacht for Colum Cille
co nnóebaib Alban alla,
for anmain Adamnán áin
rolá cáin forsna clanna.[1]

Colman's Hymn.

A.—SAINTS OF WEST AND EAST

ADAMNAN, Sept. 23, was ninth abbot of Hí, and died in 704. The oldest spelling of his name is Adomnán, which is not a diminutive of Adam, but, as Kuno Meyer has pointed out, a diminutive of the name Adomnae, 'great terror,' found as a common noun *adamnae* in Colman's *Hymn; adomnae*, AU 825. In the *Life* of St. Serf it is Edheunanus, Odau'ðdanus, and in place-names it is regularly Eódhnán.

On the west there are Killeonan, near Campbeltown in Kintyre in 1481 Killewnane (Orig. Paroch.); a cell and sanctuary of St. Adamnan in the isle of Sanda off the Mull of Kintyre; *Croit Eódhnain*, Adamnan's croft, in Glenfalloch; Rowardennan on Loch Lomond probably means 'point of Adamnan's cape.' *Crois Adhamhnáin* is in North Uist near Dùn Rosail; it is also called *Clach na h-Ulaidh*, 'stone of the grave,' or 'of the praying-station.' In the east Adamnan is the patron of Dull; *Muileann Eódhnain*, 'Adamnan's mill,' and *Magh Eódhnain*, his plain, are near Bridge of Balgie in Glen Lyon; *Ard-Eódhnaig*, Ardeonaig, in 1494 Ardewnan (RMS), is on the south side of Loch Tay; *Fuaran Eódhrain* (dialectic for Eódhnain) is near Grantully, in Strath Tay; the church of Forglen in Banffshire was formerly called Teunan Kirk, *i.e.* Saunct

[1] 'Blessing on Colum Cille, with the saints of Alba on the other side; (blessing) on the soul of glorious Adamnan, who imposed a law on the clans.'—*Thes. Pal.*, ii. p. 306.

Eunan's Kirk (Coll. A. and B., 508); in Aboyne there were St. Eunan's well and tree (*ib.*, 633); St. Adamnan's chapel was at Furvie, Aberdeenshire (Old and New Stat. Acc.); in Tannadice parish, Forfar, was Sanct Eunendi's Seit, now St. Arnold's Seat (Skene), but in 1744 the minister of Tannadyce calls it St. Ernan's Seat (Macfarlane); in Campsie there was St. Adamnan's Acre, 1567 (RMS); in Badenoch the knoll on which the church of Insh stands is called *Tom Eódhnain*;[1] the Croft of St. Adampnan in Glen Urquhart appears in 1556.[2] There were a chapel and altar of Adamnan at Dalmeny, and according to Bower Adamnan was abbot in Inchkeith in the Firth of Forth—which of course is wrong; there may be confusion here with Adamnan of Coldingham.

How Adamnan stayed a great plague is still told in Glen Lyon, and the spot where he planted his crozier, thus setting a limit to the plague, is pointed out at Craigianie.[3] Near this spot, on the south side of the public road, is a small upright slab with a cross on either side, one of them very small; the knoll on which the cross stands is *Tom a' Mhóid*, 'the moot-knoll.'

Áibind, daughter of Mane, of Cluain Draignech, was one of the holy maidens subject to Brigit (LL 353 b; Lismore Lives, p. 336). Her name, which means 'delightful,' Lat. *amoena*, would be now *Éibhinn* or *Aoibhinn*, and she may be commemorated in Killevin on Loch Fyne. The old parish church of Calder in Nairnshire was at Barevan, for *Barr Éibhinn*, now contracted into *B'réibhinn* in Gaelic, as in 'the church of Breven,' 1665 (RMS). In 1275 (Theiner) it is 'ecclesia de Ewen.' The church of Inch in Badenoch appears to have been hers.[4]

---

[1] The old bell of Insh, says the local legend, was stolen once on a time and taken south of the Grampians, but it returned of its own accord, ringing out as it crossed Druim-uachdair 'Tom Eódhnain, Tom Eódhnain!'

[2] Dr. William Mackay, *Urquhart and Glenmoriston*, p. 335.

[3] For the miraculous powers of a saint's bachall see C. Plummer, *Vitae Sanctorum Hiberniae*, p. clxxv. Mochua transferred the Yellow Plague to his bachall (*Lismore Lives*, p. 287).

[4] Shaw, *Hist. of Moray*; he writes the name 'Ewan,' as he does also in the case of Barevan.

Angus was the saint of Balquhidder. *Clach Aonghuis*, 'Angus' stone,' is a slab on which marriages and baptisms were performed. The site of a chapel near the church was called *Oirrinn Aonghuis*, 'Angus' offering,' *i.e.* 'the offering made to Angus,' and a spot where he was believed to have preached is *Beannachd Aonghuis*, also styled *Beannachd Aonghuis 'san Oirrinn*, 'Angus' blessing in the Offering.'[1] Gorman and Mart. Don. commemorate at Aug. 11 a saint called Mac Cridhe, otherwise Mac Cridhe Mochta Lughmhaigh, 'the dear one (lit. "heart's son") of Mochta of Louth'; O'Donnell calls him MacRith, Mochta's attendant. Gorman and Mart. Don. add 'Aonghus was his first name.'[2] As the *Féill Aonghuis*, 'Angus' fair,' in Balquhidder was held on the first Wednesday after the second Tuesday of August, *i.e.* on or about Aug. 11, there is reason to identify the saint with this disciple of Mochta; Mochta himself died in 535 (AU). His name seems to occur in Fetter Angus near Old Deer, where there was a chapel which once served a parish of that name. In Knapdale there is *Cill Aonghuis*, Killanaish.

Barre, Barra, or Bairre of Cork, Sept. 25, died about 610. His name is short for Findbarr, 'white-crown.' *Eilean Da-Bharr*, 'insula de Sanctbarr,' 1507 (RMS), 'isle of thy-Barr,' is at Campbeltown. At *Cill Bharr*, Kilbarr in Barra, there was an image of the saint in Martin's time (*fl.* 1700); here his anniversary was observed on Sept. 27. The proceedings were conducted on horseback, and concluded by three turns round St. Barr's church; the 'three races of a gathering' (*trí graifne oenaich*) may be said to be common form.[3] Chapel Barr at Mid-Geanies in

---

[1] *Clach* (or *Leac*) *Aonghuis* used to be recumbent in front of the pulpit of the old church; it is now set upright inside the present church. It bears the figure of an ecclesiastic in his robes, holding a chalice. The Balquhidder market was held at *Tom Aonghuis*, 'Angus' knoll,' almost opposite the gate of the present manse.—(Note by Rev. David Cameron, minister of Balquhidder).

[2] Another saint called Mac Creiche was known also as Mac Croide Ailbe or Mac Ochta Ailbe, 'Ailbe's dear one' (Plummer, *Misc. Hag. Hib.*, p. 8.)

[3] Angels ran three races for Colmán mac Luacháin (Todd Lecture, xvii. p. 86).

Ross-shire is said to have been built by Thomas MacCulloch, abbot of Fearn (1486/1516—Orig. Paroch.). Barr was the patron of Dornoch in Sutherland, where was St. Barr's Croft, and where his fair was held on Sept. 25.[1]

A saint named Da-Bhì, 'thy-Bì,' is commemorated in *Cill Da-Bhì*, Kildavie in Mull, Kildawy 1630 (RMS), and in Kintyre, Kildavie 1545, *ib.* Another Cill Da-Bhì is in Flodigarry, Skye; *Cladh Da-Bhì*, 'thy-Bì's graveyard,' is at Morenish on Loch Tay, and *Cill Da-Bhì*, with a cemetery, was near Stix (*na Stuiceannan*) at Taymouth. The name is a variant of Mo-Bhì or of Mo-Bhiu. Two saints of the former name are among the twelve apostles of Ireland, Mo-Bhì mac Natfraich and Mo-Bhì mac Beoain, styled *Clárenech*, 'flat-faced,' otherwise called Berchan, Oct. 12 (Oengus, Gorman, etc.), who died in A.D. 545 (AU). Mo-Bhiu was abbot of Inis Cuscraid. It is just possible that in the Scottish commemorations Da-Bhì or Do-Bhì is a short form of Berach, Feb. 15, abbot of Cluain Coirpthi (p. 301).

Bláán, Aug. 10, is commemorated in Kilblane in the Southend of Kintyre, in Kilblain near Kingarth in Bute (Blaeu), Kilblaan near Inveraray, and in Camas Bhlathain on Loch Shiel in Moidart, near St. Finan's church. In the east his chief monastery (*prím chathair*) was Dunblane; *Caibeal Bhlathain*, St. Blaan's chapel, is at Lochearnhead; Kilblain is near Old Meldrum in Aberdeenshire. It is unlikely that the saint's name is found in Strathblane, which is spelled Strachblachan, etc., in the thirteenth century.

Cill nam Bràithrean, 'the brethren's church,' was the old name of the church of Lochgoilhead (Old Stat. Acc.). *Cill nam Bràthair*, with the same meaning, Kilnabraar 1548 (RMS), is one of the seven churches of Strathbrora. The brethren may be Peter and Andrew, or James and John. The other churches in Strathbrora are *Cill Pheadair Mhór*, *Cill Pheadair Bheag*, *Cill Chaluim Chille* (St. Columba's),

[1] Macfarlane, *Geog. Coll.*, ii. p. 104. St. Barr's genealogy is given in Rawl. B 502, 90 g; LL 352 f; BB 223 c, 231 a; LBr. 20 a.

Cill Eathain (St. John's), *Cill Mhearáin* (St. Mirren's), and *Cill Ach-Breanaidh.*

Of the seventeen saints named Brénaind, Brénainn, the most famous are Brénaind Mocu Alti, son of Findlug, of Clonfert, May 16, who died in 577 (AU); and Brénaind, son of Neman, of Birr, Nov. 29 who died in 573 (Tighernach) or in 565 (Tighernach, AU). The former is the 'Voyager,' who founded the monastery of Ailech in *na h-Eileacha Naomha* (p. 81). There are besides on the west *Cill Bhrianainn,* Kilbrandon in Lorn, in Mull, and in Islay. In the east he was the patron of Boyndie in Banffshire, where his feast was kept on May 16, and his fair in Brannan Howe on May 26, new style. 'The Brannan Stanes' are in this parish. He appears to have been the special patron of Bute, for the Butemen were of old called the 'Brandans,' in Wyntoun the 'Brandanys of Bute' (viii. 4327-4360). Kilbrennan Sound was given me in Arran as *an Caol Srandanach,* probably a corruption of *an Caolas Brandanach*; here Kil- therefore is for *caol,* a strait, but the explanation of the latter part is obscure.[1]

The name *Brigid* was borne by fifteen female saints, including Brigid of Kildare, Feb. 1, who died in 525, and occurs often with us. Examples in the west are: Kilbride in the Isle of Seil; Kilbride joined with Kilmore in Lorn; Kilbride in Islay, in Coll, in Tiree (p. 92), in Bute, in Arran, in North Uist, and in Skye—Strath and Trotternish. On the east there are Kilbryd in Menteith 1610 (Ret.), Blarbrid 1275 (Boiamund), apparently St. Bride of Blair in Atholl; Logynbrid, *ib.*, in the diocese of Dunkeld, now included in Auchtergaven; 'ecclesia Sancte Brigide de Kilbrigde' 1219 (Chart. Inch.), is Kilbride in Strathearn; Panbride in Forfarshire; St. Bride's chapel at Kildrummy (Ant. A. and B., i. p. 589); St. Bride of Skene in Aberdeenshire (*ib.*, p. 279); St. Bride's Well, near the old church of Cushnie; Lanbryde, Lamnabride, Lamanbride, 1215 (RM), Lambride, *c.* 1391,

[1] Brendan's genealogy is in LL 349 a; BB 218 f; LBr. 13 e; Oeng. *Fél.*, p. 132.

ib., in Morayshire; Duninbride, *c.* 1200 (RM), a chapel or church in Inveravon parish; 'Cromdale Bhrid,' St. Bride's Camdale near Ruthven, Tomintoul (Ant. A. and B., ii. p. 308); St. Bride's chapel at Benchar in Badenoch; St. Bride's chapel at Conon, Ross-shire, whence Logiebride, the old name for the parish of Urquhart or Ferintosh. *Allt Brìghde*, 'St. Bride's burn,' is near Conon House. There was also a chapel of St. Bride in Tarbat parish.

The Pictish Chronicle (p. 6) tells how Nechtan Morbet, king of all the provinces of the Picts, being driven from his kingdom by his brother Drust, sought St. Brigit in Ireland and begged her to pray God for him. He was restored to his kingdom, and in the third year of his reign Darlugdach, abbess of Kildare, came from Ireland to Britain, an exile for Christ's sake, and in the second year after her coming Nechtan offered up Abernethy to God and to St. Brigit in presence of Darlugdach, who sang Alleluia over that offering. This King Nechtan reigned, according to Skene, from about A.D. 457 to 481. Brigit was born about 452 (AU), and died in 525 (FM), when she was succeeded by Darlugdach; she was therefore about twenty-nine when Nechtan died. The tale is thus open to criticism, but it is of importance as giving the tradition of very early activity of the Irish Church in the eastern midlands of Scotland.[1] We may note Mallebride, *i.e. Mael Brighde*, priest at Abernethy *c.* 1100 (RPSA).

The *Libellus de Nativitate S. Cuthberti*, a tract of little historical value, mentions a St. Brigit of Dunkeld, who was probably the same as Brigit of Kildare.

Sir David Lindsay mentions the image of 'Sanct Bryde, weill carvit with ane kow.' He says also that people sought 'to Sanct Bryde to keip calf and know.' See also *Carmina Gadelica*.

Cainer, daughter of Cruthnechan, in Gorman at Jan. 28, Cainer, daughter of Airmend, and Cainer, daughter of Fergna, were among the holy maidens who were subject to

[1] The little anachronism as to Darlugdach is of no consequence: the writer naturally styles her abbess of Kildare, which she was, though not then.

Brigit (LL 353 b; Lismore Lives, p. 336). Another saint of the same name, Cainer, daughter of Caelan, etc., is mentioned in LL 354 c;[1] her day in Gorman is Nov. 5. The name is also Cainder, Cainner. It is probably one of them who is commemorated in *Gleann Cainneir*, Glenkannyr or Glenkenner, now Glencannell, at the head of Loch Ba in Mull, where there was a chapel (Blaeu), and in the church of Bothkenner in Stirlingshire; see Kirkinner.

Cainnech Mocu Dalon of Achadh Bó, Oct. 11, the friend and helper of Columba, died in 599 or 600. It is certain that he spent a considerable time in Scotland, though it is more than doubtful whether he visited the eastern midlands, as Skene thought. *Innis Choinnich* at the mouth of Loch nan Ceall, 'loch of the churches,' in Mull, is named after him; in Fordun it is 'insula sancti Kenethy.' His churches are Kilchenich in Tiree, Kilchenzie for *Cill Choinnigh* in Kintyre, Coll, and South Uist. The parish of Laggan in Inverness-shire is *Lagan Choinnich*, 'Cainnech's Laggan,' or hollow. The site of the old church of Laggan, with its cemetery, near the head of Loch Laggan, is not in a hollow, but on a pleasant sunny plateau: the original *lagan* was somewhere near this. The *Life* of the saint says that he dwelt in a desert place at the foot of a certain mountain in Britain, but the sun did not shine on it owing to the hill coming between; it is doubtful whether this description can apply to any spot near the old church. In the east, a note on Oengus states that he had a *reclés* in Cell Rígmonaid, *i.e.* in the church of St. Andrews. From this it has been inferred that Cainnech founded a monastery there,[2] but the usage of the term *reclés* is against such an inference. The usual meaning is 'cell' or 'oratory'; in the Irish translation of the Latin *Life* of St. Martin of Tours it is used to translate *cellula*. There also we are told that Martin's *reclés* was a little cell attached to the church, and that his monks seldom came forth from their *reclésa* except to go to the church. In Ireland it was a common custom to have a

---

[1] But 'Cainder ingen Fáelán áin,' BB 230 b 10.

[2] *E.g.* Skene, *Celtic Scotland*, ii. p. 137.

reclés or chapel in or attached to a church and bearing the name of a saint, *e.g.* at Armagh there was a *reclés* of Brigid and another of Colum Cille. All, therefore, that can be inferred from the note on Oengus is that in the church of St. Andrews there was an oratory or chapel named in honour of Cainnech. The Aberdeen Breviary gives the festival 'of St. Caynicus, abbot, who is held as patron of Kennoquhy in the diocese of St. Andrews.' Kennoquhy is now Kennoway, in RPSA Kennachin, and the statement of the Breviary is probably based on a mistaken derivation of the name from Cainnech, for there seems to be no confirmation of it. In Cambuskenneth, Stirling, Cambuskynneth 1147 (Cart. Camb.), the second part may be *Cinaetha*, gen. of *Cinaed*, a different name from *Cainnech*, and really the origin of Kenneth in English.[1]

Catán, Dec. 12 and Feb. 1 (Gorman), Bláán's preceptor, was connected with Bute. *Cill Chatáin*, Kilchattan, occurs in Bute, Colonsay, Islay, Gigha, Luing, and Kintyre. *Cladh Chatáin* is in Ardnamurchan, and Ardchattan on Loch Etive was the seat of a priory. He was also connected with Stornoway (Orig. Paroch., ii. 381). In the east, he is the saint of Aberruthven, 'ecclesia Sancti Catani de Aberruadeuien,' 1198 (Chart. Inch.).[2] Gille Catain is the eponymous ancestor of Clann Chatain, the Mackintoshes and Macphersons of Badenoch.

Cessóc, *Ceaság*, March 10, is stated by the Aberdeen Breviary to have been born in Cashel in Munster, of royal family, a statement which would be of much interest if it were confirmed. He is perhaps the same as *Cessán*, who is styled 'son of the king of Alba, and a chaplain of Patrick.'[3] His name is a reduced form of some compound beginning with *cess*, a spear. He is specially connected with Luss on

[1] Genealogies of Cainneach are given in LL 348 i; BB 218 e; LBr. 16 a.

[2] The genealogy of a saint of this name is given in LL 348 f; BB 218 a; LBr. 15 c—Catan m. Matain, m. Braccain, m. Caelbaid, m. Cruind ba drui. 'Crund who was a druid' was of the Dál nAraide of Ulster, who were reckoned Cruithne.

[3] Is annsin atá fer gráda dom muinntir-se .i. Cessán mac rig Alban, 7 sacart méisi damsa hé': 'it is there, *i.e.* in Ros meic Treoin, that an official of my household dwells, namely Cessán,' etc. *Acall.* l. 486.

Loch Lomond. In 1566 (RMS) mention is made of 'the lands of Buchquhannan, with the bell and alms of St. Cassog' (*cum campana et elimosina S. Cassogi*). On the east he is the patron of Auchterarder, 'ecclesia Sancti Mechesseoc de Eohterardeuar,' 1200 (Chart. Inch.), 'Sancti Mahessoc de huctherardouer,' 1211, *ib.* *Tom mo Cheasaig*, 'my-Kessock's knoll,' and *Féill mo Cheasaig*, 'St. Kessock's fair,' are at Comrie and Callander in Perthshire. *Cladh nan Ceasanach* in Glenfinglas, near Glen Main Burn, seems to mean 'graveyard of St. Kessock's people.' In Strathearn there is 'Barrnakillis (Church-hill) with the chapel and holy bell of St. Kessog,' 1538, 1542 (RMS). *Port Cheiseig*, Kessock Ferry on the Beauly Firth, is probably named after the saint. In 1270 (Reg. Pasl.) 'Mauricius filius Gilmekesseoch' was witness at Paisley.

The Ciarán who is commemorated in Scotland is probably Ciarán of Cluain mac Nois, styled 'Mac in t-Sáir,' Sept. 9, who died in 549 (AU). In the west there are *Cill Chiaráin*, Kilchieran or Kilcherran, near Kilchoman in Islay, in Lismore, and in Kintyre, where Kilkerran was the old name of Campbeltown parish; the site of the town is *Ceann Loch Cille Chiaráin*. There was a church of St. Ciaran at Lianishader in Barvas, Lewis. *Caibeal Chiaráin*, his chapel, is at the north-east end of Loch Awe. In the east, *Caibeal Chiaráin* once stood on Boreland farm, Loch Tay, near the gate on the road to the farm-house, and a meadow there is called *Dail Chiaráin*.[1] Another *Dail Chiaráin* is on Duneaves, between Duneaves House and the river Lyon. The saint is also connected with Fetteresso in Forfarshire, and his well was in Glenbervie. He had also a chapel in Strathmore in Caithness.

Colmán, a name borne by 218 saints, is a diminutive of *colum*, 'a dove,' from Lat. *columba*, so that this, the most popular of all Irish names, is not a native term. The Martyrology of Aberdeen has Colman at Feb. 18 as patron

[1] The font of St. Ciaran's chapel is extant, and has been placed by Mr. Alexander Campbell of Boreland, with a suitable inscription, near the site of the chapel. Ciaran's genealogy is in LL 348 i; LBr. 16 a; Oeng. *Fél.*, p. 204.

of Terbert, *i.e.* Tarbat in Ross-shire, where St. Colman's chapel was. His name survives in *Port mo-Cholmáig*, Portmahomack, 'my-Colmóc's haven,' and probably in *Cill mo-Chalmáig*, Kilmachalmaig, on the Kyle of Sutherland, where there are an old cemetery and remains of a chapel. St. Colman's Well is, or was, at Teinleod above Fowlis Castle. Féill mo-Chalmáig, 'my-Colmóc's Fair,' was held at Moulin in Perthshire at the end of February (Old Stat. Acc.). On him the note to Oengus' *Félire*, Feb. 18, has 'Colmán from Ard Bó in Cinél Eoghain (on Loch Neagh), or in Alba north of Monadh (the Grampians).'[1]

At June 7 Mart. Aberd. has Colmoc of Inchmahome, *i.e.* *innis mo-Cholmáig*, 'my-Colmóc's isle,' in the Lake of Menteith. This is the day of Colmán or mo-Cholmóc of Druim Mór in Ulster. Kilmachalmaig in Bute is supposed to commemorate the same saint.

St. Colm's Fair in Badenoch was on Jan. 15; Gorman and Mart. Don. have two Colmans at Jan. 13. The *Féill Choluim* at Dingwall was on the last Tuesday but one of July; compare Mo-Cholmóg, July 19, Mart. Don.

In North Uist is Kilchalman, 'Colman's church.'

Colum Cille, St. Columba, June 9, was born in or about 521, on the day of the death of Buite, son of Brónach. In 563 he left Ireland for Alba, and died in Iona in 597. His descent and the main facts of his life are well known, and his numerous commemorations in Scotland have been collected in great detail.

Adamnan states that monasteries of Columba founded among both the Picts and the Scots were held in great honour in his own time (*c.* A.D. 700), but unfortunately he gives no list of them. Those which he mentions, in addition to Iona, are Campus Luinge and Artchāin in Tiree, one in the island called Elena, and one in Hinba. Cella Diuni was on Loch Awe, and Kaille au inde was probably in Morvern. For the east our information is later and fragmentary. The

---

[1] 'Colmán ó Ard Bó i nCinél Eogain, nó i nAlbain fri Monaid (*read* Monad) atuaid.' His genealogy is given in LL 347 f, BB 216 d; LBr. 14 c. He was eighth in descent from Colla Uais.

legend of Deer claims that this monastery was founded by Colum Cille and by Drostan, and that its early grants of land were made to God and to Colum Cille and to Drostan. Written about five hundred years or more after the event it describes, this account has to be taken with due caution. But whether Colum Cille visited Buchan or not—and there is nothing really against it—the clerics of Deer could not have been mistaken in claiming that their monastery was a Columban foundation, in the sense that it was founded from Iona. There is also reason to believe that Banchory-Ternan was an early and important Columban centre of which Torannán was abbot. The position of Aberdeen is less clear, owing to the uncertainty as to the identity of Do-Chonna with St. Machar.

Churches styled *Cill Chaluim Chille* are found on the west in the parishes of Kildalton and Killarrow in Islay, in Kilninian parish in Mull, in South Kintyre, in Morvern, at Kiel in Ardchattan, in Duror of Appin, in Canna, in St. Kilda, in South Uist at Howmore, in Benbecula, in North Uist, on Eilean Choluim Chille in Loch Erisort on the east side of Lewis, on the Eye peninsula near Stornoway, and at Snizort in Skye. He was commemorated in Iceland (p. 149). The only site in the east called *Cill Chaluim Chille* is in Strath Brora in Sutherland. He was the patron saint of Auldearn, of Petty, of Kingussie, and of Daviot and Belhelvie in Aberdeenshire, and in most of these places fairs were held on his day. He was also the patron of Dunkeld, to which some of his relics are said to have been taken about A.D. 849. He was the patron of Arryngrosk, now Arngask (Reg. Cambus., p. 35), and of Dollar (Lib. Plusc., p. 281). He was commemorated also in Burness and Hoy in Orkney, and at Olrick in Caithness.[1]

The lands of Forglen in Aberdeenshire were long held in early times by the immediate custodians of the reliquary of Columba called the 'Bracbennach, first heard of in

[1] For further commemorations see Skene, Adamnan's *Life of Columba*, and Forbes, *Calendars of Scottish Saints*. In cases where the saint's day is not known from fairs or otherwise, commemorations of some other Colum or Colman are liable to be confused with those of Colum Cille.

connection with William the Lyon's grant of it and the lands to the monks of Arbroath between the years 1204 and 1211, but the tenure was probably of much older date. The reliquary was borne in battle as a *vexillum*, and may well have been present at the battle of Harlaw. From about 1420 its custodians were the Irvines of Drum, the first being apparently a son of Sir Alexander who was killed at Harlaw. It is a beautiful work of art, considered to belong to the eighth century, and is now at Monymusk. Its name is a compound of *breac*, variegated, and *beannach*, peaked, lit. 'the variegated peaked one,' with reference to its ornaments and its sharply sloping top, like the ridge of a house.

Comgan, Oct. 13, is mentioned by Oengus, and is styled 'of Cluain Connaidh in Cuircne' in Roscommon by Gorman and Mart. Don. In LL 363 h he is called 'Comgan Céle Dé,' 'the Culdee,' but Gorman and Mart. Don. put Comgan Céle Dé at Aug. 2. The Comgan commemorated in the Scottish Calendars on Oct. 13 is said in the Aberdeen Breviary to have been the brother of Kentigerna, daughter of Cellach Cualann of Leinster, who died in 734, but I find no details of his life in the Irish authorities.

With us he is commemorated in *Cill Chomhghain*, Kilchoan, in Islay, and in Ardnamurchan and on Loch Melfort in Argyll. In Skye there are *Teampull Chomhghain* on Loch Eisheort and *Cill Chomhghain* in Glendale. There is also Kilchoan in Knoydart, and *Cladh Chomhghain*, 'Comgan's cemetery,' is in North Uist. He is the patron of Loch Alsh in Wester Ross, where he is said to have come from Ireland with Kentigerna and Fillan. In the eastern part of Ross there is *Cill Chomhghain* in Kiltearn parish, now called Mountrich. He is the saint of Turriff in Aberdeenshire; in 1273 Alexander Cummyn, Earl of Buchan, makes grants to God and to Blessed Mary and to S. Cunganus of Turreth in respect of an almshouse which he had built there. Turriff was the seat of an early religious house; its *fer-léighinn* is mentioned in the Book of Deer. Gille Comgain, mormaer of Moray, died in 1032 (AU). The name Mac Gille Chomhghain, 'son of St. Comgan's servant,' was long found near Dingwall, and also in Argyll.

Connán of Tech Connáin is in Gorman at June 29, and another Connán at July 1. There is nothing to show which of these, if either, is commemorated in *Cill Chonnáin*, Killiechonan, on Loch Rannoch, or in *Dìseart Chonnáin*, 'Connan's hermitage,' at Dalmally. Connan's fair at Glen Orchy was held on the third Wednesday of March (New Stat. Acc.). In Waternish, Skye, there is *Cill Chonnáin* at Trumpan Head, and the Conon River at Uig in Skye is *Abhainn Chonnáin*, probably from the saint. I have met no commemorations of Conán of Eigg.

Mo-Chonóc of Cell Mucroisse, Dec. 19, is stated to have been a son of Brachan of Brecheiniauc (p. 162), which may have been the northern district now represented by Brechin. Kilmachonock appears as a parish church, apparently in North Knapdale, in 1664 (RMS). According to Skene, Mo-Chonóc was the patron saint of Inverkeilor in Forfarshire.[1] Cell Mucroisse is supposed to be Kilmocrish or Kilmuckridge in Co. Wexford, but it is to be noted that Mucross was the name of a district near St. Andrews, also that Mucrosin, now Muckersie, was the name of an old parish now part of Forteviot near Perth. It is perhaps impossible to settle the point definitely, but Mo-Chonóc's church may have been in Scotland. His brother Iast was connected with Lennox. The Welsh form of his name is Cinauc, Cynauc, Cynog, and we have Lan Cinauc, now Llangunnock, in the Book of Llandaf; Merthyr Cynog is in Brecon. In Cornwall there is Bo-Conock.[2]

Cormac is commemorated in *Cill Mo-Charmáig* at Ardeonaig on Loch Tay, and *Coire Charmáig* in Glen Lochay at the head of the loch may be named after him, as *Coire Chunna* on the north side of Loch Tay is named after Cunna of Glen Lyon. There are no data to determine which Cormac is commemorated.

Cill Mo-Charmáig was at Keills in North Knapdale, in Blaeu Kilmacharmick. A few miles south by west of this is the islet called Eilean Mór, in Fordun 'Helant Macarmaig, where is also a sanctuary.' In 1579 (RMS) it is 'insula de

[1] *Celtic Scotland*, ii. p. 36, note. [2] J. Loth, *Rev. Celt.*, xxix. p. 247.

Sanct Mackchormick.' In 1507 King James IV. confirmed a series of nine charters produced by the bishop of Lismore, one of which, granted by Roderic son of Reginald, lord of Kintyre, was witnessed *inter alios* by Maurice, parson of Cillmacdachormes. The charter is undated, but Reginald died in 1207. The writer of the New Stat. Acc. gives the island as Ellan-more-vic-O'Charmaig and the mainland church as Kilvic O'Charmaig (Cill mhic O'C.), and they are still so called in Gaelic. As the ancient Irish gentilic term *moccu*, *maccu*, was regularly made later into *mac úi*, *mac O'*, it is fairly certain that the saint's name contained this element, and both churches may be perhaps ascribed either to Báetán Maccu Cormaic, abbot of Cluain mac Nois, who died on March 1, A.D. 664, or to Abbán Maccu Cormaic, Oct. 27, of Magh Arnaide.

Donnán of Eigg, April 17, is commemorated in the west in *Cill Donnáin*, Kildonan, in Kintyre, Eigg, Skye, Arran, Uist, Loch Broom, and in *Seipeil Dhonnáin*, 'Donnán's chapel,' in Kishorn, probably also in *Eilean Donnáin*, Ellandonan, Lochalsh. In the east there is Kildonan in Sutherland, with *Cathair Dhonnáin*, 'Donnan's chair'; his Fair was held there. He was the patron of Auchterless in Aberdeenshire, where his staff is said to have been kept.[1] A saying, probably of some antiquity, is

'Cill Fhinn 's Cill Duinn 's Cill Donnáin,
Na trì cilltean as sine an Albainn.'

'Killin, Kildun, and Kildonan—the three oldest churches in Alba.' I have met no genealogy of Donnán.

Dubhthach of Tain is commemorated on March 8 (Mart. Aberd.), and Tain is *Baile Dhubhthaich*, 'Dubhthach's town.' He is usually identified with *Dubhthach Albanach*, 'of Alba,' who died in 1065 at Armagh, and is described as 'chief soul-friend of Ireland and Scotland' (AU). The epithet *Albanach* does not necessarily imply more than that his work, or part of it, was in Scotland. It is difficult, however, to understand how a religious centre so important as Tain undoubtedly was could have been founded at so

[1] *Antiquities of Aberdeen and Banff*, i. p. 505.

late a period in the history of the Scoto-Irish Church; one would rather expect it to belong to the pre-Norse period, when that Church was at the height of its vigour and activity. On the west, the church of Kintail is *Cill Dubhthaich* or *Clachán Dubhthaich*, 'Dubhthach's stone-cell,' on *Loch Dubhthaich*, Loch Duich. *Cadha Dhubhthaich*, 'Dubhthach's path,' is a pass from the glen of the river Conag to Glen Affric. On the east, there was a chapel of the saint, 'Divi Duthasi,' at Forres (1611, RMS), and an altar in St. Nicholas' Church in Aberdeen (Ant. A. and B., i. 206). The prebendary of St. Duthac in Orkney is mentioned in 1545 (RMS). Kildduthie in Kincardineshire may be for *Cill Dubhthaigh*. Lethen in Nairnshire is in Gaelic *Leathan Dubhthaich*; the Laird of Lethen is *Tighearn Leathan Dubhthaich*, 'lord of Dubhthach's broad (slope)'; this may or may not be the saint.

Kilmeny in Fife and in Islay is probably for *Cill M'Eithne*, 'my Eithne's church'; the Islay name is now in Gaelic *Cill Mheinidh* for *Cill Mh'Eithne*. Forgandenny in Perthshire is probably 'Eithne's Forgan' (p. 381). There is no indication as to which saint of the name is commemorated.

Fáelán, later Faolán, 'little wolf'—a reduced form of Fáelchú—was the name of sixteen saints (Gorman). The Martyrology of Aberdeen has Fillan of Strathfillan at Jan. 9, which is the day of Fáelán of Cluain Moescna in Fir Tulach, West Meath. *Féill Faoláin*, Fillan's Fair, at Killin, was held on that day. The Aberdeen Breviary makes him the son of Kentigerna, daughter of Ceallach Cualann of Leinster, who died in 734 (AU), and states that he came to Loch Alsh with his mother and his uncle Comgan. He is the saint of *Cill Fhaoláin*, Kilillan, in Kintail, where the Kintail people claim that he is buried.[1] Comgan is the saint of Loch Alsh and Kentigerna's *cill* is on the south side of Loch Duich. Irish records, so far as known to me, give

[1] Local tradition has it that St. Fillan's body was taken in a galley (*birlinn*), which lay for a night at Camus Longart on Loch Long and then sailed up to near Kilillan. Hence, the name Loch Long (Loch of Ships). Columba, it is added, sent a sod from Iona to cover his grave, which is placed north and south.

no genealogy of Fáelán of Cluain Moescna, nor do they state that he was the son of Kentigerna.

Fáelán styled *amlabar . i . nemlabar*, 'the dumb,' is commemorated on June 20 (Oengus), and a note in LBr. states that he was a son of Oengus son of Natfraech, who was a son of Corc of Cashel, and that he was of Ráth Érenn in Alba and of Cell Fáelán in Laiges or Lecks. His period would thus be round about A.D. 500. Ráth Érenn, 'Ireland's rath,' I believe to be Rottearns in Ardoch parish (p. 227). St. Fillan's village, Dùn Fhaoláin, and the saint's chair and well, all between Comrie and Loch Earn, have been mentioned (p. 227). There were apparently two chapels of St. Fillan at or near Doune Castle.[1] 'Sanct Phillanis burn' was at Luncarty, 1597 (RMS). Kilellan in Kintyre and near *Tollard*, Toward Point, in Inverchaolain parish, Argyll, commemorates a saint of this name, but as to which there are no data. *Gleann Fhaoláin*, with *Làirig Fhaoláin* at its head, is off Glen Etive; its stream is *Allt Fhaoláin*, which joins Etive at *Inbhir Fhaoláin*.

The fact that the Battle of Bannockburn was fought on Midsummer Day suggests that the saint to whom Bruce made his orisons on the night before, and whose hand played so great a part in the battle, was St. Fillan of Rath Érenn. Sir David Lindsay in his *Complaynt to the King* swears 'be Sanct Phillane.'

Fínán, styled *lobur*, 'the infirm,' has his day on March 16 (Oengus), which serves to identify him with Finnianus of March 18 in Mart. Aberd. He was of the race of Oilill Ólum of Munster, and his genealogy is given among those of the other saints.[2] He is said to have been placed over Swords by Colum Cille (Reeves, Adamnan, p. 279). In the west he is commemorated in *Cill Fhìonáin*, Kilennan,

[1] Capellania de Sanct Phillane infra castrum de Doune et capellania Sancti Phillane extra idem castrum situata 1602 (*Ret.*). This seems to have been 'St. Fillan's chapel in Menteith,' 1536 (RMS).

[2] Fínán mac Conaill maic Echach maic Thaidc maic Céin maic Oilella Óluim—LL 350 c; BB 220 e; LBr. 18 a. This genealogy is evidently defective; it would place Fínán in the fourth or early fifth century. On the confusion between Fínán Lobur and Fínán of Cenn Eitigh, see Plummer, *Vitae SS. Hib.*, p. lxix.

in Islay; in *Eilean Fhìonain*, 'Finan's isle,' in Loch Shiel, Moidart, where his fair was held on March 18, being the day after St. Patrick's Day; in *Cill Fhìonain*, Kilfinane, 1637 (RMS), at the north-east end of Loch Lochy in Inverness-shire; in Kintyre: S. Fynnanus of Killenane, 1592 (RMS); in Kilfinan in Cowal: 'ecclesia Sancti Finani in Kethromecongal,' 1253 (Reg. Pasl.); Killenane in Islay, 1588 (RMS). In the east his churches are *Cill Fhìonain* on Loch Ness at Abriachan; *Seipeil Fhìonain*, 'Finan's chapel,' at Foynesfield, Nairn [1]; Migvie in Marr: 'ecclesia Sancti Finnani de Miggeveth' (RPSA); here Finzean's fair was held in March or April (Forbes) [2]; Abersnithack in Monymusk. In 1643 it was reported to the Presbytery of Inverness 'that there was in the Paroch of Dunlichitie ane Idolatrous Image called St. Finane, keepit in a private house obscurely.' The image was duly burned at the Market Cross of Inverness.[3]

Skene supposed Lumphanan in south Aberdeenshire to be 'a corruption of Llanffinan,' and compared Llanffinan in Anglesey, suggesting a Welsh origin.[4] The old forms are against this: Lufanan, Lunfanen, Lunphanan, Lunfanin (all from PS), Lunfannan, 1357 (Ant. A. and B.), Lynphannane, 1480, *ib.*, Lumpquhannan, 1487 (RMS). The term *lann* was used freely in Ireland in connection with churches, and in Scotland it is common in certain districts in the sense of 'enclosure, field,' *e.g.* *Lainn Chat*, Lynchat in Badenoch, 'wildcats' field.' Lumphinnans in Ballingry parish, Fife, may therefore be 'St. Finan's field'; compare the common 'croft' of a saint. Who St. Finan of Anglesey was is not certain; he was probably different from our saint.

Findoca is described in Mart. Aberd. at Oct. 13 as 'virgo de qua est ecclesia in dyocesi Dunblanensi,' 'a virgin who has a church in the diocese of Dunblane.' Gorman and

---

[1] Foynesfield is *Seipeil Fhìonain* in Gaelic; perhaps it is for 'Fínan's field.'

[2] The form 'Finzean' suggests rather *Finnén*, but no saint of that name is commemorated about that date.

[3] *Inverness and Dingwall Presbytery Records* (Scottish Hist. Soc.), p. 1.

[4] *Celtic Scotland*, ii. p. 193.

Mart. Don. give at that day ' Findsech, a virgin from Sliabh Guaire in Gailenga,' now part of Meath. *Findsech*, from *find* with the fem. suffix *-sech*, means 'fair lady,' and would naturally be made *Findóc*, whence the Latin *Findoca*. She was the saint of Findogask (Lib. Scon) in Perthshire. The church of Innishail in Loch Awe is 'ecclesia Sancte Findoce de Inchealt,' 1275 (Chart. Inch.). She is also commemorated in *Cill Fhionnáig*, Killinaig, in Mull, and in *Cill Fhionnáig* in Coll.

Maol Rubha of Applecross, whose day is April 21, but in Scotland Aug. 27,[1] is next to Colum Cille the most famous saint of the Scoto-Irish Church. He was born in A.D. 640, and became abbot of Bangor in Ulster. In 671 he crossed to Alba, founded the monastery of Applecross (*Aporchrosan*) in 673, and died there in 722 at the age of eighty. His grave is still revered, and the person who takes earth from it is ensured safety in travelling and return to Applecross. According to the Aberdeen Breviary he was slain by Norsemen at Urquhart, near Conon Bridge, but in Maol Rubha's time there were no Norsemen in Scotland, nor do the Irish authorities give any hint of his having died other than a natural death. His name is given in two forms, both represented in Gaelic now. One form is Máil Rubai (AU 671, 673), Máel Rubai (*ib.*, 722), Máel Rubae (Rawl. B 502, 94 b 9), Máel Rubha (BB 213 a 4), etc. whence the form *Cill-á-Rubha* in Islay. The other is Máel Rubi (LL 359 b), Máel Ruibe (BB 228 e 27), Máeli Ruibi, in gen. (BB 213 b 34), whence the northern form seen in *Loch Ma-Ruibhe*, Loch Maree, etc.[2] Máel Ruibi

---

[1] This being St. Rufus' day, with whom he was evidently confused in Scotland.

[2] Maele Rubi abad Bennchoir, LL 359 b; Mail Rubai in Britanniam nauigat, AU 671; Mail Rubai fundauit aecclesiam Apor-croosan, AU 673; Mael Rubai in Apur-chroson anno LXXX etatis, AU 722. His genealogy is given in LL 347 3 :—Mael Rubae son of Elganach s. o. Garb s. o. (F)oirballach s. o. Conbairenn s. o. Cremthann s. o. Binnech s. o. Eogan s. o. Niall of the Nine Hostages; also in BB 216 a and LBr. 13 e. His father's name is also given as Elgad (BB 23 a 43) and as Mael-gonaich. His mother was Suaibsech or Subtan, *e.g.* Suaibsech mathair Mael Rubha maic Mael-gonaich—BB 213 a 4; Subtan ingen Setna siur Comgaill mathair Maeli Ruibi—BB 213 b 34. She was sister of Comgall of Bangor.

occurs also as the name of another man.[1] The same variation is found in *suba*, delight, gen. *subai*, and *suibi* (Ml. 47 d 2). The second part is the gen. sg. of *ruba*, a cape or point, also a copse or wood, and *Máel Rubai* means literally 'servant of the cape' or 'of the copse'; such names were given doubtless from some connection of birth or childhood, *e.g. Máel Gemrid*, 'winter's servant,' *Máel Snechta*, 'snow's servant.' *Máel Rubha* is exactly the same in meaning as the personal name *Máel Ruiss* or *Máel Roiss*, where *ross* also has the double meaning of 'cape' and 'wood.' In such combinations the original notion of 'servitude' is in abeyance; *máel* is here practically equivalent to *gille*, 'lad.'

In the west Maol Rubha is commemorated in Applecross, where is his girth (p. 125); in Loch Carron, where *Clachán Ma-Ruibhe* appears to have been the site of the parish church (Blaeu); in Gairloch, where Maol Rubha's chapel stood in *Eilean Ma-Ruibhe*, an islet in *Loch Ma-Ruibhe*, Loch Maree; in *Cill Ma-Ruibhe*, written by MacVuirich *Cill Maoilridhe*, in Arasaig; in Kilmolrow or Kilmarow, 1603 (RMS) in Muckairn parish, Argyll; in Kilmolrow a church and parish in Kintyre, 1599 (RMS); in Kilmolroy, Craignish, 1580 (RMS); in Kilarrow, Kilmolrow 1500 (Orig. Paroch.), in Islay, now in Gaelic *Cill-á-Rubha*; in *Cill Ma-Ruibhe*, Kilmarie, in Strath, Skye; where is also *Aiseag Ma-Ruibhe*, Askimulruby 1505 (Orig. Paroch.), 'Maol Rubha's ferry'; in *Cill Ma-Ruibhe*, Kilmolruy (Blaeu), in Bracadale, Skye. *Aird Ma-Ruibhe*, with an old cemetery near it, is in Bernera, Harris. The asseveration '*Ma-Ruibhe!*' is often heard in Harris.[2] In the east there is *Preas Ma-Ruibhe*, 'Maol Rubha's copse,' near Strathpeffer in Contin, of which parish he was patron. The 'copse' is now a private burial-ground. His fair, *Féill Ma-Ruibhe*, used to be held at Contin, then at Dingwall. In Sutherland he was the saint of Lairg, where there is *Eilean Ma-Ruibhe*, Yl Mulruy (Blaeu), in Loch Shin. Keith

[1] *Celt. Zeit.*, xiv. p. 111. Ui Maelrubai were a sept in Leinster.

[2] For this I am indebted to Rev. Malcolm Maclean, Applecross, a Harris-man.

in Banffshire is Kethmalruf in *c.* 1214 (RM); his fair here was called 'Summareve's Fair,' *i.e.* 'Saunct Ma-Reve's.' A piece of land near Pethnick was 'Sanct Malrubus stryp,' 1576 (RMS). In Fife there was a chaplainry of 'Sanct Maruiff' under the castle of Craill, 1620 (RMS). In Perthshire there is *Ath Maol Ruibhe,* Amulree, 'Maol Rubha's ford,' on the river Braan, with an old church site called *Cill Ma-Ruibhe.* One of the two Amulree markets used to be held on the first Tuesday and Wednesday of May.[1]

Cill Mhaodháin, Kilmodan or Kilmadan in Glendaruel, Cowal, is Kilmodan, 1250 (Reg. Pasl.), 'ecclesia Sancti Modani,' 1299, *ib.* The saint is *M'Aodhán,* earlier *M'Aedán,* or *M'Aedóc,* 'my-Aedán.' Sixteen saints bore this name, including Aedán of Lindisfarne, Aug. 31, and Aed or Maedóc of Ferns, Jan. 31. Another is Maedóc, March 23, styled by Oengus 'Mo Maedóc mind nAlban,' 'my Maedóc, Alba's diadem.' He was son of Midgna and so on to Catháir Mór, and some authorities make him of Fid Dúin in Kilkenny,[2] but this is not certain. In view of Oengus' designation he should be commemorated in Scotland, and this may be one of his churches. The Glendaruel market used to be held on the Saturday before the last Wednesday of October.

Modan was also the saint of Rosneath, older Neveth; Michael Gilmodyn was parson of Neveth in 1199 (Reg. Pasl.); Gilmothan, son of the sacrist of Neveth, and John MakGilmothan appear in 1294, *ib.* The Aberdeen Breviary assigns this church to a Modan of Feb. 4, who appears in none of the Irish Calendars; he may be the same as Modan of Glendaruel; the Gaelic pronunciation of his name is probably irrecoverable now, but the *th* of the personal names indicates *dh.*

The Scottish Calendars mention another Modan at Nov. 14, 15, specially connected with Falkirk (Mart. Aberd.) and

---

[1] In the modern forms, *Ma-Rubha, Ma-Ruibhe,* the *l* of *maol* has become assimilated to *r* following.

[2] So all the genealogies—LL 351 d; BB 221 f; LBr. 18 b; also LL 357 d (Mart. Taml.). But LL 360 f, 361 d (Mart. Taml.) and notes on Oengus put Maedóc of Fid Dúin at May 18 and Aug. 13; a note on Gorman puts him at May 18, so also Mart. Don.

Filorth, now Fraserburgh, in Buchan. Nothing further is known of this saint, except that his arm was long kept at Falkirk, and that a silver image of his head is traditionally said to have been kept at Fintray in Aberdeenshire, and used to be carried in procession in order to influence the weather as required.[1] Oengus has Aed mac Bricc of Cell Air in Meath at Nov. 10.

There is nothing to determine the saint of St. Maddan's chapel at Freswick in Caithness. St. Madden's Well is near Kirkton of Airlie, Forfar.

A saint named Màillidh, who is not mentioned in the Calendars, is commemorated in *Cill Mhàillidh*, Kilmalde 1372 (RMS), Kilmallie on Loch Eil in Inverness-shire, the name of the largest parish in Scotland. About eight miles from Kilmallie is Glen Mallie, with the river Mallie and at its mouth Invermallie on Loch Arkaig, all named from the saint of Kilmallie. 'Air Malie,' 'by Malie,' was a form of asseveration (Old Stat. Acc., viii. p. 407). In Sutherland, the old name of Golspie parish is Kilmaly, Culmaliun, Culmalyn, 1275 (Theiner), Culmalin 1471, Kilmaly 1536 (Orig. Paroch.). This may represent *cùil Mhàillidh*, 'Maillie's retreat' (*secessus*); compare *Cùil Bhrianainn*.[2] There is also *Dail Mhàillidh*, Dalmally in Argyll, with a burn called *Allt Mhàillidh*. On the east, 'Westir and Estir Eglismaldiis,' 'church of Màillie,' in Kincardineshire appear in 1503 (RMS), now Inglismaldie on North Esk. In Fife there is 'the half carucate of land near the church of Sanct Maling now called Inchekkerie, together with the chapel of Buchadlach now called Eglismaly, 1611 (RMS); in 1668 (Ret.) there is 'the church of Sanct Maring now called Inchkiery, with the chapel of Buchadlach now called Eglismaldie.' Here *ld* represents *ll*, as in Clochfoldich, Logierait, now in Gaelic *Cloich Phollaich*. Eglismaly was in the barony of Grange, 1611 (RMS), and Inchskyrrey is placed west of Grange House, Burntisland, on Blaeu's map.

[1] Forbes, *Calendars*, p. 402.

[2] E. W. B. Nicholson in his *Golspie* supposed that the saint of Kilmaly in Sutherland is Mo-ling; the fact that Mo-ling is stressed on the second part makes this impossible.

Martin of Tours, Nov. 11, is styled by Oengus 'sliab óir iarthair domuin,' 'the mount of gold of the western world.' *Cill Mhàrtuinn*, Kilmartin, is in Glassary, and in Skye. St. Martin's is a Perthshire parish; *Sgìre Mhàrtuinn* was the Gaelic name of the old parish of Kilmartin in the Black Isle, now Culicudden. Kyref Martin *c.* 1200, Kynnef Martyn *c.* 1230 (RM), was a place in the region of Strathavon in Banffshire with a church or chapel of Martin. A chapel of St. Martin stood near the house of Ulbster in Caithness; it was the burial-place of the family of Ulbster. Isle Martin is in Loch Sween, with a chapel site on the mainland. As to Isle Martin in Loch Broom, *Talamh Mhàrtuinn*, 'Martin's land,' and Balmartin in Uist, and Strathmartin in Forfarshire, I have no proof that they are connected with the saint. For Formartin, see p. 118.

Mernóc is for *Mo Ernóc*, 'my-Ernoc,' an affectionate form of Ernáin, Ernín, or Ernén. The Mart. Aberd. states that the saint of Kilmernoch (Kilmarnock) is Mernocus of Oct. 25, which, as we have seen, suggests Ernáin of Midluachair, otherwise known as T'Ernóc (thy-Ernóc) of Cell na Sacart, whose day is Oct. 26 (Gorman), and who died in 714 (FM). Whether this is the saint commemorated elsewhere in the west we have no means of knowing. Kilmernach in Cowal, 1472 (RMS), survives in Kilmarnock on the east side of Loch Striven. Ardmarnock in Kilfinan parish, Cowal, had a chapel of St. Marnock with a churchyard and near it the remains of what seems to have been a *clochán*.[1] Inchmarnock off Bute is written Inchemernok by Fordun, who states that it had a cell of monks (*cella monachorum*). This was called later St. Marnock's chapel.

A saint of the same name was commemorated in the east on March 1, identified by Forbes with M'Ernóc mac Creseni (Oengus), a young contemporary of Columba, called by Adamnan 'Erneneus filius Creseni,' who died in 635 (AU), and whose day is Aug. 18, but there is no proof to support the equation. In the Irish calendar Ernín, daughter of Archenn, is commemorated on Feb. 28 (Gorman), but our

---

[1] *Old Stat. Account*, xiv. p. 258.

Mernóc was a man, if the records are correct. He may well have been a Scottish saint not recorded by the Irish writers. In Mart. Aberd. he is Mernanus, in the Breviary Marnanus. He was specially connected with Aberchirder in Banffshire, where he is buried ; the parish is now called Marnoch after him. In connection with a perambulation of lands belonging to the church of Aberchirder in 1493 certain men were to be sworn 'apone Sanct Marnoyss ferteris' (*i.e.* portable shrine, from Lat. *feretrum*). In the same proceedings a solemn oath was sworn in presence of St. Marnan's head (*capite Sancti Marnani presenti*).[1] 'Sanct Marnoys well' was on these lands.[2] 'Sanctus Marnocus de Loychel' (now part of Lochel-Cushnie) is mentioned *c.* 1200. St. Marny's well is at Benholm in Kincardineshire. Botmernok, 1359 (Reg. Brech.) in Panmure parish is now Boath ; the chapel of Bothe is mentioned in a charter of David II. (RMS). The Mass of St. Marnocus was appointed to be celebrated yearly by the vicar of Monikie (Reg. Brech.). Inchmarnock was the name of an old parish in Aberdeenshire now part of Aboyne and Glen Tanar. Dalmarnock is in Little Dunkeld parish. Compare Killearnan, p. 321 ; also p. 83.

Mo-Luóc of Lismore, June 25, died in 592 (AU). According to the Aberdeen Breviary he is buried at Rosmarkie in Ross-shire. His name is an affectionate form of *Lugaid*, *Lughaidh*, gen. *Lugdach*, *Lughdhach* (Ogham *Lugudeccas*).[3] In the west, his churches are *Cill Mo-Luáig*,[4] Kilmaluag, in Lismore, where there is also *Port Mo-Luáig* ; in Raasay ; in Tiree ; in Trotternish, Skye, where Kilmaluag is the old name of Kilmuir parish ; in Mull, Kilmalwage in Treshnish, 1638 (RMS) ; in Kilberry, Knapdale ; *Teampull Mo-Luigh*, (for *Mo-Luae*, gen. of *Mo-Lua*, another affectionate form of *Lughaidh*), called by Martin 'the Church of *St. Mulvay*,' at Eoropie in Lewis ; elsewhere Martin writes

---

[1] *Antiquities of Aberdeen and Banff*, ii. pp. 212, 213.

[2] *Ib.*, p. 215.

[3] His genealogy is—Mo-Luóc mac Luchtai maic Findchada m. Feidlimthe m. Sogain Salbuide m. Fiachach Araidi—LL 348 h ; BB 218 d 43 ; LBr. 15 e ; LL omits Fiacha Araide, BB and LBr. omit Fedelmid.

[4] *Luág* is pronounced as two syllables ; it might be written *Lughág*.

Muluy;[1] in the isle of Pabbay, where a chapel was 'dedicated to St. Muluag' (Martin).

In the east—*Cill Mo-Luáig* and *Croit Mo-Luáig* at Ballaggan near Inverfarigaig on Loch Ness; *Dabhach Mo-Luáig*, Dochmaluag, in Strathpeffer, 'my-Luag's davoch'; Croftmaluag, now Chapelpark, at Raitts, Badenoch, whence Gyllemallouock, a native in Laggan *c.* 1224 (RM); Sammiluak's chapel at Kildrummie for 'Sanct Mo-Luag'; 'S. M'huluoche de Tharuelund,' *i.e.* Tarland in Aberdeenshire, appears in 1165/71 (RPSA); 'ecclesia Sancti Muluoch de Taruelund,' 1207/28 (Ant. A. and B., ii. p. 17); Moloch was the saint of Mortlach, also of Clatt, where St. Molloch's fair was held. Gillemelooc was an Aberdeenshire name, *c.* 1200 (RPSA). A well called Simmerluak was close to the old church of Clova. 'Molouach' was the saint of Alyth, where his fair was held. According to Lachlan Shaw, the historian of Moray, he was patron of Cromdale on Spey.

The saint of *Cill Mhuirich*, Kilmorich in Cowal, now part of Lochgoilhead, is Muireadhach. Gorman and Mart. Don. have at Aug. 12 Muireadhach bishop of Cell Alad; Mart. Taml. (LL 361 d) has on that day Murchad of Cell Alaid. Our saint may be Muireadhach, abbot of Hí, who died in 1011 (AU). Near Dunkeld are Kilmorich and St. Muireach's well.

Ninian appears in the Scottish Calendars at Sept. 16. In Bede he is Nynia (ablative); Alcuin, in a letter to the monks of Whithorn written before A.D. 804, calls him Nyniga (genitive); in the same document the latinized form Ninianus appears for the first time. Nynia, Nyniga probably represent the British name Nynnyaw.[2] Later we have in vernacular Scots Ringan, and in Gaelic Truinnean. An early instance of the former is Rineyan in 1301 (Bain's

---

[1] *Western Isles*, p. 225, where he gives an account of the 'green stone, much like a globe in figure, about the bigness of a goose egg,' called '*Baul Muluy, i.e., Molingus* his stone globe.'

[2] 'The two horned oxen, one of which is beyond, and the other this side of Mynydd Banawc . . . and these are Nynnyaw and Peibaw whom God turned into oxen on account of their sins.'—*Mabinogion*, 'Kulwch and Olwen.'

Cal);[1] still earlier is the Norse Rinan, if Rinansey, now North Ronaldsay in the Orkneys, is rightly explained as Ninian's Isle.[2] The change of initial *n* to *r* is usually ascribed to Gaelic influence, on the analogy of *e.g. Mac Réill* for *Mac Néill*, but the explanation seems doubtful. The change in question occurs in Gaelic only after *c* of a closely related word, and I have met no certain instance of it which is really early.[3] If, however, this explanation is accepted, the change could have arisen only through the adoption into Gaelic of the phrase 'Sanct Ninian' pronounced 'Sanc' Ninian,' and as 'sanct' is never found in purely Gaelic commemorations, while it occurs regularly in Scots, the phrase must have been borrowed from Scots vernacular. There seems no reason, however, why the form 'Ringan' should not have developed in Scots. The Gaelic form *Truinnein* appears in *Cill an Truinnein*, *Teampull an Truinnein*, *Slios an Truinnein*, in Glen Urquhart, and in Killantringan of Galloway and Ayrshire; it is evidently for *Cill Shant Rinnein*, 'Sant Ringan's church,' taken over into Gaelic from Scots.

Oengus' *Félire* has at Sept. 16 'Moninn nuall cech gena,' 'Moninn the cry of every mouth,' with the variants Moenenn, Moinenn, Moinend, Monenn; a note adds, '.i. Moinenn Cluana Conaire,' 'this is Moinenn of Cluain Conaire,' now Cloncurry in Kildare. The Martyrology of Tamlacht has 'Monenn Cluana Conaire' (LL 362 f). Gorman has 'Moinend (altered into Moenend) epscop Cluana Conaire.' The Martyrology of Donegal has 'Maoineann epscop Cluana Conaire.' The forms of the name taken together indicate that it was *Móinenn* or *Móenenn*, later *Maoineann* (two syllables); it is in fact the same as that of the saint of March 1, in Oengus *Moinend*, *Moinen*, *Monenn*, in LL 373 d

---

[1] Edward is informed that the Scots heard that 'my lord your son' was on pilgrimage to St. Rineyan, and they removed the image (of the saint) to New Abbey, and on the next morning . . . it had gone back to St. Rineyan.—Bain, ii. p. 311.

[2] *Heimskringla* (before A.D. 1237); *Orkneyinga Saga*.

[3] The change of *n* to *r* after *c* does not seem to appear in the phonetic Gaelic of the Dean of Lismore (early sixteenth century).

Moinain (accusative), in Gorman Mōenenn, in Mart. Don. *Maoineann*, in AU *Moenu*, bishop and abbot of Clonfert, our St. Monan (d. 572). The true nominative is *Moenu*—saints are often commemorated in the genitive case after a word for 'death' or 'festival' understood or expressed.

The bishop of Cluain Conaire has been identified with Ninian. A marginal note on Gorman's *Félire*, referring to him, states, 'Monenn . i. id est Ninnianus episcopus Candidae Casae,' 'this is Monenn, that is to say Ninian of Candida Casa.' This statement is probably based on the *Life of Ninian* mentioned by Ussher as extant among the Irish in his time (1581 to 1656). It related how Ninian left Candida Casa on account of a vision, and went to Ireland; there he obtained from the king a suitable place called Cluayn-coner, where he established a great monastery and where he died after spending many years in Ireland. This *Life* is now lost, but the Bollandists, who had a translation of it, thought very poorly of it and say that it was full of falsehoods.[1] This appears to be all the evidence for the identification. Stokes in his second edition of Oengus lends it some countenance by adopting the reading *Moninn*, which he divides as *mo Ninn* and translates 'my Ninn,' adding in his index, 'has been identified with Ninian.' But the variants given above are against Stokes' assumption that the name contains *mo*; the strong probability is that Móenenn of Cluain Conaire and Ninian are two different persons, and that the writer of the *Life* referred to by Ussher took advantage of the superficial resemblance of the names to connect Ninian with Ireland.

In the numerous commemorations of Ninian[2] his name never appears in its native form; what appears is either the latinized form or a Gaelic form derived therefrom through Scots vernacular. He is thus a notable exception

---

[1] See Forbes's *Introduction to the Life of St. Ninian*. It explains Wytterna (Whithorn) as from Terna and Wyt, being the names of a certain blacksmith and his son. Bede's account says that Ninian was buried at Whithorn.

[2] Collected by Bishop Forbes. Their number and range alone show that they cannot be all ascribed to Ninian's period.

to the rule that though the name of a native saint may be found latinized in a Latin document, it is the native name, handed down by tradition, that appears in commemorations. It is also notable that we have no record of any personal name formed from his name with *maol* or *gille* prefixed; nor have I met an instance of a fair being named after him. All this points to a tradition broken and subsequently revived, and I have already suggested that the revival of the Ninian cult took place in the twelfth century, and that it was then revived for the purpose of lending a sanction in the eyes of the people to the changes introduced by David. When Ailred states that 'the holy pontiff (Ninian) began to ordain presbyters, consecrate bishops, distribute the other dignities of the ecclesiastical ranks, and divide the whole land into definite parishes,' he is ascribing to Ninian exactly what was being accomplished in David's time, and at the same time insinuating that David's changes, instead of being innovations, were in reality a restoration of the ancient and pure system of Ninian: the monastic Scoto-Irish Church was regarded as an unauthorized and discredited interlude. That Candida Casa bore the name of St. Martin of Tours, who may have been alive when it was founded, indicates (1) Gaulish influence on early Christianity in Scotland, (2) that Ninian regarded his monastery as a daughter of Tours (p. 148).

According to the tradition preserved by Bede, Ninian converted the southern Picts, who dwelt on the side next Bede (*i.e.* the south side) of the lofty mountain ranges which separated them from the Northern Picts, or as we should say, south of the Grampians.[1] This statement of course does not affect Ninian's work in Galloway; what Bede is concerned to make clear is that Ninian's labours left the

[1] Bede says that Columba went to Britain 'predicaturus verbum Dei provinciis septentrionalium Pictorum, hoc est eis quae arduis atque horrentibus montium iugis ab australibus eorum sunt regionibus sequestratae. Namque ipsi australes Picti, qui intra eosdem montes habent sedes, multo ante tempore, ut perhibent, relicto errore idolatriae, fidem veritatis acceperant, praedicante eis verbum Nynia episcopo reverentissimo et sanctissimo viro de natione Brittonum, qui erat Romae regulariter fidem

northern Picts untouched: their apostle, says Bede, was Columba. That the southern Picts were converted much earlier than those in the north is supported, among other things, by the legends regarding their king, Nectan, who reigned in the latter half of the fifth century.

On the west, in addition to the commemorations in Wigtownshire and Ayrshire, there is, as recorded by Jocelin, the cemetery consecrated by St. Ninian 'at Cathures which is now called Glasgu,' in which the first to be buried was Fergus, a contemporary of Kentigern. That a cemetery consecrated *c.* A.D. 400 should lie unoccupied for nearly two hundred years is incredible; Jocelin's statement, however, is evidence that the commemoration of Ninian at Glasgow is earlier than the twelfth century. The New Stat. Account of Argyll (p. 429) states that 'in the island of Sanda (off the Mull of Kintyre) are situated the ruins of a chapel dedicated to St. Ninian, together with two crosses of very rude design.' The lands of St. Ninian of Kintyre comprised the six merklands of 'Machriria and Garnacapbeck' (Ret.), situated apparently near Machrahanish. There is a chapel of St. Ninian in Bute. In the east there are commemorations of Ninian in almost every county, the farthest north being St. Ninian's Isle in Shetland. Some of the northern commemorations may be due to the influence of the Abbey of Fearn whose first abbot was a canon of Whithorn and died at Fearn (now Easter Fearn in Edderton parish) *c.* 1236. Ninian has been wrongly associated with Nonakiln in Rosskeen parish, which is for *Neimheadh na Cille* (p. 249), and with *Inbhir Inneoin* at the mouth of *Allt Inneoin* in Glen Lyon, where *inneoin* is 'anvil.'

Mo-Thatha is the form assumed in Scottish Gaelic by the Irish name *Tua*, 'the silent one,' for an earlier *Tóe.* The only Tua in the Irish records is at Dec. 22, where

---

et mysteria veritatis edoctus; cuius sedem episcopatus, Sancti Martini episcopi nomine et ecclesia insignem, ubi ipse etiam corpore una cum pluribus sanctis requiescit, jam nunc Anglorum gens obtinet. Qui locus, ad provinciam Berniciorum pertinens, vulgo vocatur Ad Candidam Casam, eo quod ibi ecclesiam de lapide, insolito Brettonibus more, fecerit.'—*Hist. Eccl.*, iii. p. 4. Compare Plummer's *Baedae Opera*, ii. p. 128.

Oengus has 'ronn-ain itge Tuae .. nád labrae,' 'may Tua's prayer, which is not speech, protect us.' *Cill Mo-Thatha* is on Loch Awe in Argyll. Balmaha, with St. Maha's well near it, is in Buchanan parish on Loch Lomond. *Loch Mo-Thatháig*, anglicized Loch Mahaick, is in the Braes of Doune, east of Callander, and a little to the west of the loch is the Annat Burn. *Cill Mo-Thatha* is the old name of Glengairn church in Aberdeenshire, where a fair was held called *Féill Mo-Thatha*.

Talorcán, Oct. 30, bears a name that was common among the Picts, but nothing is known of his period or history. He does not appear in the Irish lists, unless he is to be connected with Mac Talairc, one of the *nóeb macrad* or holy youths in LL 369 g; Gorman has Meic Thalairc, perhaps for Mac Talairc, at Sept. 8. Adamnan mentions a certain Irishman 'Baitanus nepos Niath Taloirc,' 'Báetan descendant of Nia Taloirc,' *i.e.* 'of Talorc's champion'; here Talorc is probably the name of a divinity. I have not met the name elsewhere in Ireland. The church and cemetery of *Cill Taraghláin* were at one end of a plain lying above the rocks on the north side of Portree in Skye (Old Stat. Acc.). In Inverness-shire there is the parish of *Cill Taraghláin*, Kiltarlity, Kyltalargy *c.* 1224 (RM), Kyntalargyn *c.* 1226, *ib.*, etc. Logyn-talargy in the deanery of Buchan, 1275 (Theiner), is now Logie-Buchan. Talorgán was patron saint of Fordyce, where an annual fair was held 'in festo S. Tallericani'; here too was St. Tarkin's well, in a burn close to the church.[1]

Torannán, June 12, was one of the seven sons of Oengus son of Aed, son of Erc, son of Eochaidh Munremar,[2] which places him in the same generation as Gabrán son of Domangart, son of Fergus, son of Erc, etc., who died in 559. He was therefore contemporary with Colum Cille,

[1] It has been asserted that Talorcan was one of Donnan of Eigg's monks, probably through a misreading and misunderstanding of the name given in the genitive in the list in LL 359 a: *Iarloga*, nom. *Iarlug*, later *Iarlogha*, nom. *Iarlugh*. This is of course a totally different name; Talorcan does not appear among Donnan's monks.

[2] LL 350 f; BB 220 g 52; LBr. 18 c.

but older. A short but important account of the seven brothers is given in a poem of nine quatrains in the LL copy of the Martyrology of Tamlacht, ascribed to Colum Cille.[1] They were born in Alba, and crossed to Ireland. The first to go were Mo-Thrianóc, Itharnaisc, and Eoganán or Eogan. Thereafter the others crossed the sea to Ireland on a slab of stone,[2] namely Torannán, Agatán, Mo-Chullian or Mo-Chuille, and Troscán. These four prayed that they should die on the same day, and obtained their wish, dying all on June 12. The sons of Oengus were first cousins of Berach (p. 301).

Torannán is described by Oengus as—

'Torannán buan bandach
tar ler lethan longach.'

'Torannán of lasting fame and of many a deed, over the wide ship-abounding sea.' His name has many forms. Gorman gives Tarannan; in the genealogy given in BB and LBr. he is Mo-Dairen; in LL Mo-Thairen; in notes to Oengus he is Mo-Thoren, Mo-Thoria, Mo-Thairea. His churches in Ireland were Tulach Fortchirn in Ui Cennselaig in Leinster and Druim Cliab or Drumcliff in Sligo. The Irish *Life* of Colum Cille says 'Colum Cille left Mo-Thoria in Drumcliff, and he left with them a crozier which he had made himself.'[3] O'Donnell's version here has Mo-Thairen. This took place before Colum Cille left Ireland for Alba, and accords with the data of the genealogies.

A note on Oengus states that Torannán 'is in Alba,' *i.e.* he has churches there, and goes on to equate him with Palladius, a confusion which probably arose from the tradition that he had been rejected in Ireland and had returned to Alba. This tradition survives, or survived recently, in Uist, where his memory is revered.

Torannán had an oratory on an islet, now a peninsula, in Loch Caluim Cille near Baile a' Mhanaich, Balivanich,

[1] LL 354 d e—'Secht maic áille Oengusa/lotar co iath nÉrend,' etc.
[2] This method of sailing was adopted by other saints, *e.g.* Cuthbert.
[3] LBr. 33 a.

in Benbecula.[1] Taransay, an island off West Loch Tarbert, Harris, is a Gaelic-Norse hybrid meaning 'Taran's isle,' as Barra means 'Barr's isle.' It has the remains—foundations merely—of a small church called *Teampull Tharáin*, with a burial-ground called *Cladh Tharáin*; both temple and cemetery were exclusively for the use of women.[2] This form of the saint's name corresponds to that in Tarannan, Mo-thoren, Mo-Thairen. In the Scottish Calendars he is commemorated on June 12 as Mo-thoria (Kal. Drum.), Terrenanus (Kal. Arbuth. and Mart. Aberd.), Ternanus (Kal. Brev. Aberd. and Camerarius), Tarnane (King and Dempster). In the east he is specially connected with Banchory-Ternan on the Dee in Aberdeenshire, and this is the reason why he is styled *abb Benncair*, 'abbot of Bennchar,' in the note in Gorman and in Mart. Don. Here his bell, called the Ronnecht, was kept till the Reformation; here too his head was preserved, and the skin of the part that had been tonsured and anointed was seen by the compiler of the Aberdeen Martyrology about 1530. He was also connected with the parishes of Slains and Arbuthnot, and a chapel and well at Findon in Banchory-Devenick, near Aberdeen, bore his name (Forbes). For the vowel change in *Torannán* : *Ternan* compare *Solomon* : W. *Selyf*.

B.—SAINTS OF THE WEST

Báetan, later Baodán, was the name of five saints commemorated by Gorman on Jan. 14, Feb. 5, March 1, March 23, Nov. 29. We have, however, no evidence as to which of these, if any, is found in the west. *Cill Bhaodâin*, Kilbedane 1508, Kilbodane 1536 (RMS), is the church of Ardgour, 'Kilboyden in Morvern,' 1613 (RMS); compare Kilvoydan in Clare (Joyce, iii.). The parish church of Ardchattan 'is above the bigg church a litle on the syd of ane hill in a pleasant place, where the sun uses daylie to rise upone when it ryseth upone one pairt of the country,

[1] The traditions relating to Torannan are given in Dr. Alexander Carmichael's *Carmina Gadelica*, ii. pp. 78-83.

[2] Martin, *Western Isles*, p. 49; *Carmina Gadelica*, ii. p. 82.

and this is called Kilbedan.'[1] The church has been long a ruin, and the place is now called *Baile Bhaodáin*, Balliebodane 1603, Ballebadin 1631, Ballibodan 1697 (Orig. Paroch.). *Suidhe Bhaodáin*, 'Baodan's seat,' was in Gleann Salach (p. 262).

Kildavanan in Bute is Kyldavanan 1429 (RMS), Kilmavanane 1476 (Orig. Paroch.), Kildovannane 1588 (RMS), etc. Here *da*, *do*, are for *do*, 'thy,' with a variant *ma*, 'my.' The saint is probably the same as in Kilvannan in Uist, in Blaeu Kiluanen, which Martin (p. 88) calls St. Bannan's chapel. I have not heard the Gaelic pronunciation of either name, but Kilbannon near Tuam in Ireland is 'Benén's church,' from Benignus, the disciple and successor of St. Patrick, Nov. 9 (Gorman and Mart. Don.).

Berach, Feb. 15, was the son of Nemnann and abbot of Cluain Coirpthi. He stayed for some time in Hí with Columba, and Adamnan relates his adventure with a sea monster in crossing from Hí to Tiree. His *Life* states that he visited Aedán at Aberfoyle (p. 225). He is the saint of *Cill Bhearaigh*, Kilberry in Argyll, an ancient parish now joined to Kilcalmonell, where his bell is said to have been preserved. His grandfather was Aed, son of Erc, son of Eochaidh Muinremar (LL 351 c); Berach was therefore a first cousin of Torannán and his brothers.

Broc, May 1, was the saint of Rothesay Parish, formerly called in Gaelic *Cill Bhruic* or *Sgìreachd Bhruic*; the annual fair was *Féill Bhruic*,[2] held on the first Wednesday of May, styled St. Broc's day. 'Sanct Broc' was one of the seven daughters of Dalbrónach of Dál Chonchobair in the Deisi of Bregia (LL 354 c 3; LBr. 23 a); her sister Bróicsech was the mother of Brigit (Thes. Pal., ii. p. 325). She has been confused with St. Brioc, a disciple of St. Germanus of Auxerre.[3]

Cáintighearnd, Kentigerna, Jan. 7, daughter of Cellach Cualann of Leinster, died in 734 (AU). Her church, *Cill*

---

[1] Macfarlane, *Geog. Coll.*, ii. pp. 153, 515.

[2] *Old Stat. Account*, i. p. 301; *New Stat. Account*.

[3] Forbes, *Calendars*, p. 291.

Chaointeoirn or *Cill Chaointeord* (so locally), Kilchintorn in a rental of Forfeited Estates, is on the south side of Loch Duich. She is said to have died an anchorite on Inch Cailleach, 'the nuns' isle,' in Loch Lomond, the church of which was dedicated to her.

Cill Chaomháin, Kilkivan in Kintyre, may commemorate one of the saints named Cóemán, later Caomhán, nine of whom are mentioned by Gorman and fourteen in Mart. Don.

Cill Mo-Cheallaig in Islay is 'my-Cellóc's church'; Gorman has four saints of this name; there are six in Rawl. B 502, 93 h 42, etc. The saint commemorated may be Cellach, abbot of Hí from 802 to 815.

Kilwhipnach, Kilcobenauch 1607 (RMS), in south Kintyre, suggests *Coidbenach*, bishop of Ard Sratha, who died in 707 (AU), but the phonetics are doubtful.

For Kilmahoe in Kintyre see p. 162.

Cill Choireil in Lochaber, Kilkarall 1476 (RMS), where Iain Lom is buried, is supposed to commemorate St. Cyril of Alexandria, Jan. 28, given by Gorman as *Cirill*, but one of the saints named *Cairell* or *Coirell* is more likely.

Colmán Eala is the saint of *Cill Cholman Eala*, Kilcolmanel in Knapdale (p. 187). The Irish *Life* of the saint says that he visited Kintyre and obtained a monastery from the king for ridding the district of a monster (*péist*).[1]

Cill Chomáin, Kilchoman in Islay, is 'Comman's church'; ten saints of the name are in Mart. Don. The saint commemorated here is most probably the 'presbyter honorabilis' of that name mentioned in Adamnan's *Life* of Columba. He was a son of the sister of Virgnous or Fergno, fourth abbot of Hí, and was of the race of Conall Gulban.[2]

Cill Chonaid, Killiehonnet in Lochaber, commemorates *Conat*, 'little hound,' feminine in form, but unknown otherwise; compare *Conóc*, *Mo-Chonóc* in Gorman and Mart. Don., of the same meaning; also Kilquohonedy (p. 166).

---

[1] Plummer, *Lives of Irish Saints*, i. pp. 175, 176. His genealogy is in Rawl. B 502, 91 b; LL 352 f; BB 223 b; LBr. 21 a.

[2] Skene, *Life of Columba*, pp. 92, 207, 297.

Kilchousland in Kintyre commemorates Constantin (p. 188).

Cill Chrèanan, Kilchrenan on Loch Awe, is Kildachmanan or Church of St. Peter the Deacon, 1361 (Orig. Paroch.), later Kildachymanan, Kildachrenan, *ib.*; Kildeknanane, Kildechrannane, 1594 (RMS); Kildechranane, 1614, *ib.*; Kildachranane, Kildochranane, 1629, *ib.* Here *da, de, do* is the honorific *do*, 'thy,' which has dropped in the present form of the name; the modern *cr* is for an older *cn*. In Ireland *Cell mac n-Énain* became Kilmacrenan, *Cell mhic Creunain* (for *Cell mhac C.*) (Hogan). The name is obscure to me.

Kilmaronag on Loch Etive is in Gaelic *Cill mo-Chrònaig*, on record Kilmacronag 1532, Kilmachronage 1633 (Orig. Paroch.). The saint is therefore Crónóc, parallel to Crónán, both being from *crón*, saffron, yellow-coloured. Thirty saints named Crónán are mentioned by Gorman and Mart. Don. Killichronan in Mull is in Gaelic *Coille Chrònain*, 'wood of the murmuring noise.'[1]

Cummein, abbot of Iona, Feb. 24, mentioned by Adamnan as Cummeneus Albus, and who died in 669, is the saint of *Cill Chuimein*, Fort Augustus, near which is his *suidhe* or seat.

Cill Chamáig, Kilchammaig in Knapdale, is 'Camóc's church.' Gorman has Mo-Cammóc of Irris Cáin at Ap. 13.

Cill Mo-Chumáig, Kilmochumaig at Crinan, Argyll, commemorates a saint *Cummóc*, diminutive of Cumma; Mo-Chumma of Druim Ailche is in Gorman at Jan. 4, another at June 13. Our saint may be Do-Chumma son of Aedán, son of Eochaidh, son of Muiredach Muinderg (LL 349 d). A corrupt Gaelic form is *Cill Ma-Thunnáig*.

Kildalwin, 1525 (RMS), Kildalvane 1542, *ib.*, in Cowal, is on the west side of Glendaruel, between Kilbride Beag and Auchenellait.

Kildallok 1481 (RMS), Kildallage 1586, in South Kintyre, is 'Dallóc's church'; the name is a variant of *Dallán*, from *dall*, blind; Dallán mac Forgail, Jan. 21, was the author of *Amra Coluim Cille*; Mart. Don. has Dallán of Aelmagh at Dec. 14.

---

[1] My authority is Mr. John M'Cormick.

Kildavaig near Ardlamont Point, with a *clachán*, is 'Damhóc's church'; compare Damnat of Sliab Betha, a female saint, fron *dam*, *damh*, ox, with the diminutive suffix *-nat* (Oengus, note on June 13).

Cill Fhearchair, a disused burial-ground opposite Shiel school in Glenshiel, Ross-shire, may commemorate a saint named Ferchar who does not appear in the Calendars.

Kilfinnichen, *Cill Fhionnchάin*, in Mull commemorates Findchan, a contemporary of Columba, who founded the monastery of Artcháin in Tiree. Mart. Don. has five saints of that name.

'Insula S. Finlagani in Yle,' 1427 (RMS), 'the isle of St. Finlagan in Islay,' has the remains of a chapel built and equipped by John of Islay, Lord of the Isles (d. 1386). In this isle, which was called *Eilean na Comhairle*,' 'Isle of Council,' the Lord of the Isles was wont to hold his council round a table of stone.[1] The name is a diminutive of Findlug, formed like Flannagan, etc. (p. 166).

Fintóc, a variant of Fintán, is seen in *Cill Fhionntâig*, Kilintag in Morvern. Hence *Mac Gille Fhionntáig*, reckoned as equivalent to 'Lindsay,' and anglicized 'M'Lintock.'

Fintán gives *Cill Fhionntáin*, Killundine, in Morvern (p. 93). Mart. Don. has twenty-two saints of this name.

Flannán of Cell da Lua, Dec. 18, may be the saint of the Flannan Isles; his genealogy is given in LL 351 b. In the *Life* of St. Abban, he is made contemporary with Brendan (d. 577), Moling (d. 697), and Munnu (d. 635), but of course all these were not contemporaries.[2] Gorman has two other saints of the name, which is from *flann*, red.

Cill mo Liubha, Kilmalieu, at Inveraray is 'my-Liubha's church,' *Liubha* being the modern form of *Liba*. Gorman has Mo-Liba of Gleann dá Locha, three saints called Liba, and two called Do-Liba; Mart. Don. has four saints called Mo-Libha. The mother of Mo-Liba, son of Colamda, was Cailltigern, sister of Cóemgen, and one of his brothers was Dagan, who died in 639 (FM). Another *Cill mo-Liubha* is

[1] Compare Martin, *Western Isles*, p. 240.
[2] Plummer, *Vitae SS. Hib.*, i. p. 25.

on the west side of Loch Linnhe in Kingairloch parish; 'Kilmalew in the lordship of Morvern,' 1508 (RMS). On the north side of Loch Eil is *Achadh da-Liubha*, Achdaliew, 'thy-Liubha's field.' Killmalive in Skye is on record in 1666 (RMS).

Cill Mhic Eoghain, Kilviceun in Mull, is Kilmakewin 1587 (RMS), 'the church of the son (*mhic*) or of the sons (*mhac*) of Eoghan.' The latter is probably correct, for Gorman has 'trí meic Eoghain,' 'the three sons of Eoghan,' at May 19. Reeves refers it to Ernan mac Eoghain, St. Columba's nephew (Adam., p. 415, n.). Compare, however, *Cell mac nEogain* in Connacht (Hogan).

Máeldubh, probably for *Máldubh*, 'black prince,' is commemorated in Kylmalduff, Inverary, 1304 (Orig. Paroch.). Five saints are so named in Mart. Don. Two quatrains in praise of Maeldubh son of Aed Findliath are in LL 348 (footnote).

Kilmaglass, Strachur, appears to mean 'church of the sons of Glas'; compare *Dùn mac Glais*, Dunmaglas, at the head of Strath Nairn.

Mo-Laisse, Mo-Laise, is a reduced form of Laisrén with *mo*, my, prefixed. Of several saints who bore that name, the one commemorated in Scotland is probably Mo-Laisse, given as son of Cairell and Maithgemm, 'noble jewel,' daughter of Aedán mac Gabráin, king of Dál Riata. He is also called Do-Laisse, and he was abbot of Lethglend; in recording his death in 639, AU gives his father as Cuinide. In a memorial quatrain he is styled 'abbot of *rathchell*,' which has been taken to be the name of a place not identified; the first part of *rathchell*, however, as shown by the rime with *Maithgemm*, is *rath*, grace, and the term means 'grace-church,' 'church of grace.'[1] At April 18 Oengus has 'Laissrén

[1] Maithgemm ingen Aedáin maic Gabráin rig Alban máthair Mo-Lassi maic Cairill; is é congab Lethglend, unde—

'Mo-Lasse lassar di thenid/cona chlasaib comaid
abb rathchille, rí in tsenaid/mac Maithgeme Monaid.'

'Maithgemm daughter of Aedán mac Gabráin, king of Alba, was mother of Mo-Laisse son of Cairell, and it is he who founded Lethglend, whence (the quatrain)—Mo-Laisse, a flame of fire, with his choirs in

lassar buadach/abb Lethglinne lígach,' 'Laisren, a victorious flame (? a flame of virtue), gentle abbot of Lethglenn.' He is the saint of Holy Island at the entrance to Lamlash Bay in Arran, where his cave is. Lamlash is 'Helantinlaysche' in Fordun, for which should be read 'Helantmlaysche,' *i.e. Eilean M'Laise*, 'Mo-Laise's isle.'[1] The town of Lamlash is still called in Gaelic *an t-Eilean*, 'the island,' and *an t-Eilean ard*, 'the high island.' It is therefore clear that Lamlash is for *'lean-M'Laise*, the first syllable of *eilean* having been elided in unstressed position, and that the name was primarily that of the island now called Holy Island. MacVurich says of Raghnall son of Somerled, who died in 1207, that it was he who endowed the monastic order of Mo-Laise.[2]

From consideration of the dates it is clear that the account of Mo-Laisse's parentage given above and in LL, BB, LBr. as noted below is incorrect. Muiredach Muinderg, first Christian king of Ulster, died in 479 (FM). His son Cairell, king of Ulster, died in 526 (FM). As Aedán mac Gabráin was born *c.* 533, a daughter of his could not have been wife of Cairell. Nor could Mo-Laisse, who died in 639, have been son of Cairell, who died 113 years before. Mo-Laisse, however, may have been a grandson of Cairell, as stated in the note on Oengus, and son of Maithgemm. If so, it is of interest to observe that Báetán, son of Cairell, the famous king of Ulster, to whom Aedán mac Gabráin is stated to have made submission,[3] was Mo-Laisse's uncle. There

---

harmony, abbot of a church of grace, king of the synod, son of Maithgemm of Monadh.'—LL 372 b; so also BB 212 b 37, omitting the reference to Lethglend; compare Oengus' *Félire*, p. 116 (note), where Stokes prints Ráthchell, wrongly, and (p. 117, note 5) gives a reference to the genealogy of Mo-Laisse son of Natfraech, not Mo-Laisse of Lethglend. The genealogy of the latter is given in LL 349 d, BB 219 c 8, LBr. 16 d:—Mo-Laisse Lethglinde m. Cairill Chruaid (LL; Cruind LBr., maic Cruind BB) m. Muredaig Muindeirg m. Forgo (m. Dallain m. Dubhthaig m. Miann, LBr.). The note to Oengus' *Félire*, p. 116, has:—Mo-Laissi mac mic Cairill m. Ardgail m. Muiredaig Muindeirg.

[1] Fordun regularly writes *eilean* as *Helant*, *e.g.* 'Helant Macarmyk,' etc.

[2] As é fós do chumhdaigh ord riaghailt Mho-Laisi.—*Rel. Celt.*, ii. p. 156.

[3] Skene, *P.S.*, p. 127; see above p. 130.

remains, however, the difficulty that in AU Mo-Laisse's father is said to have been Cuinide.

A saint named Lassair is commemorated in *Cill Lasrach*, otherwise *Cill Tobar Lasrach*, near Port Ellen in Islay. There is, however, nothing to show which of the saints of this name is meant; see Killasser.

Mundu or Munnu is for *Mo-Findu*, an affectionate form of Fintén, Fintán, or Finntáin, Fionntáin. The great saint of this name was Fintén Mocumoie son of Talchan, to whom Adamnan devotes a section of his *Life of Columba*. He was the founder of Tech Munnu in Wexford, and also abbot of Cluain Eidhnigh (Gorman). He died in 635 (AU), and his day is Oct. 21. He is commemorated in *Cill Mhunna*, Kilmun, on *an Loch Sèanta*, Holy Loch, in Cowal. Finnart near it probably means here 'holy cape,' from *fionn*, white, then 'holy.' There are also Kilmun on Loch Avich and Kilmun near Inveraray. *Eilean Mhunna* in Loch Leven, Ballachulish, is 'Munn's isle,' with an ancient burial-ground and remains of St. Munn's church. In 1497 (RMS) mention is made of part of the lands of Inverquhapil 'called Pordewry, with the staff of St. Munde called in Gaelic (*Scotice*) *Deowray*,' *i.e. deòradh*, 'the pilgrim'; compare St. Fillan's staff called *an Coigreach*. Inverchapel is at the foot of Loch Eck. 'Forum S. Moindi,' St. Munn's fair, is mentioned at Earls Ruthven, Forfar, 1562 (RMS).

Kilnave in Islay is *Cill Nèimh;* Ardnave is *Aird Nèimh*; *Eilean Nèimh*, with a desecrated sanctuary, is off Ardnave. Gorman has four saints named *Nem*, with *e* uncertain as to quantity; he has, however, three saints called *Némán*, which is a diminutive of *Ném*, and *Cill Nèimh* must be 'Ném's church.'

In *Cill Mo-Naomhaig*, Kilmonivaig, in Lochaber, the saint is *Náemóc*, now *Naomhág*, 'little saint'; compare Náemhán mac Ua Duibh, given by Mart. Don. at Sept. 13, latinized Sanctanus. Compare *Dùn Naomháig*, Dunnivaig, the ancient seat of the Macdonalds of Islay. *Cill Mo-Naoi'in* in Iona is probably for *Cill Mo-Naoimhín*.

Cill Naoinein, anglicized Kilninian, in Mull is Keilnoening 1561 (Orig. Paroch.), Kilninane 1642 (RMS). Skene suggests that the name is from Nennidius, son of Eochaidh, of Mull (*de partibus Mula*), who visited St. Brigit shortly before her death.[1] The Irish Life of St. Brigit in LBr. 66 a says that she received the Communion from Ninded Lámidan ('pure-handed'), after he had come from Rome. But it is doubtful whether Ninded or Ninnidh (Mart. Don.) would become Naoincin.

'Nechtán nár de Albae,' 'noble Nechtan of Alba,' is in Oengus at Jan. 8; a note in LBr. has 'anair de Albain,' 'from the east from Alba.' Gorman has at that day 'Nechtain nóebóg,' 'Nechtain a holy virgin.' In Ireland, Nechtán or Nechtain was of Dún Gemin, Dungiven in Co. Derry; he is probably the Nechtan whose death is recorded in 679 (AU). This Scottish saint is commemorated in *Cill Neachdáin*, Kilnaughton in Islay, and in *Cill Mo-Neachdáin* in Iona.[2] He appears in the east as Nathalan.

Nessán or Mo-Nessóc appears in *Clach Mo-Neasáig*, a boulder on Loch Etive side at Taynuilt. Gorman has seven saints of this name. Nessán of the sept Corco Ché flourished about A.D. 550.[3]

---

[1] *Celtic Scotland*, ii. p. 34.

[2] The Rev. R. L. Ritchie of Creich, Sutherland, informs me of a saying known to him as a native of Iona: 'thiodhlaic mi mo naoi nighnean mar sheachdnar an Cill Mo-Neachdain ann an I'; 'I buried my nine daughters as seven (*i.e.* in seven burials) in Cill Mo-Neachdain in Iona.' The meaning is that there were two double burials or one triple burial. There is a similar saying with regard to Cill Da-Bhì, Kildavee, in Taymouth Castle grounds:—

'Cladh Cill Da-Bhì,
Thiodhlaic mi mo naoinear ann mar thrì;
A nochd is buidhe Doirbhean
An déidh na cloinne caoimhe.'

This is translated in Cameron's *Guide to Aberfeldy* (p. 30):—

'The burying ground of Kildavee!
'Twas there I laid my nine as three;
Dear this night is the cross-grained one
After the rest so loved are gone.'

The writer explains the quatrain as 'the wail of a mother of ten children, who, during a plague, lost nine of them, and buried them three at a time, the least attractive of all being left her.' For the Gaelic and for the reference I am indebted to the Rev. C. M. Robertson, Islay, who got the tradition from the late Mr. Alex. Gow.

[3] *Laud Genealogies* in *Celt. Zeit.*, viii. p. 315.

Odrán, later Odhrán, Columba's kinsman and follower, who died and was buried in Iona during the founder's lifetime, is given by Oengus at Oct. 27, 'Odrán sab sóer snámach,' 'Odrán, a post noble and buoyant.' He is commemorated in *Reilig Odhrán,* Odhrán's cemetery,' in Iona, in *Cill Odhrán* in Colonsay, and in *Cladh Odhrán,* 'Odhrán's burial-ground,' in Tiree.[1]

A quatrain of the *Náemsenchas* (BB 230 a 9) says of him—

'Odrán Iea in chrábhaidh chruaidh
mac Aingein co mórbuaidh,
maic Boguine in gaisgidh ghér
maic Conaill Gulbain maic Néill.'

'Odrán of Hí, of severe piety, son of triumphant Aingen, son of Boguine of keen valour, son of Conall Gulban, son of Niall (of the Nine Hostages).' Odhran and Colum Cille were both great-grandsons of Conall Gulban.

Rónán, abbot of Cenn Garadh, Kingarth in Bute, died in 737 (AU), but I find no data to determine his day. Twelve saints of this name are given by Gorman. It is probably Rónán of Kingarth who is commemorated in *Cill Rónáin* in Islay and Kilmaronock near Dumbarton. *Teampull Rónaig* is believed to have been the old parish church of Iona; a church at Eoropie in Lewis is also so called. A tiny chapel in the remote isle of Rona is named after him.[2] An islet off the west coast of Mainland in Shetland is called St. Ronan's Isle. The island of Rona off Raasay is, however, Norse *hraun-ey,* 'rough isle.'

Slébhine, abbot of Hí, who died on Mar. 2, 767, is commemorated in Kilslevan, in Gaelic *Ci' Sléibheainn* (foı *Cill S.*), near Port Askaig in Islay.

Senchán of Kilmahanachan, Kilmosenchane 1609 (RMS), in Kintyre, may be Senchán of Imlech Ibair, Dec. 11, called Mo-Shenóc by Oengus, or he may be one of the saints named Senach. Senach mac Cairill was one of the

---

[1] How Odhran was buried alive is told in the Irish *Lives* of Colum Cille; compare the modern version given in *Celt. Rev.*, v. p. 107, from the papers of Father Allan MacDonald; here Odhran is called Dobhran.

[2] See the very interesting account of Rona in Martin's *Western Islands*. p. 19.

three seniors of the seed of Conaire (*tri senóire Síl Conaire*), LL 353 a.

According to Martin there was in the island of Vallay a chapel of 'St. Ulton,' but the statement cannot be verified now. Ultán of Ard Breccáin is in Oengus at Sept. 4. 'One of the habits of Ultán was to feed with his own hands every child who had no support in Erinn, so that he often had fifty and thrice fifty with him together' (Mart. Don.), whence Oengus says :—

'I márflaith cen etail
indat bláithe beccáin,
agait mór a maccáin
imm Ultán Aird Breccáin.'

'In the great sinless kingdom, wherein little ones are blooming, greatly play his children round Ultán of Ard Breccáin.' He died in 657 or 663 (AU).[1] He was descended from Tadg son of Cian, son of Oilill Olom, king of Munster (*Fél.*, p. 202).

C.—SAINTS OF THE EAST

Cill Mo-Bheónáig, Kilmaveonaig, in Blair Athol, is Kilmeuenoc, Kylmevenet (read -ec), 1275 (Theiner), Kilmawewinok, 1595 (RMS); the saint is probably Beoghna, second abbot of Bangor, who died in 606, and whose day is Aug. 22 (Gorman and Mart. Don.). *Clach na h-Iobairt*, 'the stone of the offering,' behind Bridge of Tilt Hotel, perhaps commemorates an offering of land to this church.[2]

Oengus has at Oct. 26 Beóán, conjoined with Nassad and Mellan, three saints of British origin in the church of Tamlachta Menann in Ulster; Beóán is commemorated

---

[1] A note on Oengus (p. 201) says, 'after the plague called Buide Connaill every babe without maintenance was brought to Ultan, so that often fifty, or a hundred and fifty, were with him at the same time.' According to AU this plague began in 664.

[2] Kilmaveonaig has been regarded as a commemoration of Adamnan, on the analogy of Ardeonaig on Loch Tay (Johnston). Ardeonaig, however, is for an older Ardewnan, and has arisen from the analogy of the neighbouring Ardtalnaig and Ardradnaig. No early form of Adamnan's name ends in *-óc*, nor does his name appear with the prefix *mo*, which in this case would require to be doubled—*Cill Mo Mh'Eodhnaig*.

also by Gorman, Mart. Don., and the Scottish Calendar of Drummond. At this day the Aberdeen Breviary has St. Bean, placed by Mart. Aberd. at Foulis in Strathearn. Bean is G. *Beathán*, a Scottish form of *Beóán*; compare *beathach*, *beothach*, an animal; in the Life of St. Cadroe *Beanus* is written alternatively *Beoanus* (*P.S.*, p. 111). The meaning is 'lively person.'

At Dec. 16 Oengus has Mo-Beóóc of Ard Chainroiss, in Gorman and Mart. Don. Mo-Pioc of Ard Camrois and Ross Caoin. A later hand in Mart. Don. writes 'Mo-bheóg ag Aongas . i . Beanus,' 'this is Mo-bheóg in Oengus, that is to say Bean.'

Both these saints lived before A.D. 800, for Oengus composed his *Félire* about that year. They have, however, been confused with a real or fictitious Beyn or Bean, styled bishop of Mortlach and first bishop of Aberdeen, whose date is put by Forbes at A.D. 1012.[1]

Our commemorations in Scotland are probably of Beóán, Oct. 26. The Charters of Inchaffray have 'ecclesia sancti Beani de Kinkelle,' A.D. 1200, etc., and 'ecclesia sancti Beani de Foulis' in A.D. 1210, etc., parishes in Perthshire. A document of 1329 (*ib.*, p. 111) is dated 'in festo beati Beani episcopi et confessoris.' One of Pont's MS. maps has 'Haly Mill,' *i.e.* holy mill, on the river Almond near a bridge above Kinchreigan near Buchanty; this is doubtless 'Sanct Mavane's mill,' 1542 (RMS), corresponding to the present mill at Buchanty near the Fendoch Burn, called in 1542 'torrens de Connachane.' A writer in the Old Stat. Acc. (vol. xv. p. 252) mentions M'Bean's chapel near M'Bean's bridge on Almond at Buchanty, obviously for Ma-Bean's (*Mo-Bheathan*) chapel, etc. T. Pont has also 'Suy Maven,' *i.e.* *Suidhe Ma-Bheathain*, 'my-Bean's seat,' between the head of Glen Almond and Glen Quaich, apparently about the head of Stuck Chapel Burn (*i.e.* *Stùc a' Chaibeil*, 'rock (or pinnacle) of the chapel').

[1] Adam King and Dempster put Bean of Mortlach and Aberdeen at Oct. 26; Camerarius, Dempster, and Usuard put him at Dec. 16, as also Forbes. See Reeves' note in Mart. Don., p. 337.

Inchmacbany, the old name of Inch parish in Aberdeenshire, is probably 'my-Bean's haugh.' Balveny in Mortlach parish is supposed to mean 'Bean's stead,' with reference to Beyn of Mortlach. Near Inverness is *Torr Bheathain*, 'Bean's knoll,' while 'the Bught' at Inverness was *Cill Bheathain*.[1] At Erne-frear (Arnpryor) in Menteith there is said to have been a chapel of St. Bean, and another at Fasslane in Rosneath—a western commemoration. Sir David Lindsay in his 'Satire of the Three Estates' has 'I mak ane vow to Sanct Mavane.'

'The vicar of Kynbethot' (read -oc) in the deanery of Mar is mentioned in 1275 (Theiner); Kynbethoc appears as a parish in the Taxation of benefices *c.* 1366 (RPSA); in 1507 (RMS) it is Kilbethok. *Bethóc*, now *Beathág*, is a woman's name; here it may be the name of a saint who is not known otherwise. Dolbethoc, 'Bethóc's meadow,' was granted to the Culdees of Monymusk by Gilcrist, earl of Mar (Ant. A. and B.).

Cladh Bhranno is an ancient graveyard near Bridge of Balgie in Glen Lyon, where there was once a chapel. The font is still there, and the bell is in the custody of the minister of Glen Lyon. The two merklands 'de le Brandvoy' appear in 1502 (RMS). *Cladh Bhranno* is 'graveyard of Branub,' a personal name with gen. *Branboth*, formed from *bran*, a raven, like *Cathub*, gen. *Cathboth*, *Fianub*, gen. *Fianboth*, *Cóelub*, gen. *Cóelboth*.

Cladh Mo-Bhrìgh, the churchyard of the old church of Lemlair by the seaside near Waterloo, east of Dingwall, commemorates one of the saints named *Bríg*, *Brígh*, explained as 'uigorosa uel uirtuosa,' 'vigorous or full of virtue.'[2] Bríg, daughter of Amalgad, of Achad Aeda and Bríg, daughter of Fergus, of Cell Brígi, were among the holy maidens who were subject to Brigit (LL 353 b; Lism. Lives, p. 336). Other saints of the same name are mentioned in Gorman and in the note to Oengus' *Félire*, Jan. 31.

Cill Chasaidh, Killiechassie, near Aberfeldy, is Kelcassin

---

[1] Fraser-Mackintosh, *Invernessiana*, p. 112.

[2] Plummer, *Vitae SS. Hib.*, ii. p. 348.

c. 1200 (Lib. Scon). The saint's name is formed from Cass, as Barre, Barri, is from Barr; compare the neighbouring *Cill Chaomhaidh.* Cass and the derivatives Cassán, Cassín or Caissín, are names of saints in Gorman and Mart. Don.; Mo-Chasóc was of Tech na Comairge in Donegal (LL 349 g). *Cass* means curly, twisted; also quick, rapid; as a personal name it is probably a reduced form of a compound. There was a chapel at Killiechassie, and a graveyard where descendants of the Wolf of Badenoch are buried. West of Killiechassie House is *Tom a' Chanoin*, 'the canon's knoll,' and *Poll a' Chanoin*, 'the canon's pool,' is in Tay opposite Aberfeldy.

Buite, Dec. 7, son of Brónach, died on the day of Colum Cille's birth. His lineage is traced to Tadg son of Cian, son of Oilill Olum, king of Munster.[1] Buite was for some time with St. Teilo of Llandaf in Wales;[2] thereafter he came to the land of the Picts, where he raised King Nechtan from the dead, and received from the king the fort (*castrum*) in in which the miracle took place. Buite thereupon consecrated it as a church.[3] The king was Nechtan Morbet, who reigned from about 457 to 481. Skene suggested that Buite's name is found in Kirkbuddo, Carbuddo, in Guthrie parish, Forfarshire,[4] on record as Kirkboutho 1463, Kerbutho 1471, Kyrkbotho 1474, Kirkbuddo 1511 (all RMS). About three and a half miles therefrom is Dunnichen, the supposed site of the battle of Dun Nechtain in 686 (AU), Dunnichtin, 1575 (RMS), 'Nechtan's fort.' While all this hangs together well, the second part of Carbuddo, Kirkbuddo, cannot represent Buite, which would appear as -buddie, -buthie. This difficulty might be overcome by supposing a diminutive form Buiteóc or Buitheóc;[5] compare Balcormok, later Balcormo. The tradition as to Buite's visit to Nechtan's court connects with that of the offering of Abernethy to Brigit, and forms another indica-

---

1 LL 351 b; Rawl. B 502, 90 h; LBr. 19 a; *Fél. Oeng.*, p. 256.

2 See Plummer, *Vitae SS. Hib.*, i. p. xxxv; *Celt. Zeit.*, vi. p. 447.

3 Plummer, *Vitae SS. Hib.*, i. p. 88.

4 *Celtic Scotland*, i. p. 134.

5 In AU and FM the saint's name is Buithe.

tion of very early Irish Church influence in the midlands. Buite was of the royal family of Munster.[1]

Coeddi, bishop of Iona, who died in 712, was a signatory to Cáin Adamnáin, 'Adamnan's Law,' or the 'Law of the Innocents,' promulgated in 697 (AU), where he is *Ceti epscop*, 'Ceti the bishop.' His day in Gorman and Mart. Don. is Oct. 24. His commemorations seem confined to upper Strath Tay. *Dail Mo-Choid*, 'my-Coeddi's meadow,' is a field on Duneaves farm, on the Lyon in Fortingal. *Féill Mo-Choid* used to be held at *Cois a' Bhile*, Coshieville, on Aug. 9 (old style). Logierait is Login mahedd, Logy mehedd, Logy Mached in the Book of Scone, for *Lagan Mo-Choid*, 'my-Coeddi's hollow'; compare *Lagan Choinnich*, p. 276. He seems to appear also in Inchcad, conjoined with Clony (Clunie in Stormont), 1275 (Theiner).

Cill Chaomhaidh, Kilchemi and Kylchemy in Book of Scone, Kylkeve, 1451 (RMS), Gillumquhambie 1549, *ib.*, is now anglicised Killiehangie; it is on the right bank of Tummel above Logierait. The site of the church is now cultivated, but two cross slabs remain by the roadside. On Nov. 2 Gorman commemorates 'Cóemhi . i . Albannach ó chill Cóeimhi,' 'Cóemhi, a man of Alba, from Cell Cóemhi'—in all probability the saint and the place now dealt with.

Kilconquhar, Kylconchare 1480, in Fife, is 'Conchobar's church'; for *Conchobar* becomes *Conchar* in later Gaelic. The saint does not occur in the Irish calendars, but Hogan gives *Cell Conchubair* from the Book of Fermoy.

Do-Chunne, given at Sept. 6 by Gorman and Mart. Don., is commemorated in *Cladh Chunna*, a small cemetery near Invervar in Glen Lyon. Over the hill, on the Loch Tay side, is *Coire Chunna*, Carwhin, 'Cunne's corrie' (cf. p. 326).

---

[1] The Latin *Life* of Buite or Boecius states that on one occasion he went to the Cianacht of Bregia, which was his paternal soil, and adds 'et, cum regem adiret, eum, quia gentilis erat, non admisit.' Skene, referring to this passage in the index to his *Picts and Scots*, has 'goes to the Kyanactei, but is repelled as a foreigner.' The real meaning of the passage is, of course, 'when he came to the king, he refused to receive him, for he (the king) was a pagan.' In this curious error Skene has been followed by Forbes and others.

Cill Mo-Chùg, Kilmahog near Callander, contains the name *Cùg* which is anglicized as Cook, an Islay surname. The commemoration can hardly be one of Mo-Chúóc, a saint given by Gorman and Mart. Don. at Oct. 21, with the note 'ua Liathain.' It may be one of Cuaca, Jan. 8 (Gorman), of Cell Chuaca, later Cell Choca, on the border of Meath. Compare Kyrkecok, p. 167. Kilmahog fair was held on Nov. 15.

Curitan, March 16, is styled 'bishop and abbot of Ross Meinn,' in a note to Gorman; in Mart. of Tamlacht he is 'of Ross maic Bairend,' in the LL copy he is 'of Ross Mend Bairend' (p. 357 b). 'Curetan epscop' follows 'Ceti epscop' as a surety for *Cáin Adamnáin* in 697. He was certainly connected with *Ros Mairenidh*, Rosmarkie, in Ross-shire, and as his other designations cannot be identified, the probability is that it is *Ros Mairenidh* that was really meant. *Cladh Churadáin* is at the church of Bona, at the lower end of Loch Ness, and between Loch Ness and Caiplich is *Suidhe Churadáin*, his seat. At the church of Corrimony, in Glen Urquhart, there is *Cladh Churadáin*, and not far from it are *Croit Churadáin*, 'Curitan's croft,' and *Tobar Churadáin*, his well, on Buntait. The saint of the old church of Farnua, north of Inverness, is called Corridon by the Minister of Wardlaw. At Struy in Strathglass is another *Cladh Churadáin*; a fourth is a small rectangular graveyard north of the farmhouse of Assynt, Novar. *Cnoc Churadáir*, for *Churadáin*, is a hillock north of Ardoch, Alness. 'Ecclesia de Ciridan' in the deanery of Buchan, 1275 (Theiner) is an interesting trace of influence from Rosmarkie in this region subsequent to the time of Mo-Luag (p. 292).

'St. Cuthbert's croft' at Peterculter in Aberdeenshire is mentioned in Antiq. of Aberdeen and Banff, p. 355. A seventeenth-century writer says, 'one mile to the S.W. of the town of Wick stands the chappel of Haulster called St. Cuthbert's Church. The common people bury their dead about it.'[1]

Tobar Chaoibeirt near Clachan of Kilmaluag, Skye, is supposed to be 'Cuthbert's Well.'

---

[1] Macfarlane, *Geog. Coll.*, i. p. 160.

Devenic, Nov. 13, is the patron of Banchory-Devenick or Lower Banchory near Aberdeen, and of Methlick. The Aberdeen Breviary makes him contemporary with Columba and Mauricius or Machar, and says that he went to Caithness, while Mauricius went to the Picts. When dying he asked that his body should be taken to some one of Mauricius' churches, and it was accordingly taken to Banquhory Devynik, where a church was raised in his honour. This rather confused legend is of no authority, but that the saint was known on the north side of the Moray Firth appears from the fact that St. Teavneck's fair was held at Creich in Sutherland.[1] St. Devenach is among the abbots invoked in the Dunkeld Litany.

St. Deuenac of Methlic is mentioned in 1365 (Reg. Ep. Aberd.). Banchory-Devenick is Banchry Deveny 1244 (Reg. Vet. Aberbr.), Banchry Deuenech 1256, *ib.*; Benchori de Neveth 1275 (Theiner), Banchry Deueny 1333 (Ant. A. and B.), Banquhory Deuiny 1346, Banchory Deueny 1438, *ib.*; forms with Devenik, etc., become regular after 1600. Landewednack or Landewennoc in Cornwall (Forbes), and Landevenech in Brittany (Skene) have been compared, both from St. Guenoc or Guenec with prefixed *de*: the difficulty is that in these the stress falls on *ven*, whereas, now at least, Devenick is stressed on the first syllable.

The name, which may be from Lat. *Dominicus*, is British, and is to be compared with *Dyfnig*, the name of a saint who accompanied Cadfan from Armorica to Britain about A.D. 520.[2] We may also compare the Welsh saint *Deunauc* or *Dyfnawg*, son of Medraut, son of Kaurdaf, son of Kradauc, whose day is Feb. 13,[3] and who is found at that day in Oengus, etc., as *Mo-Domnóc*; it was he, says Oengus, who in a little boat (*curchán*), from the east over the sea, brought the race of bees (*síl mbech*) to Ireland.

Drostan, Dec. 14, is not found in the Irish Calendars.[4]

---

[1] Macfarlane, *Geog. Coll.*, iii. p. 104.

[2] Rees, *Essay on Welsh Saints*, p. 224.

[3] Rees, *Cambro-British Saints*, p. 268; *Essay*, p. 295; *Archiv f. Celt. Lexik.*, ii. pp. 176, 194.

[4] At Dec. 14, Oengus has 'Drusus (Drursus, Trursus) cona thriúr,' Drusus with his three,' namely, Zosimus, Theodorus, and Nicasius.

The Book of Deer, in an entry which was probably written in the early part of the eleventh century, says that his father was Cosgrach, that he was a pupil of Colum Cille, and that Colum Cille and he came from Hí to Aberdour in Buchan, where Bede the Pict granted them the *cathair* of Aberdour and another *cathair*, namely Deer. The Aberdeen Breviary says that he was of the royal stock of the Scoti, that his parents handed him over to his uncle Columba, then in Ireland, to be educated, and that he assumed the monastic habit at Dalquhongale, of which he became abbot. Thereafter he became a hermit and built the church of Glenesk in Forfarshire. His bones are in a stone tomb at Aberdour. Fordun makes Drostan a great-grandson of Aedán mac Gabráin; his father was a son of the king of Demetia or South Wales. This would place him in the latter part of the seventh century.

Of these accounts, that in the Book of Deer is by far the oldest, and in the absence of other early information may be accepted as nearest to the truth as to Drostan's period and associations. No proof exists, so far as known to me, for the assertion sometimes made that Drostan was earlier than Columba and independent of him or of Iona.

The well-known inscription at St. Vigeans in Forfarshire, 'Drosten ipe Uoret ett Forcus,' most probably commemorates three clerics, one of whom may be this saint. It is assigned to the eighth century, which was probably the period of the Forcus commemorated, but that does not necessarily imply that Drosten was his contemporary.[1]

Drostán is a Gaelic form of E.Celt. *Drustagnos*, the genitive of which appears in the inscription *Drustagni hic jacit*

[1] A good photograph of the stone and inscription is reproduced in Nicholson's *Keltic Researches*,' p. 74. I accept Mr. F. C. Diack's explanation of *ipe* as a contraction for *in pace*. The inscription thus reads, 'Drosten . in peace . Uoret . and . Forcus.' The form suggests that Drosten was not contemporary with the others, but earlier. With Uoret may be compared names of kings of the Picts, namely, Wrad, Uurad, Feret, Feradacus, etc., son of Bargot, and Elpin filius Wroid, Uuroid, Ferat, Feradach, etc., in Skene's *P.S.*, and Welsh Guerith in Lib. Land. Uoret may be represented by St. Terott or Tirot, Teorot, whose chapel was in Fordel, Fife, 1570, 1611 (RMS).

Cunomori filius, found in Cornwall, and means primarily 'sprung from Drust.' Drust was a name common among the Picts, and becomes in Gaelic Drost; compare Lat. *furnus*, an oven, whence G. *sorn*, a kiln, W. *ffwrn*, an oven.

A grant to 'ecclesia Modhrusti de Markinge,' 'the church of my-Drust of Markinch' in Fife (RPSA) indicates that the saint was also known as Drust; Dempster connects him with St. Andrews. *Ard Trostáin* is on Loch Earn. Near Lochlee in Forfarshire are 'Droustie's well' and 'Droustie's meadow,' and he is said to be the patron of the church of Edzell. In Aberdeenshire he is the saint of Aberdour and of Deer; 'the church of St. Drostan of Inchemabani,' now Insh in Aberdeenshire, appears c. 1230 (Chart. Lind.); he is also patron of Skirdurstan, 'Drostan's parish,' now part of Aberlour, and of Rothiemay in Banffshire. St. Drostan's chapel stood near Dunnachton in Badenoch. Urquhart in Inverness-shire is in Gaelic *Urchardan Mo-Chrostáin*; just west of Balmacaan House in Glen Urquhart is *Croit Mo-Chrostáin* for *Mo-Dhrostáin*, 'my-Drostan's croft.' Here too were his relics, under charge of a *deoir* or 'dewar,' who had a croft called *Croit an Deoir*. In Caithness he is commemorated at Westfield in Halkirk parish and at Brabster in Canisbay; the church of Canisbay was also probably his. An islet off North Uist is called *Eilean Trostáin*.

Kilduncan, Kyldonquhane, 1378 (RMS), Kyldinechane, 1375, *ib.*, in Fife, may be for *Cill Donnchon*, 'Donnchú's church,' but no saint of that name is in the Irish calendars.

Lachlan Shaw in his History of Moray states that a chapel at Kinrara in Alvie parish, Badenoch, and another at Achnahatnich in Rothiemurchus were dedicated to St. Eata. Whether this was Eata, one of the twelve English pupils of Aedan and then abbot of Lindisfarne and of Melrose, or whether it is the Irish saint Ite, Ide, is not certain.

Englacius or Englatius, Nov. 3, is a fictitious saint with a curious history. A note on Oengus' *Félire* says of a saint named Murdebur, 'in Darbais (*i Darbais*), in the south-east of Scotland he is, *i.e.* in Ferann Martain near Buchan, and Oenglais is now the name of the place.'[1] 'South-east' is an obvious error for 'north-east.' 'I Darbais' is for 'i nTarbhais,' 'in Tarves,' for *i nTarbhais* is sounded *i Darbhais*. Tarves is a parish of Formartin, and just over its present boundary, in the adjoining parish of Methlick, there is Andet, which is for O.Ir. *andóit*, later *annáid*, 'a patron saint's church,' always a sure sign of an ancient religious foundation. *Oenglais* means 'unique brook,' 'choice brook,' and it was doubtless the name of the streamlet which is close to Andet; the name would be given with reference to its sanctity. It is this stream which was elevated into 'Beatus Englacius abbas' of the Aberdeen Breviary, with regard to whom Forbes remarks, 'no details of the life of this saint are known. . . . All the lists associate him with the parish of Tarves,[2] where his local name is Tanglan. There is a Tanglan's well at the village, and Tanglan's ford on the Ythan.' In these names *t* is the *t* of *Sanct*; for *Oeng-* into *Ang-*, compare *Oengus* into *Angus*. 'Sanct Anglas' well' became 'St. Tanglan's well'; 'Tanglandford' is a little way below the spot where the brook *Oenglais* falls into Ythan (Ordnance Survey Map).[3]

Murdebur is commemorated at Nov. 3 by Oengus, Gorman, and Mart. Don., along with his brother Corcunutan. The Litany ascribed to Oengus invokes 'Murdebur sapiens et scriba' with the gloss 'frater Caemain,' 'Murdebur the sage and scribe, brother of Caeman.' It also invokes Corconutan 'frater Murdebuir' (LL 373 b). Cáeman or Cóeman is identified by other references: 'Cóeman Enaig Thruim 7 a dá bráthair . i . Murdebur 7 Corconutan,' 'Cóeman of Enach Truim and his two brothers, Murdebur and Corconutan' (LL 350 a; 385 a). He was contemporary with Mo-Chóemóc of Liath Mo-Chóemóc, who died in 656

[1] 'Murdobur . i . sapiens . . . i Darbais a n-airther deiscert (*read* tuascert) Alban ata i fail Bucan, 7 Oenglais nomen loci eius nunc (*Fél. Oeng.*, pp. 240, 241).

[2] This is not correct. Brev. Aberd. and Mart. Aberd. say he is of Tarves; Camerarius makes him reverenced in Strathbogy. Other lists either omit him or make no reference to Tarves.

[3] Compare Plummer, *Bede*, ii. p. 26, for other fictitious saints.

(AU). Murdebur's period was therefore about A.D. 590 to 660. He was of the Dál Meissi Corb of Leinster (BB 220 a b). The historical importance of a saint's day is brought out strongly by the fact that the correct day of the real Murdebur is ascribed in our Calendars to the fictitious Englacius.[1]

At Andet in Methlick there was in later times a chapel of St. Ninian.

Along with Englacius goes Glascianus, confessor and bishop, at Jan. 30 in the Aberdeen Breviary, of whose life, as Forbes remarks, we have no details. He is connected with Kinglassie in Fife. In RPSA and Reg. Dunferm. old forms of Kinglassie are Kilglassi, Kilglassin,[2] for *Cill Glaise*, 'church of the brook,' like *Cell Glaissi* in Ireland (Hogan). A well near the church is known as St. Glass's well or St. Finglassin's well. Here again the stream (*glais*) on which the church stands has been made into a saint; Findglassin is *find glaisín*, 'holy streamlet.' Near the church is an eminence called Finmont, for *finn monad*, later *fionn mhonadh*, 'white hill,' here probably 'holy hill.' The real saint of Kinglassie is unknown.

Erchard, Yrchard, in Gaelic *M'Eircheard*, Aug. 24, is the saint of Kincardine O'Neill, and of Glen Moriston in Inverness-shire. According to the Breviary of Aberdeen, the only source for his life, he was a disciple of St. Ternan, and was born and died in Kincardine O'Neill; this is of course of little real value. On the north side of the village, we are informed by a writer in Macfarlane's Collections (i. p. 102), there is a St. Erchan's well. In Glen Moriston there are *Clachán Mheircheird*, *Suidhe Mheircheird*, and *Fuarán Mheircheird*, the saint's 'kirktoun,' seat, and well. Here too within living memory was his bell, which possessed wonderful virtues.[3] No other commemorations of him are known,

---

[1] Dempster's note is interesting: 'Englatius the bishop, who saw the destruction (*halosin*) of the Picts, and bewailed it before the war.'

[2] But 'ecclesia de Kinglassin,' 1224 (Theiner); Kinglassi, 1275, *ib.*

[3] It floated on water, but there was an injunction against testing this quality for the purpose of a wager. See Dr. William Mackay's *Urquhart and Glen Moriston*, p. 322; *Transactions of the Gaelic Society of Inverness*, xxvii. p. 149.

nor does he appear in any Irish record. His name may be explained as *air-cheard*, ' over-artificer,' ' skilled artificer ' ; compare *Daigcherd*, ' good artificer,' the name of an Irish saint (LL 348 a), who was St. Ciaran's artificer (LBr. 15 c).

Killearnan in the Black Isle, Ross-shire, is in Gaelic *Cill Iurnáin* ; there is also *Carn Iurnáin* in the same parish. Killearnan in Kildonan parish, Sutherland, is also *Cill Iurnáin* now in Gaelic, on O.S. map Kilournan, in Blaeu Kilirnan. *Iurnán* results from *Iarnán* owing to the strong palatalization characteristic of the Gaelic of Easter Ross and Sutherland, and *Iarnán* represents the earlier Ernán. In 1744 the minister of Tannadyce in Forfarshire stated that the church of Tannadyce was formerly called St. Ernan's church, and that north fron the church is a very high mountain called St. Ernan's Seat.[1] On Moll's map (1745) it is St. Ann's Seat, now St. Arnold's Seat. There is nothing to show which of the saints named Ernán, Ernín, is commemorated here (compare p. 291).

Ethernan or Itharnan, Dec. 2, is said in the Breviary of Aberdeen to have founded the church of Rathin in Buchan ; on the east side of Mormond Hill ' there is a large solitary den called St. Ethernen's Slack,'[2] where his hermitage is supposed to have been. Madderty in Perthshire is ' ecclesia Sancti Ethernani de Madernin,' 1200 (Chart. Inch.) ; in 1219 his name is spelled ' Ithernanus,' in 1220 ' Ydarnasius,' *ib.* He was also the saint of the Isle of May in the Firth of Forth, and ' St. Tuetheren's fair ' was held at Forfar (Forbes). Bishop Ethirn son of Laitbe, of Domnach Mór maic Laitbe, is at May 27 in Gorman, but our saint is probably Itarnan or Itharnan who died among the Picts in 669 (AU), and who is not mentioned in the Irish Calendars.

Féchín of Fabhar or Fore in West Meath, Jan. 20, died of the plague in 665 or 668 (AU). The affectionate form of his name is Mo Écu, for Mo Fhécu ; in the Latin *Life* Féchín is made Fekinus, Fechinus. In Scotland St. Vigeans in Forfarshire is said to be from a latinized form of Féchín,

---

[1] Macfarlane, *Geog. Coll.*, i. pp. 286, 287.
[2] *Ib.*, i. p. 57.

but I have met no direct proof of this except the fact that St. Vigeans fair used to be held on Jan. 20.[1] In evidence, however, that the saint was revered in the east we have the personal name Malæchín, for Máel Fhéchín, 'Féchín's servant' in the Book of Deer; Malechí, which occurs in an earlier section of the book, is probably the same.

Fergna Britt, 'the Briton,' March 2, was fourth abbot of Iona and died in 623, in Adamnan Virgnous and (once) Fergnous. His style of 'Britt' must have been derived from some connection with the Britons, for his descent was purely Irish; similarly the epithet *Albannach* is applied to Irish saints who worked in Scotland. It is probably his name that appears in *Tobar Fheargáin*, 'Fergan's well' at Pitlochry and near Tomintoul—both famous wells, especially the latter. *Tìr Fheargáin* is the present day Gaelic for Tir Arragan in Mull. M'Lergain, for *Mac Gille Fheargáin*, 'son of Fergan's servant,' is an Islay name.[2]

The legend of St. Fergus, Nov. 15 or 18, given in the Aberdeen Breviary appears to be founded on historical fact. He was for many years a bishop in Ireland, and then came to the west of Scotland and to Strogeth, where he founded three churches. Then he went to Caithness. Thereafter he visited Buchan, resting in a place called Lungley, where he built a basilica, which to this day exists, dedicated to himself. Then he came to Glammis, where he consecrated a tabernacle to the God of Jacob, and died full of years. His bones were placed in a marble shrine, and his head was taken to the monastery of Scone.

A council held at Rome in 721 passed certain canons, and among the bishops who signed them was 'Fergus a Pict, a bishop of Scotia,' [3] *i.e.* of Ireland. This indicates that he was a native of the north of Scotland who had come to

---

[1] Forbes, *Calendars*, from Miller's *History of Arbroath and its Abbey*. Féchín's genealogy is in Rawl. B 502, 90 g; LL 352 g; BB 223 c; LBr. 21 a. He was descended from Cairbre Nia, son of Cormac, ancestor of St. Brigit of Kildare.

[2] His genealogy is in *Félire*, p. 87.

[3] Fergustus Episcopus Scotiae Pictus huic constituto a nobis promulgato subscripsi.—Haddan's *Councils*, vol. ii. pt. I, p. 7.

Ireland, like Iogenan the Pict, a presbyter whom Adamnan mentions as being in Leinster. He is doubtless the Fergus Cruithnech commemorated in LL 362 d, Gorman and Mart. Don. at Sept. 8; his church in Ireland is not mentioned. The church of Strageath in Strathearn and the neighbouring churches of Blackford and Dolpatrick are dedicated to Patrick, which, as Forbes remarks, shows an Irish connection. In Caithness his churches are Wick and Halkirk. The *Féill Fhearghuis*, 'Fergus' fair,' was held at Wick on the fourth Tuesday of November. Lungley or Longley in Buchan is now St. Fergus. The church of Dyce was also his. At Glammis are St. Fergus' cave and well. Dalarossie in Strathdearn is in Gaelic now *Dail Fhearghuis*, but we have no evidence as to the dedication of the church there.

It is more than likely that we have St. Fergus' name in the St. Vigeans inscription, 'Drosten ipe Uoret ett Forcus,' ascribed on good authority to the eighth century (p. 317).

Cill Fhinn, Killin, is found near Garve in Ross-shire, on Loch Brora in Sutherland, at the head of Loch Tay, in Stratherrick, and on Loch Freuchie in Glen Quaich, Amulree. It has been taken to mean 'white church' (dative-locative), which is, of course, possible; but the fact that Loch Garve is still called *Loch Maol-Fhinn*, 'Fionn's servant's loch'—a name of old type—suggests a Saint Fionn, though none of the name appears on record.[1] *Cill Duinn*, Kildun near Dingwall, can hardly be other than the dative of *cell donn*, 'brown church.'

Gabréin, June 24, is said in Mart. Don., quoting the *Life* of Mo-Chua of Balla, in Mayo, to have been a bishop of the Britons and a fellow-student of Mo-Chua, to whom he offered his church of Gael in Fir Rois (in Louth or Meath). Mo-Chua was fostered by Comgall of Bangor, who died in 602, whence Gabréin's period can be inferred. He may be the saint of Kilgourie, Kylgoverin 1275 (Theiner), in Fife.

Ath Mo-Ghriam, 'my-Griam's ford,' a little below Bridge

---

[1] Compare Gillebert mac Gillefin c. 1166, who witnessed a charter of Uchtred son of Fergus of Galloway (Bain); Cudbert mac Gilguyn (of Galloway), 1296, *ib.* Here Fin and Guyn seem to be short forms of Finnén and Guinjn respectively.

of Lyon in Fortingal, may contain the name of a saint. It was a dangerous ford just above a deep pool called *Linne Lonaidh*.

Eglesgreig, now St. Cyrus in Forfarshire, is *ecclesia Cyrici*, 'St. Cyric's church.' Cyricus, martyr in Antioch, is commemorated as Giric by Oengus at July 16; in his Prologue he calls him Ciric. The Pictish Chronicle records an eclipse of the sun which took place in the ninth year of the reign of Ciricius or Girg 'in ipso die Cirici,' *i.e.* on St. Cyric's Day, A.D. 885. In Ireland Mael Giricc, abbot of Fore, died in 932 (AU). In Scotland Malgirc mac Tralin appears in the Book of Deer, A.D. 1131/2; Kilegirge son of Malis is witness, *c.* 1197 (Chart. Lind.); Malgirhe, a canon, witnesses in *c.* 1190 (Chart. Inch.), probably the same as Malgirk of Mothel (Muthil), c. 1198, and again *c.* 1200 *ib.* Greig is a common surname in the east of Scotland north of Forth. Tulachgrig in Tarves, now Tullygreig, belonged to the bishop and church of Aberdeen (Coll. A. and B., p. 338).

Itharnaisc, Dec. 22, was the saint of Lathrisk in Fife; the church of St. John the Evangelist and of St. Atherniscus of Losceresch was dedicated on July 28, 1243.[1] He was one of the seven sons of Oengus (p. 299), and his church in Ireland was Claenad, now Clane, on the left bank of the Liffey.

Kentigern or Mungo, Nov. 30, was patron of Glengairn in Aberdeenshire where the parishioners met on Jan. 13. This, however, appears to be a secondary dedication, for Glengairn church was called *Cill Mo-Thatha* (p. 298). The old church of Kinnoir, now included in Huntly parish, is also claimed for him, and there are here St. Mungo's hill and well. Early religious influence is suggested by the name Anatswell on the east side of St. Mungo's hill, but here also the dedication to Mungo may be secondary, as in the case of Muirdebur's Annat at Methlick, subsequently the site of a chapel of St. Ninian.

Of Lolan, Sept. 22, nothing authentic is known, except that he was the saint of Kincardine in Menteith. A toft

[1] A. O. Anderson, *Early Sources*, ii. p. 521; RPSA.

and garden there 'for St. Lolan's bell' are on record *c.* 1199 (Reg. Cambus.). His holy bell goes with the mansion and manor of Kincardine, 1542, 1582 (RMS), and St. Lolan's croft goes with the church land of Kincardine, 1620 (RMS).

Clach Mo-Lucháig, a large boulder in a garden by the roadside at Fortingal, may contain the affectionate form of a name such as Luchar or Luchta or Lochein, all names of saints; but Luch, 'mouse,' was itself a woman's name.[1] At one time, as I was informed, scolds were fastened to this stone; part of the irons used for fastening them remain in it.

Machar or Machorius is at Nov. 12 in the Scottish Calendars; he does not appear, under that name at least, in the Irish Calendars. He was the saint of Aberdeen, where his name survives in the parish and cathedral of Old Machar, while north of Don is New Machar. The church of Old Aberdeen is styled 'ecclesia Sancti Machorii' in Pope Adrian's Bull of 1157. The formula of early grants to the Church, both in the documents that are reckoned spurious and in those that are admitted to be authentic, is 'to God and to Blessed Mary and to Blessed Machorius and to the bishop of Aberdeen,' *e.g.* in William the Lyon's charter confirming David's grants, and his later charter of 1170 regarding the lands of Brass (Birse).

The legend of St. Machar, as given at some length by O'Donnell in his *Life* of Colum Cille (A.D. 1532) from sources that cannot now be checked, is in brief as follows. Mo Chonda, otherwise Macarius, was the son of an Irish prince named Fiachna and of Findchaem his wife. He was fostered by the king of Connacht and instructed by Colum Cille, who was his kinsman. With him he went to Alba, and was sent by him along with twelve others, whose names are not given, to the province of the Picts, with instructions to make his abode in a place where he should find a river shaped like a staff (*bachall*). He did so, and

---

[1] 'Croidh, cuius mater Luch uocabatur, quod sonat latine mus.'—Plummer, *Vitae SS. Hib.*, ii. p. 10.

after performing many miracles there, bringing hosts to the faith, and building churches, he went with Colùm Cille to Rome, where the Pope gave him the new name of Mauricius and made him bishop of Tours. At Tours he died three and a half years thereafter.

Reeves was doubtless right in identifying Mo Chonda or Mo Chonna with ToChannu Mocu Fircetea, who is named in Codex B of Adamnan's *Life of Columba* as one of the twelve who accompanied Colum Cille to Alba. It seems likely, though Reeves does not mention this point, that he is the saint whose genealogy is given as DaChonna of Ess mac nEirc son of Eochaidh son of Illand son of Eogan son of Niall Nóigiallach [1]: as Colum Cille was the great-grandson of Conall Gulban son of Niall Nóigiallach, he and DaChonna were both kinsmen and contemporaries. DaChonna of Ess mac nEirc is given by Gorman and Mart. Don. at March 8, where Oengus has Conandil. The note calls him Connadil 'at Ess mac nEirc in Connacht,' explaining that his name was Conna, and that *dil*, 'dear,' was added by his mother. Further than this we cannot go: as to the identity of ToChannu or DaChonna with Machorius or Mauricius there is no independent evidence. It is of course certain that Colum Cille never visited Rome, and this part of the legend, together with that relating to Tours, may be safely discredited. To doubt the existence of a saint called Machar or Machorius would, however, be going too far: the early and authoritative mention of him as patron of the church of Old Aberdeen is sufficient.

Kilmaichlie in Inveraven parish, Banffshire, is Kynimachlo, 1544 (Ant. A. and B.), Kinmachroun, 1624 (RMS), Kinmachlein, 1632 (Ret.), Kinmachlone, 1649, *ib.*; Blaeu has Kilmachlie; in Strathspey Gaelic now it is *Cill Mèichlidh*. As a chapel stood here, the forms with Kin- are probably due to the common confusion of Kin- and Kil-; if the second part is a saint's name, the saint is unknown: the various suggestions made as to his identity

[1] LL 348 a; LBr. 14 a.

all ignore the fact that in *Maichlie* the stress is on the first syllable.[1]

Kilmadock near Callander, Kylmadoc, 1275 (Theiner), being stressed on the last syllable cannot be referred to M'Aedóc, which is stressed on the first. It is 'my-Doc's church,' and the saint is that Docus of whom we are told that 'the second or monastic order of saints received a Mass from Bishop David and Gillas and Docus the Briton.'[2] Docus is a shortened form of Cadog of Llancarvan, the eminent Welsh saint of the sixth century;[3] the commemoration is therefore an indication of British influence. The church stood at the junction of the Annat Burn with Teith.

The same saint is commemorated in St. Madoes in the Carse of Gowrie, in the vernacular called Semmidoes or Semmidores,[4] and on record often as 'ecclesia S. Medoci,' 1517 (RMS), 'ecclesia de Sanct Madois,' 1525, *ib.*, etc. Glendoick, north by east of St. Madoes church, is not connected, being Glendovok 1484 (RMS), etc.

Kilmalemnock in St. Andrews-Lanbryde parish, Morayshire, is Kilmalaman 1426, Kilmalaymak 1497, -lamage 1499, -lannak 1531, -lenno 1532, -lemnoch 1547, -lemnok 1633, -lemocke 1661 (all RMS). In the numerous other

---

[1] Forbes gives Machalus, April 25, but this is for Mac Caille, 'lad of the veil' (*pallium*), who invested Brigit with the veil. Mackinlay gives Maughold, who is commemorated in Man, and is equated with Macc Cuill, an Ulster robber converted by Patrick. Another conjecture is Mac Loig (*Celt. Zeit.*, ix. p. 364). All these are stressed on the second part and are therefore impossible.

[2] Skene, *Celtic Scotland*, ii. p. 49.

[3] Compare 'espoc Sanctan 7 espoc Liathan 7 Mádoc oilither tri mic Channtoin rig Bretan'; 'bishop Sanctan and bishop Liathan and Madoc the pilgrim, three sons of Cannton king of the Britons' (LBr. 22 a); the corresponding entries in LL 253 b and BB 224 b are obviously corrupt. The preface to Sanctan's Hymn has 'Bishop Sanctán made this hymn . . . as he went from Clonard westward to Inis Matóc. He was brother to Matóc, and they were both of the Britons, and Matóc came into Ireland before bishop Sanctán' (*Thes. Pal.*, ii. p. 350). The brothers are said to have been grandsons of Muiredach Muinderg, king of Ulster, who died A.D. 479.—*ib.*, p. xxxix.

[4] Mackinlay, *Pre-Reformation Church*, p. 21.

instances in which it appears in RMS the spellings are all variants of -lamok with occasional instances of -lennok. L. Shaw writes: 'at Forrester's Seat stood the church of Kil-ma-Lemnoc.' The ordinary pronunciation now is said to be Kil-Molymock.[1] The commemoration is very doubtful; it may be one of Mo-Lomma (Mart. Don.) or Lommán (Oengus, etc.).

Mayota or Mazota, Dec. 23, is connected with the tradition of the gift of Abernethy to St. Brigit. According to the Aberdeen Breviary Brigit came across to Scotland on that occasion with nine maidens, of whom Mayota was the chief. She served in the church of Abernethy, and is commemorated in Dulmaok on the Dee in Aberdeenshire. Boece makes her one of the Nine Maidens, daughters of a St. Dovenald who lived in the Den of Ogilvie in Forfarshire. Dulmaok is Dulmayok, 1157 (Reg. Ep. Aberd.), later Dulmaok, Dalmaock, Dalmayock, etc., pronounced Dalmaik (Old Stat. Acc.). The lands and forest of Drum were next to Dulmayok, hence the name Drum-mayok, now Drumoak.

The record forms and the pronunciation show that the name ended in *-oc*, not in *-ot*, and that the stress fell on the first syllable; the *z* of Mazota should indicate *gh* or *dh*. All this suggests a possible M'Aedóc, M'Aodhóg; compare Cell M'Aedhog in Kilkenny, now Kilmogue, also Cell M'Aedhog in Kildare, now Kilmeague (Hogan), corresponding exactly to Drumoak and Dalmaik respectively. M'Aedóc of Lismore is in Gorman and Mart. Don. at Dec. 29, and he is probably the real saint. Mayota or Mazota is a fiction of the same kind as Kevoca (p. 189); it may be noted that the name does not occur among the holy maidens invoked in the Dunkeld Litany.

Portmoak, Portemuoch, *c.* 1152 (RPSA), at Lochleven in Kinross, may contain the saint's name.

The saint of St. Monans in Fife is probably Móinenn given by Oengus at March 1, in Mart. Don. Maoineann, in AU Mōenu. He was bishop of Brendan's monastery of

[1] D. Matheson, *Place-Names of Elginshire*, p. 185.

Clonfert, and died in 572 (AU). The Litany of Oengus invokes 'fifty men of the Britons who were with Mōinu in Land Lére,'[1] considered to be now Dunleer in the diocese of Armagh. A charter of David II. grants certain lands in Fife 'Deo et Beate Marie Virgini et Beato Monano' and to 'the chaplains in our chapel of St. Monan, which we have founded anew,' 1369 (RMS).

The Scottish Calendars give Moroc at Nov. 8. According to Mart. Aberd. he was patron of Lekraw (Lecropt), and was buried there. Camerarius makes him abbot of Dunkeld. Forbes states that in Lecropt he appears as Maworrock. I have failed to verify this, but it indicates Mo-Bharróc, the affectionate diminutive of Barrfind, and at Nov. 8 Oengus and the others commemorate Barrfind son of Aed, of Achadh Chaillten in Leinster, who was brother of Findbarr, July 4, of Inis Doimle or Inis Teimle near Waterford.[2] Kilmorich near Dunkeld, with St. Muireach's well, can have no connection with this saint.[3] *Cill Mhóraig*, Kilmorack, near Beauly, has been explained by Macbain as for *Cill Mo-Bharraig*,[4] but the position of the stress makes this impossible: *Mo-Bharrag* could not become *Mórag*. Kilmorack is Kilmorok 1437 (Orig. Paroch.), Kilmoricht 1521, *ib.*, and commemorates a Saint Móróc who may or may not be the saint intended by the Calendars. *Móróc* is a diminutive of the female name *Mór*.

A group of churches—Tullich and Bethelnie in Aberdeenshire and Cowie in Kincardineshire—have as patron a saint called in the Scottish Calendars Nothlan, Nathalan, or Nethalen, in the vernacular Nauchlan or Nachlan. The Aberdeen Breviary says that he was born and buried at

---

[1] Cóecu fer do Brénaind la Moinain i Laind Lére (LBr. 23 b). Cóica fer de Breth*naib* la mac Móinain i lLaind Léri (LL 373 d). I have translated LL, omitting *mac*.

[2] Oengus, *Félire*, p. 167; genealogies of Findbarr are given in LL 347 b, BB 215 e, LBr. 13 d, and *Félire*, p. 167.

[3] Equated by Forbes, *Calendars*, p. 414; Forbes also says 'in Blaeu's Atlas is Kilnamoraik, near Loch Lochy,' but what Blaeu has is Kilmanevach, *i.e.* Kilmonivaig.

[4] *Place-Names*, p. 154.

Tullich; his well is at the old church of Bethelnie in Meldrum parish. An old rime that used to be current at Cowie says:—

'Between the kirk and the kirk ford
There lies St. Nachlan's hoard.'

The spot indicated is now part of Cowie graveyard. This curious rime has been understood to imply that the saint was rich, not to say miserly; the true explanation, however, as it seems to me, is to be found by supposing that the couplet represents a Gaelic original. In Scottish Gaelic the term for a treasure or hoard has long been *ulaidh*; in earlier Gaelic, however, this word meant a tomb, especially a tomb of stone, also a station in the shape of a stone altar on which bodies were placed before burial; a praying-station (p. 260). We need only suppose that the English couplet was formed at a time when the old meaning was forgotten, and that the original reference was to the saint's tomb or possibly to a station such as that mentioned.[1] Such 'grave-verses' are well known in old Gaelic and in Welsh.

The saint's day is Jan. 8, coinciding with that of Nechtan, who died in 679 and who undoubtedly laboured in Scotland (p. 308), and the coincidence is not accidental. His name is a compound, the first part of which is E. Celt. *nectos*, pure; the second may be the not uncommon *launos*, joyful: *Necto-launos*, 'pure-rejoicing' or possibly 'rejoicing in purity,' would become *Neithlaun*, *Naithlon* in British; compare Ythan (p. 210). The inference is that Nathalan was a British saint who, like Fergus and Iogenan, went across to Ireland, where he was known as Nechtan, stayed there for some time, and came back to Alba. We may compare also the unknown saint of Cambusnethan in Lanarkshire (p. 202).

---

[1] *Ulaidh* is dative of *uladh*, earlier *elad*, etc., used as nominative. For its use in connection with the grave of a saint, compare:

'A chloch thall for elaid úair/Buite búain maic Brónaig báin';

'thou stone yonder upon the cold tomb of ever-famous Buite, the son of Brónach.'—Kuno Meyer, *Todd Lecture* xiv. p. 18.

The Gaelic couplet on St. Nathalan might run thus:—

'Eadar an chill is an áth/tá ulaidh naomhtha Nathlán.'

A saint called Nechtan was honoured in Cornwall and Devon; in Brittany he is Neizan.[1]

It has been supposed that Bethelnie, Bothelny 1452 (RMS), means 'Nathalan's hut,' but this is clearly impossible.

The Culdees of Monymusk possessed a half carucate of land named Eglismenythok 1211 (RPSA), Eglismeneyttok 1245, Eglismenigcott 1260, *ib.*; this seems to be the same, in name at least, as Eglismonichto 1482 (RMS), Eglismonitho 1511, *ib.*, Eglismonichto and the salmon fishings called Palmanichto on the north side of Tay under the barony of Eglismonichto, 1619 (Ret.), Eglismonichtone, etc., in the parish of Monifieth, 1672, *ib.*, in Forfarshire. The name here appears to be Mo-Nechtán or Mo-Nechtóc; for the ending compare Balcormow, 1450 (RMS), for Balcormoc, and for the vowel Dunnichtin, 1575, Dunnichen, for *Dun Neachtain*. For the forms we may compare Bennethy, etc., from *bendachtain* (p. 264).

'Muchrieha's well' is about a mile and a half north-west of the church of Aboyne, near a stone with a cross cut on it. Near this cross stood a stone with a hollow cut in it, called 'Muchrieha's chair'; it was broken up soon after 1800.[2] This indicates a saint named *Mo-chridhe*, 'my dear one,' lit. 'my heart,' an appellative rather than a real name. The diminutive form *Mo-chridóc* [3] appears as the name of sixteen Irish saints; we may compare also *Mac Craidhe Mochta*, 'Mochta's dear one,' otherwise Aengus (p. 272), and *Mac Croide Ailbe*, 'Ailbe's dear one,' otherwise Mac Creiche.[4] There is nothing by which we can identify this Aberdeen saint. *Baile Mo-chridhe*, Balmachree, is the name of a farm in Petty near Inverness.

Sciath, Sept. 6 (Oengus), Sept. 6, Jan. 1 (Gorman and Mart. Don.), Sept. 4, 'adventus reliquiarum Scethi filiae Méchi ad Tamlacht,' LL 362 c (Mart. Taml.) In the other authorities her father is Mechar. She was of Feart Scéithe,

---

[1] J. Loth, *Rev. Celt.*, xxx. p. 150.

[2] *New Stat. Account*, p. 1059.

[3] Rawl. B 502, 94 a 10.

[4] Plummer, *Miscellanea Hagiographica Hibernica*, p. 8.

now Ardskeagh, in Co. Cork, and is made contemporary with Ailbhe of Imlech (Emly) who died between 527 and 542.[1] It is this early Munster saint who is commemorated in St. Skay's chapel in the parish of Craig in Forfarshire. This parish includes the old parish of Dunninald otherwise called the parish of St. Skaochy, 1539 (RMS) or of St. Skeoche, 1604, *ib.*; these forms of the saint's name may be compared with that in Ardskeagh. They seem due to confusion between *sciath*, gen. *scéithe*, a wing, shield, and *scé*, gen. *sciach*, hawthorn. Sciath was one of the 'Three Maidens' of Síl Conaire of Munster (LL 353 a).

Servanus or Serf, July 1, though mentioned by Irish writers as the saint of Culross in Fife, has no place in the Irish Calendars. He was a British saint, and it would appear that his work was mainly in the province of Fortriu or Strathearn. He has been connected with Palladius (*c.* 400), with Kentigern, and with Adamnan; Skene places him round about A.D. 700, and he may be right. His name may be compared with that of Seruan son of Kedic son of Dumngual Hen; this Seruan flourished in the latter part of the sixth century and had a son Mordaf.[2]

The church of St. Servanus of Dunin (Dunning) in Perthshire appears in 1219 (Chart. Inch.); his church of Alueth (Alva), near Stirling in 1272 (Reg. Cambusk.), and St. Serf's well was on the glebe near it. He was patron of Moydeuard (Monzievaird), 1219 (Chart. Inch.), and of Tulliedene, 1220, *ib.* 'Lie pait myre de Sanct Serf,' 1591 (RMS), was in Auchtermuchty. A notice in RPSA records that 'Brude son of Dergard, according to tradition the last king of the Picts, gave the Isle of Loch Leven to God and to St. Serf and the Culdee hermits dwelling there.' This Brude is equated by Skene with Brude son of Derili, who died in 706; Reeves preferred to make him Brude son of Feredach, who died in 843, and was in fact the last king of the Picts. The tale of St. Serf's argument with the Devil is told in his

[1] Plummer, *Vitae SS. Hib.*, i. p. 58 and n.

[2] Mordaf and Beli, who died in 627 (Ann. Camb.), were great-grandsons of Dumngual Hen.

Latin *Life* and by Wyntoun. Lorin or Lorne Mac Gil serf witnessed a charter at Kenmore in 1258 and again at Crieff in 1266 (Chart. Inch.). About 1143 King David greets his bishops, earls, sheriffs, servants, and gilleserfis of Clackmannan, but the exact position held by those 'servants of St. Serf' is not clear beyond that they had an interest in the subject of the grant—the common of the wood of Clackmannan.

St. Syth, who is not mentioned in any of the Calendars, must have had some vogue in pre-Reformation times, for Sir David Lindsay says of the people of his own time—

> 'Thay ryn, quhen thay haif jowellis tynte,
> To seik Sanct Syith or ever thay stynte';

i.e. 'when they have lost jewels, they make no stay till they reach St. Syth.' The patronage of St. Syths belonged to the Earl of Linlithgow in 1696 (Ret.). Kilsayth, apparently in Islay, is on record in 1665 (RMS). Kilsyth in Stirlingshire appears in Gaelic as *Cill Saoif* in MacVurich; in a poem composed about 1650 and printed in 1786 (Gillies, p. 86) it is *Cill Saithe*; in a *crosantacht* composed about 1650 it is *Cinn Saighde* (for *Cill Saighdhe*); [1] it is now known as *Cill Saidh(e)*, pronounced exactly like English 'sigh.' In record it is Kelvesyth 1210, Kelnasydhe (read Kelua-) 1217 (Reg. Glas.); here *Kel* is for *cell*, church, and *ve*, *ua* are for *mhe*, *mha*, representing aspirated *mo*, my, in unstressed position; 'Kelvesythe' is 'church of my-Syth.' What name is represented by Syth is doubtful; MacVurich's *Cill Saoif* might be for *Cill Saidhbhe*, 'Sadhbh's church,' but the other forms do not agree with his.

St. Tears chapel is in Caithness near Castle Sinclair in Wick parish. The minister of Wick, writing in 1726, describes it as 'ane old chappel called by the common people St. Tears, but thought to be in remembrance of

---

[1] This composition in verse and prose deals with the doings of the Gael in Montrose's campaigns of 1645/6. The words are 'dochum an mhachaire dheighenuidhsin Chinn Saighde,' 'up to that last field of Kilsyth,' Montrose's crowning victory in 1645. For a transcript of the whole document (TCD H 3. 18,791) I am indebted to Miss E. Knott.

Innocent day, the commons frequenting that Chappell having their recreation and pastime on the third day of Christmass' (Dec. 28).[1] It was also customary on Innocents' Day to leave bread and cheese in the chapel as an offering to the souls of the children slain by Herod.[2] The name appears to be a translation of G. *Cill nan Deur*, 'church of tears'; compare *Cell na nDér*, 'cella lacrimarum,' St. Beccan's church of Cluain Aird in Co. Tipperary.[3]

Triduana or Treduana, Oct. 8, is mentioned as one of the two abbesses who came with St. Boniface to Pictland;[4] a saint of that name is also mentioned as one of three ladies who came with Regulus when he brought the relics of St. Andrew to Scotland, and who were buried at the church of Anaglas.[5] Her legend in the Aberdeen Breviary tells how she took out her eyes in order to avoid the attentions of a certain prince. Practically the same tale is told of the Irish saint Cranat, and a similar one is told of Brigit.[6] Thereafter Triduana devoted herself to fasting and prayer in Lestalrig in Lothian, where she died and was buried. In Sir David Lindsay's time people resorted to St. Tredwell at Lestalrig 'to mend their ene.' The Orkneyinga Saga tells how Bishop John of Caithness, having his tongue and his eyes cut out at the instigation of Earl Harald in 1201, had them miraculously restored by a visit to the tomb of Tröllhæna, the Norse form of Triduana.

The *triduanum ieiunium*, 'three days' fast,' was a well-known institution in the Celtic Church,[7] whence O.Irish

[1] Macfarlane, *Geog. Coll.*, i. p. 159.

[2] *Old Stat. Account*, x. pp. 27, 29; *New Stat. Account*, Sutherland, p. 201; Caithness, pp. 133, 160; *Orig. Paroch*, p. 772.

[3] Plummer, *Vitae SS. Hib.*, i. pp. 17, 18.

[4] Skene, *P.S.*, p. 423.

[5] Skene, *P.S.*, p. 187; the text has *Keduana*. The church of St. Anaglas was at St. Andrews.—*P.S.*, p. 187.

[6] Plummer, *Miscellanea Hagiographica Hibernica*, pp. 158, 160, *seqq.*; Stokes, *Lismore Lives*, pp. 40, 188, 322. Cranat put out both her eyes, Brigit put out one; both had their sight miraculously restored.

[7] See Plummer, *Bede*, ii. p. 78. Beccán of Cluain Aird spent his whole life 'in lacrimis et ieiuniis triduanis,' 'in tears and fasts of three days.'—Plummer, *Vitae SS. Hib.*, i. p. 17.

tredan, a fast of three days. The name Triduana, 'lady of the three days' fast,' was probably not the saint's real name but given her on account of the rigour of her fasting. She may have been one of the 'bevy of virgin-girls' commemorated by Oengus at Oct. 8, the *septem filiae* of Cell na nóebingen, 'church of the holy girls,' in the precinct of Armagh.[1]

The several parts of the district of Clyne near Brora, Sutherland, are distinguished as Clynelish, Clynekirkton, Clynemilton, and Kintradwell. This last is Clyntredwane, 1566 (Orig. Paroch.), later Clyntraddel, in Gaelic *Clìn Trolla*, 'Triduana's Clyne,' or 'Tröllhæna's Clyne';[2] we may compare *Lagán Choinnich*, 'St. Cainnech's Laggan,' Forgandenny, 'St. Eithne's Forgan,' etc. In Caithness there is *Croit Trolla*, 'Tröllhæna's croft,' and in Papa Westray, Orkney, the saint's name appears in Trallyo. How Triduana came to be Tröllhæna in Norse is not clear to me. In Aberdeenshire, near the church of Kinellar, is Cairntradlin, Cartralzeane 1478 (RMS), Carnetralzeane 1511, *ib.*, Cartrylzer 1547, *ib.*, Carnetradleane 1611, *ib.*, Cartrilzear 1617, *ib.*, Carndradlane 1625, *ib.*, 'Triduana's cairn.' There are standing stones near the spot, and there is a sculptured stone at the church. In Banffshire there appears on record Cartrilzour, 1573 (RMS), Cartrilzear, 1617, *ib.*, Mynone and Cartrylezear, 1615, *ib.*, apparently Draidland, 1625, *ib.*; this seems to be the same name as that in Kinellar. Her fair was held at Rescobie in Forfarshire, where she is said to have lived an eremetical life before she went to Lestalrig. The variety of forms of her name is due to the fact that it is not native.

Volocus or Makwoloch, Jan. 29, gave name to the old parish of St. Walach, now part of Glass in Aberdeenshire. He was also patron of Logie in Mar, where his fair was held :—

> 'Wala fair in Logie Mar
> the thirtieth day of Januar.'

[1] Oengus, *Félire*, p. 222.

[2] *Clìn*, as the Rev. C. M. Robertson has pointed out, is dialectic from *claon*, a slope; Clyne parish is *Sgìre Chlìn*, for *Sgìre Chlaoin*.

His legend in Aberdeen Breviary conveys no information except that he came at a time when the faith was not yet received in all parts of Scotland. Camerarius connects him with Candida Casa, Balveny, Strathdon, and Mar, placing his death in 733. In the Litany of Dunkeld he is invoked among the bishops as Makknoloch (presumably for Makkuoloch). In fact nothing certain is known of this apparently local saint. His name suggests the rather rare Irish name Uallach, 'haughty,' diminutive Uallachán (FM).

The introduction of Christianity into the north and the history of its progress in the various districts will always be matters of great interest and not a little difficulty. In this chapter I have tried to give documented facts with regard to the saints commemorated in so far as such facts are available, and for these we have to depend in the main on Irish sources supplemented to some extent by Welsh authorities. Some general remarks may be made in conclusion.

1. In order to appreciate the status and the authority of the early clerics it is well to bear in mind that they were usually, or at least often, men of high birth, to whom rule came naturally. The reason why the genealogies of so many of them were known and preserved is that they belonged to ruling families. The political influence of such men must have been by no means negligible, though we hear little about it in Scotland except in the case of Columba.

2. The earliest purely Irish missions to Scotland of which we have tradition are those of Fáelán, son of Oengus, the first Christian king of Munster, and of Buite, who was descended from the same royal family as Fáelán, to the eastern Midlands, with which may be taken the traditional gift of Abernethy to Brigit. These events took place, if the tradition is accepted, before Columba was born, and about fifty years after the death of Ninian, who, according to Bede's tradition, had turned to the true faith the Picts who lived south of the Grampians.

3. The tradition as to Fáelán and Buite, together with the commemoration of Sciath—a very famous saint of Munster—in Forfarshire, agrees with and supports the other

evidences for early influence from Munster in the eastern Midlands. Fínán Lobur, who was also ultimately of Munster origin, is widely commemorated in west and east, but his period was much later.

4. The facts that Torannán or Ternan was abbot of Banchory-Ternan and that he is commemorated elsewhere in the east, *e.g.* at Arbuthnot, and that his brother Itharnaisc was commemorated in Fife, are proof of the presence of what may be properly called the Columban church in the east during the sixth century. This agrees with the statement in the *Amra* that Columba 'taught the tribes of Tay,' and it supports the tradition as to the connection of Deer with Iona.

5. Influence of the British church is seen in the commemorations of Cadoc of Llancarvan and possibly of Kentigern. Devenic's name is British; he appears to have been an important cleric. Nidan of Midmar, whose name is not preserved in any place-name, is said to have been a disciple of Kentigern (Aberd. Brev.), and has been identified with Nidan of Llanidan in Anglesey. Our Nidan may have been Welsh, but his day is Nov. 3, while that of the other is Sept. 30.

6. Among the natives of Pictland who were either trained in Ireland or were connected with the Irish Church are Nathalan or Nechtan, Fergus, and Eoganan. It is uncertain whether the sons of Brachan were Picts or Welshmen; Mo-Chonóc, whose name is gaelicized, is commemorated with us, but I find no trace of his brother Iast, who is said to have been connected with Lennox.

7. The connections between the district between Dee and Spey on the one hand and Sutherland and Caithness on the other are worth noting: Drostan, Fergus, and Devenic are common to both. Mo-Luag and Curitan of Rosmarkie are both commemorated between Dee and Spey as well as in Ross and Inverness-shire. On the west coast the absence of commemorations between Loch Broom and Cape Wrath indicates probably a very sparse population in early times.

It may be useful to add a list of the saints whose period is known or can be determined approximately.

A.D. 400-500.
Ninian, d. *c.* 432.
Broc.
Patrick, d. *c.* 460.
Mo-Choe, d. 497.

A.D. 500-550.
Áibind.
Brigit, d. 525.
Buite, d. 521.
Cadóc.
Cainer.
Ciarán of Cluain Mac Nois, d. 549.
Darerca or Mo-ninne, d. 517 or 519.
Eogan of Ard-sratha.
Fáelán mac Oengusa.
Sciath.

A.D. 550-600.
Aed mac Bric, d. 589 or 595.
Aengus.
Báithene, d. 600.
Berach.
Brendan of Cluain Ferta, d. 577.
Cainnech, d. 599 or 600.
Catán.
Colum Cille, d. 597.
Comán.
Mo-Chonóc.
Constantin.
Cormac úa Liatháin.
Dallán Forgaill.
Draigne.
Drostán.
Ernán, uncle of Colum Cille.
Ernán mac Eoghain, nephew of Colum Cille.
Fínán Lobur.
Findbarr of Magh Bhile, d. 579.
Findchan.
Fintán, Fintén.
Itharnaisc.
Mo-Luóc, d. 592.
Móenu, Móinenn, d. 572.
Odhrán.
Torannán.

A.D. 600-650.
Barre of Cork, d. 610.
Beogna, d. 606.
Bláán.
Colmán Eala, d. 611.
Congual, Connel.
Mo-Chutu, d. 637.
Donnán, d. 617.
Ernín mac Creseni (Mernoc), d. 635.
Fergna Britt (Virgnous), d. 623.
Kentigern, d. 612.
Mo-Laisse, d. 639.
Machan.
Munnu (Fintán), d. 635.
Murdebur.
Oswald, d. 641.

A.D. 650-700.
Báetán Maccu Cormaic, d. 664.
Mo-Chóemóc (St. Quivoc), d. 656.
Cummein, d. 669.
Cuthbert, d. 687.
Féchín, d. 665 or 668.
Itharnan, d. 669.
Nechtán or Nathalan, d. 679:
Ultán, d. 657 or 663.

A.D. 700-750.
Adamnán, d. 704.
Caintigern, d. 734.
Coeddi, d. 712.
Comgán.
Curitan.
M'Ernóc of Midluachair, d. 714.
Fáelán, son of Caintigernd.
Fergus.
Máelruba, d. 722.
Rónán, d. 737.

A.D. 750-
Slébhíne, d. 767.
Muireadhach, d. 1011.
Dubhthach, d. 1065 (?).

CHAPTER XI

BRITISH NAMES

MANY names of British origin, especially those preserved by classical writers, have been noticed in the earlier chapters. I now propose to give some further account of this important element.

In the old Welsh Laws it is stated that an ancient and famous king of Britain, Dyfnwal Moelmud, caused the island to be measured from Penrhyn Blathaon to the promontory of Pengwaed in Cornwall, which was found to be 900 miles, and that is the length of the island. A tale of the Mabinogion says of Drem, son of Dremydid, that when the gnat arose in the morning with the sun, he could see it from Gelli Wic in Cornwall as far off as Penrhyn Blathaon in North Britain. This interesting name ought to correspond to Dunnet Head. Blathaon occurs in the Mabinogion as the name of a man connected with Arthur. It is worth noting that the Britons of old had a native equivalent of 'from Land's End to John o' Groat's.' Another interesting reference in the Mabinogion is that regarding the Arthurian warrior Osla Gyllellvawr (Osla of the big dagger, *cultelli magni*), whose sheathed dagger, when laid across a stream, would serve as a bridge sufficient for the armies of the three Isles of Britain and of the three Isles adjacent. The latter three Isles, as we learn from Nennius, were Inis Gueith, the Isle of Wight (*Vectis*); Eubonia or Manau, *i.e.* Man; and 'that situated in the extreme limit of Britain beyond the Picts, and it is called Orc,' *i.e.* Orkney. To which Nennius adds, 'so it is said in an ancient proverb, when men speak of judges or kings, He judged Britain with its three Isles.' With this goes the claim that 'the Britons at first filled the whole island from the sea of Icht[1]—the English Channel—to the sea of Orcs' or the Pentland Firth. These traditions, ancient in the time of Nennius, date from, or at least refer to, a period before the Gaelic conquest.

The names that occur in the older Welsh literature relating to *Gwyr y Gogledd*, 'the men of the North,' have been collected by Sir E. Anwyl.[2] Some of them belong to places south of the Border, such as *Caer Weir*, Durham; *Caer Lliwelydd*, Carlisle; *Catraeth*, too, which Skene equated with Calatria and placed in the south-east of Stirlingshire, is now considered to be Catterick in Yorkshire.[3] We may now mention the more important of those not already noted.

First of these is *Eidyn*. This, as Sir J. Morris-Jones has established, is the correct spelling in modern Welsh, not *Eiddyn* or *Eiddin*. He points out further that the final *n* rhymes with words that have final *-nn* in modern Welsh. The term is used alone and in the combinations *Din Eidyn*, *Dinas Eidyn*, *Minit Eidyn*. On the Gaelic side we have 'obsessio Etin' in 638 (AU), where *Etin* represents O.Welsh *Eitin*. In his *Four Ancient Books* Skene identified *Etin* with Carriden in Linlithgowshire,[4] but later in his *Celtic Scotland*, he says, 'that Etin here is Edinburgh need not be doubted.'[5] The Pictish Chronicle records that in the reign of Indulfus (*Indolb*) 'oppidum Eden' was evacuated by the Angles and left to the Scots. This is the modern *Dùn-éideann*, Dunedin or Edinburgh, which correctly represents O.Welsh *Eitin*, and O.Ir. *Etin*. The modern Gaelic form, it may be added, shows that *Etain* of Tigernach (A.D. 638) is wrong, for it would result in a modern *Dùn-éadain(n)*; the chronicler probably thought of O.Ir. *étan*, gen. *étain* (now *éadann*, gen. *éadainn*), a face, which is not uncommon in names of places. In the twelfth century

---

[1] *Muir n-Icht* cannot mean 'sea of Wight,' as it is sometimes translated; it must be 'sea of Ictis,' an island mentioned by Diodorus and identified by Dr. Rice Holmes with St. Michael's Mount off the coast of Cornwall.

[2] *Celtic Review*, iv. pp. 125, 249.

[3] Morris-Jones, *Taliesin*, p. 67.

[4] Vol. i. p. 178, etc.

[5] *Celtic Scotland*, i. p. 249, note 4.

Symeon of Durham wrote *Edwinesburch*; the adjective *Edwinesburgensis* occurs in King David's charter to Holyrood; *dinas etwin* occurs in a Welsh MS. assigned to the early fourteenth century.[1] Here folk-etymology has connected the name with the English personal name Eadwine, now Edwin, presumably the king of Northumbria who was killed in A.D. 633. It is extremely unlikely that this king had any connection with Edinburgh. The genuine anglicized form is seen in 'Dunedene, which in English is Edineburg';[2] in King David's mandate of 1126 'apud Edenburge'; 1142, Edeneburg (Hol. Chart.); Edeneburgum, 1143 (Lawrie, p. 120); Edinburg, *c.* 1143 (*ib.*, p. 121), etc., and in the modern Edinburgh. The Old Welsh forms could not result from Eadwine, which would yield *Etguin* nor could the O.Irish *Eitin*, nor the modern *-éideann*.[3]

The meaning of *Eidyn*, *Dùn-éideann*, is quite obscure. It is not unlikely that we have the same word in Etin's Ha', the name of a broch on Cockburnlaw in Berwickshire, which reminds one of the Welsh *kynted Eidyn*, 'the hall of Eidin'; Duneaton in Lanarkshire may also be compared, and somewhere near the foot of Lauderdale there was a fort called Dunedin (Lib. Melr.) 'The Reid Etin' is mentioned in 'the Complaynt of Scotland' (*c.* 1550) as a popular story of a giant with three heads; it is referred to by Sir David Lindsay.

Minit Eidin, 'the upland (*mynydd*) of Eidyn,' may be the old name of the Braid Hills, where 'Braid' stands for Gaelic *bràghaid*, upland, as in Breadalbane. If Eidyn was the name of a district, as it may have been, this would be parallel to *sliabh Manann*, now Slamannan. Sir E. Anwyl thought that Minit Eidyn was Arthur's Seat. Skene considered Mynydd Agned, the scene of one of Arthur's battles, to be an alternative name for Mynydd Eidyn, but it would seem that no proof of this can be found.

In the early charters of Holyrood and elsewhere the

---

[1] *Archiv. f. Celt. Lexik.*, ii. p. 168.

[2] Life of Monenna, *Celtic Scotland*, ii. p. 37, note.

[3] Compare Sir J. Morris-Jones, *Taliesin*, 77-81; Egerton Phillimore in *Y Cymmrodor*, vol. xi.: both consider Eidyn to be a district.

Castle of Edinburgh is often called 'Castellum Puellarum,' 'the Maidens' Castle.' There has been a good deal of speculation as to the origin of this designation, but there can be little doubt that *puellae* is used here as it is in Adamnan's *Life of Columba*, 'in aliquo puellarum monasterio,' 'in a monastery of maidens,' *i.e.* of nuns. Similarly Bede mentions 'virginum monasterium' in the same sense.[1] The name is first applied to the Castle in the reign of David I., and must refer to the legend of St. Monenna and her maidens, according to which she founded seven churches in Scotland, one of which was in Edinburgh, on the top of the rock, in honour of St. Michael.[2]

Another name evidently connected with the North is *Aeron*. In a poem in the Book of Taliesin it comes after *Bretrwyn*, 'promontory of Troon,' *i.e.* 'of the nose'; next after it is mentioned 'the wood of Beit' (*coet beit*), which Skene made Beith in Cunningham. Elsewhere Aeron is associated with *Clut*, Clyde.[3] In Wales Aeron is a river name, representing, as Sir E. Anwyl pointed out, an early *Agrona*, 'Slayer,' a goddess name, from *agro-*, Gaelic *àr*, slaughter, Welsh *aer*. The name reminds us of the old Scots rime :—

'Till said to Tweed:
Though ye rin wi' speed,
And I rin slaw,
For ae man ye kill,
I kill twa.'

Sir J. Morris-Jones would equate our Aeron with the river Ayr,[4] whose genitive case is seen in *Inbhir-àir*, the Gaelic for Ayr town, meaning possibly 'estuary of slaughter.' This implies a change of name involving a sort of translation into Gaelic, but we shall see that translation of Welsh names was not uncommon. The only alternative is the

---

[1] Adamnan's *Life*, ii. 42; Bede, *Eccles. Hist.*

[2] *Celtic Scotland*, ii. pp. 37, 38. One *Life* has 'quinta vero (ecclesia) in Dunedene, quae Anglice lingua dicta Edenburg.' Another says, 'apud Edenburgh in montis cacumine in honorem Sancti Michaelis alteram edificavit ecclesiam.'

[3] *Taliesin*, p. 76.

[4] *Ib.* p. 77.

Earn of Renfrewshire; it answers better phonetically, and it is possible as regards position, but one would hardly expect so small a stream to be mentioned along with Clyde, or to have given its name to a district.

The Tract on the descent of the men of the North mentions *Catrawt Calchuynid*, 'Catrawt of Calchvynydd,' and the same place is mentioned in the Book of Taliesin. Catrawt was a brother of Clydno of Eidyn, of the race of Coel, and flourished in the sixth century. *Calchvynydd* means 'chalk hill,' 'lime hill,' and Skene correctly identified it with Kelso, which, says Chalmers, 'derived its ancient name of Calchow from a calcareous eminence, which appears conspicuous in the middle of the town, and which is still called the Chalkheugh.' The Chalkheugh overlooks the river Tweed. Here, then, we have Welsh *mynydd* translated into *heugh*, 'height.'

Taliesin, the bard of Urien of Rheged, mentions a district in the north called Goddeu, a term which means 'trees,' and as a place-name, 'forest.' Fflamddwyn—who is probably Deotric, son of Ida of Northumberland—marched in four hosts against Goddeu and Rheged. He came from Argoed to Arfynydd. He demanded hostages: were they ready? Owein and Urien defied him: he should have no hostages. On their refusal followed the battle of Argoed Llwyfein.[1] Elsewhere Taliesin speaks of 'protecting Goddeu and Rheged.'[2] The Angles came from the east; the seat of their kings was at Bamborough. Goddeu may be taken to have been adjacent to Rheged, and the order of the words suggests that it was nearer the Angles. Rheged, as we have seen, included Carlisle and the region westward along the northern shore of the Solway. It is practically certain that Goddeu was north of Rheged, and it appears in fact to represent the district known later as the Forest, now Selkirkshire. 'This shire was called the Forrest because it was wholly covered with woods except the tops

---

[1] Morris-Jones, *Taliesin*, p. 156; Skene, *Four Ancient Books*, i. p. 365; ii. p. 189.

[2] Skene, *ib.*, i. p. 351; ii. p. 190.

of the mountains, which are covered with heath.'[1] Like Manau on Forth, it was within the bounds of the Guotodin.[2]

Argoed is a compound of W. *ar* and *coed*, meaning 'on-wood,' 'near-wood,' corresponding to G. *Urchoill*, from *air* and *coill*. *Llwyfein* means 'elmwood'; *Arfynydd* means 'near (the) mountain.' The position of these places is unknown, but it may be conjectured that the two latter were not very far from the head of the Solway. Taliesin also mentions Llwyfenydd: 'the lands of Llwyfenidd, mine is their wealth, mine is their courtesy, mine is their bounteousness.'[3] They were, as Sir J. Morris-Jones remarks, the home lands of Urien, Taliesin's patron. The name is for an earlier *Leimanion*, 'elmwood,' corresponding to G. *Leamhnacht*, Lennox, with which Skene identified it. It, must however, have been in Rheged, and a trace of it may survive in Leuin, a streamlet placed on Blaeu's map near the head of Solway.

Another of Urien's battles was that of Gwen Ystrad. 'The men of Prydyn,' says Taliesin, 'advanced in hosts to Gwen Ystrad.' After the defeat of the enemy by Urien—

> 'At the gate of the ford I saw blood-stained men laying down their arms before the hoary weirs. They made peace . . . with hand on cross on the shingle of Garanwynion. The leaders named their hostages—the waves washed the tails of their horses.'[4]

Gwen Ystrad means 'white strath,' possibly 'holy strath.' Skene identified it with the Gala valley; Sir J. Morris-Jones suggests Wensleydale in Yorkshire. But the scene is the seashore, and the enemy are men of Prydyn, that is to say Picts. During the sixth century, Urien's period, the Picts were the scourge of the west coast as far south as the

---

[1] Macfarlane, *Geog. Coll.*, iii. pp. 168, 169.

[2] Morris-Jones says, 'Goddeu seems to mean the country between the two walls' (*Taliesin*, p. 73); 'Manaw either is or is in Goddeu' (*ib.* p. 74). Skene put it in Lanarkshire, equating the name with Cadyow or Cadzow.

[3] *Taliesin*, p. 182; compare p. 176; Skene, *Four Ancient Books*, i. pp. 352, 346.

[4] *Taliesin*, p. 162; Skene, *F.A.B.*, i. p. 343.

Severn estuary, and what Taliesin describes has all the appearance of a Pictish raid on the Solway coast of Rheged.

One name which we might have expected to find, but do not find, in the Welsh literature is Dunpelder, now called Traprain Law, the important stronghold in East Lothian, whose excavation has shed such a flood of light on the civilization of this district in the early centuries. 'Mons Dunpelder' was one of the seven sites on which St. Monenna planted her churches. It was from the lofty hill called Dunpelder that St. Kentigern's mother, according to Jocelin's account, was precipitated by her father's order.[1] About a mile to the south of Mount Dunpelder the lady's father was killed by a swineherd, and the spot was marked by the erection of a royal stone (*lapis regalis*, the Irish *ríglia*), with a smaller carved stone on the top of it.[2] The hill is marked 'Dunpendyrlaw' in Blaeu. The Old Stat. Account of the parish of Prestonkirk (1794) says, 'the only considerable hill in the parish is Traprane Law, formerly called Dun-pender.' The second part of the name is O.Welsh *paladyr*, now *paladr*, pl. *pelydr*, a spear-shaft, and the meaning is 'Fort of Spear-shafts'; for the idea we may compare *Dún na Sciath*, 'fort of shields,' which occurs more than once in Ireland. It is of interest to note Geoffrey of Monmouth's 'fortress of Mount Paladur, which is now called Shaftesbury,' founded, he says, by Hudibras. There was another fort of the same name near Glasgow, in 1545 (RMS) Dunpelder; 1602 Drumpendare, 1607 Dumpelder, 1608, *ib.* Dunpelder, etc., *passim*, now Drumpellier. In *paladyr* the stress fell on the penultimate syllable; *-pelder*, later *-pellier*, on the other hand, follows the Gaelic system of stressing the first syllable.

We now come to some further hints which are of importance for understanding the nature of the process that British names underwent through the influence of Gaelic,

---

[1] *Life of St. Kentigern*, c. iii.; the *Fragment of the Life of St. Kentigern* says that she was thrown down from Kepduf, now Kilduff, about four and a half miles north-west of Dunpelder.

[2] *Frag.*, c. vii.; there is a standing stone on the flat a little to the south-west of Dunpelder.

and it may be desirable to mention first one or two general considerations. The process of displacing one language by another usually takes considerable time. There is a transition period during which a good deal of bilingualism is inevitable. The length of this period will vary according to circumstances ; after a conquest by violence accompanied by settlement it may be expected to be shorter than when the change is due to what is called peaceful penetration. An instance of the latter is the introduction of English into Easter Ross, a district which has been more or less bilingual for at least two centuries, the position being that the common man spoke Gaelic and a little English or often none, while the better classes used English freely and could also speak Gaelic. In that district the change from Gaelic to English has proceeded rapidly within the last fifty or sixty years. There are other considerations which cannot be mentioned here, but one thing is obvious, namely. that the change from one language to another is greatly helped when the two languages are closely akin and have a certain amount of common vocabulary. Modern Gaelic and modern Welsh, though they differ so much that the Welshman and the Gael are quite unintelligible to each other, have nevertheless a large common element which makes Welsh very much easier for a Gael than for an Englishman, and *vice versa* ; in early times the similarity was greater. As regards the place-names, some elements of ordinary occurrence are the same in both languages, with perhaps a shade of difference in meaning ; some others are so similar that the Gaelic speaker, who understood them quite well, readily turned them into their Gaelic forms. These are general considerations which, I think, may be usefully kept in view.

Bede, writing before 733, says of the Wall between Forth and Clyde that it begins two miles west of the monastery of Aebbercurnig (Abercorn), at a place called in the Pictish language *Peanfahel*, but in English *Penneltun*, and that, running westwards, it ends near the city of Alcluith (Dumbarton Rock). Less than a hundred years later Nennius (*c.* 800) says that the Wall 'is called in the British language

Guaul,' and a marginal note adds that it runs from Penguaul, which in Gaelic (*Scottice*) is called *Cenail,* but in English *Peneltun,* as far as the mouth of the river Cluth and Caerpentaloch, where it ends in a rustic work. This note is probably about two hundred years later than Bede's account. Bede says nothing of a Gaelic form, but of course that does not prove that such a form did not exist in his time. The important thing is that here we have no less than four different forms of the same name: (1) *Peanfahel,* which is 'Pictish'; (2) *Penguaul,* which is British, *i.e.* Early Welsh; (3) *Cenail,* which is Gaelic; (4) *Penneltun,* which is English, or, rather, an anglicized form of (1), with *tun* added. The clash of languages in this district about the Wall is reflected in a very remarkable way.

The Wall in British is *guaul,* modern Welsh, *gwal,* from Lat. *vallum*: *Pen-guaul* is 'head of Wall,' 'end of Wall.' The 'Pictish' form is *Pean-fahel.* To Bede, of course, Pictish meant the language spoken in his own time by the people who were then, in his day, called Picts, and in this particular case by those 'Picts' who lived near the Wall, *i.e.* the old Caledonians. Now it ought to be obvious that, whatever value the term *Peanfahel* may possess as a specimen of the old Caledonian language, it is valueless as a specimen of the language of the ancient Picts. It is, however, of much importance in another way. The first part *pean* is British *pen,* earlier *pennos,* head. As Welsh or British initial *gu* (*gw*) corresponds to Gaelic initial *f,* both standing for an earlier *v, guaul,* later *gwal,* corresponds to Gaelic *fāl,* a rampart, and it is *fāl,* or rather its genitive *fāil,* that is written by Bede as *fahel.* His spelling simply denotes that the vowel is long; compare *Mahelgun* for *Maelgun; Dahal* for *Dāl* in John Major; Bede's own *Daal* for *Dal*; Adamnan's *cuul* for *cūl. Pean-fahel* is in fact a British name which has been half turned into Gaelic. Its form indicates that already in Bede's time the 'Picts' of the district had been subjected to Gaelic influence, and were well on the way to becoming Gaelic speakers. By A.D. 900 or thereby there is a further change: W. *pen,* head, is displaced by the Gaelic *cenn,* and we get the full-blown Gaelic name *Cenail* for *cenn-f(h)áil*, where aspirated *f*, being silent, is not written. It may be added that in O.Ir. compounds *cen* for *cenn* is not uncommon. The place is now Kinneil, stressed on the second part, as it ought to be. Here then we have a clear instance of a British compound name becoming first half Gaelic, then wholly Gaelic. The gloss on Nennius contains a closely parallel instance of the same procedure, for *Caer-pen-taloch* can be no other than *Kir-kin-tilloch*, in old spellings Kerkintalloch, Kirkintolach, etc. (RMS I, index), meaning 'fort at (the) head of (the) eminences' (G. *tulach, tilach*). In its earlier form *pen-taloch* would have been *pen-bryn* or such. The people who used the names knew perfectly well the meaning of such terms as *pen, cenn*; *bryn, tulach*, in both languages. Another example of *pen* becoming *cenn* is *Kin-pont* in Linlithgowshire, as against Penpont in Dumfriesshire; had the name been completely turned into Gaelic, it would have been Kin-drochet. The Welsh *pont*, bridge, from Latin *pons, pont-is*, seems to appear also in Petponti (RPSA), now Pitpointie, Auchterhouse, Forfar; and in Pontheugh, Berwickshire (Ret.). Doubtless many a British *pen* has been displaced by Gaelic *cenn*,[1] and what is true in this case is true of other British terms, though the nature of the case seldom admits of proof.

The same process can be illustrated from Gaelic and English. *Drochaid Charra*, 'bridge of the rock-ledge,' becomes Carr-bridge. *Allt na Frìthe*, 'burn of the forest,' becomes Free-burn, near Tomatin. *Tom Nochta*, 'naked knoll,' in south-east Perthshire, has become Naked Tam. *Tom Chàtha*, 'knoll of husks,' in Muthil, is now Shilling Hill, *i.e.* the knoll where corn used to be husked or 'shilled.' In 1150 (Chart. Cambus.) Torwood near Stirling is *nemus de Keltor*; here 'wood' translates 'Kel,' *i.e.* G. *coille.*

---

[1] The change was not confined to place-names. Samuil Pennisel, 'Hanghead,' son of Pabo Post Priten, becomes in Irish Samuel Cendisel. His wife was Deichtir, daughter of Muiredach Muinderg, the first Christian king of Ulster, who died in 479 (FM). Sons of theirs were Bishop Sanctan and Matocc the Pilgrim (*ailithir*). *Arch. f. Celt. Lexik.*, i. p. 210; BB 213 b 20; 214 a 23; LL 372 c.

Dunedin into Edin-burgh is on the same principle. These are a few of the cases of half translation or quasi-translation. Whole translation is seen, for instance, in *Allt nan Albannach* in Ross-shire becoming Scotsburn ; *Cnoc an Arbha* in Ross-shire becoming Cornhill ; *Leth-chlamhaig* in Ross-shire into Gledfield ; *an t-Ath Leathan* in Skye into Broadford ; *Baile Meadhonach* into Middleton ; Dippool in Carrick into Black Burn ; Falkirk, properly Fawkirk or such 'Varia Capella,' is a translation of *an Eaglais Bhreac*, which again may represent a British form, for Symeon of Durham has Egglesbreth. In these examples, and many more might be added, the English speakers knew the meaning of the Gaelic terms which they translated, as in a much earlier time the Gaelic speakers knew the meaning of the British terms.

British names were taken over into Gaelic, and when this took place they were sometimes given a Gaelic colouring. At the present day the Sutherland man, and others also, will say for 'I was calling on him,' 'bha mi *cailigeadh* air' : he has thus appropriated into Gaelic the English 'calling' by making *-ing* into *-ig* and adding the usual Gaelic ending for verbal nouns. This colouring is now difficult to detect, for unfortunately the Gaelic pronunciation of names south of the isthmus of Forth and Clyde is, as a rule, quite lost, and the same is largely true of the region from Spey to Forth, and even beyond Spey. We happen to know the Gaelic of Renfrew ; it occurs in the Dean of Lismore's Book, written *ryn frewi^t^*. and as it rimes with *riú*, 'to them,' this is to be read *Rinn-friú* or *Rinn-friúth.* Alexander MacDonald spells it *Renfriu*, stressed on the second part and riming with *tùs*, 'beginning.'[1] This is clearly the Welsh *rhyn-frwd*, 'point of current' ; compare 'inter burgum de Reynfru et le Nesc del Ren,' 'between the burgh of Renfrew and the point called the Ren'—at the junction of Gryfe and Clyde. Here the name has been taken over without change, and that at an early period. Another instance is 'the fords of Frew' on Forth in Menteith, called in English 'the Frews,'

[1] Dean of Lismore in *Rel. Celt.*, i. p. 91, l. 4 ; MacDonald, first edition, p. 118.

and in Gaelic *na Friùthachan.*[1] The name was applied to the shallow places where the water ran swiftly and was fordable. The fords of Frew were the first place above Stirling where the Forth was fordable; they were thus of great importance in the time before bridges, and I have heard them mentioned as one of the seven wonders of Scotland. For the present purpose the form *Friùth-ach-an*, formed like *Tròisichean*, illustrates Gaelic colouring. The vowel here is long, while that of Gaelic *sruth* is short. We may therefore suspect Welsh influence in *Srùthán*, Struan, at the junction of the rivers Garry and Erichdie; the meaning is not 'streamlet,' as in *an Sruthán*, Struan in Skye, but 'current-place,' 'stream-place.' The Perthshire *Srùthán* is not accompanied by the article.

Welsh *traws*, O.W. *tros*, across, occurs in Trostrie, Wigtown, for *tros-tref*, 'cross-stead,' 'thwart-stead'; Troustrie in Crail parish, Fife; Troston in Glencairn parish, Dumfriesshire, where W. *tref* has been translated by English *tun*. With these compare *y Drostre* in Brecon, Trostref and Trostre in L. Land., corresponding to the Gaelic *Baile-tarsuinn*, Baltersan. Trously in Stow parish, Midlothian, represents W. *trawsle* or *trosle*, 'thwart-place.' Near Bridge of Lyon in Fortingal there is *Ard-tràsgairt*, 'height of thwart-field'; the place so called is at the upper end of a big flat bounded on two sides by a bend of the Lyon, and this flat, between the two limbs of the bend is the *tràsgart*; compare *trosgardi*, L. Land., p. 145. Here *gart*, a field, is common to Welsh and Gaelic, as *rinn* was in the case of Renfrew. *Traws* appears in Gaelic dress in *na Tròsaichean*, also *na Tròiseachan*, the Trossachs, meaning 'the cross-places,' a name which applied originally to the small hilly region that stretches crosswise between Loch Katrine and Loch Achray. It is probably a Gaelic adaptation of *trawsfynydd*, 'cross-hill,' which is so common in Wales. I have been told that Crossmount, on the south side of

[1] Compare 'benorth the Frew'—Macfarlane's *Geog. Coll.*, i. p. 339.

Loch Rannoch, is *Tròscraig* in Gaelic, but I have not been able to verify this.

Another instance of the *-ach* ending is *Tèadhaich*, already noted as the Gaelic name of Menteith.

W. *pren*, G. *cran*, a tree, occurs over a large part of Scotland. It is found by itself in Pirn, formerly Pren, near Innerleithen, and Pirn in Stow parish, Midlothian, forming a parallel to G. *Craoibh*, Crieff. Pren becoming Pirn is an instance of the metathesis of *r* after a consonant which is so common in Lowland Scots, *e.g.* Burntisland for earlier Bruntiland; Glencorse for Glencrosk; Birse for Brass; Pitgerso for *Peit* (*an*) *ghréasaighe*, 'the shoemaker's share'; 'girse' for 'grass,' and so on. It seems to be a fairly modern feature. The plural of *pren* is *prenau*, which is perhaps seen in Pirnie near Maxton, Roxburgh. In composition with another term *prenn* appears in Pirn-taiton, Pryntaytoun 1598 (RMS), Stow, Midlothian; the second part is not clear, possibly *tiddyn*, 'a measure of land, a small farm,' literally 'house-land.'[1] Primside, Roxburgh, is Prenwensete in Chart. Melr., which seems to be for *pren-gwyn*, 'white-tree' (? hawthorn), with English *sete*, 'seat' or 'set'; compare Forestaresete, Forestarissete, which in Lanarkshire becomes Fortisset. The form Prenwen suggests *gwen*, fem., rather than *gwyn*, masc., but *pren* is masculine in Welsh; it may, however, have been originally neuter like O.Ir. *crann* (later masc.), in which case it might well have been fem. dialectically. Printonan, in Berwickshire, may be 'tree of the bog,' from W. *tonnen*, sward, bog; in Blaeu it is Primtanno; in Retours Prentonen. Primrose, in the counties of Midlothian, Berwick, and Fife (north-east of Denino) may be 'tree of the moor,' *rhos*. Barnbougle, in Linlithgowshire, is Prenbowgall, Pronbogalle, Pronbugele, in RMS I, for *pren bugail*, 'herdsman's tree'; 'The cattle of the tref,' says Professor J. E. Lloyd, 'grazed together in the wide pasture which surrounded it, under the eye of the village herdsman or

[1] Morris-Jones, *Welsh Grammar*, p. 146.

bugail.' [1] With this we may compare Barncluith, near Hamilton, probably 'Clyde tree.' Pirncader, Stow, appears in Armstrong's map of Berwickshire (1771); the Retours have 'Easterton, Middleton, and Netherton of Princadoes,' apparently near Fountainhall Junction; the second part is doubtful. Nithbren, now Newburn in Fife, has been noted (p. 54). *Pren* is probably seen also in Prinlaws, 1440 (RMS) Prenlas, in Leslie parish, Fife, for *prenglas*, 'green tree'; but here again the old spelling suggests that *pren* was feminine, in which case the name would be *prenlas*. There is also Pirnhill in Perthshire. Pairney in Perthshire, and Kinpurnie in Forfar may be from the plural of *prenn*, or they may be locatives from a gaelicized *prenach*, corresponding to G. *crannaich*, locative of *crannach*, 'tree-place.' Prenteineth, which appears in connection with Rutherglen in the Acts of Parliament of David I., seems to be of the Peanfahel type; the second part seems to be *tened* (*teineadh*), the genitive of O.Ir. *tene*, fire. It is at any rate the exact equivalent of the Irish *bile tened*, 'fire tree,' the ancient name of a place in Moynalty, Meath, now 'Billywood.' [2] Joyce suggests that this name takes its origin from the fact that Beltane fires were lighted under or near the tree, and this would explain why Prenteineth was so well known as to be a suitable point, together with seven others, for fixing an extensive boundary.[3] Traprain in Haddington, Trepren, 1335 (Bain) is for *tref-bren*, 'tree-stead.'

Another British term of wide range and of rather common occurrence is 'carden,' W. *cardden*, 'thicket, brake.' The oldest instance of it is in Adamnan's *Airchartdan*, now *Urchardan*, in Glen Urquhart. There is a parish of Logie-Urquhart in the Black Isle, and there is Urquhart in Morayshire, both probably the same as Urquhart in Inverness-shire. But the Urquharts and Leden Urquhart of Strathmiglo in Fife, and Urquhart near Dunfermline

---

[1] *History of Wales*, i. p. 296.
[2] Joyce, *Irish Names of Places*, i.
[3] MS. p. 14.

are of quite different origin, being, as the old spellings show, from G. *urchar*, a cast, a shot, perhaps with reference to some real or mythical feat of casting.[1] The combination *Cinn-chardain*, Kincardine, 'copse-end, wood-end,' occurs in north-east Ross-shire; in Badenoch; in Menteith; in Kincardine-on-Forth, Perthshire; in Fordun parish, Kincardineshire, whence the county name; in Blackford parish, Perthshire; and in Kincardine O'Neil, Aberdeenshire. The old form of Cardross, in Dumbartonshire, is Cardinross, 1208-33 (Orig. Paroch.), meaning 'copse point,' or perhaps rather 'copse moor'; the latter is the meaning of Cardross in Menteith. Drumchardine is in Kirkhill parish, Inverness-shire; Pluscarden, near Elgin, is now, at any rate, stressed on the first part. Cardenden in Fife means the den or hollow at or near Carden. Fettercairn in Forfar appears on record as Fothercardine, Fettercardin (RMS I, index). Carden appears with the suffix *-ach* in Cardenauch (RMS I) in Buchan, now Cardno, 'copse place,' and in Cardnye, Cardenys (pl.), near Loch of Skene, Aberdeen, now Cairnie. Cairney in Forteviot parish, Perthshire, is Cardenay, 1318; Cardny, 1313-14; Cardenai, 1444-5. The parish of Cairnie in Strathbogie was of old Cardeny. These facts indicate that our present 'cairn' and 'cairnie' do not necessarily come from *carn*, a heap of stones. The barony of 'Cardeny betuix the lowes' (*inter lacus*) was near Dunkeld, now Cardny. It is thus clear that the Gaelic *-ach* (locative *-aigh*) was added freely to this British term.

The Gaulish and O.British word for 'head,' 'end' was *pennos*, as in *Penno-vindos*, 'white-headed,' a personal name. The Welsh is *pen*, formerly *penn*; O.Irish *cenn*, now *ceann*. I have no sure instance of *pen* north of the isthmus of Forth and Clyde, nor from Galloway or Ayrshire; in these regions it must have been displaced by Gaelic *ceann* before our records begin. The instances that occur may be given according to counties.

[1] The reference may be to a spur or offshoot of rising ground.

Roxburgh

Pennygant Hill, at the head of Hermitage Water, may be compared with Pennigant in the north-west of Yorkshire; it may stand for *pen y gwant*, 'head of the mark,' or 'of the butt.' Peniel Heugh, *i.e.* Peniel Height, Crailing, near a fort or forts, may be compared with Bede's Penneltun, where Pennel is for *Pen-gwal*, 'walls' end.' Pennymure is a commanding eminence which was the site of a Roman camp on the left bank of the river Kale, and doubtless represents *pen y mûr*, 'head of the wall.' It is of special interest as one of the few names which can be connected with the Roman occupation, and may be compared with the Welsh *Tomen y Mûr*, 'the mound of the wall,' once an important Roman station. Penchrise, south of Hawick, is placed by Blaeu—misprinted Peneress for Pencress—close to 'Pen Hill'; old spellings are Pencriz, 1380 (Bain's Cal.); Penercerys, 1368 (Lib. Melr.); Penchrist (Ret.). Pennango appears in Lib. Melr. 1153-65; it seems to have been near the junction of Teviot and Allan Water, and it was in the barony of Cavers; Penangoshope was among the lands claimed by England in 1380 (Bain's Cal.).

Selkirk

The Pen of Ettrick is on the Dumfries border. Penmanscore, mentioned by Chalmers, is *Pen-maen*, 'head of (the) stone,' with English 'score.' Penmaen is a fairly common name in Wales; two places so called are near Dolgelly. *Penfaen*, with mutated *m*, would mean 'head-stone.' Penistone Knowes, south of Loch of the Lowes, may be a corruption by folk-etymology of *Pen-ystum*, 'head of the bend'; compare 'the Pennystone' in the parish of Kirkmabreck, 'under which money is fancied to be.' [1]

Berwick

Penmanshiel is *pen-maen*, 'head of (the) stone,' with English 'shiel.'

Peebles

Peniacob, 'James' head,' is the old name of Eddleston, also Penteiacob, 'head of James' house.' Penvalla is a hill in Stobo parish, and south-east of it is Penveny.

Haddington

Pencraig, in East Linton, is in Blaeu Pencraick; it is the same as Penncreic of L. Land., p. 136, meaning 'rock-end,' equivalent to the common Gaelic name *Ceann na Creige*,

[1] Macfarlane's *Geog. Coll.*, ii. p. 67.

Kincraig. Pencaitland is spelled Penketland, 1296 (Bain's Cal.); David of Penkatlond signs the Ragman Roll; John de Penkatelen, 1296 (Bain's Cal.); Pentkateland, Petkatelande, 1315-21 (RMS.). A Life of Gildas mentions 'mons Coetlann, quod sonat intrepretatum monasterium nemoris,' 'the hill of Coetlann, which means the monastery (*lann*) of the grove (*coet*, now *coed*).' Before *lann* came to mean 'monastery,' 'church,' it meant an enclosure, a court; thus *coetlann* would naturally mean 'an enclosure, or clearing, in or near a wood,' and Pencaitland probably means 'head,' or 'end,' of such a clearing. A glen off Glen Etive is called *Gleann Ceitlein*, or *Ceitilein*, 'Glenketland,' but though *Ceit(i)lein* may be a British name, it would be rash to equate it with Pen-caitland. Penratho (Ret.), now obsolete, is probably for *Pen-rhathau*, 'head of (the) raths,' from *rhath*, a circular fortified place; compare Ratho in Midlothian.

Penicuik, in Reg. Dunf. Penycok, is probably for *Pen y Gog*, 'Cuckoo-head,' from *cog* (earlier *coc*), a cuckoo. As *cog* is feminine, its *c* should here change to *g*, but our anglicized forms of Welsh names often, one may say even usually, do not show the mutation and the same applies to the corresponding 'aspiration' or 'lenition' of Gaelic names. The name of the cuckoo is not uncommon in Welsh names, *e.g. Llwyn y Gog*, 'the cuckoo's grove,' in Cardigan; with us 'gowk,' which seems to be a loan from Welsh, appears in 'Net Whowaig or Geks seit' (Blaeu) north of Glengennet in Ayrshire, where 'Net Whowaig' is for G. *nead chuthaig*, 'cuckoo's nest.' Gowk's Hill is near Gorebridge, and 'Cukoueburn' was near Primside, Roxburgh (Lib. Melr.). The cuckoo appears in Gaelic names also, as Torchuaig, 'cuckoo's knoll,' a rather high hill near the foot of Strath Braan, Perthshire. Plenploth, in Stow parish, is Plenploif, 1593 (RMS); Plenploff, 1598, *ib.*; Plamploch (Ret.); it appears to be for *Pen-plwyf*, with which may be compared *Blaen-plwyf* on Nant Wysg, a tributary of Aeron, Cardigan. *Plwyf* is from Lat. *plēb-em*, and means 'people, community, parish,' and our name probably means 'head of (the) hamlet.' Blaeu has Pendourik north-east of Dalhousie

Midlothian

Castle. Cockpen may be for *coch ben*, 'red head,' without mutation of *p*; the stress, however, is on -pen.

Dumfries

Penpont, says the Old Stat. Account, 'is probably derived from *pendens pons*, an arched bridge, there being a bridge of one semicircular arch, supported by two steep rocks, over the river Scar.' The meaning is of course 'Bridge-end.' Pennersaughs, in Annandale, is often on record, *e.g.* Richard de Penresax 1194-1214 (Bain's Cal.), Pennyrsax with its mill (RMS, i.). This appears to be for *Pen yr Sax*, 'Saxon's head'; in modern Welsh the article is *y* before a consonant, *yr* before a vowel or *h*, but in O.Welsh the *r* stood before a consonant. *Saxo* is in modern Welsh *Sais*, pl. *Saeson*. We may compare Glensaxon in Westerkirk parish, Dumfries, Glensax in Peebles, and *Gleann Sasunn*, 'glen of (the) Saxons,' near Kinlochrannoch.

Renfrew

Pennel is Penuld, *post* 1177 (Reg. Pasl.); with it goes Barpennald, *ib.*, later Foulton (*ib.* p. 412); this is *pen-allt*, 'head-hill,' 'head-cliff,' like G. *Ceannchnoc*.

Lanark

Penty, in Shotts parish, may mean 'head of house'; compare Penteiacob.

Llanerch, a clear space, a glade, occurs in Lanark of Lanarkshire, pronounced locally Lainrick, and in Barrlanark near Glasgow. Further north we have Lanrick near Loch Vennachar, in 1507 Lanark, Estir Lanarkie (RMS), and Lanrick in Kilmadock parish near Callander, both in Gaelic *Laraig* for *Lanraig*. There are also Lendrick in Kinross, and Lendrick near Kirriemuir, Forfar. All these show metathesis.

Instances of *uchel*, high, have been noted (p. 209).

Perta, 'wood,' 'copse,' was the old name of several places in Gaul, now *Perthes* (plural, for *Pertae*); it is also given in Holder as the name of a grove goddess. Derivatives are *Pertacus*, 'Forrester,' a man's name found on an inscription in Britain, and *Perticus saltus*, a forest now *le Perche* in France. In Welsh *perth* is a bush, brake, copse, found with us in the name of the town and county of Perth, in Gaelic *Peart*, gen. *Peairt*. In Forfar there are Pert, Muir of Pert, Brae of Pert, Mill of Pert, and the parish of

Logie Pert, all in the north-east, near the river North Esk. In Aberdeenshire there are on record Perthok and Ludcarn *alias* Perteoc (Exch. Rolls, iv.); Ludcarn or Ludquharn is in Longside parish. The land of Perthbeg in the parish of Mortlach appears on record often down to 1658 (RMS), and this is the furthest north instance. Larbert in Stirlingshire is Lethberth, 1195 (Johnston); Lethberd, Lethbert (Antiq. Tax.); Lethberde 1370, etc. (RMS); it seems to be for *lled-berth*, 'half-wood'; compare Narberth in Pembroke, c. 1248 Nerberd.[1] Pappert Hill in Dumbarton and Lanarkshires, and Pappert Law in Selkirk, probably have *perth* as their latter part. The Perter Burn, a tributary of Tarras, Dumfries, shows an *-r*-suffix, perhaps collective.

An early Celtic *trebo-*, 'dwelling,' perhaps also 'village,' is seen in the tribal name *Atrebates* for *Ad-trebates*, 'nigh-dwellers.' With *Contrebia*, 'joint-settlement,' we may compare G. *caidreabh*, 'society, intimacy,' earlier *co-treb*, 'dwelling together,' 'fellowship.' In Welsh *tref* is a homestead, hamlet; technically in the Welsh Laws it was a division consisting of four 'gauaels,' each of 64 acres, and four trefs made a 'maynaul'; *cantref* is the largest division of land in a lordship or dominion, a 'hundred.' The Irish cognate *treb*, *treabh*, means 'place of abode, home, region, family.' In Sc. Gaelic *treabh* is used only as a verb, 'to plough'; *treabhar* means houses collectively, especially farm-buildings:[2] the uninhabited and untilled waste is *dìthreabh*, 'wilderness.'

Treabh is extremely rare in Irish place-names, the only instances known to me being *Oentreb*, 'single stead,' now Antrim, and *Seantruibh*, 'Shantry,' 'old stead,' the name of a village north of Dublin, and these may have been influenced by British usage. In Wales, on the other hand, *tref* is as common in names of places as *baile* is with us and in Ireland.

---

[1] J. B. Johnston, *Place-Names of England and Wales.*

[2] *Treabhar*, pronounced *treo'r*, is used freely in the Gaelic of Easter Ross, but I have never met an Islesman who knew the term.

In the southern part of Scotland this term is found all over; north of the Forth and Clyde isthmus it occurs fairly often on the eastern side of the country; in the west two possible instances occur in the Kintyre region. The Gaelic-speaking people of Galloway and Ayrshire evidently retained it as a legacy from Welsh. In view of this it may be inferred that in Scotland it is to be classed with the British element rather than with the Gaelic element. Several of the formations in which it occurs are obviously purely British; others show Gaelic colouring.

In the south it appears by itself in at least three instances. A charter confirmed in 1440-1 was given by the Countess of Douglas 'apud castrum de Trefe,' *i.e.* the Douglas stronghold of Threave on an islet in the Dee; it is described in Macfarlane as 'an Island called ye Threave.'[1] Another charter by the same lady was granted 'apud le Treffe,' 'at the Tref.' The other two are in Carrick (Kirkoswald and Kirkmichael), now also Threave, but Treyf in 1504 (RMS).

We may take next a group in which the old spellings begin with Traver-, Trever-, as Travernent, now Tranent. Skene took this to be a Gaelic word *treabhar*, meaning, as he thought, 'a naked side,' but this is an error into which he was led by O'Reilly's dictionary: the word has no such meaning.[2] There is, however, as we have seen, a word *treabhar*, 'farm-buildings,' and it actually occurs in *Treabhar*

---

[1] Macfarlane's *Geog. Coll.*, ii. p. 63.

[2] O'Reilly says '*treabhar*, the side,' and refers to a quatrain which gives the meanings of some terms not in common use. The first couplet is:—

'Greann ainm d'ulchain, líth nach locht;
fec fiacail treabhar taobhnocht.'

The metre is Deibhidhe, which requires that the last word of the second line of the couplet shall have one syllable more than the last word of the first line; therefore *taobhnocht* is one word, and it means 'bare-sided.' The translation is: '*Greann* is a name for a beard—a faultless lesson (?); *fec* is a firm bare-sided tooth.' O'Reilly made the mistake of thinking that the second line meant '*fec* is a tooth; *treabhar* is a bare side,' whereas both *treabhar* and *taobhnocht* are adjectives qualifying *fiacail*, a tooth; the whole line is explanatory of *fec*. The expression 'líth nach locht' is a cheville of doubtful meaning, as chevilles often are; Ir. *líth* means 'a festival'; I have taken *líth* here to be Welsh *llith*, a lesson; Lat. *lect-io*.

nam Preas, Tornapress, the name of a farm in Lochcarron parish. This is the only instance known to me north of Forth. It may, however, be said without hesitation that 'Trever' is Welsh: it is either *tref yr*, 'stead of the,' or *tref ar*, 'stead near,' followed by the defining noun. In modern Welsh the article is *yr* before a vowel or *h*; *y* before a consonant; besides these there is *'r*, used before both vowels and consonants when the preceding word ends on a vowel.[1] Old Welsh, however, had only the two forms *ir* (now *yr*) and *r*; *ir* is therefore possible before consonants in the old forms of our place-names, while before vowels *ir* (*yr*) is regular in any case. The preposition *ar*, 'on, near,' is also possible, but if and when it occurs, we cannot certainly distinguish it from the article. Names of this class are not confined to Scotland, for instance there are on record Treverman in Cumberland, which seems to be for *tref ir maen*, 'stead of the stone'; Trewarmeneth, 1282 (Close Rolls), 'stead on the hill' in Cornwall.

In Kirkcudbright there is Terregles, Travereglys 1365 (RMS), for *tref yr eglwys*, 'church-stead,' corresponding to G. *Baile na Cille* or *Baile na h-Eaglaise*; compare Trefeglwys in Montgomeryshire.

In Dumfries, Trailtrow is considered to be Trevertrold of King David's Inquisition, *c.* 1124. Trailflat, on the water of Ae, is Traverflet, 1165-1214 (Lib. Calch.); it was once a parish, but is now included in Tinwald. Boiamund's list of churches which contributed to the Crusades (1275) contains the names of six 'in the Deanery of Glenken,' five of which are in Kirkcudbright. The sixth is Trevercarcou, and the only place extant or on record which this suggests is Carco on Crawick Water in Kirkconnel parish, Dumfries.

Trabroun in Lauderdale is Treuerbrun, *c.* 1170 (Reg. Dryb.). It is on a hillside, and stands for *tref yr bryn*, 'hill-stead,' like G. *Baile a' Chnuic*, Hilton. Skene places it in Roxburghshire, but if there is a Trabroun there, I have failed to find it.

In Haddington there is another Trabroun, also fairly

---

[1] See further, Morris-Jones, *A Welsh Grammar*, p. 192.

high. Tranent is Trauernent, *c.* 1127 (Chart. Hol.), Treuernent, 1144, 1150, *ib.*; Trauyrnent, 1266 (Exch. Rolls and Reg. Dunf.); Travernant once in Chart. Hol. (William I.). The last part is *nant*, pl. *neint*, a brook, dingle, valley, and in view of the persistence of the spelling *-nent* the meaning is probably 'stead of the brooks,' or 'of the dells'; compare *Nantu-ates*, 'valleymen,' the name of a Gaulish tribe bordering on the Lake of Geneva. *Trefnant* is common in Wales.

Traquair in Peeblesshire is Treverquyrd, *c.* 1124 in King David's Inquisition; Treuequor, a. 1153 (Chart. Melr.); Trauequayr, *c.* 1150-1242 (Reg. Glasg.); Trauercuer, 1174, *ib.*; Trefquer, *c.* 1200-16, *ib.*; Trafquair, 1186, *ib.*, etc. The church and hamlet stood at the junction of the river Quair with the Kirkhouse or Kilhouse burn, and the name means doubtless 'stead on Quair,' or 'of the Quair.' 'Quair' itself may perhaps be compared with *Gweir*, earlier *Gueir*, the river Weir, from a still earlier *Vedra*, which is supposed to mean 'the clear one.'

In Ayrshire there was Trevercrageis (an English plural) or Trevercraig in Carrick (RMS, i.), for *tref yr graig*,' rock-stead,' corresponding to G. *Baile na Creige*, Balnacraig, and to the Welsh *Tre 'r graig*. It is the place which on Blaeu's map appears a little to the north-west of Girvan called Trochchraig, and is on record as Trewchreg, 1498 (RMS); Trochreg, 1523, *ib.* Tarelgin in Kyle is Trarelgin, 1449 (RMS), which may be another instance. Another name in Kyle, which seems to be now obsolete, appears as Trarynzeane, 1437 (Great Chamberlain's Accounts); Trarinzeane, 1467, 1487 (RMS); Tarrinzane, 1487, *ib.*; Terringzane, 1511, *ib.*; 'the common of Crawfuirdstone alias Terringzeane in the parish of Cumnock,' 1647 (Ret.); perhaps for *tref yr fynnon*, 'well-stead,' like Auchintibber.

Some other instances of this formation are difficult to locate. King David I. confirmed to the Abbey of Kelso 'Treuenlene . . . as Vineth held it,' also the rock of Treuenlene (Lib. Calch.). Sir A. C. Lawrie considered this charter spurious. The name appears as Traverlen four times in the Book of Kelso between 1165 and 1214; Treuerlene, 1245-54. One might suggest—somewhat doubtfully—

a comparison with Pethferlen or Petenderlyn, granted to the Hospital of St. Peter in Old Aberdeen (the Spittal) between 1172 and 1179. This is for *peit an fhir-léighinn*, 'portion of the *fer-léighinn*,' or lector, an important official of a Celtic monastery, who had charge of instruction. In modern Welsh the corresponding term is *gŵr lên*; 'the *gŵr lên*'s tref' would be *tref 'wr lên*, with the middle part unstressed.

Trevergylt occurs in King David's Inquisition as pertaining to the church of Glasgow; its position and meaning are uncertain.

Another place which occurs in the same document is Treueronum, going along with Ancrum (Alnecrumba), Lilliesleaf (Lillescliva), Ashkirk (Asheschyrc). Later it is Trauerenni, 1170 (Reg. Glasg.), Treveranni, *ib.*, etc. It has been identified with 'Tryorne in Roxburghshire,'[1] which appears often on record in the connection 'the lands of St. Leonards, Snawdone, Tryorne, Cannomunt croft,' 1637 (Ret.). All the rest of these places are in the parish of Lauder, and I can find no place in Roxburgh called Tryorne. In the indexes to the Register of the Great Seal Tryorne is identified with Trearn in Cunningham, and this place is Triorn and Triarn in Blaeu, also Triorne (?) (*sic*) in 1655 (RMS). The Ayrshire place, however,—when one can be sure that it is meant—does not occur in connection with Snawdone, etc.

A much larger number of names have *tref* followed directly by a specific or qualifying term; some have the qualifying term first, like Ochiltree.

In Ayrshire, Ochiltree, the two Threaves, and Travercraigs have been noted. In addition there are about ten instances in Carrick, three in Kyle, and one in Cunningham.

Tralorg is Trewlorg, 1459; Trolorg, 1523 (RMS). Tralodden is Troloddan (Ret.). Trowier is Trowere, 1430; Traver, 1505 (RMS). Traboyack (two) is Trabuyage, 1413 (RMS). Tranew is Trownaw, 1504 (RMS; also in index Treunewr). Tradunnock is Trodonag, 1492 (RMS); also Carrick

[1] See Lawrie, *Early Scottish Charters*, p. 303.

Trodonag-Makcowben, *ib.*; compare 'ecclesia de Tredenauk' in the Book of Llandaf, now Tredunnock in Montgomeryshire. Troquhain is Treuchane, 1371 (RMS); Treuechane, *ib.*; Troquhan, 1511, *ib.* Troax is Trowag, 1549 (RMS); the present form is therefore an English plural. Trogart (Ret.) seems to be now obsolete. Giltre and Giltre Makgrane appear on record, 1511 (RMS).

It may not be without historical significance that these names tend to go in groups. There is a group of five near Girvan, with a sixth a few miles south-east of them. Another group of five is round about Barbrethan, 'the Britons' height.'

Kyle — Trabboch is Trebathe, 1303 (Bain's Cal.); Trabeathe, 1451 (RMS); other spellings, from the index of RMS, i., are Trabeache, Trebach, Trabeche, Tirbeth. Treesmax is Treyvinax, 1511 (RMS); Trevenox, Trenemax (Ret.). Tarelgin is Trarelgin, 1449 (RMS). All these, and also Ochiltree, lie in the same region.

Cunningham — Trearne, east of Beith; I have not met this name on record.

The explanation of these names, or of most of them, would involve so much pure guesswork that I prefer not to attempt it. The second part of some of them may be Gaelic; they have all been influenced by Gaelic. It is quite possible that some do not contain *tref*, but most seem to fall properly under this head.

Lanark — Troloss at the head of Powtrail Water, is apparently from *los*, Corn. *lost*, a tail, now obsolete in Welsh; the Gaelic *gasg*, a tail, is often applied to a 'tail' of land, *i.e.* a place where a plateau ends in an acute angle and narrows down to the vanishing point.

Wigtown — The only instances noted are Ochiltree and Threave (Ret.) which latter may be one of the Ayrshire Threaves.

Kirkcudbright — Terregles, Threave, and Trostrie have been noted. Troqueer, the name of a parish, is Trequere, 1372-4 (RMS); it may be the same as Traquair, and may contain an older name for the Cargen Water, which flows through the parish. Tregallan, in this parish, does not seem to be on record. Troquhain, in Balmaclellan parish, is Trechanis, 1467 (RMS), with English plural; compare Troquhain in Carrick. Tralallan or Trolallan, on record as in Parton parish, seems obsolete. Rattra in the parish of Borgue will be discussed later.

The instances already given—Trailtrow, Trailflat, Trevercarcou—seem to be the only certain cases. Dumfries

I have none from Roxburgh or Selkirk.

In addition to Traquair there is Trahenna, on the ordnance survey map Trahennanna. Dreva on Tweed, east of Broughton, is Draway, 1649 (Ret.); Drevay, 1688, *ib.*; it may be for (*y*) *dref-fa*, 'the tref-place.' Near it on an eminence there are what seem to be traces of an old settlement. Peebles

Trabroun, and, probably, Tryorne, are both in Lauderdale. Berwick

Trabroun and Tranent have been noted. Longniddry is Langnudre, 1424 (RMS). Without the prefix, which is presumably English, it is Nodref (twice), *c.* 1315 (RMS). There are also Niddrie near Edinburgh and Niddrie in Linlithgow; old spellings are Nodereyf, 1266 (Exch. Rolls); Nudref, 1290 (Theiner); Nudreff, 1296 (Bain's Cal.); Nodref 1335, *ib.*; Ballengrenagh in the tenement of Nodreff, 1335, *ib.*; Nodref, 1336, *ib.*; Nodrefe, 1337, *ib.*, etc. The first part may be *newydd*, 'new,' influenced by Ir. *nuadh*, older *nue*; in Sc.G. regularly *nodha*; the meaning would be thus 'Newstead.' Haddington

Traprain appears in a 'chartour of Traprene and Dunpeldare by the Erle of Merche to Adam Hyeburn, to be hald in blench for ane paire of gilt spurris or 2 sol. sterling for them' (RMS, i.). In the body of the charter, which was granted in David II.'s time, it is 'Trepprane unacum monte de Dumpeldar'; also Trepprene, 'tree-stead,' strictly *tref-bren*.

Besides Niddrie there is Soutra, the name of an old parish, once in Haddingtonshire, now joined to Fala. The hospital of Soltre was founded by Malcolm IV., *c.* 1164. The early spellings are regularly Soltre. Midlothian

Niddrie and Ochiltree have been mentioned. Linlithgow

We have noted the old division of Fothrif, 'sub-settlement,' and Troustrie, 'thwart-stead.' Capledrae, near Lochore, is Capildray (RPSA); it may be from G. *capull*, Fife a horse; W. *ceffyl*; compare *Mochdre*, 'swine-stead,' in Wales.

Clackmannan — Menstrie is Mestryn, 1261 (Bain's Cal.); Mestreth, 1263, Mestry, 1315 (RMS); Menstry, 1392 (RMS). The *n* seems to be intrusive, and the name is probably *maes-dref*, 'hamlet on the plain'; in Wales *Trefaes* is common, *faes* being the mutated form of *maes*, a plain, a field.

Stirling — Fintry is Fyntrif, a. 1225 (Chart. Lev.); Fyntryf, 1225-70, *ib.*; Fyntrie, *ib.*, where Fin- stands for G. *finn*, white, fair; the name represents W. *gwendref*, 'pleasant stead,' which was, as I consider, the original.

Perth — Rattray, in Blairgowrie, is Rotrefe, 1291 (Bain's Cal.); Rettref, 1296, *ib.*; Rothtref, 1305, *ib.*; Rotreffe, Rettreff in rolls of Robert I. (RMS, i.). The first part seems to be for G. *ràth*, a circular fort, usually of earth; W. *rath*, a circular fortified spot; a mound, hill; for the *o* compare Rothiemurchus, Rothie-may, etc. The ancient castle of Rattray stood on a rising ground close by the modern village, and was probably preceded by a still more ancient fort. Similarly Rattra in Borgue parish is close to a moat.

Forfar — Fintry, near Dundee, is doubtless the same as Fintry in Stirlingshire. There is also Trostrie, 'thwart-stead.' Drumhendrie, in Marykirk parish, is Drumhenrie 1588 (RMS); here the latter part is probably G. *seanruigh* (aspirated), 'old slope,' 'old reach.'

Aberdeen — Fintray parish is Fintreth, Fintrith 1178 and *passim* (Chart Lind.); Fintrefe *c.* 1219 *ib.*; Syntref (for Fyntref) in Theiner; Fyntreff 1316 (Reg. Epis. Aberd.). It is very fertile. There is also Fintray in the parish of Kingedward; 'de maiore Fyntrie' 1375 (A. & B.); Meikle Fyntre 1505 *ib.*; 'the barony of Cantress commonly called Fintries in the parish of Kingedward' 1625, 1634 (RMS and Ret.). The alternative form 'Cantress,'—an English plural—suggests that in this case the original may have been *can-dref*, for *canto-treb-*, 'white stead.' Rattray, in Buchan, is Retref in Theiner (several times); Rettref, Rettreffis in rolls of Robert, I. (RMS, i.); Rethtre in rolls of David II. *ib.* Close by is the ancient castle of Rattray, and the name is evidently the same as Rattray in Blairgowrie. Another name on record is Clyntre: duae Clintreis 1368 (RMS); 'le Crag de Clentrethy' 1316 (Reg. Episc. Aberd.); 'litil Clyntree,' *ib.* It may be compared with Clentry in Auchertool parish, Fife; the first part seems to be G. *claon*, awry, sloping. Fortrie, which occurs several times in Aberdeen and Banffshires, appears to be a Gaelic form of *gor-dref*, 'big-stead,' the opposite of Fothref. Trefor Hill near Kirk of Rathen in Buchan, Tramaud and Trefynie (Milne) may be further instances.

The only instance here is the ancient name Moray itself. Moray

Cantray and Cantraydoun are in Gaelic *Cantra* and *Cantra an Dùin*, 'Cantray of the Fort'; Cantradoun 1468 (RMS). This is probably for an early *canto-treb-*, 'white stead.' Inverness

Muchtre in Knapdale appears in 1554 (RMS). Cuilghailtro in Kilberry parish, Kintyre, is Coulgaltreif 1511 (Orig. Paroch.), in Gaelic *cùil Ghailltreo* (or *Ghalltro*), with which may perhaps be compared Giltre in Ayrshire. If these are genuine instances of *tref*, the term must have come to Kintyre from Ayrshire. Argyll

Caer is an entrenched or stone-girt fort; in Wales it very often indicates the sites of Roman camps. It has been regarded as a loan from Lat. *castra*, through a supposed British-Latin *casera*.[1] It has also been referred to a late Latin *quadrum*, which became *cadrum*, then in Welsh *caer*, as *Vedra*, the Wear, became *Gweir*, and *cathedra* yields *cadeir*.[2] To the latter derivation it may be objected that *cadrum* would rather yield *caeir* than *caer*. In any case the term is of British development, for there is nothing in continental Celtic from which it could have come. The great Roman camp at Ardoch was called in Gaelic *Cathair Mhaothail*, 'fort of Muthil,' and here G. *cathair* may well have been for an older British *caer*. A fair number of native forts are called *caer*, but more might have been expected. The numerous native strongholds between Forth and Tweed

---

[1] Thurneysen, *Handbuch des Alt-Irischen*, p. 517.
[2] Professor J. Glyn Davies, *Y Cymmrodor*, xxvii. p. 148.

are usually styled Chester or Castle; sometimes Rings or Faulds, *i.e.* folds, corresponding to G. *buaile*. The ramparts of these are usually of earth, circular in form, with deep ditches between them; the area enclosed is large enough to contain structures for living in—their sites are often visible—and also a large number of cattle. Thus they answer rather to the Irish *ráth*. The Chesters, it may be noted, are found in Fife, *e.g.* south-west of Denino and east of Newport; Bochastle, near Callander, is Mochastir 1452 etc. (RMS), (p. 240). As *caer* is feminine, a following mutable consonant should undergo mutation, but in the forms as we now have them this is not the case. Mr. Egerton Phillimore observed the same feature in Cornish names, and inferred that *caer* was sometimes masculine, but, as has been seen, this absence of mutation is rather the rule in the forms of our names as preserved in English.

It is often difficult in a given instance to determine whether the first part is *caer* or something else. For example, G. *ceathramh*, a fourth part, appears as *Ker-*, *Keir-*, *Keire-*; *carn*, a cairn, is occasionally *Car-*; *cor*, a rounded hill; *coire*, a cauldron; *carr*, a rock ledge, projecting rock, have also to be considered. If the remains of an ancient fort are known to exist at the place, the question becomes one of a choice between *caer* and G. *cathair*, earlier *cathir*, gen. *cathrach*, and is sometimes settled by old spellings or by the occurrence of the genitive case, as in Strath-cathro for *Srath Cathrach*, where the inflection shows that the name is Gaelic.

Ayr Carwinning near Dalry, is a hill with remains of a fort contained by three circular walls, covering two acres. The name is from the saint of Kilwinning, the adjoining parish. A grant of a fort to a saint was not uncommon. Monenna or Darerca, according to her legend, was specially addicted to choosing hill-forts as the sites of her religious establishments, and we may compare Dunblane, named after Bláán.

Renfrew Cathcart is Kerkert 1158 (Acts of Parl. I). Katkert 1165-1173 (Reg. Pasl.), also Ketkert in the same period; William de Kathkerke 1296 (Ragman Roll); his seal bears 'de Chatkert' (Bain's Cal.); Alane of Kerthkert 1451 *ib.*; a writer in Macfarlane's Collections says 'the Carcarth water runs out of Whitloch,'[1] indicating a seventeenth century pronunciation on the same lines as the earliest spelling. Both forms make good sense: Kerkert would represent *caer gert*, 'fort on Cart'; Ketkert would represent *coet cert*, 'wood of Cart.' The explanation of the double form may be that there were two different names.

Cardonald, in Abbey parish, may be 'Donald's fort,' but I have no proof. Donald is the English form of G. *Domhnall*, O.W. *Dumnagual* or *Dumngual* is a name which occurs several times in the royal family of Strathclyde.

Carfin near Bothwell is Carnefyn 1489 (RMS), 'white cairn.' Carmyle, Old Monkland, is Kermill 1240; it may be for *caer-foel*, 'bare fort,' with *m* of *moel* not mutated. Carmunnock is Cormannoc 1177 (Reg. Pasl.), which may mean 'monks' close,' but it is not a 'caer.' Carmaben in Dolphinton is on record in 1573 and 1617 (RMS), and looks like 'Mabon's fort,' with unmutated *m*; compare Lochmaben. Carmichael, Kermichel 1179 (Reg. Glas.), Karemigal *c.* 1250 (Acts of Parl.) is 'Michael's *caer*.' Lanark

Caer Rian was supposed by Skene to be connected with Loch Ryan, as it probably is. In view of the adjectival form *Rhionydd*, the ancient *rigonios*, we should have expected *Caer Rion*, 'fort of (the) king.' With Cargen Water, in Troqueer parish, we may perhaps compare *Caergein*, a fort in Wales seemingly.[2] Carruthtre, on record in 1487 (RMS) may be 'Uchtred's fort.' The name Uchtryd occurs several times in the Mabinogion: 'Uchtryd Varyf Draws, (of the Cross-beard), would spread his red untrimmed beard over the eight and forty rafters which were in Arthur's hall.' Galloway

Caerlaverock is to be compared with Carlaverock in Tranent parish, and Carlavirick near Cramalt on Megget Water, Selkirk; there are also Laverick Stone, south-west of Kirkbampton, Cumberland, and Laverock Law in Dumfries

---

1 *Geog. Coll.*, ii. 591.

2 Skene, *Four Ancient Books*, i. 479.

Gladsmuir, Haddington, which latter may mean simply 'Larkhill.' The old spellings are all like the present, *e.g.* Carlaverok 1371 (RMS). The second part is most probably the personal name which is in O.W. *Limarch* (L. Land.), then *Lifarch*, in the Welsh genealogies *Llywarch*; for the vowel developed after *r* may be compared O.Ir. *erelc*, an ambush, later *elerc*, now *eleirg*, Elrick; the form was probably influenced by the analogy of A.S. *láwerce*, M.E. *laverock*, a lark. Carruthers, in the eastern section of Middlebie parish, has an earthwork adjacent. It is Caer Ruther 1350 (Johnston), and may be compared with Carruderes, conjectured to be in Berwickshire.[1] The second part is probably a personal name; compare Rotri map Mermin 877 (Ann. Camb.), who is in Gaelic *Ruaidhri mac Muirminn* in 878 (AU). Keir, a parish name, probably represents *caer*; the spot primarily so called appears to have been south of Penpont, with Courthill near it. Sanquhar, like the other places of that name, means 'old fort,' and is Gaelic in form. The site, at the junction of Crawick Water with Nith—whence Conrick, 'meeting'—must have always been important, as is also indicated by the Gaelic name Knockenhair, 'watch hill,' a little way up the Crawick valley. The name of the 'old fort' may be preserved in Carco or Carcow on Crawick Water opposite Knockenhair, but Carco is stressed on the first syllable. Skene thought that the name Crawick stands for *Caer Rywc* of the Book of Taliesin, on the analogy of Cramond for *Caer Aman*.

Roxburgh

Caerlanrig on Teviot is Carlanerik 1610 (RMS). There seems to have been a fort at the spot,[2] and the name is for *caer lanerch*, 'fort in the glade.' This appears to be the place given as Carlaverik 1511 (RMS); chapel of Carlavrock 1662 *ib.*; in Blaeu it is Carlanryik, and in the Old Stat. Account it is Carlenrigg.

---

[1] Bain's *North Durham*, App. 39; J. B. Johnston, *Place-Names of Scotland*, p. 67.

[2] *Old Statistical Account of Cavers*, xvii. p. 92: 'At a place called Carlenrigg, a number of Roman urns were dug up about five years ago; but when these camps were formed, or the urns deposited, the present incumbent has never been able to discover.'

Cardrona in Traquair parish has a hill fort close by. It Peebles
is Cardronow *c.* 1500 (Johnston), probably for *caer dronau*, 'fort of circles,' from W. *tron*, a circle, throne, with mutation of *t*. The curious name 'Pharaoh's Throne,' on the west border of Tongland parish, Kirkcudbright, may perhaps be a corruption of a Welsh name containing *tron*.

Carfrae in Lauderdale is *caer fre*, 'hill fort,' from W. *bre*, Berwick
a hill, mount; compare *Moelfre*, 'bare hill,' common in Wales. There is a fine large fort here.

Carfrae, south-west of Garvald Mains, is the same in Haddington
meaning as the last mentioned.

Carcant on a tributary of Heriot Water may be from Midlothian
O.W. *cant*, white, now *can*, or from W. *cant*, orb, rim of a circle. The position is a circular opening at the head of a *cul de sac*. Carkettyll appears in 1543 (RMS) and *passim*; Blaeu places Karkettill near the right bank of Esk, a little above Roslyn; it is for *caer Gatell*, 'Catell's fort'; the personal name Catell or Catel was not uncommon among the Britons. Caerketton Hill, a summit of the Pentlands about three miles or less to the north-west, is probably a corruption of the older form. Carnethy, another summit of the Pentlands, may be, as Sir John Rhys suggested, for *carneddau*, the plural of *carnedd*, a heap of stones; the surface of the hill is full of stony accumulations. The term occurs in Wales, *e.g.* in *Maes y Carneddau*, 'plain of the cairns,' near Lake Vyrnwy, Montgomery; Carneddau near Erwood Station, Brecon. The stress is on the penult, as it is in Car-nethy.

Cramond, near the mouth of the river Almond, is Caramonth 1178 (Johnston); Karramunt in the reign of William I. (Chart. Hol.); Karamunde 1293 (Johnston). The river name is the same as that of the Perthshire Almond which is *Aman* in Gaelic, and Caramonth means 'fort of Almond.' The modern form, Cramond, is due to contraction of the first syllable, arising from the stress on the first syllable of Almond; similarly Caraile becomes Crail, and in Welsh *caled-dwfr*, 'hard-water,' becomes *Clettwr*, a river name.

Carriden is stressed on the first syllable. The *Capitula* Linlithgow prefixed to Gildas' *Excidium* have 'Kair Eden, a very ancient city, about two miles from the monastery of Abercurnig, which is now called Abercorn.' Other early spellings are Karreden *c.* 1148 (Chart. Hol.); Karedene 1336 (Bain's Cal.); 'baronia de Carden in constabularia de Lithcu,' 1336 *ib.* (vol. iii. p. 340); Caridyn 1337 *ib.* The Old Stat. Account states that Carriden was then (*i.e.* 1794) pronounced Carrin. 'Kair Eden' is usually equated with 'Caer Eidyn' of the old Welsh poems, meaning 'fort of Eidyn,' and if this is correct, the name should now be 'Creden,' as 'Karramunt' became 'Cramond,' etc. Carriden is the parish next to Abercorn on the west; its church is right on the seashore; there is neither trace nor tradition of a town or village called Carriden. Carlowrie near Dalmeny is Carlouri 1336 (Bain's Cal.). There is no proof that Car- here is *caer*; the second part may be a man's name, compare Lowrie's Den, near Penicuik; Lowrie's Knowes, near Fast Castle, Berwick.

Dumbarton The only instance is Kir-kin-tulloch, already discussed.

Stirling Castlecary is Castelcary 1450 (RMS); the second part may be *caerydd*, the plural of *caer*, with reference to the important Roman fortifications there.[1] Carmuirs near Falkirk is 'duas Carmuris' 1457 (RMS); Westir Carrmure, Eister Carrmure . . . the Cairnmure-mylne 1632 (RMS). As there appears to have been a Roman fort here, the first part may be taken as *caer*; the second part is fairly certain to be W. *mûr*, a wall, from Lat. *murus*; compare Pennymure.

Perth The Pictish Chronicle records a place named *Ceirfuill* near Abernethy; this must be now Carpow, 'fort on the Pow,' or 'sluggish stream,' W. *pwll*, G. *poll*, as in *Obarphuill*, Aberfoyle. This specialized meaning of *pwll*, *poll*, obtained great vogue in the south-west, and occurs several times in Perthshire. The old spelling *Ceir*, reflected elsewhere in the English form 'Keir,' may be compared with *Kêr*, the form assumed by *caer* in Brittany, *e.g.* *Kêr Leon*, near Brest.[2] It is specially common in the Vale of Menteith

[1] *The Roman Wall in Scotland*, p. 207 ff.

[2] Pedersen, *Vergl. Grammatik*, i. 197.

from Keir near Dunblane westwards. An eighteenth-century writer says: 'Kier is one of a chain of rude forts (which are all called Kiers) that run along the north face of the Strath or Valley of Menteith. There are Kiers at Achansalt, 'the field of the good prospect,' at Borland, at Balinackader, 'the fuller's town,' at Tar, 'the groin,' and in many other places in that direction, all similar to one another in respect of situation, construction, prospect, and materials.'[1] The Keir of Achansalt is still in part extant, and it belongs to the circular stone structures that are found in North Perthshire and westwards as far as the valley of the Nant at Taynuilt. Another writer in the Old Stat. Account states that in the parish of Kippen in Stirlingshire—south of Forth—there are several small heights called Keirs, and that on the summit of each there is 'a plain of an oval figure surrounded with a rampart.'

Carpoway in the parish of Abdie is Carpullie 1625 (RMS), Fife for *-pollaigh*, old genitive of *pollach*, 'a puddly place,' or 'place of a hollow, or hollows.' It is to be compared with Carpow above. Early spellings of Kirkcaldy are Kircalathin 1150, also Kircaladinit, Kircaladin, Kirkaldin. The second part contains W. *caled*, hard, probably in composition with *din*, a fort. This supposes an early *Caletodūnon*, 'hard fort,' to which *caer* was prefixed in later times.

Carbuddo has been noted (p. 313). Carmylie may be Forfar 'warrior's fort,' from *mìlidh*; compare Kinmylies near Inverness, in Gaelic *Ceann a' Mhìlidh*.

In the parish of Skene there are the remains of a circular Aberdeen fort with broken-down stone walls on the summit of 'the Keir hill.'[2]

The most common term for a fort in Gaulish is *dūnon*. On the continent it is wide-spread, ranging as far as Carrodūnon on the Oder, identified with modern Krappitz, and

[1] Dr. Robertson of Callander, in *Old Stat. Account* of Lecropt. Dr. Robertson's explanations of the meanings are correct. 'Achinsalt' is in Gaelic, *Ach an t-seobhalt*, dialectic for *Achadh an t-Seallaidh*, 'field of prospect.'

[2] *Old Stat. Account*, xvii. p. 58.

Noviodūnon near the mouth of the Danube; Singidūnon was the old name of Belgrade. In Welsh *dūnon* becomes *din*, in Gaelic *dùn*. Owing to Gaelic colouring the British *din* can seldom now be distinguished from Gaelic *dùn*; for instance, Dunpelder was certainly Din- before it became Dun-, though the former never appears on record.

Din appears in Din Fell on Hermitage Water. Dinmont is 'Fort-hill.' Dinwoody or Dinwiddie occurs in Applegarth parish, Dumfries, and in Ettleton parish, Roxburgh. The latter is Dunwedy 1504 (Orig. Paroch.). With the suffix *-le*, 'place of,' there are Dinley on Hermitage Water, and Dinla-byre in upper Liddesdale.

A derivative of *din* is *dinas*, a fort. It has been distinguished from *caer* as 'a camp of refuge,' while *caer* is a permanently inhabited stronghold.[1] But near Dunpelder, which certainly was permanently inhabited, there is Cairn-dinnis, 'cairn of the fortress,' so that the rule is not universal. The form *dinis*, it may be noted, is that prevalent in Cornwall. With us it appears as Tinnis in Tinnis or Tennis near Yarrow Church, Tinnis Burn, a tributary of Liddel; and Tennis Castle, Drummelzier. There are forts near all these places. This change of *d* to *t* is seen in Tintern for *Din Dirn*, and in Tin-tagel.

An interesting and difficult group consists of names that begin with Par-, Per-; this seems to be, in most cases at least, the same term as occurs often in certain parts of Brittany in the sense of a plain, a parcel of land.[2] In Welsh it is found in *parlas*, a grass-plat, for *par(r)* and *glas*, green, where it is feminine. Pardivan in Haddingtonshire seems to occur in Pardauarneburne 1144 (Reg. Newb.), a tributary of the Esk. Parduvine is near Gorebridge. Pardovan in Linlithgowshire is Purduuyn 1282 (RPSA); Perdovin 1542, etc. (RMS). With these may be compared Perdovingishill in Renfrewshire 1478 (RMS).

[1] Professor J. E. Lloyd, *Y Cymmrodor*, xi.

[2] J. Loth, *Mélanges H. D'Arbois de Jubainville*, p. 226.

They seem to be for W. *par-ddwfn*, 'deep-field,' with reference either to soil or to position.

Parbroath in Forfarshire is Perbroith 1512 (RMS); Perbroth 1538 *ib.*; Parbroth 1581, 1592, *ib.*; it is probably to be compared with Arbroath; Brodie, in Gaelic *Brothaigh*; and Dulbrothoc. Parconnen was in St. Vigeans parish (Macfarlane).

Pardew, now Broomfield near Dunfermline, is Pardusin in (RPSA). Purgavie in Forfarshire is Purgevy 1406 (RMS); Pargavie is east of the Loch of Lintrathen in Perthshire. In the latter two, the second part may be *gàbhaidh*, gen. of *gàbhadh*, danger. Parwheys is near Maybole in Ayrshire.

Perwinnes, north of Bridge of Don in Aberdeenshire, is Parwenneis 1583 (RMS); Perveines marresium, 'moor of Perwinnes,' 1592 *ib.*; its second part appears to contain *gwen*, the fem. form of *gwyn*, white. The place is now called also 'Moor of Scotstown.'

William I. granted to the monks of Cupar the lands of Parthesin, now Persey at the foot of Glen Shee in Perthshire, in Gaelic *Parsaidh* and *Parasaidh*. Percie in Birse, Aberdeen, is Parci 1170 (Reg. Aberd.). Pearsie in Forfarshire is 'Myddil Perse commonly called Bagraw' 1529 (RMS). There is also Persie near Mulben in Banffshire, which, in the absence of old forms, may be a derivative of *preas*, a copse, like Terpersie (p. 420). With the first we may perhaps compare Pardusin above.

The Welsh term *pant*, a valley, hollow, appears in Pant in Stair parish, Ayrshire. Final *t* is not uncommonly dropped in Welsh, as in *nan* for *nant*, a valley, and it appears to have dropped with us in the case of *pant*, and that this explains certain names whose first part is 'Pan.' Panbride and Panmure in Forfarshire are Pannebride, Pannemor 1261 (Acts of Parl.); in William the Lyon's reign the latter is Pannomor, *ib.* The parish of Panbride has two hollows or dens with streams which unite about a mile from the sea. The church and manse are close to the larger of these, and this site, with the houses near by, is Panbride proper, meaning 'St. Brigid's hollow.' Panbryd in Colmonell parish, Carrick, appears in 1574 (RMS). Panmure House stands near the head of the larger den, whose stream flows through the policies; Panmure is therefore most probably for *pant mawr*, 'big hollow.' In the smaller den is the farmhouse of Panlathy Mill; the farmhouse of Panlathy is a little distance off, and not in the den; 'Panlethy with the mill' 1528 (RMS). Pannanich in Glenmuick parish, Braemar, famed for its wells, may be a derivative with Gaelic terminations. Panlaurig 1509 (RMS) 'in the territory of Duns,' Berwickshire, is probably for 'Panlanrig,' in which case it would mean 'hollow in the glade.' Panbart Hill 1573 (RMS) was part of the lands of Hetherwick in Haddingtonshire, and may be for *pantbert* 'hollow-wood,' 'wood in the hollow.'

One of the terms for a bog or quagmire in Welsh is *mig-*, *mign*, pl. *mignoedd*; there are also *mignen*, *mignwern*, *migwern*, the latter two having as their second part *gwern* in the sense of 'swamp'; *migwyn* is 'white moss on bogs.' Hence in Welsh place-names we find, for instance, *Migneint*, 'bog valleys,' a hilly district on the borders of Denby and Carnarvon. A similar term occurs widely in Scotland on the east side from Sutherland southwards.

Midmar in Aberdeenshire is regularly Migmar in the older records, and is still so called in Gaelic, meaning 'bog of Mar'; it is the lowest part of Mar, the middle division of Mar being Cromar.

Midstrath in Birse parish, near Aboyne, is Migstrath 1170 (Reg. Aberd.), later Megstrath, and is still Migstra in vernacular Scots; it means 'bog strath,' and is boggy still; compare the Welsh *Migneint*.

With these go Craigie Meg in Glen Prosen, and Craig Mekie in Glen Isla, both in Forfarshire; in the latter 'Mekie' is probably the genitive of a Gaelic formation in *-ach*, 'bog-place.'

Migvie in Stratherrick is in Gaelic *Migeaghaidh*; it lies low by Loch Garth. Dalmigavie in Strathdearn is *Dail Mhigeaghaidh* and *Dail Mhigeachaidh*, a low-lying flat under a steep hill-face. In Badenoch, near Feshie Bridge, is

Creag Mhigeachaidh, in Macfarlane Craig Megevie. Miggewethe 1275 (Orig. Paroch.) is the old form of the modern Migdale in Creich parish, Sutherland, part of which is very boggy. Migvie in Mar is 'ecclesia Sancti Finnini de Miggeuth' *c.* 1160 (Ant. A. & B.), in the title 'Migaueth'; Miggeueth *c.* 1180, Migaueth *c.* 1210 (RPSA), etc. There is also Migvie near Kirriemuir.

Meckfen in Perthshire is Meggefen, Mekfen, *c.* 1230, Mekven 1443 (Chart. Inch.); here the second part is probably *faen*, the mutated form of *maen*, a stone (p. 387).

Meigle in Perthshire is Migdele in the Legend of St. Andrew;[1] Mygghil 1378 (RMS); Mygille 1378-89 *ib.*; here the second part is *dol*, as in Dolket, now Dalkeith; compare Welsh and Cornish *dol*, a meadow. For *dol* becoming *del* in unstressed position compare *Sestel, Sestill,*[2] now 'Chesthill,' G. *Seasdul*, in Glen Lyon. Megdale in Dumfriesshire, and Craigow-meigle (*i.e.* Meigle of Craigow) in Orwell parish, Kinross, are doubtless the same.

Megdale in Dumfriesshire is on the Water of Meggat, and there is also Megget Water in Selkirkshire, 'aqua de Megot,' 1509 (RMS). The ending may be compared with the Mid. Welsh abstract and collective ending *-et*, later *-ed*, as in *ciwed*, a rabble (Lat. *civitas*), *colled*, loss;[3] the meaning is '(water of) bogginess.'

Strathmiglo in Fife is Stradmigeloch (RPSA). The church and village are on the upper part of the river Eden, called here Miglo. The marshy nature of the valley, of old at least, is indicated by the names Pitgorno and Gornogrove to the west of the village, and by Myres Castle and Nethermyres to the east in the adjoining parish. There is a small loch near the stream a mile to the south-west of the village, and westwards of it are Nether Pitlochie and Westfield of Pitlochie. The second part of *miglo* is most probably G. *loch*, borrowed into Welsh as *llwch*, giving the meaning 'strath of bog-loch.'

Migger in Strathearn, near Comrie, in Megour 1474 (RMS)

---

[1] Skene, *P.S.*, 188. [2] Macfarlane, *Geog. Coll.*, ii. 562; RMS, ii.
[3] J. Morris-Jones, *A Welsh Grammar*, p. 231.

in G. *Migear*; it is still a boggy, marshy place. The *-r* suffix appears to be collective, like Ir. *salch-ar*, dirt, from *salach*, dirty. About the middle of Glen Lyon is *Migear-naidh*, now a pleasant meadow, flat and low by the river side; it is Meggarne 1502, now 'Meggernie.' The latter part here may be suffixes like Gaelic *Muc-arnaich*, *Muc-arnaigh*, but more probably we should compare W. *migwern*, boggy meadow, pl. *migwerni* or *migwernydd*.

All these names appear to be derivatives or compounds of *mig*. There are besides some which seem to represent *mign*. Migdale in Creich, Miggeweth in 1275, is now in Gaelic *Migein*, with *g* as in English *rig*, whereas slender *g* in Gaelic is regularly a forward *k*, as in *rick*. Migdale Rock is *Creag Mhigein*. Here, then, we have three forms with a common element. The second part of Migdale is probably Norse *dalr*, a dale; Miggeweth (Migvie) is a compound. With Migein are to be compared Craig Meggen in Glen Muick, Aberdeenshire, and Megen Burn, a tributary of Girnock, Ballater.

Another widely spread term is that seen terminally in Pit-four, Doch-four, etc. Here *f* represents *ph*, and the unaspirated *p* appears in the closely related Pōrin, G. *Pórainn*, in Strath Conon. Dr. A. Macbain pointed out long ago that this is to be compared with W. *pawr*, pasture, grazing; Breton *peuri*, to pasture. This term survived in Gaelic as *pór*, gen. *pùir*, and the fact that it is used with the article in Gaelic shows that it was treated as a Gaelic word. Its derivative *Pórainn* is not found with the article, and does not seem to have been retained in Gaelic speech. The Book of Deer records that 'Comgeall son of Aed gave from Orti till thou reach *Fúrené* to God and to Drostan.' The words are *níce Fúrené*, which is for *gonice Fúréne*, and here *Fúréne*, being the object of *gonice*, 'till thou reach,' has its initial consonant aspirated; *i.e.* the nominative un-aspirated is *Púréne*. This may be compared with Mid. Irish *arténe*, a pebble, the diminutive of *art*, a stone. *Púréne* is probably represented elsewhere, but the Strath Conon *Pórainn* is a different formation, corresponding rather to W. *poriant*, pasture, or to the plural of *pawr*, namely *porion*, which suggests an original *n*-stem. We have no data for certainty, but *Verturion-es* yielded in Gaelic a nominative *Foirtriu*, whence gen. *Foirtreann*, dat. *Foirtrinn*. Similarly *Pórainn* may be a dative form of an *n*-stem. The instances of the various forms which I have noted are as follows :—

Baile Phùir, Pitfour, in Rogart parish. Sutherland

Pórainn, Porin; *Innis a' Phiuir*, 'Inchfuir, 'pasture Ross meadow,' formerly Pitfour, Kilmuir Easter. This is an example of the tendency to palatalization in this and other districts, heard even in English, *e.g. smyooth* for 'smooth.' *Baile Phùir*, Pitfour, Avoch.

Do'ach Phùir, Dochfour, 'davach of pasture,' near Inverness Inverness. *Dail a' Phùir*, Delfour, Kincraig, 'dale of pasture.'

Delfour in Cromdale parish. Moray

Furene as above, now represented by Pitfour in Old Deer Aberdeen parish; Petfoure 1382 (RMS). There are also Balfour in Tullynessle; Tillifour in Monymusk, 'pasture hill'; Tillyfourie in Cluny, where the *-ie* ending may perhaps be for *-aigh*, the dative of a derivative in *-ach*; Pourin also occurs.

Balfour, Edzell. Kincardine

Pourie, in Murroes parish; compare Tillyfourie. Forfar

Purin in Falkland parish, Pourane 1450 (RMS); Balfour in Fife Markinch; Pitfirrane, Petfurane 1474 (RMS), Dunfermline.

Trian a' Phùir, Trinafour, Struan, 'third of the pasture'; Perth *Tom a' Phùir*, Glen Artney forest, 'knoll of the pasture'; Pitfour, Petfur 1357 (RMS); Pethfoure 1358 *ib.*, in St. Madoes parish.

Tìr a' Phùir, Tirafuir, 'land of pasture,' in Lismore, Argyll with a Pictish broch; *Peighinn a' Phùir*, Pennyfuir, 'pennyland of the pasture,' near Oban.

W. *maes*, Corn. *maes*, *mes*, Bret. *meaz*, all meaning 'an open field, a plain,' are from E. Celt. *magos*, stem *mages-*, a plain, with *-st-* suffix. This is probably the first part of Menstrie (p. 364). It forms the second part of Pol-maise in Stirling, 'pool (or hollow) of the plain.' Rothmaise in Rayne parish, Aberdeen, *c.* 1175 Rothemase, 1333 Rotmase (Reg. Ep. Aberd.), is 'fort on the plain,' corresponding to Rothiemay in Banffshire, where the second part is the genitive of G. *magh*, a plain. In Linlithgow there is on record Okelfas, Ogelfas *c.* 1200 (Ch. Hol.), Ogleface 1392, Ogilface 1450 (RMS), etc., possibly for *uchelfaes*, 'high-field'; compare Okeltre for Ochiltrie (RPSA). In some cases at least *maes* appears as *moss* when it is the generic term in a compound and therefore unstressed. Moss-fennon in Glenholm parish, Peebles, is Mesfennon 1296 (Ragman Roll), Mospennoc 1250 (Lib. Melr.) where *p* is probably for *f*; the second part may be W. *fynnon*, a fountain; there is a spring at the place. Mospebill 1506 (RMS), now Mosspeeble in Ewesdale, is probably for *maes pebyll*, 'tent-field,' like the Gaelic Achpopuli, for *achadh poible*, at Abriachan, Inverness. Other possible examples are Mosmynning 1595 (RMS), now Mossminning in Lesmahagow; Moscow near Kilmarnock, apparently from G. *coll*, W, *collen*, hazel; Moscolly in the constabulary of Haddington 1463 (RMS), like *Bealach Collaigh*, 'hazel pass,' west of Wyvis; Mosgavill 1588 (RMS), now Mossgiel near Mauchline, from G. *gabhail*, W. *gafael*, a holding; perhaps also Moss Maud and Moss Brodie in Aberdeenshire.

E.Celt. *magos*, a plain, is *ma* in Welsh, now used only terminally in composition in the mutated form *fa*, as in *porfa*, 'pasture-place.' This seems to form the second part of Ogilvie, Ogeluin 1239, Oggoueli *circ.* 1172 (Ch. Inch.), in Blackford parish, Perthshire, also in Banffshire 1655 (RMS); compare Ogilface above. Glen Ogle at the head of Loch Tay is in Gaelic *Gleann Òguil*, which is obscure to me; there is another Glen Ogle or Glen Ogil in Forfarshire. Morphie, earlier Morfy, is on the coast at the border of Kincardine and Forfarshires, an elevated plain on the north side of Esk, low on the south side.[1] This seems to be identical with W. *morfa*, a sea-plain corresponding to G. *mormhoich*. In Migvie (p. 374) the second part is probably as in Ogilvie; see further p. 500.

---

[1] Flat piece of land extending along the sea, or salt marsh, would apply to the Forfarshire side of the Esk . . . plain overlooking the sea would be an exact description of Morphie.—J. Crabb Watt, *The Mearns of Old*, p. 374.

The Annals of Ulster note at A.D. 756 a battle styled 'bellum gronnae magnae,' and the same event is recorded by Tigernach with the explanation in Gaelic, '. i . mōna mōire,' *i.e.* '(battle) of the big bog.' The place is now Moneymore in Co. Derry. Thus *gronna* is the equivalent of G. *mòine*, a bog; a mossy moor. The Latin Life of St. Carthach or Mochuta tells how the saint built his cell 'in gronna deserti'; in the Irish Life the corresponding passage has 'ar móin fhásaigh,' 'on a desert bog.' Cogitosus' Life of St. Brigid mentions the construction of a road through a *gronna* which was very deep and almost impassable. Another form of the word is *gromna*, quoted by Holder. Further instances of the word are given by Ducange. Zimmer says that *gronna* is peculiar to Irish Latin of the post-Norse period, but this is not correct, for it occurs in the Book of Landaf: 'per medium gronnae,' 'through the middle of the bog.'[1] Whether it occurs in the place-names of Wales I do not know, but it was evidently vernacular with us, though it does not occur in literature.

It is found in place-names on the east side from Linlithgow to Kincardineshire. I have not observed any instance between Dee and Ness, but two cases of it seem to occur near Beauly. It is found (1) as Groan, Groam; (2) by metathesis in Scots as Gorn; (3) from *gronn* there was evidently formed *gronnach*, 'marshy,' or 'marshy place,' which appears, with metathesis, as Gorno, and in the dative (or genitive) as Gornie, for *gronnaigh*.

The instances noted are as follows:—

Gromyre, for Grom-myre or Gron-myre, in Torphichen parish, is a case where the second part is a translation of the first. Balgornie is for *Baile Gronnaigh*, 'mire-stead.' Linlithgow

Kinghorn, Kyngorn 1374 (RMS), etc., is still pronounced locally, as I am informed, Kin-gorn with stress on the second part. It means 'bog-head,' and adjacent to it is Grange-myre. In Myregornie, near Kirkcaldy, 'gornie' means 'at bog-place,' and 'myre' translates it. Pitgorno, in Strathmiglo, is for *Peit ghronnach*, 'boggy bit'; here *-ach* Fife

---

[1] *L. Land.*, p. 224 (Evans and Rhys).

becomes in Scots *-o*, as in Durno for *Dornach*, and many other instances. Gorno Grove is close to Strathmiglo village, and Myres Castle is eastward of the village. Miglo itself is from *mig*, 'bog.' Balgrummo, Balgormo 1480 (RMS) may be compared with Braegrum in Perthshire; it is from *gromach*; compare *gromna*, the other form of *gronna*.

Clack-mann Craighorn is a hill behind Alva, meaning 'myre rock'; near it is Myreton Hill.

Perth Groan, in Logie Almond, is a mossy place. Braegrum, in Methven, means 'upland of the moss'; the tract below it is mossy. It is near Meckfen, which probably means 'bog-stone.' In the same parish is Grundcruie, a mossy place with hard subsoil; the second part is G. *cruaidh*, hard, and I take the first part to be *gronn*, with developed *d*, as in Almond, etc., rather than *grunnd*, 'bottom or channel in river'

Kincardine Kinghornie, near Bervie, is for Kin-gornie, 'head of marsh-place'; compare Kinghorn.

Inverness Groam, in Kirkhill, appears also as Groan;[1] it was a very wet place, and part of it is marshy still. On the left bank of the Beauly river, not far above Beauly Bridge, is Groam of Annat, *Grom na h-Annaide*. Near the farmhouse there is a swamp now, and the place must have been always swampy. We have already noted that St. Carthach made his cell in a *gronna*, and the cell of the saint who founded this *annaid* was probably similarly situated.

It is fairly certain that this is the second part of the compound Forgrund, now Forgan. There are four places of this name: (1) Forgan in Fife, Foregrund and Foregrundseihire, 1144, 1150 (RPSA); here 'seihire' is doubtless for 'scihire,' whence G. *sgìr*, *sgìre*, parish; English 'shire.' To the grant of 1150 king Malcolm IV. added half a ploughgate of land called Chingoth, *ib.*; here the first part is G. *cinn*, dative of *ceann*, head, and the second part is *gáith*, *góith*, a marsh, as in the derivative term *góethlach*; 'Gáethlaige Meotecdai,' 'the Maeotic marshes'

[1] 'John Grant in Groan,' 1774 (*Forfeited Estates Papers*, p. 77).

was the old Gaelic name for the Palus Maeotis or Sea of Azof. The term is not uncommon with us, *e.g.* 'Bog of Gight' is in Gaelic *Bog na Gaoithe*, literally 'bog of the bog.' (2) Longforgan in the Carse of Gowrie; Forgrund, 1178-82 (Chart. Lind.); Langforgrunde, 1377-82 (RMS). (3) Monorgan, situated below the latter; Monorgrund, 1377-82 (RMS). (4) Forgandenny near Perth; Forgrund, 1215-21 (Chart. Inch.); Forgruntheny, 1382 (RMS). The second part here is the woman's name *Ethne*, the name of St. Columba's mother and of many others. Ferchard, earl of Strathearn, and his wife 'Ethen' endowed the church of St. Cattan of Aberruthven with tithes and land before 1171,[1] and it may be this lady's name which is attached to Forgan here. The personal name Gille Ethueny, 'Ethne's lad,' 1246 (Chart. Lind.) occurring in Perthshire rather indicates, however, that Ethne is here a saint's name.

Forgrund is compounded of *for*, over, on, and *gronn*, and means 'on or above the bog.' For the spelling we may compare that of Morgan, earl of Mar, whose name is spelled Morgrund, *c.* 1153.[2] The relation of the Carse of Gowrie pair of names, Forgan and Monorgan, is interesting, Forgan is the place 'above the gronn'; the *gronn*, that is to say, occupied the site of what is now Monorgan. In later times, when the meaning of *gronn* was forgotten, this moss or bog was styled 'the bog of Forgan,' *i.e.* in Gaelic *Móin-Fhorgruinn*, Monorgrund.

E.Celtic *cēto-* (for *caito*), as in *Cēto-briga*, means 'wood,' O.W. *coet*, now *coed*. Cormac's Glossary has '*coit coill isin chombric*,' '*coit* is Welsh for wood,' and explains the Irish name Sailchoit as 'willow-wood.' The modern form is probably seen in Knockcoid in Kirkcudbrightshire. When taken over at an earlier stage it appears with fair certainty as *-cet* in Pencaitland (p. 355). Bathgate is Bathchet, Batchet, Batkeht, Bathketh, Bathcat (Ch. Hol.), Bathket, 1306, Bathkethe, 1337 (Bain's Cal.); it may be for

---

[1] *Charters of Inchaffray*, p. lviii.

[2] Lawrie, p. 427.

baeddgoed, 'boar-wood.' Rothket 1195, Rothketh 1199 (Ch. Lind.), in Aberdeenshire may be 'fort-wood.' In other cases it seems to be now represented by *keith*, as in some of the old spellings just given. Dalkeith is Dalkeid *c.* 1142 (Reg. Newb.), Dolchet *c.* 1144 (Ch. Hol.), Thomas Dalket 1238 (Bain's Cal.), Dalkethe 1336 (Bain), etc., meaning 'meadow or plateau of the wood.' Inchkeith, an island in the Firth of Forth, is 'insula Keð' in the *Life* of St. Serf; another Inchkeith is inland near Lauder. The old barony of Keith was in south-west Haddingtonshire (RMS I), where there are now Upper and Lower Keith, Keith Marischal, Keith Water, in the parish of Keith Hundeby or Humbie. The 'Forrest of Kyth' (Blaeu) in Largs parish, Ayrshire, is now 'Ferret of Keith.' Keith in Forfarshire, Keith, 1489 (RMS), is near Ledcrieff, 'tree-slope.' Keithock is near Brechin; Kethik, 1492 (RMS), is in Coupar-Angus parish; Kethak, 1636 (RMS), is in Mortlach, Banffshire. In Aberdeenshire are Keithinche, 1612 (RMS), Keithney, Kethny, 1631, *ib.* Keith in Banffshire is Ket, Keth, Kethmalruf, 1200-24, (RM) (p. 289). Keithmore is in Kirkhill, Inverness. Balkeith near Tain is in Gaelic *Baile na Coille*, 'wood-stead.' This list might be added to.

It is, however, by no means certain that every place called Keith or containing Keith as an element represents the word for 'wood,' or that all the names given above contain it. The surname Keith in Caithness is in Gaelic *Càidh*, and Balmakeith near Nairn is in Gaelic *Baile Mac Càidh*.[1] Inverkeithing in Fife is Inuerkethyin, *c.* 1120 (Lib. Scon), Inverkethin, Innerkethyn, *c.* 1150 (Reg. Dunf.), Inverkethyne, 1372, etc. (RMS), in Gaelic *Inbhir Cheitean*;[2] compare Inverkeithny in Banffshire.

In O.W. *poues* means 'rest, repose,' 'quies' (Zeuss, p. 1053), whence 'Powys paradwys Cymry,' 'Powys the

[1] I do not know the origin of *Càidh*; that it represents a short form of *Mac Dha'idh*, 'David's son,' is phonetically possible but otherwise unlikely, for this form of shortening does not occur in any other northern name.

[2] *Blàr Inbhir Cheitean*, the battle of Inverkeithing, disastrous to the Macleans, was fought in 1651.

Eden of Wales.' T. Pont has Pouis on Liddel Water near Castleton. Compounded with *-ma*, it gives O.W. *pouisua*, *poguisma* (L. Land., pp. 149, 249), 'station, settlement,' whence probably Posso, a notably pleasant place on Manor Water, Peebles, near the Castlehill of Manor; its situation and that of Pouis suggests that both were ancient seats of British chiefs. Possil, Glasgow, may be from *poues* with the suffix *-el*, as in *presel*, meaning much the same as *pouisua*.

Manor, Peebles, is Maineure, Menwire (Orig. Paroch.), Mener, 1323 (Acts Parl.), and appears to be the same as W. *maenor*, spelled Mainavre in Domesday Book, and said to mean 'the stone-built residence of the chief of a district.' [1] This fits in with Castlehill of Manor.

W. *pebyll*, a tent, pavilion, occurs with English plural in Peebles, Pobles *c.* 1124 (Reg. Glasg.), Pebles *c.* 1126 (Lawrie), Pebbles, Pebles, *c.* 1141 (RPSA), Peples *ante* 1136 (Lawrie). The G. form is *pobull*, and Pobles suggests Gaelic influence. There is another Peebles near St. Vigeans, Forfar. Dalfibble in Dumfriesshire shows Gaelic aspiration of *p* to *f(ph)*; Pibble occurs in Galloway. The term corresponds to the common English 'shiel, shiels.'

In Lindifferon, Fife, the second part is probably the same as O.W. *dyffrynn*, a watercourse, valley, 'vallis' (Zeuss, p. 138), the Dyffryn of Merioneth, etc., explained as for *dyfr-hynt*, 'water-way'; [2] *hynt* represents an early *sento-*, path, way, as in *Gabro-sentum*, 'goat-path.' The G. equivalent is *séad*, a journey, M.Ir. *sét*, a way. The first part may be *lann*, a field, enclosure: Landifferoun, 1540 (RMS).

A little west of Borrowstoun Ness and right in the Firth of Forth, Blaeu's map has 'Ruines of Cast Karig Lion.' Castle Lyon is said to have stood between Kinneil House and the sea, but has now totally disappeared. From the situation, near the eastern end of the Roman Wall, it is possible that Lyon here stands for Lat. *legion-*, as in W. *Caer*

---

[1] Professor J. E. Lloyd, *Y Cymmrodor*, ix.

[2] Egerton Phillimore in Owen's *Pembrokeshire*, p. 358.

Lleon, Chester, for *Castra Legionum*, and that 'Karig Lion' means 'the rock of the legions,' or 'of the legion.' Other names connected with the Wall are Kinneil, Cleddans, Kintocher for *cinn tóchair*, '(at) end of causeway.'

Linlithgow is Linlidcu, 1138, 1144 (RPSA), Linlitcu (Chart. Hol., David I.), Linlithcu, 1150 (Reg. Dunf.); Thomas de Linnithuc, *c.* 1150 (Reg. Glas.), Lithcu 1336, etc. (Bain's Cal.), Linlithcu, Linlithqu, etc., 1384-92 (RMS), Lithgeo 1399, *ib.*, William de Lythcu 1375, *ib.*, Pel de Lithcu (Fordun), Johannes de Lvihtgow, 1433.[1] Linlithgow was well known in Gaelic; its wells, reckoned as one of the marvels of Scotland, are *tobraichean Ghlinn Iucha* (or *Iuch*), and the Linlithgow measure was *tomhas Ghlinn Iucha* (or *Iuch*).[2] A MS. of about 1800 gives it in Gaelic as *Gleann Iucho*. The Dean of Lismore records 'bellum prope Glenvchow *alias* Lithkow percussum,' 'a battle stricken near Glenvcho,' etc.[3] The Wardlaw MS. has 'Glenugh, Lithgow.'[4] With this is to be compared the Latin form *Limnuchum* used by George and David Buchanan, Andrew Melvin, and others.

The first part is W. *llyn*, a lake; 'Linlithgow' is 'lake of Lithgow.' Lithgow, Lithcu, Litcu, is a compound of which the second part is most probably W. *cau*, a hollow, like Ir. *cua*, O.Ir. *cue*, a cup, a hollow, as in the not uncommon Irish name *Senchue*, later *Senchua*, 'old hollow,' anglicized as Shanco, Shancough, Shankough. The first part is an adjective, and is probably for W. *llaith*, damp, moist, as in *Llaithnant*, 'wet valley,' at the head of Afon Dyfi, or possibly O.W. *luit*, now *llwyd*, grey, as in *Litgarth* for *Luitgarth*, 'grey garth,' in Lib. Land. (index).

The Gaelic form may be explained on the supposition that 'Lithgow' was understood to mean 'wet hollow,' and translated into *fliuch chua*. 'Linlithgow' would then be *linn fhliuchua*, pronounced *linn liucho*;[5] in the combination

---

1 *Sheriffdom of Lanark and Renfrew*, p. 127 (Maitland Club).

2 Compare Gregorson Campbell, *Witchcraft and Second Sight*, p. 279.

3 Adv. Lib. MS., XXXVII.

4 *Wardlaw MS.*, p. 64.

5 The final vowel (unstressed) is indefinite in sound.

tobraichean Linn Liucha and such, *linn* was readily confused with *ghlinne* genitive of *gleann*, which latter was assumed as a new nominative. The second *l* was dropped.[1]

Glasgow is Glasgu 1136, etc. (Lawrie), Glasguensis episcopus 1122, etc., Glasgo, -gu, -cu, -ku (RMS I). Jocelin, in his *Life* of St. Kentigern, written after 1175, mentions 'Cathures, which is now called Glasgu,' and 'Glesgu, which means *cara familia*, now called Glasgu'; [2] elsewhere his form is Glasgu. By *Glesgu*, explained as 'dear community,' he means that the name is W. *clas cu*, from *clas*, a close, cloister, used in the sense of a religious community, and *cu*, dear. But, apart from the fact that *clas*, being masculine, would not mutate the following *c* to *g*, the stress on the first part proves *glas* to be adjectival, qualifying the second part. There is the further fact that the name occurs elsewhere without any religious association. Glascogrene, 1538 (RMS), was in the barony of Glenbervie in Kincardineshire; in Aberdeenshire, Glasgow (Glasco) Forest and Glasgow (Glasco) Ego are in Kinellar parish—Glasgoego 1478, Glaschawe 1490-1500, Glasgow 1511, Glasco

[1] For the substance of the following notes on the Burgh Seals of Linlithgow I am indebted to Mr. James Russell, Town Clerk. The earliest is from the Ragman Roll, 1296, and bears 'a Hound, collared, passant dexter, through water in base wavy: tail curved, hairy: legend—(Gothic capitals)—Sigil. Commune de Linlithqu.' The second, of 1375, has on the obverse 'a Hound passant to dexter, through a stream, with open mouth, pointed ears, and curved tail, collared and chained to a ring on pole. Legend (Gothic capitals)—Sigillum Commune B'gi de Lilithcu.' The third, of 1661, has on the obverse on a Mount surrounded by water a Hound passant to dexter, collared, and chained to a Tree with three branches. Legend (Roman capitals)—as in No. 2. The seal of 1721 is similar to No. 3, with some difference in details. In Matriculation of Burgh Arms, July 16, 1673, the second coat is given, Or, a Greyhound bitch sable, chained to an oak-tree, within a Loch proper. The *Old Stat. Account* (vol. xiv. p. 548) says of Lithgow, 'it has been chiefly traced to the Erse language, in which *gow* expresses a *dog*, and *lith* a *twig*; and supposed to allude to a black bitch, which, according to tradition, was found fastened to a tree in the small island on the E. side of the loch.' The hound on the Burgh Seal may have been suggested by *cu* of *Litcu*.

[2] 'Cathedralem sedem suam in villa dicta Glesgu (B. M. Glaschu), quod interpretatur Cara Familia, que nunc vocatur Glasgu (B. M. Deschu, *leg.* Cleschu), constituit.'—*Life of St. Kentigern*, Forbes, p. 182.

1524 (Macdonald), Mekle Glasgow *alias* Glasgowforest, 1642 (RMS).

In Gaelic, Glasgow is *Glaschú*, *Glaschu* in MacVurich and often later, also often *Glascho*, which is the pronunciation now. There are therefore two forms: (1) Welsh *Glasgu*, with *c* mutated to *g* after the adjective, (2) Gaelic *Glaschu*, *Glascho*, with *c* aspirated after the adjective; compare the spelling *Glaschawe* above of the Aberdeen place. The first part, *glas*, green, is common to Welsh and Gaelic; the second part is the same as in Lithgow, and the name most probably means 'green hollow,' which was doubtless descriptive of the ancient site on the Mellendonor Burn, now Molendinar.

Partick is Perdeyc *c.* 1136, Perdehic 1172, Pertheic 1174, 1179, Perthec 1186 (Reg. Glas.), Perthwyk-Scott 1452 (RMS), etc. In Gaelic it is *Pearraig*, for *Pearthaig*: 'cho lùthmhor ri muileann Phearraig,' 'as active as Partick mill,' used to be applied to an active and restless child.[1] A prophecy of Thomas the Rimer says 'you may walk across the Clyde on men's bodies, and the miller of Partick Mill (*muileann Phearraig*), who is to be a man with seven fingers, will grind for two hours with blood instead of water.'[2] The phonetics of *Pearraig*, *Pearthaig*, are regular, like G. *parras*, M.Ir. *parthas*, O.Ir. *pardus*, from *paradisus*; they suggest an earlier British *Peredic*, taken over into Gaelic with stress on the first syllable. The name may be compared with *Tom-pearrain*, for *-pearthain*, at Comrie, but both are obscure to me.

Carnwath in Lanarkshire is Karnewid, 1179, Karnewic (? -wit), 1172 (Reg. Glas.), Carnewithe, 1315, Carnwythe, 1424, Carnewith, 1451 (RMS), evidently for W. *carn gwydd*, 'cairn of (the) wood.'

Carstairs is Casteltarras, 1172 (Reg. Glas.), Castrotharis, *c.* 1250 (Acts of Parl.), Castalstaris, 1540 (Johnston), Carstaris, 1579 (RMS); the 1250 form may be latinized.

---

[1] For this saying I am indebted to the Rev. C. M. Robertson.

[2] Gregorson Campbell, *Superstitions of the Scottish Highlands*, p. 271, where the name is misspelled *Pearaig*.

The second part may be the same as Tarras Water in Eskdale.

In A.D. 717 a battle was fought between the British and the men of Dál Riata at a stone called Minuirc (AU ; Tigern.), supposed by Skene to be Clach nam Breatan in Glen Falloch. Here *min* may perhaps correspond to W. *maen,* a stone, while *uirc* may be compared with W. *iwrch,* a roebuck. Keating makes it *Cloch Mhionnuirc,* but this form cannot be taken as authoritative.

Methven in Perthshire is Methfen, 1211 (Chart. Inch.), so also in 1371, 1376 (RMS), and in Reg. Arbr. 'Methven Wood' is *coill Mheadhoin* in MacVurich (Rel. Celt., ii. p. 198), which agrees with the Old Stat. Account; 'in Gaelic said to signify the Middle' (*i.e. meadhon*).[1] It was given me as *Meadhanaidh,* and from an older source as *Meadhainnidh,*[2] probably better written *-aigh,* from a gaelicized nominative in *-ach.* The name seems to be for W. *meddfaen,* 'mead stone.'

In the neighbourhood is Meckfen, situated between Braegrum (p. 380) and Methven Moss, Meggefen, Mekfen, *c.* 1226, Mekven, 1443 (Chart. Inch.), Mecven, 1376 (RMS). The first part may be compared with Megget, etc. (p. 375), giving the meaning 'bog-stone,' like the Gaelic *Clach na Bogaraigh* in Edderton, Ross-shire.

Ruthven occurs often north of Forth. Instances are Ruthven in Strath Nairn; at the foot of Strathdearn: Rothuan in Stratheren, 1236 (RM); near Kingussie; near Perth—Ruthven Castle or Huntingtower: Rotheuen, *c.* 1233, etc. (Chart. Lind.); near Tomintoul in Banffshire; in the Enzie, Banffshire: Rothfan *c.* 1224 (Ant. A. & B.); an old parish of Strathbogie, now Cairnie; Rothuan, *c.* 1208 (Ant. A. & B.); Logie-Ruthven, now Logie-Coldstone, in Mar: Logyrothman twice in 1200 (Reg. Aberd.); a parish of Forfarshire: Rothuan, Rotheuen, Rothuen (Reg. Arbr.);

---

[1] *O.S.A.*, x. p. 609; the writer adds, 'the adjective *dow*, or black, is commonly added in speaking of it.'

[2] This latter form was given me about 1900 by Mr. John Whyte, who had it from his father. In Atholl it was given me as *Mèinntigh, i.e. Meadhainntidh,* with developed *t.*

Aberruthven in Auchterarder: Aberruadeuien, *c.* 1198 (Chart. Inch.)—in the margin Aberrotawin, Aberrotheuin, *c.* 1199, Aberrotheuin *c.* 1200, Aberrotheuin *c.* 1211 (Chart. Inch.). The first four are well known in Gaelic as *Ruadhainn*, pronounced almost as one syllable—*Rua'inn*. Ruthven near Perth seems to occur in the Book of the Dean of Lismore as *roywone* riming with *zwine*, being end words in a Deibhidhe couplet probably for *ghuin/Ruadh-mhuin*. In the case of all the places of this name with regard to which I have got information there is red stone or red earth or ferruginous deposit, and while we might compare W. *rhuddfaen*, 'red stone,' the name is almost certainly G. *ruadh-mhaighin*, 'red spot,' 'red place' (O.Ir. *maigen*). Hogan gives five instances of *Maigen* in Ireland, anglicized as Moyne. Mayen, in Rothiemay parish, may mean 'little plain' (*maighin*). I am informed that Ruthven near Coldstream, Berwick, is a fairly recent importation.

CHAPTER XII

British-Gaelic Names

The generic terms which we have been considering show no signs of having remained effective in the language after the period of transition from British to Gaelic, nor do any of them, with the exception of *pór* and *gronn*, appear to have been adopted into Gaelic in the earlier stages of that transition. Certain other terms, however, remained effective, being retained in the Gaelic of Scotland, though they are not found in the Gaelic of Ireland. The process that took place may be illustrated by what went on and is still going on in course of the change from Gaelic to English. When the Gaelic speakers began to take to speaking English, they made up their deficiencies in that language by using Gaelic words freely instead of the English terms with which they were not yet familiar. This was a custom with which I was very familiar in my own native district a good many years ago. At that time the English of Easter Ross was full of Gaelic words for which we had no handy English equivalents. We said, for instance, 'I have a *meanmhainn* in my nose,' *i.e.* a premonitory tickling sensation; a child who would not eat his porridge was told 'you 'll be a *taidhbhse*,' *i.e.* literally, 'a phantom'; of a pithless man it was said there was 'nothing in him but a *blianach*,' *i.e.* 'a meagre creature'; and so on almost *ad infinitum*. As time went on the Gaelic terms became fewer; the more proficient in English were inclined to make merry over those who were less proficient. The man who went to Invergordon and demanded in a shop some pounds of 'beef-*uain*'—*i.e.* 'beef of lamb'—instead of mutton, was known as 'beef-uain' ever afterwards; another worthy man earned the name of 'twenty-ten' because he used that expression for the more regular 'thirty.' At the present day the number of Gaelic terms in the English of that district is much less than it was forty years ago, but some survive still. Nowadays the influence of the public school and the newspaper is decisive against their becoming permanent, but in older times, when conditions were more favourable, some of these Gaelic words became so firmly established that they are still current in the Lowland Scots of the districts that were once Gaelic-speaking. The Garisch classic *Johnny Gibb of Gushetneuk* contains several such survivors, *e.g.* 'ablach,' 'bourach,' 'clossach'; in Fife they still use 'car-haun'' for 'left hand,' the first part being Gaelic *cearr*, left. In Galloway corn was at one time separated from the husks by rubbing it with the bare feet: 'this they call Lomeing of the corne'; here 'lomeing' is Gaelic *lomadh*, 'stripping.'[1] These examples may be sufficient to illustrate what takes place. The process is, of course, not peculiar to Scotland: it is well known in Ireland, and something of the same nature must have happened wherever one language has been displaced by another. From one point of view the terms which survive from the earlier language may be said to be borrowed into the later, but they are not really loans in the ordinary sense. What was borrowed in this case, to put it somewhat paradoxically, was not the Gaelic but the whole of the English which the people gradually came to use.

The few British terms which remain in Gaelic are to be accounted for in the same way, not as loans, but as survivals of the older speech. Their survival was due doubtless to a certain fitness which may not have been the same in all cases. One condition which must have been effective was the presence or absence in Gaelic of a handy synonym conveying the same shade of meaning. Thus British *obar*, 'confluence,' had no chance against Gaelic *inbhear*, for both had exactly the same meaning. On the other hand, *preas*, a copse, was a useful general term for which it was

[1] Macfarlane's *Geog. Coll.*, ii. p. 101; several other terms there recorded are Gaelic: 'awell,' for 'ath-bhuaile'; 'fay' for 'faithche,' which is 'Foy' in certain place-names; 'lene' for 'léan'; 'cork' for 'corcar'; 'glassons' from 'glas,' 'grey.'

difficult to find an exact Gaelic equivalent; in the east *tom* has its British sense of 'rounded hill.'

One of the most important of the survivals is *monadh*, hill ground; O.Breton *-monid*; Cornish *menedh*, older *menit*, a mountain; Welsh *mynydd*, older *minit*, a mountain, from an early *monijo-*; compare Lat. *mon-īle*, necklace; G. *muin*, neck, back; *muintorc*, a neck-collar. This postulated early form would yield *muine* in Gaelic, and O'Reilly and O'Brien give for *muine* 'mountain,' but no example of this meaning appears to exist in literature, the regular meaning being 'brake,' 'shrubbery.' The form of *monadh* shows that the word came into Gaelic from a form *monid*, *i.e.* at a stage when the original *o* had not as yet been affected, but when the terminal *-ijo-* had become *id*. This latter change took place early, for it is not found in the loans from Latin into Welsh, the inference being that by the time these loans were established the process was either complete or well advanced.[1] The reduction of *o* is due to the position of the stress. In modern Welsh the stress is regularly on the penult, but in earlier times it was on the last syllable of the word as it stood after dropping the inflexional ending.[2] When the reduction took place is not ascertained, but O.Breton *-monid* shows *o* unchanged. Old forms are: bellum monith Carno 729 AU; bellum montis Carno 728 Ann. Cambr.; ryal (al. brwydyr) Mynyd Carn, Bruts, *PS.*, p. 123; Dubtholargg rex Pictorum citra Monoth, 782 (AU); montana scilicet Moneth transiens, *PS.*, p. 186.

In modern Gaelic *monadh* means hill ground, hilly region; in Perthshire it is *mon*, and this short form was apparently used elsewhere in districts which are not now Gaelic-speaking.

In the older literature, especially in poetry, *monadh* is used as the name of a district or territory denoting originally, it would seem, the mountainous part of Scotland, but later applied in a wider sense. Eoghan, son of Niall of the Nine Hostages, who flourished in the first half of the

---

[1] Pedersen, *Vergl. Grammatik*, i. p. 241.

[2] Strachan, *Introduction to Early Welsh*, p. 6.

fifth century, is styled 'great Bear of Monadh,' with reference to the part he played in his father's expeditions to Britain.[1] Gabrán, king of Dál Riata, who died in A.D. 560, is 'king of Monadh.'[2] His son Aedán is styled king of Alba (*rí Alban*), but Aedán's daughter, the mother of Molaise, is 'Maithgemm of Monadh.' Mongan mac Fiachna, who was slain in Alba—apparently in Kintyre—in A.D. 624, is referred to as 'Mongan of Monadh,' from his connection with Dál Riata in Alba, in the same way as other Irishmen who stayed long in Scotland are styled 'Albannach.'[3] Maol Coluim mac Cinaetha, the victor of Carham, is styled 'king of Monadh.'[4] One of the ancient 'chief tales' was entitled *Táin Monaid i nAlbain*, 'the driving of the cattle of Monadh in Alba.' Among the battles of the Fian are 'the battle of Monadh, the battle of Kintyre, the fortunate battle of Islay.'[5] The hero Goll slew 'great Donn of Monadh, son of Ruadh of the coastland of Alba.'[6] Wyntoun, who knew the old tradition, says of Fergus Mór, son of Erc :—

'Oure all the hychtis evyrilkane
As thai ly fra Drwmalbane
Tyll Stanmore and Inchegall,
Kyng he mad him oure thaim all.'[7]

Here 'kyng of the hychtis' is equivalent to 'king of Monadh.' In the Gododdin poetry relating to the battle of Catraeth, contained in the Book of Aneurin, repeated reference is made to a king of the North who is styled *Mynydawc*, 'of the mountain,' 'of the Highlands.' Skene

---

[1] 'Eogan mac Néill art mór Monaid'; *Celt. Zeit.*, viii. p. 299.

[2] 'a rí Monaid in marggaid,' 'thou king of Monadh of the mart' (Rawl. B 502, 86 b 16). In BB 231 a 29, certain saints are described as 'do síl Loairn ríg Alban/maic Erca na mór margadh,' 'of the seed of Loarn, king of Alba, son of Erc of the great marts.'

[3] 'Mongan mac Fiachna Lurgan ab Artur filio Bicoir Pretene lapide percussus interiit'; 624 Ann. Tiger.; 625 AU; LL 204 b 3.

[4] 'Ba rí Monaid Maol Colaim'; *Duan Albanach* in *P.S.*, p. 63.

[5] 'cath Monaid cath Chinn-tíre/agus cath ághmhor Ile'; *Duanaire Finn*, p. 37.

[6] 'Donn mór Monaid, Echtcolla/dá mac Ruaidh Oirir Alban'; *Duanaire Finn*, p. 11.

[7] *Orygynale Cronykil*, bk. iv. l. 1117.

in his edition of the *Four Ancient Books*, considered that the king referred to was Aedán mac Gabráin, and this may be correct, though he does not repeat the suggestion in his *Celtic Scotland*. In any case it is clear that the Welsh *Mynydawc* is the equivalent of the descriptive genitive *Monaidh* in the passages I have quoted.

Other expressions are *Magh Monaidh* and *Clár Monaidh*, 'the plain of Monadh,' 'the surface of Monadh,' *i.e.* of Scotland north of the isthmus between Forth and Clyde. A sixteenth-century poem claims for MacCailin, Earl of Argyll, 'the headship of the Gael of the plain of Monadh.' [1] Another styles MacCailin 'bulwark of the plain of Monadh.' [2] Ranald of ClanRanald, who died in A.D. 1514, is styled 'a warrior by whom was preserved the surface of Monadh'; MacLeod is 'the forest tree over the surface of Monadh.' [3]

There are also references to *sliabh Monaidh*, 'the mountain of Monadh,' which may occasionally mean the Grampian Range. Of Domhnall Breac, king of Dál Riata, it is said 'though Sliabh Monaidh were of gold, he would distribute it in gifts at one time.' [4] One of the ancient tales relates that a hero named Sean Garman took to Ireland certain cattle from the *Sídh* or Fairy Hill of Findchad in Sliabh Monaidh in Alba.[5] This *Sídh* is also referred to as *Sídh Monaidh* (Hogan). A poem addressed to Eoin mac Suibhne on his setting out from Ulster in 1310 to claim his ancestral lands of Knapdale has 'hail at the streams of Sliabh Monaidh to MacSuibhne of Sliabh Mis.' [6]

---

[1] 'ceannas Gaoidheal mhoighe Monuidh'; Adv. Lib. MS., LII, 3a.

[2] 'mór-dtimchell muighe Monaidh'; *ib.*, 11a.

[3] 'laoch lér cothuigheadh clár Monaidh'; *Rel. Celt.*, ii. p. 218. 'an coillbhil ós chlár Monaidh'; Adv. Lib. MS., XXXIX p. 31a.

[4] 'gemad ór Sliab Monaidh, no-s-fodail (Domhnall Breac) fri h-óen uair'; *Fleadh Dúin na nGedh*, p. 56.

[5] 'tuc Garman leis a buar co Mag Mesca ingine Buidb iarna breith dósom a sídh Findchaid a Sléib Monaid a nAlbain'; 'Garman took his cattle with him to Mag Mesca of the daughter of Bodb, after they had been brought by him from Findchad's Fairy Hill in Sliabh Monaidh in Alba'; YBL (fcs.) 446 b 11; compare BB 193 b.

[6] 'Fáilte ag srothaibh Sléibhe Monaidh/do MacSuibhne Sléibhe Mis'; Dean of Lismore in *Rel. Celt.*, i. p. 103.

Dún Monaidh, 'the fortress of Monadh,' is usually further described as 'baile rígh Alban,' 'the king of Alba's stead.' Congal, king of Ulster, who rebelled against the king of Ireland, went for aid to Alba and came to Dún Monaidh, where the king of Alba dwelt. He got the aid he desired and fought the battle of Magh Rath in A.D. 637, where he was killed.[1] One of the Fenian tales tells how four young warriors went to Dún Monaidh in Alba, where they met the king.[2] The hero of another tale announces that on the previous night he had slept in Dún Monaidh, the king of Alba's stead.[3] Cuchulainn goes to Dún Monaidh in Alba in quest of Emir, his wife.[4] The sons of Usnech, who lived in Alba in exile, are styled as 'from Dún Monaidh.[5] One of the exploits of the Fian was the taking of Dún Monaidh.[6] John Carswell, writing in 1565, states on his title-page that Dún Monaidh is another name for Edinburgh,[7] by which he probably means simply that Edinburgh was in his time the seat of the king of Scotland. In the Dean of Lismore's Book, Eoin MacGriogóir, chief of Clan Gregor, who died in 1519, is twice styled 'of Dún Monaidh,' possibly with reference to his descent from the kings of Scotland.[8] The old literature, it is to be noted, gives no hint of the position of Dún Monaidh. Skene identified it with Dun Add in the moss of Crinan, not far from Lochgilphead, which he says is called in Gaelic 'Monadhmor.' 'Dun Add,' he says, 'was called from the moss which surrounds it Dunmonaidh.'[9] Here, however,

---

1 *Battle of Magh Rath*, p. 46.

2 Is and sin luidsim reomaind in cethrar soer ochlach sin co Dún Monaid in Albain . . . 7 is annsin doriacht Mogdhurnn rí Alban isteach'; 'then we, the four noble warriors afoiesaid, went on our way to Dun Monaidh in Alba . . . and then Mogdurnn the king of Alba came in'; *Acall. na Senorach*, l. 3069. (The king's daughter Aine was married to Finn.)

3 *Silva Gadelica*, p. 276.

4 *Rev. Celt.*, vi. p. 184; Mackinnon, *Catalogue*, p. 218.

5 *Celt. Rev.*, i. p. 151, 152; ii. p. 445.

6 'robhrisedar Dún Monaidh'; *Duanaire Finn*, p. 68.

7 'dún Edin darab comhainm dún monaidh.'

8 Adv. Lib. MS., XXXVII., p. 304.

9 *Celt. Scot.*, i. p. 229; iii. p. 129.

Skene has gone astray. The moss of Crinan is not 'Monadh-mor'—which would mean 'big hill'—but *a' Mhòin Mhór*, 'big moss,' 'big bog'; he has confused two different words, *monadh*, mas., and *mòin*, fem. Nor is there any indication whatever that the fort has at any time been known by any other name than Dún Add. Dún Add, however, was certainly an important strónghold, and may very well have been used by the kings of Dál Riata. Further, the remarkable combination of the footprint in the rock, the cup or *ballán* hewn out of the rock, and the sculptured figure of a boar, all close together near the top of the fort, suggest rather strongly that here may have taken place the ceremonial for the inauguration of a king. That, however, does not prove it to have been Dún Monaidh. Native tradition makes Dunstaffnage the ancient royal seat, and though Dunstaffnage cannot be proved to have been Dún Monaidh, it has in its favour that its present name dates from a time subsequent to the Norse invasion. But the probability is that Dún Monaidh was used loosely to denote the seat of the Gaelic kings of Scotland wherever it might be placed.

The chief *monadh* in Scotland is, of course, the range of mountains now known as the Grampians, formerly the Mounth, which divides Scotland north of Forth into two divisions, the northern of which was 'ultra Moneth,' 'ultra Muneth,'[1] the southern was 'citra Monoth.'[2] Wyntoun makes frequent mention of 'the Mownth,' sometimes of 'the Mwnthis,' as—

> 'The Kyng Alysandyre in Elgyne
> Held his Yhule; and come oure syne
> The Mwnthis, passand till Mwnros.'[3]

By 'the Drum,' on the other hand, he always means the range which runs from north to south, near the west coast, Adamnan's 'Dorsum Britanniae.' The supersession of the old term is unfortunate, but it still clings to various parts

---

[1] 1198 *Chart. Lind.*

[2] A.D. 782 *AU.*

[3] *Oryg. Cron.*, vii. l. 2823.

or sections of the system. West of Dalnaspidal the range is called in Gaelic *Monadh Dhruim-uachdair*, 'mountains of upper-ridge,' more commonly, I think, *Druim-uachdar*, simply, as in the old dirge :—

'Tha mi sgìth is mi siubhal
Leaca dubha Dhruim-uachdair;
'S beag an t-iongnadh dhomh féin sud:
Chaill mi deagh mhac an duine uasail,
Is e sìnte 'n a bhreacán,
'S a chasan 'san luachair.'

'I am weary traversing the black flags of Druim-uachdar; small wonder that same: I have lost the good son of a noble; he lies stretched in his plaid, and his feet in the rushes.'

East of Dalnaspidal, and bounding Atholl on the north, is *Monadh Miongaig*, called by T. Pont 'the mountayn of Minegeg,' in Blaeu 'Mountains of the Minigeg,' meaning 'mount of little-cleft' (*gàg*). On the Kingussie side there is the forest of Gaick, *i.e. Gàig*, dative of *gàg*, cleft. Further east, separate parts of the Mounth are distinguished as *Mon' Chollaigh*, Mount Cowie, 'hazel mount'; *Mon' Chapull*, Moncaple, 'mount of horses'; *an Tulmon*, from *tul*, 'brow,' or *tul* given by Cormac: 'is tul gach nocht.' 'everything naked is *tul*.' *Mon' Chainb* is in English 'Camp Hill'; its meaning is obscure to me. *Mon' Caoin*, Mount Keen, is 'beautiful mount' or more probably 'smooth mount,' for G. *caoin* has both meanings.[1]

Another *monadh* which deserves special notice owing to its historical associations is the 'Royal Mount' or 'Montreal' of Scotland, near St. Andrews in Fife. The *Life* of St. Cadroe says that the Chorisci—with whom we are not concerned—made conquest of Rigmonath and Bellethor, two towns far apart from each other.[2] The earlier of the two versions of the Legend of St. Andrew, printed by Skene, makes Regulus come to the summit of the King's Hill, that is Rigmund.[3] The other version says 'there

[1] I owe the Gaelic forms of these eastern names to Mr. F. C. Diack.
[2] Skene, *P.S.*, p. 108. [3] *Ib.*, p. 139.

was a royal town Rymont, called "regius mons," "the royal hill," which King Hungus (Oengus) gave to the holy apostle Andrew.'[1] The Early Irish form of the name is *Rígmonad*, 'Royal Hill,' and the name survives in Balrimund 1144 (RPSA), Easter and Wester Balrymonth, Balrymundis, 1471 (RMS); Balrymont Estyr, Balrymont Westyr, 1480 (RMS). The places are about two miles apart, at the easter and wester extremities respectively of an elevation little more than a mile south of St. Andrews, which reaches 339 feet. Southwards in Crail parish this high ground continues into the King's Muir, a name on record in the reign of David II. (RMS I). Somewhere hereabout probably was the 'Kingissete' mentioned in RPSA. Connected with this height is the name *Cennrígmonaid*, 'head of the royal mount.' The later Legend of St. Andrew states that 'Regulus came to the land of the Picts, and put in at the place called Muckros, but now Kylrimont.'[2] A note to the Félire of Oengus says of St. Cainnech of Achadh Bó that 'he has a *reiclēs* in *Cell Rīgmonaig*' (read *-aid*). This name survives in a slightly corrupt form as *Cill Rìbhinn*, for *Cill Rìghmhuin*, the Gaelic name of St. Andrews, meaning 'church of the royal mount.' Muckros is explained as 'nemus porcorum,' 'the swine's wood,' and a little along the coast south of St. Andrews there are Kinkell Braes and Kinkell, standing for G. *Ceann na Coille*, 'wood-end.'

The Legend of St. Andrew has it that King Hungus (Oengus) granted *Cursus Apri*, 'the boar's course,' to God and to St. Andrew, which gift was renewed by Alexander I. and confirmed by his brother David I.[3] Wyntoun puts it—

'The Barys Rayk in regale
To the kirk the king gave hale.'[4]

The district meant is near St. Andrews, and is reckoned to extend from east to west about eight miles in length, and in breadth two, three, four, or even five miles in some

---

[1] Skene, *P.S.*, p. 188.
[2] *Ib.*, p. 185.
[3] *Ib.*, pp. 190, 193.
[4] *Oryg. Cron.*, bk. vii. 681, 916.

places.[1] The name doubtless refers to some famous boar-hunt, such as that of the Twrch Trwyth in the Mabinogion or the Gaelic tale of Diarmaid's hunt. 'Rayk' corresponds exactly to the Gaelic *sgrìob*, and in the Tongue district of Sutherland, one of the places in which Diarmaid's hunt is located, the course taken by the boar is still known as *Sgrìob an Tuirc*, 'the boar's rayk.' Whether the name Boar-hills in St. Andrews parish is a genuine survival of the old name seems rather doubtful; Blaeu has it as Byer-hills.[2]

Rìgmonad is most probably the gaelicized form of a British name; the place was doubtless the seat, or near the seat, of the British rulers of Fife. We may compare Ptolemy's *Rerīgonion*.

Mr. Egerton Phillimore, in his edition of Owen's *Pembrokeshire*, remarks: 'The "Mountain," so common in the place-names of English Pembrokeshire means, as the Welsh *mynydd*, from which it is translated, often does, merely a common or wild unenclosed land, without any necessary reference to its hilliness. The "Mount" is also a common place-name in parts of English Pembrokeshire. . . . In Welsh Wales the name Mount is rare, and . . . only seems to occur in districts strongly occupied by the *advenae*.'

In Scotland also 'Mount' is sometimes a translation, as, for instance, 'Mount Lothian'; but in most cases it is not a translation but an anglicization of *monadh*, the form assumed by the old British term when taken over into Gaelic. As to height, *monadh* with us is applied to quite low elevations, but in every instance there is a height of some sort, so far as I have observed. There are, however, many cases in which it is difficult or impossible to distinguish the short form *mon* from *mòin*, a moss.

South of the isthmus of Forth and Clyde the instances are fewer than might have been expected in a region of which great part is hilly. The old term has been displaced by Gaelic terms and by the Teutonic *law* and *fell*, sometimes, too, by *heugh*, as in the case of Kelso.

---

[1] Lawrie, *Early Scottish Charters*, p. 391.

[2] J. B. Johnston, *Place-Names of Scotland*, 'Boarhills.'

Minto in Roxburghshire is near Minto Hills, 905 feet in height and conspicuous. Old spellings are Roger de Munethov, 1166 (Bain's Cal., i.); Myntowe, 1296 (Orig. Paroch., i. p. 321); Minthov, 1306-29 (RMS). This seems to be a compound of which the first part is O.Welsh *minit*; the second part may be Scots *how*, a hollow, a low hill; or *heugh*, a hill, as in Kelso. Roxburgh

Glentenmont, near the head of Kirtle Water on the border of Dumfries and Roxburgh, is probably 'fire-hill glen'; W. *tan*, G. *teine*, fire, compare Prenteineth. Mountbenger on Yarrow is Montberneger, -berngear, -bernger, 1563-73 (RMS).

Fordun states that Marchemond Castle was the name of the Castle of Roxburgh.[1] A writer in Macfarlane's Collections says that Roxburgh Castle was 'of old called the Castle of Marchmonth, as Stirling Castle was called Wester Snodoun, whence two of our Heraulds receive their denominations, to this day being called Marchmonth and Snodoun Heralds.'[2] There is also Marchmont in Polwarth parish, Berwickshire, bordering on the Lammermoor Hills and the Merse; the church of Polwarth was repaired in 1703 at the expense of Lord Patrick Hume, Earl of Marchmont, and of his Countess.[3] The first part is W. *march*, a horse; compare *Meall Greagh*, 'mount of horse-studs,' near Ben Lawers, where horses were pastured.

Pressmennan is Presmunet, *c.* 1160 (Lib. Melr.), meaning 'copse of the hill,' with reference to a ridge close by. Monynut, on the border of Berwickshire, is Monimet (for Moninet) in the reign of William I. (Lib. Melr.); Maninet, *ib.*; Monynett in Blaeu; Manegnut in the Old Stat. Account. It is a difficult name, and its first part is doubtful. We may perhaps compare Minnigaff in Kirkcudbright, Monygof, 1548 (RMS), and Munmaban, 1186 (Reg. Glas.), the site of the 'chapel of Horde,' dependent on the church of Kirkurd in Peeblesshire: in the former the second part appears to be W. *y gof*, 'of the smith'; Haddington

---

[1] *Scotichronicon*, v. p. 32; *Annals*, p. 1. [2] *Geog. Coll.*, iii. p. 157.
[3] *Old Stat. Acc.*, xvii. p. 96.

in the latter it is W. *Mabon*, a personal name, or *maban*, a babe.

Peebles — Mendick, a hill in West Linton, is Menedicte (a Latin genitive), 1165-90 (Chart. Hol.); Mynedicht, Mynidicht, after 1210 (Orig. Paroch., i. p. 516). The first part looks like W. *mynydd*, but it bears the stress.

Minitiuallach, *c.* 1370 (Orig. Paroch., i. p. 517) appears also to have been in West Linton.

Linlithgow — Dechmont and Dechmont Hill, Degmethe 1337 (Bain's Cal.); Dechmont, -ment (RMS I), is the same as Dechmont and Dechmont Hill in Cambuslang, Lanarkshire. The former hill is 686 feet, and 'commands a very extensive prospect'; the latter is 602 feet, and 'commands a magnificent view.'[1] The first part was probably O.W. *dag*, now *da*, good, displaced by O.Ir. *dag-*, *deg-*, now *deagh*, good; compare the hill *Deagh-choimhead*, 'good prospect,' in Muckairn parish, Argyll. In the second part -methe of 1337 is doubtless for -menthe, with which may be compared Meneted, for Menet-ted, Menteith.

Dumfries — Kinmont in Annandale, Kynmund, 1529 (RMS), being stressed on the first part means 'head hill,' like *Ceannchnoc*, 'head-hill,' in Glen Lyon and Inverness-shire.

Ayr — Montgreenan near Kilmarnock is Montgrenane, 1488 (RMS), probably 'hill of (the) sunny place'; G. *grianán*, a sunny place; a dry, hard place on which peats are dried; a sunny hill-top. Pethmont of Blaeu, Pemont 1546 (RMS), also Pemonth, Peymonth, appears to have been at or near Hawkhill, near Old Dailly, on Girvan Water. Other names now obsolete are Monediwyerge and Monemethonac, 1367 (RMS), in Carrick; the latter is apparently for G. *monadh meadhonach*, 'mid hill.'

Lanark — Dechmont in Cambuslang has been mentioned.

Stirling — Polmont, Polmunth 1319 (Johnston), means 'pool-hill,' 'hollow-hill,' from *poll*, pool, also 'hollow place'; compare *a' Chlach Phollach*, 'the stone with hollows,' a cup-marked boulder at Auchterneed, Strathpeffer.

Perth — Moncrieff near Perth is considered to have been the

[1] *Ordnance Gazetteer.*

scene of 'bellum Monid Chroibh,' 'the battle of Monad Croib,' 728 (AU), between two sections of the Picts; Tighernach has 'cath Monaigh (read *-aidh*) Craebi.' If the former stood alone, it might be possible to take *croibh* for *cruibh*, genitive of *crobh*, a hand, a claw, but *craebi* of Tighernach is certainly genitive singular of *craeb*, now *craobh*, a tree; the meaning is therefore 'hill of (the) tree,' with reference to some conspicuous tree, possibly a tribal tree, which stood there; compare Crieff, Pirn. Stormonth has been noted. Kinmont in the lordship of Methven is Kynmonth, 1578 (RMS), 'head hill,' like Kinmont in Annandale.

In A.D. 729 (AU) Nectan, king of the Picts, was defeated by Oengus, son of Fergus, in 'bellum Monith Carno iuxta stagnum Loogdae,' 'the battle of Moned Carno near Loch Loogdae,' in Annales Cambriae 'bellum montis Carno.' The site of this battle has been placed at Cairn O' Mount in the Mearns, at Loch Insh in Badenoch, and on Loch Tay,[1] but no trace of a lake so named is to be found at any of these places. *Loch Loogdae*, as has been mentioned (p. 50), is the same name as Adamnan's *Stagnum Lōchdiae*, now Loch Lochy near Fortwilliam; *Loog*, however, is British in form, like Adamnan's *Crōg* in *Crōg reth*, while *Lōch* is Gaelic in form. The battle was certainly not fought near Fortwilliam; the campaign which preceded it was in Perthshire. On the western border of that county the river Lochy rises near Tyndrum in a small loch now called *Lochán na Bì*, 'lochlet of pitch-pine.' Clifton near Tyndrum is *Achadh nan Tuirighnean*, 'field of the kings,' from O.Ir. *turigin*, explained as 'king' by Cormac and O'Clery; south-west of it, on the opposite side of the valley, is *Sròn nan Colann*, 'point of the bodies.' The name *Moned Carno* might be compared with *Ard-carna* (BB 213 b 37), which means 'height of flesh'; the second term may, however, be also compared with W. *carnau*, plural of *carn*, a cairn. Just behind Tyndrum, on the county march, is

[1] Skene, *P.S.*, index (Cairn O'Mount); *Celt. Scot.*, i. p. 288 (Loch Insh); Egerton Phillimore, *Y Cymmodor*, ix. p. 160 (Loch Tay).

Carn Droma, 'cairn of the ridge,' an ancient landmark,[1] and west of this are two cairns, while on the south side of the valley *Beinn a' Chuirn* is west of *Sròn nan Colann*. This part of Drum Alban might therefore well be styled 'hill of the cairns,' and *Mọned Carno* may be purely British. The fact appears to be that Oengus drove Nectan's forces up Strathearn and into Glen Dochart.[2]

Fife

Mount Hill in Monimail parish is '*le* Monthe in the lordship of Petblatho' (Pitbladdo), 1538 (RMS). Monimail is Monimel, 1250 (RPSA); Monymeyll, 1480 (RMS); the second part is pronounced in English 'meal,' and is for G. *mìol*, a wild animal such as a hare, or possibly for *maol*, bare. Finmont in Kinglassie parish is for *finnmhonadh*, 'white hill,' 'holy hill,' the Welsh *Gwynfynydd* in Montgomery. West of it in the same parish is Kininmonth, for *Cinn Fhinnmhonaidh*, 'head of white hill'; the English form, it is to be noted, shows neither aspiration after *finn* nor inflection—with these it would be 'Kininvonie.' There is also Kininmonth near Pitscotie. Formont Hills in Leslie parish stands either for G. *fuarmhonadh*, 'cold hill,' or, perhaps with less probability, for *formhonadh*, 'projecting hill,' or 'great hill'; with the former compare *Meall Fuarmhonaidh*, 'hill of cold moor,' Glen Urquhart. Glassmount near Kinghorn is for G. *glasmhonadh*, 'green hill.' Brackmont, Leuchars, may be from Ir. *bréch*, 'wolf,' now obsolete both with us and in Ireland. Kilbrackmont in Kilconquhar parish is Cilbrachmontin, 1550, but without older forms it is impossible to say whether the first part is *coille*, a wood, or *cill*, dat. of *ceall*, a church, or *cinn*, dat. of *ceann*, head. Here again the English form shows no inflection of *monadh*. Montrave in Largo parish is Mathriche before 1177 (Chart. Mon. North Berwick); Malthrif in rubric of same; Matheryue, before 1177, *ib.*; Mathriue, before 1228, *ib.*; Monthryve, 1587 (RMS);

---

[1] A charter of Robert II. constituted Guillaspic Campbell his lieutenant, etc., from Carndrome to Polgillippe and from Polmalfeith to Loch Long (1431, RMS).

[2] Nectan may have expected help from Dál Riata; it is rather significant that Oengus 'smote' the Dalriadic Scots in 734 and 736.

Mondthryve and Cotton of Mondthryve in Blaeu. The difference in spelling here is such as to suggest two different names in the same locality, but in the North Berwick charters 'Athernin 7 Mathriue' go together, and in the charter of 1587, which deals with the same lands of the Convent of North Berwick, 'Atherny and Monthryve' go together. The presumption therefore is that Math-, or Mal(t)-, has become assimilated to the names which begin properly with Month. The early spellings may represent British *mad-tref*, 'good *tref*.'

Names now obsolete are 'Monthquoy in the barony of Rossithe,' 1529 (RMS); Montripple apparently near Largo, 1542 (RMS); Munquhany et Strathor (Strath Ore), 1465 (RMS); Munquhane, 1459, *ib.*; Montquhanny, 1598 (Ret.).

Monameanach in Glen Isla is *monadh meadhonach*, 'mid hill.' Monthroy in Lintrathen is *monadh ruadh*, 'red mount.' Mountboy, Kinnoul, is *monadh buidhe*, 'yellow mount'; there is also Montboy in Careston parish. Kinblethmont, Inverkeillor, is Kinblathmont, 1531 (RMS), 'head of Blathmont,' *i.e.* *bláthmhonadh*, which may mean either 'smooth mount' (*bláith*), or 'flower mount' (*bláth*). Montquhir in Carmyllie parish is Muncur 1237-48 (Chart. Lind.), Munchur 1245, *ib.*, Moncur in a Roll of David II. (RMS I); Monquhir, 1540 (RMS); the second part is probably *corr*, a pit, gen. *cuirre*; compare Polmont and Curmyr, 1366 (RMS), in Kincardineshire, also Strachur. The first part may be *mòin*, moss, or *monadh*. Cunmont in the west end of Panbride parish is probably 'high mount' from W. *cwn*, E.Celt. *cunos*, high. Montreathmont Moor, Aberlemno, is Munreimund, 1325 (RMS); Monreuthmont, 1565, *ib.*; Montrewmonth, 1566, *ib.*; Montrewmont, 1580, *ib.*; Muir of Montroymont in Macfarlane (ii. p. 28). Another writer in Macfarlane (ii. p. 44), says, 'upon the Westsyde of both parishes (Farnell and Kinnaird) lyes that great and spacious forest called Mont roy mont . . . abounding in wyld foul and haires.' It is Montreathment, 1652 (RMS); Montromont, 1653; Montromonth, 1656, *ib.* The later spellings indicate W. *rhuddfynydd* or G. *ruadhmhonadh*, 'red mount,' 'Rougemont,' as the second part, and the Forfar sixteenth-century ones are consistent with this. But 'reimond' of 1325 indicates *rígmonad*, 'royal mount,' as in Balrymonth in Fife.

Kincardine

Kinmonth in Glenbervie is 'head-hill.' Mondynes in Fordoun parish is identified with Monacheden, where Duncan, son of Malcolm Canmor was slain by the Mormaer of the Mearns.[1] The *Chronicon Elegiacum* has 'in Monehedne' as the end of an hexameter, making four syllables with penultimate stress.[2] In 1214 it is Monethyn, and in Fordoun Monthechin. The second part is obscure to me.

Aberdeen

Fourman Hill near Huntly, on the Banffshire border, is doubtless *fuar mhonadh*, 'cold hill,' as in *Meall Fuarmhonaidh* and Forman Hill. Mormond Hill in Buchan is for *mórmhonadh*, 'big hill'; though the highest part is only 769 feet, it is a far-seen landmark and a massive feature in the landscape. Garmond in Monquhitter parish is probably *garbhmhonadh*, 'rough hill'; there is also Tullygarmonth, 'hill of Garmonth,' in Birse. Kininmonth, east of Strichen, is 'head of white mount,' as in Fife. Mount Medden at the head of the Cabrach, on the Banff border, is *monadh meadhon*, 'mid hill'; the representation of internal aspirated *d* (*dh*) by *d* or *dd* in English is common in this region. The range forms a watershed. Mundurno in Old Machar parish is for *monadh dornach*, 'pebbly mount'; the adjective is from G. *dorn*, W. *dwrn*, a fist, whence *dornág*, a rounded pebble like a fist. *Monadh an Àraidh*, at the head of Strathdon, is now on maps 'the Ladder Hills,' a correct translation. Montammo, Foveran, is for *monadh tomach*, 'knolly hill,' possibly 'bushy hill.' Monthammock, in Durris, is the same. Brechmount, which appears in the bounds of Maryculter,[3] is the same as Brackmont, Kil-brackmont, and probably contains the old *bréch*, wolf.

Essilmont in Ellon parish is Essilmonth, 1377 (RMS); Essilmund, 1450 (Ant. A. & B., iii. p. 7); the hill is only 219 feet. This may be direct from the British corresponding

---

[1] *Celt. Scot.*, i. p. 439.
[2] Skene, *P.S.*, p. 175.
[3] *Ant. of Aberdeen and Banff*, i. 300.

to Welsh *iselfynydd*, 'low hill'; prefixed adjectives of more than one syllable are extremely rare in Gaelic at any stage of the language, though nouns of two syllables are found not uncommonly in this position, while Welsh has no difficulty in placing a polysyllabic adjective as the first part of a compound, *e.g. ucheldre*, 'high stead.' The Gaelic word *ìosal*, 'low,' appears in Craigeazle in Galloway; Irish *íseal*, as in Gorteeshal, 'low field,' in Tipperary.[1] Crimond, the name of a parish on the Buchan coast, appears as 'the land of Creithmode (*read* -monde) in which the church of Rettref (Rattray) is situated,' 1323 (RMS); with it go 'Creichmode Nagorthe and Greichmode Belle' (*read* -monde) meaning apparently 'hungry Crimond' (*na gorta*) and 'broom Crimond' (*bealaidh*) respectively. Later there is 'Retref which is also called Crechmound'; also Creichtmont (Ant. A. & B.). The height is close to the sea and to the site of the old burgh of Rattray. There are also Crimond in Keithhall parish and in Methlick, Crechmond 1509 (Ant. A. & B.). The first part of these names is probably that found in Dul-chreichard in Glen Urquhart, where *ard*, 'height,' answers to *monadh*; compare also Crechmael, the name of a wood, but probably originally of a height (Hogan); it seems to be connected with *creachann*, the bare wind-swept top of a hill. The first part of Brimmond Hill in Newhills parish is uncertain in the absence of old forms. Wheedlemont in Auchindoir parish is Fidilmonth, 1414 (Ant. A. & B.); Fidelmonth, *ib.*; Fulzemont, 1506-7 (RMS); Fuilyement, 1610, *ib.*; Fulyemonth, 1696 (Ret.). The Old Stat. Account of Auchindoir (1794), says, 'Fulziemont, or the blood of the mountain . . . lying at the foot of a pretty high conical hill called Knock-chailich. The hill has been fortified by a double wall, and the farm has probably received its name from some bloody battle that has been fought there.' The first part of this difficult name is probably British taken over into Gaelic. The remarks of the parish minister in 1794 show the pronunciation in his time; the subsequent change

[1] Joyce, *Irish Names of Places*, vol. ii.

to Wheedlemont is probably connected with the Aberdeenshire pronunciation of initial *wh* as *f*, as in the stock example 'fa fuppit the fite fulpie' for 'who whipped the white whelp': here, on the other hand, initial *f* has been 'corrected' into *wh*; the restoration of *d*, which had been lost in pronunciation long before 1794, is further evidence that the name has been 'renovated.'

Monadh Fergie in Kirkmichael is 'hill of Fergie,' a stream. Monadh nan Eun, near the head of Aven is 'hill of the birds.' Mountblairy in Alvah is to be compared with Muieblairie in Ross-shire, in Gaelic *Muigh-bhlàraidh*; the second part is either genitive of *blàrach*, 'dappled place,' or an adjective *blàrdha* from *blàr*, 'dappled.' Montgrew in Grange is probably for *monadh na gcraobh*, 'hill of trees'; compare Bunchrew. Kelman Hill in Glass is probably for *caolmhonadh*, 'narrow hill,' which describes it. Kyninmonthe, 1407 (RMS), later Kinmonth, is 'head of white hill.'

Brightmony in Auldearn parish is Brechmond, 1507 (RMS), Brichmonye, Brightmanie, Brightmony (Ret.); probably 'wolf hill.'

Am Monadh Liath is 'the gray mountain range,' and *am Monadh Ruadh* is 'the red mountain range,' now the Cairngorms. *Meall Fuarmhonaidh* is 'lump of the cold mountain range,' a prosaic description of a far-seen mountain of shapely contour as seen from near Inverness. Balvonie, near Inverness, means 'moorstead.'

North of Inverness, though the term *monadh* is in constant use colloquially, it enters into very few names of places, and when it does, the meaning is 'moor' rather than 'hill,' *e.g. Tigh a' Mhonaidh*, 'moorhouse,' otherwise Altnamain, on the road which goes across the hill ground between Alness and Ardgay. The one exception is the old name for the Ord of Caithness, formerly called 'the Mounth' or 'Mound,' and this suggests that the term may have been displaced in other parts of the North, before our record period. In Argyll and the Isles it is rare. North of Inveraray there is *am Monadh Leacanach*, 'the flaggy range,' from *leac*, a flagstone; in Macfarlane it is Monikleaganich, 'verie dangerous to travel . . . in time of evill stormie weather, in winter especiallie, for it is ane high Mountaine.'[1] In Lorne there is *am Monadh Dubh*, better known as 'the Black Mount,' a large range of mountainous ground. The same name occurs in Mull, where there is also *am Monadh Beag*.

The generic term *Pit* occurs on the east side of Scotland from the Forth basin to Rogart in Sutherland. On the west side it is extremely rare ; the only instances known to me are Pitcon in Cunningham ; Pitmaglassy in Lochaber ; Pitalmit and Pitchalman in Glenelg ; Pitnean, now obsolete, in Lochcarron. Two or three instances occur in Lothian—Pittendriech near Lasswade, in Midlothian, and Pitcox in Haddingtonshire ; there is also Pitcox near Edinburgh. Elsewhere the distribution is approximately—Sutherland, 7 ; Ross, 17 ; Inverness-shire, 10 ; Nairnshire, 1 ; Moray, 12 ; Banffshire, 15 ; Aberdeenshire, 67 ; Kincardineshire, 25 ; Forfarshire, 31 ; Fife and Kinross, 57 ; Clackmannan, 1 ; Perthshire, 69 ; Stirlingshire, 3. It is notable further that in Sutherland all the Pits are in the south-east part of the county. In Ross they are nearly all in Easter Ross. In Perthshire they are mostly in the east and south-east, the farthest west being Pitmackie in Glenquaich, Amulree. Wherever they occur, they are found mostly on the low ground, that is to say, on the land best fitted for cultivation.

The facts of distribution show that the term had its original habitat, so to speak, in the district north of Forth, and that it was established most firmly between Forth and Spey. It was evidently current in Gaelic over a wide area ; latterly, however, it has been displaced in Gaelic by *baile*, a stead, though the old form survives in the names as anglicized. Thus, for instance, Pitfour in Rogart is in Gaelic *Baile-phùir* ; Pitkerrie in Ross is *Baile-chéirigh* ; Pitcastle near Pitlochry is *Baile a' Chaisteil* ; Pitlochry itself is *Baile Chloichrigh*. When this change took place I have not been able to determine.

---

[1] *Geog. Coll.*, ii. p. 147, 512.

Our earliest instances are in the Book of Deer: *Pett in Mulenn*, 'the *Pett* of the Mill'; *Pett Malduibh*, 'Maldubh's Pett'; *Pett meic Garnait*, 'Garnat's son's *Pett*'; *Pett meic Cobroig*, 'Cobroch's son's *Pett*'; in modern Gaelic it is *peit*. It is to be compared with Welsh *peth*, 'some, a certain quantity of something, a thing'; Cornish *peth*, 'a thing, a something, an article'; Breton *pez*, 'morceau,' 'a piece'; Low Latin *pecia*, *petia*, 'a piece'; *petia terrae*, *petium terrae*, 'a measure of land,' whence French *pièce*, Italian *pezza*. The Celtic forms *pett*, *peth*, *pez*, all postulate an earlier *petti-*, which probably existed in Gaulish, whence it came into Low Latin as *petia*. Welsh and Cornish *peth* do not occur in place-names; the term has evidently been specialized in Pictland in the sense of 'portion, share,' 'petia terrae.' In Gaelic literature it appears twice only so far as known to me, and then in the sense of 'croft': 'fuath liom droch pheit 'ga daoradh,' 'I hate to see a poor croft raised in price.'[1] That *pett* was taken over into Gaelic from British appears from its form, for Gaulish *petti-* would naturally become in Gaelic *pitt* rather than *pett*.[2]

From *peit* was formed a collective *peiteach*, 'place of petts,' the plural of which is seen in *na Peiteachán* in Killearnan parish, Ross-shire. The dative-locative is seen in *Peitigh*, Petty, the name of a parish near Inverness; there are several other places of the same name. The genitive is found in *Blàr-pheitigh*, Blairfettie, in Glen Erichdie near Struan.

Some of the 'Pits' involve personal names. Near Struan and nearly opposite Bruar is a burn called *Allt Phit'al-domhnaich* for *Allt Phit Mhaol-domhnaich*, 'the burn of the portion of Maol-domhnaich'; Petmuldonych in the barony of Strowane, 1504 (RMS). Pitmiclardie in Fife is for *Peit mhic Fhlaithbheartaigh*, 'MacLaverty's share'; the

---

[1] Adv. Lib. MS., XXXVI, 92 b 15; later in the same poem it is used in the secondary sense in which Alexander MacDonald uses *croiteag* in his *Moladh Móraig*.

[2] Irish *pit*, gen. *pite*, denotes 'an allowance of food or drink' in connection with monastic discipline; so in LBr. 10 b, 11 a; also in Stokes, *Metrical Glossaries*, Bezz. Beitr., xix. p. 102.

personal name means 'Rule-bearing.' Pitelpie in Forfar was formerly Pitalpin, 'Alpin's share,' but that is not to say, as Skene does,[1] that the name has any connection with the father of Kenneth MacAlpin. Kethirhelpie, which seems to mean 'Alpin's fort,' appears in Reg. Ep. Aberd., p. 58. Pitcarmick in Strath Ardle, Perthshire, is 'Cormac's share.' Pitewan, which occurs twice in West Forfar is 'Eoghan's share'; compare Balmackewan in Kincardine. Pitkennedy in Aberlemno parish is 'Cennéitigh's share'; the name means 'grim-headed.' Pitkenny in Fife is 'Kenneth's share,' for G. *Baile Choinnigh*, 'Kenneth's stead.' Pitcaithly near Perth is Pethkathilin, *c.* 1230 (Chart. Lind.); here the second part is probably *Cathalan*, dim. of *Cathal*; compare Balkaithley in Fife. Pitcalman in Glenelg is 'Colman's (later Calman's) share'; the local tradition is that Calman held one of the Glenelg brochs, and that his brothers held the other two, while their mother Grùgag lived at Caisteal Grùgaig, a broch near Totaig, on the south side of Loch Alsh. Pitmaduthy in Ross-shire is in Gaelic *Baile mhic Dhuibh* or *Pit mhic Dhuibh*—one of the few instances where *peit*, *pit*, survives in Gaelic—meaning 'Macduff's stead,' or 'portion'; compare Belmaduthy in the Black Isle, G. *Baile mac Duibh*, 'stead of the sons of Dubh.' Pitconnoquhoy, the old name of Rosehaugh in the Black Isle, is 'Donnchadh's (Duncan's) share.' Pitkerrald in Glen Urquhart is understood to mean 'St. Cyril's share,' but the name involved may be Cairell, not uncommon of old. Pitmurchie in Aberdeenshire is 'Murchadh's share,' for *Peit Mhurchaidh*, or in Mid. Gaelic *Pett Murchada*. Petultin, 1144 (RPSA), somewhere near St. Andrews, is 'Ultan's share'; there were seven saints so called, so that the name must have been common. Pethergus 1195 (Chart. Lind.), 'Fergus' share,' was near Mathers in Kincardineshire. Petmacduffgyl (RMS I, App. 2), in Athol, is for *Peit mac Dubhghaill*, 'share of Dugall's sons'; I do not know its position.

A few of the Pit names are connected with occupations.

[1] *Celt. Scot.*, i. p. 307.

Pitgersie in Aberdeenshire is for *Peit* (*an*) *ghréasaighe*, 'the shoemaker's share,' with metathesis of *r*, as in 'gerse' for 'grass.' The term *gréasaighe*, however, in the older language meant 'decorator,' 'embroiderer'; the modern 'shoemaker' is a specialized meaning. Pettinseir in Moray is for *Peit nan Saor*, 'the artisan's share'; *saor* is now with us specialized as 'wright.' Pitskelly in Kincardine, Forfar, and Perth-shires is 'the share of the *sgéalaighe*,' *i.e.* of the teller of tales or romances; compare Balliskilly in the Black Isle, in Gaelic *Baile Sgéalaighe*. The story-teller would be skilled in reciting the ancient tales which used to form the principal entertainment of the people in the long winter evenings. Though *bard*, 'a poet,' forms part of many names, it does not seem to occur with Pit.

A third class contains terms relating to agriculture, pasture, or stock. Pitkerrow in Forfarshire is from *ceathramh*, a fourth part. Pitcog in Kinfauns parish, Perthshire, is 'share of the fifth part' (*cóig*, *cóigeamh*); compare Pitcox in Haddington. Pitfoddels in Aberdeenshire is Badfothel, 1157 (Reg. Ep. Aberd.); Badfothal, 1359, *ib.*; Badfodall, 1390-1406 (RMS I, App. 2); Badfodalis, 1440 (RMS); Petfodellis, 1487, *ib.*; the plural form is due to the fact that there were Easter and Wester Badfodals. 'Bad' of the earlier spellings is *bad*, a spot, a clump; either it was changed to 'Pet,' or—which is perhaps more likely—there were two names close together, one in 'bad,' the other in 'pet.' The second part is *fodál*, 'a subdivision,' from *fo*, under, and *dál*, a division, a share, as in *Dál Riata*, 'Riata's share.' The term is not uncommon in connection with land; for instance, Diarmait son of Aed Slaine is recorded to have offered Tuaim Eirc with its subdivisions of land (*cona fodlaib feraind*) as a 'sod on altar' (*fód for altóir*) to God and to St. Ciaran.[1] In *fodál* the *d* is of course aspirated (*dh*), but in this region *dh* in the body of a word is often *d* in the English forms, *e.g.* Pitmedden from *medon*, later *meadhon*, middle. Pitarrow in Forfar and Fife is for *Peit arbha*, 'corn-share,' 'corn-town'; compare *Cnoc an*

---

[1] LU 115 b 39.

Arbha, 'Knockinarrow,' in Strath Carron, Ross, now 'Cornhill'; *Aird an Arbha*, 'Ardinarrow,' 'Corn-point,' in Loch Alsh. Pitbladdo in Fife is Petblatho, 1481 (RMS), 1492, *ib.*; Petblado, 1494, *ib.* It suggests comparison with *Móel-blātha*, the name of a stone in the refectory of Hí, with regard to which we are told that Columba lifted a sack of meal from off it, and that he left prosperity on all food that should be placed upon it.[1] *Móel-blātha* seems to mean 'servant of meal,' *blātha* being the genitive of a noun corresponding to Welsh *blawd*, meal. The same term probably appears in Bladebolg, 1144 (RPSA), now Blebo in Fife; which again is to be compared with Blato-bulgio(n), a place in Britain, supposed to mean 'meal-sack place.' On this basis Petblatho would mean 'meal-share'; compare *Tìr-mhine*, 'land of meal,' on Loch Awe, anglicized as Tervin. Pitmillan in Foveran parish, Aberdeenshire, is Pitmulen, *c.* 1315 (RMS), 'share of the mill,' like *Pett Mulenn* of the Book of Deer. There was also a Pethmolin in Crail parish, Fife (RPSA). Petbrain, 1564 (RMS), was in Kirkmichael parish, Perthshire: 'terrae de Petcurren *alias* Petbrain et Glengenet.' It probably means 'quern portion'; compare Auchenbrain near Mauchline. The various Pitfours have been mentioned. Pitglassie in Ross-shire is in Gaelic *Baile a' Ghlasaich*, 'Lea-stead,' but Pitmaglassy or Pitenglassy in Lochaber (on the Glengarry border) is *Bad a' Ghlaistir*, 'clump of the green land' or 'lea-land.' It also appears as Balmaglaster.

Animal names appear in several cases. *Muc*, a pig, occurs in Pettymuck in Aberdeenshire, and *capull*, horse, mare, in Pitcaple. Pittentarrow is from *tarbh*, bull; Pittendamph from *damh*, ox. Pitgaveny is from *gamhna*, genitive of *gamhain*, stirk. Pitgoberis, 1593 (RMS), 'in the barony of Mukartschyre,' is plural of Pitgober, from *gobhar*, goat. Petmarch, 1526 (RMS), in Fife is from W. *march* or G. *marc*, horse, like Markinch. Pettyvaich in Inverness and Banff-shires is from *bàitheach*, 'cow-house, byre.'

---

[1] Bernard and Atkinson, *Liber Hymnorum*, i. p. 62.

Names of this class connected with the Church have been noted (p. 267).

Other names in Pit show a variety of differentia. Some are named after trees. Pitchirn in Badenoch is for *peit a' chaorthainn*, 'rowan-tree portion.' Pitcowden in Aberdeenshire is Pitnacoldan, 1695 (Ret.), for *peit an challtuinn*, 'hazel portion.' Pitcows in Kincardineshire is from *coll*, hazel, with English plural. Pitcullen in Aberdeen and Perth-shires is from *cuilionn*, holly. Pittencrieff in Fife is for *peit na craoibhe*, 'share of the tree.' Pitcruvie or Balcruvie in Fife is *peit chraoibhe*, 'tree share,' or *peit chraobhaigh* (dative of *p. chraobhach*), 'wooded share'; Pitcruive in Perthshire may be for *peit chraoibhe*, the -*e* being lost in English; but compare Dalcruive, p. 418.

Pitcroy in Moray is for *peit chruaidh*, 'hard share.' Pitlurg in Aberdeen and Banff-shires is Petynlurg, 1226 (RM); Petnalurge, 1232, *ib.*; for *peit na luirge*, 'portion of the shank,' *i.e.* distinguished by a shank-like strip of land; compare 'the land that Forbes clemys his of Tirepressy is called Lurgyndaspok that is to say The Bischapis Leg, the whilk name war nocht likely it suld haf war it nocht the Bischapis'; 1391 (Ant. A. & B., iv. p. 379). Pitfoskie in Aberdeenshire is from *fosgadh*, gen. *fosgaidh* (now with us *fasgadh*), shelter. Other instances in this county are Pitsligo, Petslegach, 1426 (RMS), for *peit sligeach*, 'shelly portion'; compare *Sligeach*, Sligo, in the Black Isle; *Sligeachan* in Skye; Sligo in Ireland. Pitmedden is 'mid-share'; Pitmain in Badenoch is the same name, though here the medial *dh* is not represented. Pitcorthie in Fife is from Old and Mid.Gaelic *coirthe*, a pillar stone; so also are Pitforthie in Kincardine and Forfar-shires; the change of Gaelic initial *ch* into *f* in Scots is common, *e.g.* *Uachdar-chlò* in the Black Isle into Auchterflow. The presence of *coirthe* indicates a stone circle or a standing stone. Pittenweem in Fife is for *peit na h-uam(h)a*, 'share of the cave'; there are caves close by. Similarly Weem, near Aberfeldy, is in Gaelic *Uaimh* (dative) 'cave.' These indicate a pronunciation of initial *ua* as *wa*; compare 'dunnie-wassel' for *duine uasal*, a pronunciation which I have heard myself in North Perthshire. Pettinmyre, 1664 (Ret.) near Auchtermuchty is probably for *peit an mhaoir*, 'the officer's share.' Pitfoules, 1698 (Ret.) in Fife is for *peit foghlais*, 'brook share,' and Pittilloch in Falkland parish is from *tulach* or *tilach*, 'hill,' whence the surname Patullo. One of the oldest names on record in Fife is Petnaurcha, 1128 (Reg. Dunf.), granted to the church of Dunfermline by Malcolm Canmore (1070-93); this name is now represented by Urquhart near Dunfermline. It stands for *peit an urchair* (old *aurchoir*), 'the share of the cast or shot,' but what the cast was we do not know. Pitarrick near Pitlochry is in Gaelic *Baile an Tarraig*, 'stead of the pulling,' with reference to its position. The long stiff road which goes from Pitlochry to Kirkmichael ascends 'Pitarrick Brae' on the Pitlochry side, where there was a hard pull. Pitmackie in Glenquaich is now in Gaelic *a' Mhacáig*, which is obscure to me. Pitlochry is *Baile a' Chloichrigh*, from *cloichreach*, which in Irish means 'stony ground,' and that may be the meaning here also, but the reference may be to stepping stones, for which the Irish term is *clochrán*. There is another Pitlochrie on a burn in Glen Isla in Forfarshire. Pitmurthly is for *peit mhórthulaigh*, 'portion of (the) big hill.'

The name Pittendreich occurs from Midlothian to Banffshire. The furthest south instance is near Lasswade, spelled Pendendreia, *c.* 1130 (Chart. Hol.); Petendreia, *ib.* (King David's Charter); Pettenreia, *c.* 1142—all latinized forms. In Fife there is Pittendriech in Dunino parish. In Kinross, Pittendriech is north-east of Loch Leven. In Perthshire it occurs in Lethendy parish. In Stirlingshire there is Pendreich north of Bridge of Allan, of old Petendreich. In Forfarshire, Pittendriech is near Brechin. Pittendriech in Aberdeenshire was 'in the lordship of Kinmundy.' Pittendreigh in Banffshire is on Deveron in the parish of Marnoch. The old forms—apart from those of the Edinburgh one—are all alike: Pettindreche, Petindrech, etc. The places of this name appear all to be situated on slopes, usually facing the sun; *e.g.* adjacent to Pendreich, Bridge of Allan, there is Sunnylaw. Here *-drech* is simply

G. *drech*, now *dreach*, feminine in Mid.Irish, now masculine in both Irish and Scottish Gaelic, and meaning in Irish 'face, countenance,' in Scottish Gaelic 'aspect.' It usually applies to the human face, but it was also used generally, like *aodann*, as in 'in order to keep the front (*drech*) of the hostel from falling on them.'[1] In the east of Scotland the term was at one time evidently applied freely in the sense of 'hill-face.' In Welsh the cognate term is *drych*, 'aspect,' which probably accounts for Scottish Gaelic *dreach*, 'aspect,' as compared with Irish *dreach*, 'face.' It is perhaps worth noting that the early latinized forms are feminine, agreeing with the gender of *drech* in Mid.Irish.

In Wales, Cornwall, and Brittany, the term *dol*, fem., 'meadow, dale, valley,' is common in names of places. In Scotland it was also common, appearing in the older forms of our place-names as *Dol*, *Dul*, later *Dal*. In Ireland it does not occur either in place-names or in literature. It has survived in Scottish Gaelic as *dail*, fem., gen. *dalach*, on the analogy of such words as *fail*, fem., gen. *falach*, a ring; *sail*, fem., gen. *salach*, willow. There is evidence, however, of an earlier form *dol* or *doil*, gen. *dolach*. In Gaelic pronunciation of the present day, initial *Dal* is always *dail* in the west and usually in the east, but in some districts of the east *dul* is still heard—only, however, in place-names.

As to distribution, *Dal* is very common in Ayrshire, there being 83 names so beginning in the index to the Retours. It is common in Perthshire, about 46 from the same source; fairly common in Kirkcudbright, 22; Dumfries, 16; Stirling, 10; Banff, 10; less common in Inverness-shire, 15; Ross, 5; Aberdeen, 5; Dumbarton, 8; Argyll, 8. On the west coast from Kintyre northwards, and in the Isles, it is rare. The actual instances in each county are, of course, much more numerous than the instances in the index to the Retours: Milne, who ransacked

---

[1] 'ré dreich na bruighne do chongbháil ina seasamh gan tuitim orra'; *Caithréim Conghail Cláiringnigh* (Irish Texts Soc., v. p. 88).

the 6-inch Ordnance Survey Map, found 40 for Aberdeenshire. The figures, however, help to show the relative proportions. It is noticeable that the term is rare or non-existent in the regions where Norse influence was strongest—the Isles and the far North ; commonest in districts where there was little or no such influence. This is enough to show that our *dail* is not a loan from Norse *dalr*, mas., a dale, which of course occurs often in names of Norse origin, but almost always terminally, as Borrodale for *borgardalr*, 'fort-dale,' and never with the article. *Dail*, Dell, which occurs by itself as a name in Lewis, does not take the definite article and is not declined ; *Gleann Dail*, Glendale, in Skye, is 'the glen of *Dail*,' undeclined and without the article : in both these instances we have to do with Norse *dalr*, which was never adopted into Gaelic speech, though it remains in the Norse place-names.

Dull near Aberfeldy is in Gaelic *Dul*, without the article and not declined. The Irish-Latin Life of St. Cuthbert tells how he came to 'a town called Dul,' and goes on to say 'not more than a mile from it there is in the woods a high steep mountain called by the inhabitants Doilweme, and on its summit he began to lead a solitary life.' The 'mountain,' as Skene has noticed, is the Rock of Weem, *Creag Uaimhe*, the boldest of the high conspicuous bluffs on the north side of the Tay valley. Behind the village of Dull there is another bluff called *Creag Dhul*. But 'Doilweme' really means 'Dull of Weme,' *doil* being the earlier form of *dail*, and probably the dative of *dol*, fem., adopted into Gaelic. With Dull may be compared 'the Doll,' G. *an Dail*, in Strathbrora.

At the foot of Glen Lyon, just above the Pass of Lyon, there is the beautiful flat of *Seasdul*, Sestill 1502 (RMS), now Chesthill in English. Here the second part is *dul* ; the first part, which is the differentia, is doubtful, but may be compared with Cessintully in Menteith, so in 1488-9 (RMS), where *cess* may stand for G. *seas*, 'a seat, bench,' *i.e.* in this case 'a terrace, plateau.' There is a well-marked terrace or plateau not actually on the place at present called Chesthill but immediately west of it.

Dulsie on Findhorn, in Nairnshire, is in Gaelic *Dulasaidh*, for *Dulfhasaidh*; Dulsie Bridge is *Drochaid Dhulfhasaidh*. This is a compound of *dul* and *fasadh*, 'a stance.' The *fasadh* is the site of the present farmstead; immediately behind it the ground dips steeply down to a little meadow enclosed on three sides by high steep banks whose tops are on a level with the farmstead, and on the fourth by the Findhorn. This meadow is the *dul*, now called 'the Little Haugh.' Dulsie means 'haugh-stance'; compare Pitdulsie, Petdoulsie 1594 (RMS), in Aberdeenshire.

Dallas in Ross is Dollace, 1574 (RMS), in Gaelic *Daláis Bhig* and *Daláis Mhóir* (Little and Meikle D.), treated as feminine. In Nairnshire, Dallaschyle is for *Daláis na Coille*, 'of the wood.' In Moray there are Dallas, in 1232 Dolays Mychel (RM), a parish, and Dallasbrachty, formerly Dolesbrachti (RM), 'of the malthouse' (W. *bracty*).[1] The *a* of *-ais* is open, not dull. *Daláis*, older *Doláis*, is a compound of *dul*, *dol*, with most probably a term representing W. *gwas*, fem., an abode, dwelling, O.Ir. *foss*, mas., rest, act of residence, and means 'meadow-stance,' 'holm-dwelling,' or such. The Gaelic, or gaelicized, form probably appears in the term *Lethfoss*, 'half-station,' corresponding in meaning to G. *leathbhaile*, 'half-stead'; compare W. Dolassey near Bleddfa in Radnorshire.

The Pictish Chronicle records a battle 'in Dolair' fought *c.* A.D. 875, *i.e.* Dollar in Clackmannan. There are also Dollar Law and Burn in Drummelzier parish, Peebles. This name is from *dol* with *-ar* extension, as in Migear, Monar, etc. Dollerie near Crieff may be from *doilleir*, dark, opposed to Soilzarie, near Blacklunans, from *soilleir*, bright.

The old genitive singular is seen in Ballindollo, Forfarshire, Balwynddoloche, 1363 (RMS); Ballindolloch in the reign of David II. (RMS I, App. 2), which is the older form corresponding to Ballindalloch, in Gaelic *Baile na Dalach*, in Banffshire, also in Glen Lednock, Perthshire, and elsewhere, meaning 'stead of the meadow or riverholm

[1] I have not heard or seen the corresponding Gaelic term.

or haugh.' Drumdollo, 1666 (RMS) in Forgue parish, Aberdeenshire, is for *Druim Dolach*, 'haugh ridge'; '*lie sched* vocat. Abirdolo,' 1611 (RMS) in the barony of Largo, Fife—a plot of five acres—is probably 'the *abir* of (the) haugh.'

In some cases the second part is a personal name in the genitive case. Deloraine in Selkirkshire, 1486 (Ex. Rolls) is Doloriane, 1624 (RMS); Wester and Easter Dalorrens in Macfarlane (iii. p. 166). The stress is on the second syllable, though Sir Walter Scott, with his usual liberty as regards place-names of Celtic origin, put it on the last when he told of 'William of Deloraine' in the *Lay*. Sinking of Dol, Dal to Del is common in unstressed position; the second part may be the Gaelic name *Odhrán*, anglicized as Oran, but the spelling of 1624 rather suggests *Urien*. Dailly, the name of a parish in Carrick, is Dalmakerane 1323 (RMS), Dalmulkerane 1404 (Chart. of Crossraguel), for *Dail Mhaoil-Chiarán*, 'Holme of St. Ciaran's servant.' Personal names beginning with *Maol* become rare after about A.D. 1000. Dolpatrik, 1226-34 (Chart. Inch.) 'Patrick's holm' lies on the Earn a little above Inchaffray; another name for it was 'Kenandheni,' *ib.*, which is for *cenn ind enaig(h)*, 'head of the marsh,' rather than from *aonach*, gen. *aonaigh*, 'moor'; compare 'Edardoennech near the dam of the mill of Gortin,' 1208 (Chart. Inch.) in the same locality, for *eadar dà enach*, 'between two marshes.' Dalarossie in Strathdearn is Dulergus, 1208 (RM); Dulergusyn in Stratheren, 1224-42, *ib.*, representing *Dul F(h)ergusa*, 'Fergus' holme'; now in Gaelic *Dail Fhearghuis*, the name being declined in modern Gaelic as an o-stem, while in old Gaelic it was a u-stem. The church of Dalarossie stands on a fairly large flat beside the Findhorn. Lower down in Strathdearn, on the Fintag burn from Loch Moy, is Dalmagarry, in Gaelic *Dul Mac-Gearraidh*;[1] the forms Tullowch Makcarre (RM); Tullochmakerrie, 1661 (Ret.); Tulloc Smagarre in Macfarlane (ii. p. 608), must refer to a place on or connected with a ridge (*tulach*) adjacent to the

[1] *Poems of Kenneth Mackenzie* (1792), pp. 208-211.

present Dalmagarry, which is on a level haugh by the burn side, rather raised so as to form a plateau.

Dol, Dul, is found not uncommonly with names of saints, indicating an old church site or land gifted to the Church. Dolbethok, 1211 (RPSA), 'Bethoc's holme,' was granted by Gilchrist, earl of Mar, to the Culdees of Monymusk. 'Dulmernock in the thanage of Kynclevyn,' 1382 (RMS), now Kinclaven north of Perth, is 'St. Mernoc's holme.' Other instances are Dail Chiarain, Dail Mo-Choid, Dalmahoy, Dulmayok or Dulmaok, Dalmally, and probably Dalserf in Lanarkshire.

Some further examples follow :—

Kinross 'Cultcarni and Dolkoyth,' 1304 (Bain's Cal.) belonged to the bishop of Dunkeld. The former is now Cockairney, for *cuilt charnaigh*, 'rocky nook,' in Kinross parish; the latter is later Dalqueich (Ret.; RMS), 'haugh of the round hollow' (*cuach*).

Perth Dolcorachy, 1313, Dalhorochquhi, 1445 (Chart. Inch.), was in the thanage and old parish of Forteviot, north of Earn, now apparently Bankhead. The second part is for the genitive of *corrachadh*, 'odd-field,' or 'taper field.' Dulbethy and Dulcabok appear in 1490 (RMS). The former is now Dalbeathie on the left bank of Tay in the parish of Caputh, 'haugh of the birchwood' (*beitheach*, gen. *beithigh*). Dulcabok, now obsolete, is doubtless the haugh beside Tay at the church and village of Caputh, for Caputh is in Gaelic *Capaig*, in 1275 Cathbathac, -bethac (Theiner). Dulgardy, 1536 (RMS), in Glen Lochay, is now in Gaelic *Dail-ghaoirdigh*, at the foot of the mountain *Meall Ghaoirdigh*, a name obscure to me, with which may be compared Gourdie in Caputh parish. Another difficult name in Glen Lochay is Daldravaig, in Gaelic *Dul-drabháig*. Dulcrufe, 1557 (RMS), in Methven parish, is now Dalcruive or Dalcrue on Almond; the second part is *cruibh*, genitive of *crobh*, a hand or paw, from its shape; compare *Tir-ingnigh*, Tirinie, 'land of the claw-place,' near Coshieville, Aberfeldy, where there is close to the farmhouse a hollow shaped exactly like a bird's foot. Dulgus, 1587 (RMS), is now Dalguise, 'haugh of fir,' from *giús*, gen. *giúis*, now with us *giuthas*, fir.

Dulquhober 1511, Dulquhowqbeir 1554 (RMS), is now Dalfouber on North Esk near Edzell; the second part seems to be the same as Cupar and Drum-cooper in Fife and Cupar-Angus. Duldarg, 1511 (RMS), 'red haugh,' in the same district, seems to be obsolete. Dulquhorth, 1554 (RMS), now Dalforth on the left bank of Upper North Esk, is probably from *coirthe*, a pillar stone. Forfar

'Dolbrech on Duffhern,' 1255, 'le haylch which is called Dolbrech,' *c.* 1286 (RM), was near the mill of Carnousie in Forglen parish and means 'dappled haugh,' like the common *Dail Bhreac*, Dalbreck. Banff

Cromdol, *c.* 1224 (RM), is now Cromdale, in Gaelic *Cromdhail* and *Crom'ail*, 'bent haugh,' from the bend of Spey which sweeps round the glebe just below the parish church. The Gaelic term corresponding to 'the Haughs of Cromdale' appears to be forgotten in the district. Moray

Dulchreg, 1655 (RMS), in Stratherrick, is now in Gaelic *Dulchrag*, 'rock haugh,' with reference to a rock outcrop near the farmhouse. Dalcrombie on Loch Ruthven is *Dul-chrombaidh*, 'haugh of (the) bend'; it is rather elevated above the loch. Dulschangy, 1345 (RMS), now *Dul-seangaigh*, Dulshangie, in Glen Urquhart, seems to be from *seangach*, gen. *seangaigh*, 'slim place,' 'narrow place,' derived from *seang*, slender; compare *Caoilisidh* for *Caoilinsi* '(at) narrow meadow,' in Ross-shire. Duldragin of 1509 is now *Dul-dreagain*, Duldreggan, in Glen Urquhart, with which may be compared Baldragon in Forfarshire, popularly derived from *dreagan*, a dragon, O.Ir. *drac*, gen. *dracon*, a dragon. This term was used metaphorically of a valiant man, a hero: MacDougal of Dunollie is styled 'an dreagan caomh o'n Chonghail,' 'the dear dragon from Connel.'[1] The popular explanation is therefore possible in this sense. *Dul-chreichard* (p. 405) is in the same district. Inverness

In modern Scottish Gaelic the regular term for a bush is *preas*, which does not occur in Irish. It is a survival

[1] *Book of the Dean of Lismore* in *Rel. Celt.*, i. p. 98.

from British, and is to be compared with Welsh *prys*, covert, brushwood; *presel*, a brake, thicket. In place-names *preas* is used in this latter sense—'copse, thicket, covert'—as was also 'buss' (bush) in Scots, which is defined as a wood 'consisting of oake and birk.'[1] In the parish of Logie Easter in Ross-shire there was a place called 'the Bus of Preischachleif,' *i.e.* 'the bush of *preas a' chachaileith*,' 'the bush of the *preas* of the gate or hurdle.' Though *preas* is found over all the east of Scotland, it is not very common; on the west there seems to be only one instance.

Caithness *Preas a' Mhadaidh*, in Reay parish, is 'the wolf's copse'; in 1726 'Presswaddie, where there is an old decayed wood.'[2]

Ross Tornapress in Loch Carron parish is in Gaelic *Treabhar nam Preas*, 'farm-stead of the copses.'

Preas Ma-Ruibhe, 'Mael-ruba's Copse,' is near Strathpeffer; it is now the burial-place of the family of Coul.

Inverness Presmuckerach in Badenoch is 'pig-brake place'; compare *Dos-mucarain* in Ross-shire, where *dos* is the Old Gaelic equivalent of *preas*.

Moray Prescalton in Knockando parish is for *preas calltuinn*, 'hazel copse.' Pressley, in Edinkillie parish, is Presleye, 1501 (RMS).; appears to be a compound of *preas* and the suffix *le*, 'place,' found in Welsh names.

Banff Preshome may be for *preas Choluim*, 'Colum's copse.'

Aberdeen Prescoly, 1390 (RMS), is 'copse of the hazel wood' (*collach*); it seems to have been near Old Aberdeen; other spellings are Presterle, Prostoli, Prosly (RMS I, App. 2). Presquheill 1577, Pressecheild 1581 (RMS), near Inverernan in Marr, is for *preas choill*, 'hazel copse.' Terpersie, Alford, is Tirepressy, 1391 (Reg. Ep. Aberd.), for *tìr preasaigh*, either 'copsy land' (dative) or 'land of copsy place' (genitive). Dalpersie, Tullynestle, is 'copsy haugh.'

Kincardine Culperso is Coulpersauche, 1375 (RMS); Culpressache in the reign of Robert I. (RMS I, App. 2), for *cùil phreasach*, 'copsy nook.'

---

[1] *Book of Kilravock*, p. 328.

[2] Macfarlane, *Geog. Coll.*, i. p. 181.

Presnerb in Glen Isla is 'roe copse.' Bada na Bresoch in Glen Isla is for *Bad na mPreasach*, 'spot of the copsy places'; the final *n* of the article goes with the following *p*, the resulting sound being *b* (eclipsis). Pressock, Preschak 1529, 1538 (RMS), also Preciok, seems to be *preasach* with *ch* hardened in English. Forfar

Pressmuk, 1511 (RMS), 'swine copse,' was near Kinnoul. Bofressely, 1489 (RMS), Bofresle, 1511, *ib.*, was apparently near Aberfoyle; the first part is *both*, a booth, hut; the second part is the same as Pressley in Moray. Perth

Pressmennan is 'copse of the hill' (p. 399). Haddington

Press on Coldingham Moor; there is a wood there now, and Taylor and Skinner's map of 1776 shows a wood then. Berwick

The town of Dumfries is Dunfres, 1189 (Bain's Cal.); so also 1259, *ib.*; in the Ragman Roll Dunfres occurs often, Dunfrys once, Dumfres once. Elsewhere in Bain's Cal. Dunfres, Donfres, occur *passim*. R(adulph) son of Dunegal granted part of his heritage in Dronfres to the Hospital of St. Peter of York (Bain's Cal., ii. p. 421); he was a witness in 1136; the lands of Coulyn and Ruchane 'infra vicecomitatum de Drunffres' appear in 1363-4 (RMS); in the same year a charter is granted 'apud Drunfres,' *ib.*; Drunfres appears twice in a charter of 1370, *ib.*; Drumfreis, 1321, *ib.*, 1324 (thrice), *ib.*; Drumfres, 1369, *ib.*; Drumfres, 1395; in RMS I, App. 2 the forms with Drum- and Dum- occur side by side; in subsequent volumes of RMS (1424-1668) the name occurs very often and always with Drum- till 1628 when Dunfres appears; thereafter the form is Drum- usually, but not always. Wyntoun has Drumfrese, Drwmfres, Dwnfres; Fordun has Dunfrese twice. In Macfarlane, Doctour Archbald writes Drumfreis, Drumfreiss (iii. p. 185). Dumfries

There are cases where *dun* and *drum* are confused, *e.g.* Dunpelder, near Glasgow, now Drumpellier; Dunmedler is now Drummelzier. But in the case of Dumfries we have two forms from the beginning of the records—Dron-, Drun-, later Drum-, and Dun-, later Dum-, and the two forms persist. If the case here is one of confusion, the confusion cannot be cleared up; it is more likely, however, that we have to do with two names containing a common second part, not, of course, two names for the same place, but names of places close together, Dunfres and Dronfres or Drunfres. In the latter, *dron* is probably for Gaelic *dronn*, hump, a bent back, as in Dron, the name of a parish in Perthshire; in modern Irish *dronn* is pronounced *drun* according to Dinneen. Before the labial of the second part *dron, drun*, became naturally *drom, drum*, and so indistinguishable from *druim*, a back, a ridge, cognate with *dronn* and similar in meaning. The second part is assumed by Skene to indicate early settlement by Frisians, and he would make Dunfres to mean, as he says himself, 'the town of the Frisians, as Dumbarton is the town of the Britons.'[1] The difficulty here is that there is no independent ground for believing that the Frisians ever settled in Dumfries: the first Teutonic invaders were the Angles in the seventh century, and they are always called 'Saxons' both in Gaelic and in Welsh, *e.g.* Glen Saxon in Middlebie parish, etc. 'Dunfres' stands for *Dun-phreas*, 'fort of copses,' or for *Dun-phris*, 'fort of (the) copse,' as 'Duncow' in the parish of Kirkmahoe, four miles from Dumfries, is for *Dun-choll*, 'fort of hazels.' 'Dronfres,' etc., is 'copse hump,' 'copse ridge.'

In Scottish Gaelic *tom* mas., gen. *tuim*, regularly means in the east a knoll or rounded eminence, but in the west, especially in the Isles, it means a copse. In Irish the usual meaning is 'bush, tuft, thicket, grove,' though it also means 'knoll, small bank.' Welsh *tomen* is 'heap, mound, dunghill'; *tom* is 'dirt, mire, dung,' a specialized secondary meaning of 'mound,' which is found also in a well-known Scottish Gaelic idiomatic use of *tom*. Our use

---

[1] *Celt. Scot.*, iii. p. 25. The name 'Frisians' occurs in Mid.Irish as *Frési*, *e.g.* 'Frainc 7 Frési, Longbaird 7 Albanaig,' 'Franks and Frisians, Lombards and men of Alba' (LL 29 a). It is said of the hero Cuchulainn 'cim a Frēsib frithbēra,' 'he shall bring hither tribute from Frisians;' the Lombards are mentioned in the next line (*Irische Texte*, i. p. 289). 'Fort of the Frisians' would be *Dún Frés*, so that Skene's proposed derivation is sound phonetically.

of *tom* in the sense of 'knoll' has probably been influenced by British. In place-names it is not always easy to be sure whether we have to deal with the meaning of 'thicket, clump,' or that of 'knoll,' *e.g. Tom na h-Iubhraich*, Tomnahurich, at Inverness, most probably means 'knoll of the yew-wood,' rather than 'thicket of the yew-wood'; *Tom-beithe*, near Callander, Scott's 'Tombea,' probably means 'birch-knoll,' rather than 'birch-clump.' A derivative of *tom* is *tomach*, 'knoll-place,' dative-locative *tomaich*, which is the form in place-names.

Tom is most common between Forth and Spey, rare in the west. In Lewis there is *Airigh an Tuim*, 'shieling of the knoll.' *Tomaich*, Tomich, occurs in Sutherland, Easter Ross, and Kiltarlity parish, Inverness. *Tom-aitionn*, Tomatin, in Strathdearn, is 'juniper knoll'; the juniper flourishes here at a height of well over 1000 feet. *Tom Ghealagaidh* (1300 feet), probably 'knoll of the white gorge,' is at the foot of Strathdearn. In Banffshire there are *Tom an t-Sabhail*, Tomintoul, 'barn knoll,' *Toman a' Mhuilinn*, Tomnavoulin, 'little knoll of the mill,' and other such. There are many instances in Aberdeenshire, mostly simple. Perthshire has many instances also. *Tom-pearrain*, Tomperran near Comrie, has been noted. *Baile an Tuim*, Ballintuim, in Strath Tummel and Kirkmichael parish, is 'stead of the knoll.' Little Corum (1683 feet) and Meikle Corum (1955 feet), Greenloaning, stand for *corr-thom* 'taper knoll'; near them is Core Hill, a part translation of G. *corr-bheinn*, 'pointed horn or peak.' Tambeith (1279 feet), in the same district, is for *tom-beithe*, 'birch knoll.' Naked Tam (1607 feet), in Tannadice parish, Forfar, is doubtless for *tom nocht*, 'naked knoll.' All these are notable for their height. Yellow Tomach is south-east of Loch Macaterick in Carrick, and Tomluchrie, 'rushy knoll,' is in Carrick, near Knockdolian. In the south, however, instances are very rare: the term in use is *torr*.

Bad, found only in Scottish Gaelic, means 'a spot, a particular place,' then 'a clump of wood,' etc., *i.e.* a spot where trees grow; also 'a tuft of hair,' such as a man's beard. Its diminutives are *badán* and *badaidh*, often used in the sense of 'a small company' of men or animals. It seems to be simply British *bod*, 'residence,' retained in the specialized sense of 'place,' 'spot.' As to range, it occurs all along the east, but is rarely found in the south-west and Argyll.

Baad and Baads occur several times in Midlothian. Baddinasgill seems to be for *bad an asgaill*, 'spot of the armpit'; *asgall*, from Lat. *axilla*, is masculine in Sc. Gaelic, fem. in Irish. In Peebles there are Badintree Hill, Tweedsmuir, and Badlieu, the latter being probably for *am bad fhliuch* (dative), 'wet spot,' like *Bad an Fhliuchaidh*, 'spot of wetness,' near Achinalt in Ross. Bedlormie in Torphichen parish, Linlithgow, is Badlormy, 1424 (Ch. Hol.); the second part is obscure. Bedcow, Kirkintilloch, is for *bad call*, 'spot or clump of hazels,' like Badcall in West Ross and elsewhere. In Kincardineshire, Baady Craig is 'little spot or clump of the rock.' In Aberdeenshire, Baad is in Peterculter parish; Badenscoth in Auchterless is for *bad nan scoth*, 'spot of the flowers'; *scoth* is now obsolete with us. Badymichael in Dallas parish, Moray, is 'Michael's little clump.' *Bad Each* in Badenoch is 'spot of the horses,' so called because horses were apt to get bogged there. *Am Badaidh Beithe*, the Baddybae, 'the birch clumplet,' is the name of the park east of Delny Station, Ross; the same name occurs at the head of Cabrach in Banffshire in the form Bodiebae, where the *o* is notable. *Bad Sgàlaidh*, 'spooky spot,' reputed to be haunted, is beyond Kildermorie in Alness parish.[1] In Sutherland, *Bad an Donnaidh* is 'the spot of the mischance,' anglicized 'Pattergonie'; compare *Gob an Uisgich*, 'beak of the water-place,' made into Gobernuisgich, also in Sutherland. In Argyll, 'the Badd' is the name of a hill above Dunoon. *Bad nan Nathraichean* in Skye is 'the adders' spot.'

---

[1] *Sgàlaidh* represents E.Ir. *sgálda* (Cormac, No. 21), from *sgál*, a spectre.

CHAPTER XIII

River Names

The oldest version of the Cattle Raid of Cualnge tells how the hero Cuchulainn, defending Ulster alone against the host of Connacht, came to the ford of the river Cronn. Laeg, his charioteer, proposed that they should give battle to the Connacht men, who were approaching, but Cuchulainn bade him stand fast, and uttered a prayer to the river: 'I beseech the waters to help me; I beseech sky and earth and Cronn in particular. Cronn takes to fighting against them; Cronn will not let them into Muirthemne.'[1] Therewith the water rose up till it was in the tops of the trees. Here the Ulster river comes to the aid of the Ulster hero as in the plain of Troy the Scamander arose against Achilles in aid of the Trojans. Similarly, it is told of St. Ciaran of Saighir that at his prayer the river Brosnach rose against the men of the North when they proposed to invade Munster.[2] Another ancient tale tells that the hero Fráech mac Fidaig of Connacht was son of Bébinn from the Sídhe, a full sister of Boand who gave her name to the river Boyne—she was in fact the goddess of the Boyne.[3] These tales indicate definitely the way in which rivers were regarded by the pagan Gael of Ireland. Gildas says with reference to Britain that he will not cry out by name upon the mountains, wells, hills and rivers, once destructive but now made serviceable to man's uses, on which divine honour was wont to be heaped by a people who were then

[1] LU 67 a 12. The last sentences are part of the prayer.

[2] *Silva Gadelica*, p. 8 (Gaelic), p. 10 (English). The words are 'from God he (Ciaran) procured that which of proud human folk he had not gained: for . . . to bar Conn's Half the Brosnach's stream swelled over her banks so that not one dared take it.'

[3] *Celt. Zeit.*, iv. p. 32.

blind.[1] As to early Scotland, two references in the *Life of Columba* are instructive. Adamnan tells how Columba heard of a fountain famous among the heathen Picts which was worshipped as a god.[2] The other is that of the river whose name 'may be called in Latin Nigra Dea, the Black goddess.'[3] That the continental Celts regarded rivers as divine appears from the names given them. Among the Gauls a favourite ending for names of goddesses was *-ona*: Epona was the goddess of horses (*epos*); Damona was another goddess, probably of cattle (cf. Gael. *damh*, ox; Lat. *dama*, fallow deer); Nemetona was the tribal goddess of the Nemetes; Matronae were the mother goddesses. The same ending is found in names of streams: Matrona was the old name of the Marne; the fountain of Bordeaux was Divona; Ritona was a ford goddess (W. *rhyd*, ford); Axona is now the Aisne. The formation of these names proves their divinity.

In Christian times this pagan conception persisted in the form of streams and wells of virtue connected with some saint or sanctuary. Some of these were probably regarded as waters of power in pre-Christian times, and had been sained or consecrated by the Church. Traces of the old idea remain in our folklore, some of which may be mentioned. Glen Cuaich in Inverness-shire is—or was till lately—haunted by a being known as Cuachag, the river sprite. The tutelary sprite of Etive is Éiteag; a man of my acquaintance declared that he knew a man who had met her in Glen Salach—after a funeral. Dr. George Henderson has recorded that when the northern Nethy gives signs of rising in spate the people used to say, 'tha na Neithichean a' tighinn,' 'the Nethy sprites are coming.'[4] A stream in Benbecula is called *a' Ghamhnach*, 'the farrow cow,' and the custom in crossing it was to throw a wisp of grass

[1] Neque nominatim inclamitans montes ipsos aut fontes vel colles vel fluvios olim exitiabiles, nunc vero humanis usibus utiles, quibus divinus honor a caeco tunc populo cumulabatur; *De Excidio Brittanniae*, c. 4. Mommsen omits the words *fontes vel*.

[2] *Life of Columba*, ii. p. 10.

[3] *Ib.*, ii. p. 39.

[4] *Celtic Review*, i. p. 200.

into the water with the formula 'fodar do'n Ghamhnaich,' 'fodder for the Gamhnach.'[1] Within living memory river spirits persisted in the form of the *uruisg*, a half human creature who haunted streams and fords; the term is from *ar*, on, and *uisge*, water. Breadalbane was specially noted for its *uruisg* tribe, chief of whom was Peallaidh whose name is preserved in *Obar Pheallaidh*, anglicized as Aberfeldy. This sprite had a fairly wide range, for his footprint (*caslorg Pheallaidh*) is to be seen on a rock in Glen Lyon, and the wild burn of *Inbhir-inneoin*, near the foot of the glen, was a favourite haunt of his; a cataract on it is called *Eas Pheallaidh*. In Ross 'Cailleach na h-abhann,' the river hag, was dreaded at the fords of the river Orrin. Such instances of our own time help us to realize, or at least to imagine more or less dimly, the state of matters in early times and its effect on the people. It is not too much to say that the feeling of divinity pervades and colours the whole system of our ancient stream nomenclature.

As a natural consequence of this view, the river has often, as it were, a territory or sphere of influence. The lake from which it issues is regarded as belonging to the river and receives its name from it. Adamnan's expressions 'the lake of the river Nesa,' 'pool of the river Aba,' reflect the native point of view. So, too, the basin through which the stream flows and its point of confluence with river or sea are often called after the stream. In some cases the sequence is complete, as in Naver, Loch Naver, Strath Naver, Invernaver, and Tay, Loch Tay, Strath Tay, Aber Tay. More often, however, some part of the sequence is wanting, as indeed must happen when the stream does not issue from a loch. The Etive rises in *Lochán Màthair Éite*, 'the mother of Etive,' flows through Glen Etive, and falls into the salt-water Loch Etive, but there is no Inver Etive. The *Deathan*, Don, rises in *Coire Dheathain*, flows through *Srath Deathan*, and enters the sea at *Obar Dheathain*,' Aberdeen. A striking exception is the Forth, which has

[1] *Carmina Gadelica*, vol. ii. p. 83.

neither loch, strath, nor confluence named after it; nor has its tributary the Teith.

The rivers of Ireland do not, as a rule, give their names to loch or basin, and not often to their confluence. An exception is the Dea, 'goddess,' now the Avonmore, which flowed through Gleann Dea and entered the sea at Inber Dea, now Arklow. We also differ from Ireland in our use of *srath*, a strath. In O.Ir. *srath* glosses *gramen*, grass, turf, pasture; in place-names, says Joyce, 'the level, soft, meadowland or holm—often swampy and sometimes inundated—along the banks of a river or lake, is generally called *srath*.'[1] There are instances of this use with us, *e.g. Srath na h-Abhann* on the river Averon in Ross-shire, just after it leaves Loch Moire, exactly answers Joyce's description; *Srath Nìn*, Strath Noon, is a riverside holm in Strathdearn; *Srath Ghartáin*, Strath Gartney, is on Loch Katrine side. But our standard use of the term is to denote the lower ground on both sides of a stream. A strath differs from a glen in being as a rule wider and, perhaps always, smoother. In this usage, as well as in the other aspects of river nomenclature, Scotland goes with Wales rather than with Ireland.

The oldest Celtic stream names, both on the Continent and in the British Isles, are usually formed from a single stem with suffixes, as Matrona, Tamaros, Dubis. A few have no specific suffix, as Dea, 'goddess.' In this respect they differ from the personal names, which are regularly compounds such as Cingetorix. Compound names of streams, however, are not unknown in the early period: Pliny records Vernodubrum, 'alder water.'

Among our existing names of streams, Tay, Don, Forth, Naver, etc., belong to the first class; Dea is represented by Dee; compounds of E.Celt. *dubron* are numerous, as they are also in Wales, though not in Ireland. Another frequent element is *glais*, a stream, common to Britain and Ireland. In Wales *nant*, a brook, glen, forms compounds such as Mochnant, 'swine stream,' Marchnant, 'horse stream.'

---

[1] *Irish Names of Places*, ii. p. 399.

Personal names, as has been evident from the names of saints, were often used in a reduced form, arising from one or other of the compounded elements. Findbarr, for example, gives from its first part Finnén, Fintén, Fintán, Fintóc, Findia, Findu, and from its second part Barre or Bairre or Barri, Barróc ; Laisrén gives Laisse in Mo-Laisse ; Cóemgein gives Mo-Chóem, Mo-Chóemóc ; Cóemi, now Caomhaidh, is for Cóemgin or another compound of *cóem* ; Berach gives Bí in Mo-Bhí. A similar process went on in ordinary compound nouns,[1] and there can be little doubt that it accounts for certain of the forms which stream-names have assumed among Gaelic speakers.

A relatively small but very interesting section of our stream names consists of those that show the influence of the early Church, such as Oenglais, Ar-buthnot, Buadhchág, Orrin, Fyne. Some glens or straths were called after saints, such as Glen Màilie, Glen Cannel—properly Glen Cainner—Strath Fillan. Lochs similarly named are Loch Ma-Thatháig or Loch Mahaick and Loch Finlagán. The next step was to call the stream by the saint's name, whence the rivers Cainner or Cannel, Mailie, and Conon, *Abhainn Chonnáin*, in Skye, and others.

The names that follow may for purposes of convenience be arranged in groups, but the groups are by no means mutually exclusive and they are indeed to some extent arbitrary.

I. Such as show the endings (1) *-an* (early *-onā*), (2) *-ar* (early *-aros*, *-ara*), (3) *-ann* (early *-ava*), (4) *-e* (early *-ia*, *-ios*), (5) *-idh* of various origin, (6) *-ad*, *-aid*, (early *-nt-*), (7) *-nad*, *-that* (diminutive, early *-nt-*), (8) *-ág* (early *-ācos* through British), (9) *-agán* (E.Ir. *-ucán*), (10) *-án* (diminutive).

II. Names connected with colours, sound, animals.

III. Compounds of *dobur* and *glais*.

IV. Such as contain *aber*.

V. Names not included in the other groups.

---

[1] *E.g.* : *dobharchú* : *dobhrán*, otter ; *brodchu* : *brodán*, mastiff ; *síoth-bhrugh* : *síthein*, fairy knoll ; *corrmheur* : *corrág*, forefinger ; *cluasadhart* : *cluaság*, pillow ; *gearrfhiadh* : *gearr*, hare.

I. (1) The ending *-onā* which occurs in Gaulish names occurred also in early Britain. The geographer of Ravenna mentions a British river Abona whose exact position is not known. This name has as base the word which appears in O.Ir. as *ab*, gen. *aba*, a river; it is represented now by W. *afon*, a river, and by the various rivers called Avon in England and Wales. Our rivers Avon in Lanarkshire and Stirlingshire represent an early British *Abona*, not Gaelic *abhainn*, which is the dative case of Ir. *abha* used by us as nominative.

Endlicher's glossary of Gaulish terms, a short vocabulary of uncertain but early date, gives '*ambe*, rivo; *inter ambes*, inter rivos.' 'Ambis,' then, was a Gaulish word for a stream, most likely a strengthened form from the base *ab*. Hence would come *Ambona*, parallel to *Abona*, which in modern Celtic—Gaelic or Welsh—would be *Aman*. In Wales there is the river Aman with Aberaman at its mouth, and we have the river *Aman*, 'Almond,' in Perthshire, whose confluence appears in Rath-inver-amon, while its head in *Gleann Amain* is 'caput amnis Amon.'[1] The Gaelic form of the other Almond, in Linlithgow, is not known, but the names are doubtless the same. The popular derivation from G. *abhainn* (misspelled *amhainn*) is, of course, impossible.

The Don, with its two forms Deon and Doen, representing Gaelic and British tradition respectively, the Doon, and the Ythan and Nethan, have been mentioned (p. 211).

The Irvine in Ayrshire, Yrewyn, 1258 (Bain's Cal.), Irwyn, 1296 *ib.*, is probably the same as the Irfon in Cardigan.

Prosen in Forfarshire, is Glen-prostyn, 1463 (Reg. Cup.); it is probably to be compared with the Breton personal names Prost-lon, -voret, Iud-prost, and the O.Welsh On-braust (L. Land.), but the meaning of *praust*, *prost* is unknown to me.

The Conon in Ross is in Gaelic *Abhainn Chonainn*, but notwithstanding the *nn* it may be for an early *Conona*,

[1] Skene, *P.S.*, pp. 151, 179.

'hound stream,' like the Welsh *Cynon* at Aberdare and at Aberaman.

Abharan, Averon, in Alness parish, may be for *Ab-ar-ona*, based on *ab* as in *Abona*.

Lavern or Levern in Renfrewshire, is Laberane, 1539 (RMS). Laveran on Loch Lomond, is Laueran, Innerlaveran, 1225 (Ch. Lennox). Louran is a burn in Wigtownshire. These may be for *Labarona*, from *labaros*, loud, W. *llafar*, G. *labhar*, loud.

Kale in Roxburghshire, is 'aqua de Kalne," 1165-1214 (Lib. Melr.). Calneburne 1214-49 *ib.* was a tributary of Whitadder, now apparently Hazelly Burn. These may be for an early *Calona*, 'calling one.'

Caddon Water, a tributary of Tweed, appears to be the same as *Cadan* of *Inbhir Chadain*, Inverhadden near Kinlochrannoch, for an early *Catona*, from *catu-*, W. *cad*, G. *cath*, battle, 'the warring one.' The phonetics are British. Both Innerhadden and *Dail Chosnaidh*, 'meadow of fighting,' which is close by, are traditionally said to have been named after a battle fought by Robert Bruce,[1] but the tradition may be much older than his time.

(2) Some names show the ending *-aros*, *-ara*, found in *Nabaros*, Naver, *Tamaros*, Tamar.

The Tanner of Selkirkshire and the Tanar of Glen Tanar, Aberdeen, are to be compared with Tanaros, a tributary of the Padus or Po in Cisalpine Gaul. Tanaros was also the name of a Gaulish and British thunder-god, equated with Jupiter; his name is connected with Lat. *tonare*, to thunder, *tonitru*, thunder. Taranis, a Gaulish god of thunder and lightning, is probably the same name with metathesis of *n* and *r*; W. *taran*, Ir. *torann*, Sc.G. *torrunn*, *tairneanach*, thunder: Taranis was a tributary of the Garonne; Afon Tarenig is a tributary of the Wye in Wales.

Calar of Balquhidder noted for its loudness, is from *cal-*, cry, for an early *Calaros* or *Calara*.

Gamhar enters Loch Rannoch from *Loch Éigheach*, 'shouting loch'; in Macfarlane it is Gawir, with Blaeu

[1] *Old Stat. Account*, ii. p. 456.

and on O.S.M. Gaur. The meaning is 'wintry stream,' 'winter stream,' from its floods during melting snow, like Homer's χείμαρροι ποταμοὶ κατ' ὄρεσφι ῥέοντες, 'winter torrents flowing down the mountains'; Gaulish *Giamon*, 'winter month'; O.W. *gaem*, W. *gaeaf*, *gayaf*, O.Ir. *gemred*, G. *geamhradh*; O.Ir. *gam*, winter, formed on the analogy of *sam*, summer, whence *gamain*, a stirk—calves became stirks at Hallowmass—and Gamanrad, 'stirk-folk,' the name of a warrior clan of Connacht. The name is a Gaelicized form of an early *Giamaros*. Auchtergaven in Perthshire is said to have been in Gaelic *Uachdar ghamh-thir*, 'upper part of the winter land' (*O.S.A.*, xvii. p. 551); the name should be written *Uachdar Ghamhair*, for the parish church is in *Gleann Gamhair*, anglicized Glen Garry.

Bruar in Blair-Athol is noted for its deep ravine, 'spanned at intervals by natural arches and by bridges, overhung by impending rocks.'[1] The reference to 'natural arches and bridges' was possibly correct at the time of writing; at present only one remains. The E.Celtic term for bridge was *brīva*, whence the places in France now called Brives and in Britain *Dūro-brīvae*, 'fortress-bridges.' As E.Celtic *īvo* gives W. *uo* or *ua*, *e.g.* Lat. *fībula*, W. *hual*,[2] Bruar probably represents an early *Brīvaros* or *Brīvara*, 'bridge stream.'

Labhar, Lawers, is the name primarily of a stream, then of a district, on the north side of Loch Tay. The English plural is due to the three divisions, *Labhar Shìos*, East Lawers, *Labhar Shuas*, West L., and *Labhar na Craoibhe*, L. of the tree. The stream comes from *Beinn Labhar*, Ben Lawers. *Labhar* is for *Labara*, fem. of *labaros*, W. llafar loud, resounding; with us *labhar*, loud, is common in the east, *e.g. gaoth labhar*, 'a loud wind,' but seems to be quite unknown in the west. O.Ir. *labar* is 'haughty, arrogant,' but *amlabar* is 'speechless, dumb'; the word is not in Dinneen's *Dictionary of Modern Irish*. Another loud

---

[1] *Ordnance Gazeteer*; the *Imperial Gazeteer*, c. 1860, mentions 'natural arches,' and gives a picture of the arch still existing. The suggestion that Bruar may mean 'bridge-water' is due to the Rev. C. M. Robertson.

[2] Sir J. Morris-Jones, *A Welsh Grammar*, p. 110.

tumbling stream, *Uisge Labhar*, falls into Loch Oisein in Lochaber. Aberlour in Banffshire 'is derived from its situation, being situated at the mouth of a noisy burn' (*O.S.A.*, iv. p. 64); Aberlower, 1226 (Reg. Mor.). No streams of this name occur in Ireland; in Wales *Afon Llafar* is fairly common.

Labharág, 'little loud one,' is a stream in Coire Muilzie, Ross; another *Labharág* enters Loch Laggan. Their opposite is *Balbhág*, 'little dumb one,' from Balquhidder into Loch Lubnaig, in the *Lady of the Lake* stressed wrongly on the second syllable. Lavery is a tributary of Duisk in Ayrshire.

The Lugar of Ayrshire, Lugar *c.* 1200 (Lib. Melr.), is probably for **Loucara* or **Loucaros*, 'bright stream,' from *loucos*, white, W. *llug*, bright; compare Luggie. In Macfarlane's *Collections*, however, it is Lugdour (twice), which would represent an early *Louco-dubron* 'bright water.'

(3) Another group of ancient stream names has the ending *-ann*, *-unn*, the vowel being indeterminate. This is a Gaelic genitive ending, as in *Domnu*, gen. *Domnann*, *Danu*, gen. *Danann*; that Gaelic genitives of British names in *-au* (early *-ava*) were formed on this analogy appears from *Manau*, Gaelic gen. *Manann*, dat. *Manainn*. For river names in *-ava* we may compare *Anava*, now the Annan, *Amblava*, now Ambleve. Names of this class are naturally in the genitive, like many other stream names, after *srath* or *gleann*; in some instances there is a tendency to re-inflect, *i.e.* to make *-ann* into *-ainn*.

In Ross there are two rivers *Carrann*, Carron, both with rough rocky beds, from *kars-*, 'harsh, rough,' seen in W. *carreg*, stone, rock, G. *carr*, a rock shelf, *e.g.* *Drochaid Charra*, Carr Bridge. Other streams called Carron, in Nairn, Banff, Kincardine, Stirling, and Dumfries-shires, are probably of the same origin.

Lìobhunn or *Lìomhunn* of Glen Lyon is now *Lì'unn*; the Dean of Lismore's MS. spells it *leivin*, *levin*, and it rimes with *bríoghmhor*, and in Deibhidhe metre with *conn*. The base is probably as in Latin *līma*, a file, Ir. *líomhaim*, I smoothe, polish, W. *llifo*, grind, whet. The Welsh *Afon*

Llifon in Caermarthen is probably for *Limona*; compare Liman or Liuan, a tributary of Trothy in Wales (L. Land.) and *Pull Lifan* or *Lifann* (L. Land.), and 'stagnum Liuan' of the *Mirabilia* of Nennius, whose waters met the Severn at *Oper-linn-liuan*.

Gleann Comhann, Glencoe, with its river *Comhann*, is obscure to me. *Beinn Chomhainn*, Ben Chonzie, near Comrie in Perthshire, seems to be the same name, and some continental names such as *Com-ani*, a tribe of south Gaul, and the Gaulish personal name *Com-avos* may perhaps be compared, but the meaning of these is unknown. The anglicized form Coe may be from the old nominative. The record forms of Glencoe are often bad, *e.g.* Glenchomure, Glenogweris, Glenagwe (RMS I); Glencole, etc. (RMS II); comparatively good spellings are Glencowyn, RMS 1500; Glenquhoin, 1517 (Orig. Paroch.).

The river Braan in Perthshire is *Breamhainn*, from the base seen in Gr. *βρέμω*, roar; W. *brefu*, low, bleat, bray, roar. It is to be compared with *Afon Brefi*, a tributary of Teifi in Wales, which, says Rhys, 'is pronounced much the same as *brefu*, the act of lowing, bellowing, or bleating.'[1] Breamhainn is 'a turbulent and impetuous stream. . . . A fall of about 85 feet, a sheer leap at a wild chasm into a dark caldron, occurs at the Rumbling Bridge, 2½ miles from the river's mouth, and a cataract, long, tumultuous, and foaming occurs at Ossian's Hall, about a mile further down.'[2] Ptolemy's *Bremenion*, a town of the Votadini, now High Rochester, is from the same base. Bremenion is on the river Rede, described as 'rapid and often turbulent,' and it is doubtless formed from the early name of that river, meaning 'place on roaring water.' Other early names in Britain which are probably connected are Bremetennacum now Ribchester, and Bremia mentioned by the Ravenna geographer. The early British form of *Breamhainn* would be *Bremava*, which would become *Bremau*, *Bremu*, with Gaelic gen. *Bremann*, re-inflected *Breamhainn* in modern Gaelic, after *srath* and *abhainn*. It is worth mention that

---

[1] *Celtic Folklore*, p. 578

[2] *Ordnance Gazeteer.*

in parts of Easter Ross *breamhainn* is the regular term for a wheelbarrow. It may be a survival from British. Strath Braan is in Gaelic *Srath Freamhainn*, which suggests Welsh influence, for the mutation of *b* is *f* in Welsh, corresponding to Gaelic *bh*; Rhys quotes 'Llan ddewi Frefi fraith,' 'Llandewi of Brefi the spotted.' Another example of this is *Both-frac*, for *Both-bhrac*, Bolfracks, near Aberfeldy; compare Balfunning, before 1300 Buchmonyn (Johnston), in Stirlingshire, where *m* is mutated to *f* as in Welsh, not to *mh* as in Gaelic.

The river *Calann* comes from *Coire Chalainn* above Tyndrum. The haugh land by its side is *Innis Chalainn*, and the full name of the farm of Auch there is *Achadh Innis Chalainn.*[1] The base here is *cal-*, call, cry; Lat. *cal-are*, *cal-endae*. Holder cites a farm name Calonia, called after a well Calonna; Calonna is also the old form of Chalonnes on Loire. There are three rivers *Callann* in Ireland.

Peathann—so spelled by Ewen MacLachlan—anglicized Pean, enters the head of Loch Arkaig in Inverness-shire from Gleann Peathann, Glen Pean. It may stand for an early **Petava*; compare Holder's *Petavonion* and *Poetovio* or *Petovio*.

The river Nairn is in Gaelic *Abhainn Narunn*, from *Srath Narunn*, Strath Nairn, with *Inbhir Narunn*, the town of Nairn, at its mouth. It may be compared with the river *Naro* in Dalmatia with the town of *Narona* at its mouth, and the Italian *Nar* with its town *Narnia*, supposed to be connected with Lat. *nāre*, *nă-t-āre*, to swim, float.

(4) The ending *-e* represents, in some cases at least, the early ending *-ia*, as in *Boderia*, Ptolemy's name for the Forth. The actual name *Boderia* survives in Aberbothrie, south of Alyth, Perthshire, on record as Abbyrbothry, Abirbothry, etc., 1375 (RMS), where Bothry is for *boidhre*, meaning in common speech 'deafness,' but in this case, as a stream-name, 'the deaf one,' *i.e.* 'the noiseless one.' In Ireland *bodhar*, deaf, is made *bother* in English. Joyce says 'a person who is either partly or wholly deaf is said to be

[1] Auchanichalden 1537 and 1547, *Orig. Paroch.* ii., pt. I.

bothered . . . you are said to 'turn the *bothered* ear' to a person when you do not wish to hear what he says, or grant his request. . . . So well are the two words *bother* and *bodhar* understood to be identical, that in the colloquial language of the peasantry they are always used to translate each other.' [1] This is one of the many instances in which a British name has either been taken over direct into Gaelic or translated during the period of transition from British to Gaelic. A place called *Achadh Bhoidhre* is in Glen Gloy, Brae Lochaber, but I have no information as to the reason of the name.

Three streams in Inverness-shire are called *Dotha*, Doe, one west of Corpach, another nearly opposite Fort Augustus, and the third at the head of Glen Moriston—all in the same region. The name is also spelled *Dogha*, but the pronunciation shows that whether we write *th* or *gh* or *dh*, these letters are not radical, but merely used to separate the syllables, as in *nodha*, new, *ogha*, grandchild, *cnothan*, nuts. The O.Ir. form would be Dóe (two syllables), which was the name of a mythical lady who was drowned in Linn Dóe in the river Berba, now Barrow (LL 195 b); she may have been a water nymph. In the St. Gall glosses on Priscian, Lat. *tardus* is explained by *mall no doe* (Sg. 66 a 7), translated by 'slow or stupid' (Thes. Pal., ii. p. 119). The name may be for an early **Dovia*, but the meaning is doubtful.

The bell of Cill Chuimein, now Fort Augustus, was called *am Buadhach*, 'the one of virtue.' The Minister of Wardlaw relates that by reason of virtue derived from this bell 'the water, or as the vulgar calls it the wine, of Loch Ness is medicinall, and beasts carried to it or the water of the lake brought to beasts to drink, which I have often seen.' [2]

A parallel instance is supplied by the Gaelic poet Duncan Macintyre, who, in his poem on Beinn Dobhrain, lauds 'the wine of the burn of the Annat' (*fìon uillt na h-Annaid*), on the river Conghlais at the foot of the Ben; 'that is the

---

[1] *Irish Names of Places*, ii. p. 47.

[2] *Wardlaw MS.*, p. 147 (Scottish History Society).

unfailing remedy (*ìocshlainte mhaireann*) . . . good manifold was wont to be got from it without purchase.' The point of this is clearly that the water of the burn received virtue from the Annat by which it flows. The court Bards make frequent mention of 'the wine-blood' (*fìonfhuil*) of nobles: the expression refers to the potency or virtue immanent in blood of the true ruling families, on whose purity of descent the prosperity of the tribe depended. It is connected with the old belief in the divine right of rulers. Two wells of virtue in Skye are known as *Tobar an Fhìona*, 'well of wine'; one of them, it is said, used to be recommended by Fearchar Lighiche, the noted physician of the Beaton family.[1] Another *Tobar an Fhìona* is in Glen Salach on Loch Etive, connected doubtless with St. Baodan's Seat in that glen. A famous healing well at Peterhead is called the Wine Well. As wine comes from the vine, it was but natural that streams of virtue should be sometimes called *Fìne*, vine, from Lat. *vinea*, and two well-known streams of Argyll bear that name. One, near Inveraray, flows through *Gleann Fìne*, Glen Fyne, and falls into *Loch Fìne*,[2] Loch Fyne. At the foot of Glen Fyne is the church site of Kilmorich, with Clachán near it. The other flows through *Gleann Fìne*, Glen Fyne, in Cowall; near its mouth is *Aird Fhìne*, Ardyne, and east of the river is a hill called *Buachaill Fhìne*, 'shepherd of *Fìne*.' The origin of this stream's virtue was doubtless *Cill Fhaoláin*, St. Fillan's church, which stood on its bank near the river mouth. Dean Monro mentions Lochefyne in Mull, 'quherin there is a guid take of Herrings'; from his description it appears that the loch is now Loch Scridain. The name, however, is lost, unless we have a trace of it in *Aird Fìneig* on the south side of Loch Scridain near its mouth.

The Ruaidh of Glen Roy and the Gairbh of Strath Garve

[1] A. R. Forbes, *Place-Names of Skye*.

[2] 'Leomhan lonn Locha Fíne,' 'the fierce lion of Loch Fyne'; riming with *tíre*—Adv. Lib. MS., lii. 3 b; 'A cheannlaoch Locha Fíne,' 'thou chief warrior of Loch Fyne,' riming with *ríghe*—Adv. Lib. MS., xxxvi. 81 a.

are probably for *Ruaidhe* and *Gairbhe*, and may be genitive sing. fem.—'Glen of the Red,' 'Strath of the Rough,' with *abhann* understood. This seems to be the explanation of the Irish river *Garbh*: genitive 'gáir na Gairbe gainmige,' 'the roar of the sandy Garbh' (LL 298 b 9); dative 'risin Gairbh,' 'to the Garbh' (*ib.*, a 22). We may compare *Leabhar na hUidhre*, 'Book of the Dun (Cow).'

The *Duibhe*, Divie, a tributary of Findhorn, may be similarly explained, but it may mean 'black goddess' (p. 50). Glen Devon in the Ochils is latinized Glendofona, 1271 (Chart. Inch.); its stream is now in Gaelic *Duibhe*. 'Devon' is probably for an early British *Dubona* or *Dobona*; compare *Dubis*, the river Doubs in France, and its other form *Dova* (for *Doba*), 'black one.' This is a goddess name, of which *Duibhe* may be a Gaelic rendering, thus giving rise to an interesting doublet.

(5) Many stream-names, especially on the east side of Scotland north of Forth, have an ending conventionally written *-idh*, in English spelling *-ie* or *-y*. This is really not one ending but several which have fallen together, so that it is difficult to disentangle them.

In the group of streams called Lochy, Lochty, the ending, as we know from Adamnan, is *diae* (gen.), 'goddess.' The Boyndie of Banffshire may belong to this class; ? *buandea*, 'lasting goddess.'

In some cases we have to do really not with stream names but with place-names in *-ach*, gen. *-aigh*. Inverebrie in Aberdeenshire is probably for *Inbhir-eabraigh*, 'inver of the marshy place,' from *eabar*, a marsh, a puddle. Inverneny in Balquhidder is in Gaelic *Inbhir-leanntaigh*, by dissimilation from *-neanntaigh*, 'confluence of the nettly place.' The old Gaelic word for nettle was *nenaid*, gen. pl. *neannta*, whence *Abhainn Neannt(a)*, 'river of nettles,' the Nant at Taynuilt; compare Nettly Burn in Saline, Fife.[1]

Innis, a meadow, haugh, has gen. *innse* (older *inse*), and an old dative *insi*. In numerous compounds the dative is

---

[1] Invernenty may represent *Inbhir-neannta*, 'confluence of nettles,' like Erichdie, etc., below.

now *-isi* (written *-isidh*) in unstressed position at the end of names, *e.g.* *Caolaisidh*, 'at narrow haugh,' *Camaisidh*, 'at bent haugh,' *Cruaisidh*, 'at hard haugh.' Similarly *Maithisidh*, the river Mashie in Badenoch, is primarily a meadow name, 'good haugh'; *Féithisidh*, the Feshie in Badenoch, is 'boggy haugh,' from *féith*, a bog channel, a boggy place. It is hard to say whether these names are dative or genitive after *abhainn*, river.

The ending may represent an older final *-a*, from whatever cause arising. In ordinary names of places, *e.g.* Dalarossie in Strathdearn represents an old *Dul Ferguso*, later *Fergusa*, 'Fergus' Haugh'; it is now *Dail Fhearghuis*, having changed from the u- declension to the o- declension. Munlochy is for *i mBunLocha*, 'at loch's foot,' 'near the loch'; it is now *Poll Lochaidh*. Similarly Monessie in Lochaber is for *i mBun Easa*, 'near the waterfall.' Kildrummy, formerly Kindrummie, is for *Cionn Droma*, 'at ridge end.' In the same way *Gleann Eireachdaidh*, Glen Erichdie near Struan, is for *Gleann Eireachda*, 'glen of the assembly'; there is another river Erichtie in Blairgowrie. Assemblies were not infrequently held at confluences: the Lindesay's court (*curia*) was held at Inverquiech, the junction of Isla and Burn of Alyth.[1] In the present case the sites of the *eireachd* would probably have been at Struan and near Rattray.

Nochty Water in Strath Don is for *Uisge Nochta*, 'naked water,' *i.e.* bare of trees and shrubs. Lossie in Moray is for *Uisge* (or *Abhainn*) *Lossa*, 'water of herbs,' 'herbaceous stream,' from the old gen. pl. of *luss*, a plant; Kinloss, 'herbaceous head,' in the same region, has dropped the ending. Strathy river in Sutherland is *Abhainn Shrathaidh*, 'river of the strath,' from *sratho*, later *sratha*, gen. of *srath*; its valley is now *Srath Shrathaidh*, lit. 'strath of strath (-river).'

This ending may also represent final *-e*. *Abhainn Poiblidh* in north-east Ross is for *Abhainn Poible*, 'river of

[1] Ad tres curias nostras capitales de Innerkoith'—Charter of Alexander de Lindesay, Earl of Crawford, given at Dundee, June 20, 1324.

the tent,' from *poball*, borrowed from Lat. *papilio*, a butterfly; Belisarius, says Procopius, used a tent of heavy cloth, commonly called παπυλεῶν, our 'pavilion.' Near Abriachan, Inverness, is Achpopuli, for *Achadh Poible*, 'field of the tent.' The Annaty Burn near Scone is for *Allt na h-Annaide*, 'burn of the Annat.' *Coilltidh* in Glen Urquhart is probably for *Abhainn Choillte*, 'river of the woods.'

There remains a large number of stream-names in which the ending can be explained in none of the ways just mentioned. In some of these—perhaps in most of them—*-idh* represents the old Gaelic ending *-de*, *-ide*, Welsh *-eid*, later *-aidd*, used to form adjectives from nouns, as *nemde*, heavenly, *mucde*, 'suinus,' pertaining to swine. In Scottish Gaelic this ending appears as *-aidh*, *e.g.* O.Ir. *diade*, Sc.G. *diadhaidh*, godly; *tiamda*, now *tiamhaidh*, gloomy. A clear example is *Geallaidh* or *Geollaidh* in Nairnshire and Moray, Geldie twice in Aberdeenshire, with Abergeldie, *Obar Gheollaidh*; Loch Gellie in Fife; *Inbhir Gheallaidh*, Invergeldie, in Glen Lednock; Gellie near Kinross, and other Gellie burns; Dalziel near Comrie is *Dail-gheollaidh*, while Dalziel at Novar in Ross and Dalziel in Petty, Inverness, are *Dail-ghil*, dat. of *dail gheal*, 'white meadow.' In all these *Geallaidh* or *Geollaidh*, 'Geldie,' is formally identical with O.Ir. *gelde*, *geldae*, *geldai*, glossed *taitnemach*, 'shining,' in LBr. 90, and formed from O.Ir. *gel*, now *geal*, white, with *-de* suffix. This suffix corresponds to E.Celtic *-idius* in *Epidius*, which became *echde*, 'equine,' in O.Irish, and in Sc.G. would be *eachaidh*. This, however, does not necessarily imply that the stream-names in question represent directly early British names in *-idios*, *-idia*, for in that case we should expect to find stream-names of this class often in Wales and in Ireland, which as a matter of fact we do not. In all likelihood they belong to a much later stage. Again, comparison with Ireland is not in favour of regarding them as merely adjectives qualifying some term for 'stream' understood. The true explanation seems to be that they came into vogue during the period of linguistic transition, and that the practice started as a method of reducing, and incidentally gaelicizing, old compound names, and that it then led to the formation of names on the same model.

As regards reduction of compounds, we have to reckon also with the suffix *-i* found in some Irish personal names of this type. Compounds of *Cóem* and *Cass* are reduced to *Cóemi* and *Cassi*, found with us in *Cill Chaomhaidh* and *Cill Chasaidh*, now indistinguishable from formations in *-de*.

Though direct proof of the process is not available in the absence of records, there are indications of it. The farm of Callander near Crieff is at the confluence of Kelty and Shaggie, and between Kelty and Barvick Burn. Here as elsewhere, Callander, in Blaeu, etc., Kalladyr, etc., was primarily the name of a stream, in Gaelic *Caladar*, representing an early *Caleto-dubron*, 'hard-water'; Kelty, in Gaelic *Cailtidh*, is the reduced form. The name *Cailtidh*, Kelty, recurs near the town of Callander in Perthshire and in Stirlingshire; there is also Kelty in Fife. Closely connected with it is *Cailtnidh*, Keltney Burn, a tributary of Lyon.

Two streams on the north side of Loch Tay are called *Mucaidh*, whence *Eadar dà Mhucaidh*, by assimilation *Eadarra Mhucaidh*, 'between two swine-burns.' This suggests an original compound name like W. *Mochnant*, 'swine-brook,' 'swine-glen.' Similarly *Marcaidh* 'horse-stream,' occurs as a tributary of Deveron, of Spey at Crathie, of Feshie, and of Fechlin in Stratherrick, and suggests a compound like W. *Marchnant*, 'horse brook.' *Maircnidh* is the name of the stream in Rosmarkie, *Ros Maircnidh*, and also of a tributary of Fechlin; compare *Cailtidh*, *Cailtnidh*. The *n* suffix is on the analogy of such names as *Broc-n-ach*, 'place of badgers,' *Muc-n-ach*, 'place of swine,' *Samh-n-ach*, 'place of sorrel.'

Examples are: *Balgaidh*, the Bogie of Strathbogie and a river in Applecross, from *balg*, older *bolg*, a bag, with reference to bag-shaped pools; a big baglike pool on Lyon is called *Linne Bhalgaidh*. The same idea gives *Pollaidh*, the Polly in Loch Broom, from *poll*, a pool; compare Pattag (p. 447). *Banbhaidh*, the Banvie of Atholl, is from *banbh*, a pig, W. *banw*, and *Banbhaidh*, Banavie near Fort William, is probably also a stream-name. The river Banavie, which connects Loch Banavie with Loch Shin, in Sutherland, noses its way through a reedy marsh, leaving a track exactly like that of a rooting hog. The river Banw or Banwy in Montgomery has a tributary Twrch, 'boar': 'many rivers forming deep channels or holes into which they sink into the earth, and are lost for a distance, are so called in Wales. A small brook called Banw in the parish of Llan Vigan, meaning 'a little pig,' has been said to be of this family.'[1] The Atholl Banvie has dug a wild deep channel, like its neighbours Bruar and Tilt; the name is probably of British origin. Brannie on Loch Fyne side is from *bran*, a raven, O.Ir. *brandae*, ravenlike; Bran occurs several times as a river-name in Wales. *Bocaidh*, Buckie in Balquhidder, is from G. *boc*, W. *bwch*, a buck.

Cailbhidh of Glen Calvie in Ross, is from *calbh*, Ir. *colbha*, a stalk, ozier. Cattie from Glen Cat, a tributary of Dee, Aberdeen, is from *cat*, a wild cat. Connie, another tributary of Dee, may be from *conda*, canine, with assimilation of *nd* to *nn*. Crombie is from O.Ir. *cromb*, G. *crom*, W. *crwm*, bent. Cowie in Kincardineshire, is from *coll*, hazel, *collde*, 'colurnus,' hazelly; compare Hazelly Burn in Berwickshire; *oll* becomes *ow* regularly in Scots, as Towie for *Tollaigh*, 'at hole-place.'

Daimhidh, a tributary of Ardle, is from *damh*, an ox, stag. *Gobhraidh*, Gowrie, a tributary of Conon, is probably from *gobhar*, a goat, like *Bocaidh*; Gowrie of Invergowrie in Perthshire may be the same; both might, however, be compared with W. *gofer*, a rill. *Eilgnidh* in Sutherland, is from O.Ir. *ailcne*, a pebble, **ailcnide*, pebbly. Gadie in Aberdeenshire, 'at the back of Bennachie,' is from *gad*, a withe, probably from its oziers. *Gollaidh* of Glen Golly in Sutherland, is from *goll*, blind, glossed by *caoch* (Lat. *caecus*), blind. A *caochán* is a rivulet so overgrown with herbage that it cannot see out of its bed. That *Gollaidh* was wrapt

[1] *Arch. f. Celt. Lexik.*, iii. p. 45.

in foliage appears from Rob Donn's song 'Gleann Gollaidh nan Craobh,' 'Glen Golly of trees.'

Siaghaidh, on the border of Inverness and Ross, is to be compared with O.Ir. *ség*, 'oss allaidh,' a wild deer, whence comes *ségde*, later *ségda*, stately.

Tarbhaidh or *Tairbhidh* is from *tarbh*, bull, O.Ir. *tarbde*, 'taurinus,' bull-like. *Teinntidh* near Callander, is from *teine*, fire, O.Ir. *tentide*, fiery, from its rapid boiling course; compare *Eibhleag*, 'the ember,' the sparkling burn of Evelix—an English plural—near Dornoch. Taitney in Forfarshire, is for O.Ir. *taitnemde*, later *taitnemda*, glittering, brilliant. *Tromaidh* in Badenoch is from *trom*, the elder tree or bourtree; its glen is *Gleann Truim*, 'glen of the elder tree.'

Uaraidh of Strathrory (*Srath-uaraidh*) in Ross, is from *uar*, a landslip, a waterspout, still current in Sutherland in these meanings. The river has high clay banks at many parts of its course. There are also *Allt Uaraidh*, behind Abriachan, with similar banks, and at Knockan between Invercasley and Oykel Bridge in Sutherland.

Some other names may be noted under this head. *Labhaidh* in Sutherland seems to be from *lab-*, W. *llef*, voice, cry; *lab-ar-os*, loud. *Niachdaidh* in Glen Strathfarrar comes from *neikt-*, whence in Welsh the personal name Nwython in the Mabinogion. This seems to be a different grade of *nekt-*, whence O.Welsh *Naithon*, G. *Nechtan*. Cormac explains *necht* by *glan*, pure; Nuadu Necht is explained as *fer gel*, 'a white man,' 'bright man' (LL 378 b 38). The meaning of *Niachdaidh* would thus be similar to that of *Geallaidh* above. *Tealnaidh* in Sutherland may be compared doubtfully with *Telo*, stem *Telon-*, the name of a Gaulish fountain god of unknown meaning. Kirkney is probably from *cearc*, a hen, a grouse hen. Keithny of Inverkeithny is a tributary of Deveron.

Cingidh, Kingie, a tributary of Garry in Inverness-shire, is from O.Ir. *cingim*, I stride, *cingid*, *cingidh*, a champion, Gaulish *Cingeto-rix*, 'hero-king,' 'king of heroes'; compare *Ceatharnag* of *Inbhir Cheatharnaig*. It may be simply the noun *cingidh*, champion.

Luggie, a tributary of Kelvin, is probably the same as

the Llugwy of Carnarvon, Merioneth, and Anglesey, representing an early *Loucovia*, 'bright one,' from *loucos*, white, W. *llug*, bright, O.Ir. *luach-te*, 'white-hot'; compare Lugar. Goodie, from the Lake of Menteith, has many small bends and is probably from W. *gwd*, a twist, turn, *gwden*, a withe, coil, noose.

(6) The ending *-ad*, *-aid*, representing an early *-ant-*, is not uncommon in Irish names of places and streams, as *Caolaid* from *caol*, slender, narrow, *Cruaghad* from *cruach*, a stack, *Uaraid* from *uar*, cold; compare E.Celtic Noviantum, etc. The reduction of *-nt-* began before the Ogham period and has gone on ever since, *e.g.* 'parliament' becomes in modern Gaelic *parlamaid*, 'regiment' is *reisimeid*; it is therefore certain that any ancient name in *-nt-* will suffer reduction of *n* when taken over into Gaelic. In some cases the reduction doubtless had taken place in Ireland already, before Gaelic came to Scotland: O.Ir. *bélat*, 'compitum,' a cross road, a path, formed from *bél*, lip, mouth, occurring in Badenoch as *Bialaid*. O.Ir. *fidot*, from *fid*, wood, is explained by Windisch as 'aspen' and 'ash,' by Kuno Meyer as 'cudgel' (*knittel*) for driving horses; horses, however, are driven by a switch rather than by a cudgel, and the *fidot* used for this purpose was probably a hazel sapling. It occurs with us in *Buail-fhiodhaid*, Belivat in Nairnshire. My informant, who was an old man in 1906 when he gave me the names of his district, said that the old people understood *Buail-fhiodhaid* to mean 'fold of Thicket,' taking *fiodhad* to be a man's name. It probably means 'thicket-fold' or 'fold near a thicket of saplings.' It may be noted in passing that the anglicized Belivat is a good example of *dh* becoming *v*.

Examples of this ending in names of places may be given. *Treasaid*, Tressat, on Loch Tummel, means 'battle place'; W. *tres*, toil, trouble; Ir. *treas*, battle, conflict; compare Glentress in Peebles and *Treasaididh*, Tressady, in both Rogart and Lairg, Sutherland. *Dùn Turcaid*, Dunturket in Stratherrick, means 'fort of boar-place'; it is described in Macfarlane as 'a myl be east Dalnakappel, upon the east bank of Faechloyn,' but I could find no fort thereabout.

Similarly Tarvit in Fife is 'bull-place,' from *tarbh*. Lovat near Beauly is from *lobh*, rot, putrefy, with reference to the nature of the surface: Lovat is near the sea, and is now *a' Mhormhoich*, 'the sea-plain.' Another name of similar meaning is *Both-lobhach*, 'putrid booth,' in Glen Fintag, Brae Lochaber, with which may be compared *Groidich*, 'rotten place,' in Glen Finglas, Perthshire. *Noid*, for *No'aid*, Nude in Badenoch, probably represents an early British *Noviantum*, 'fresh place, green place.'

In stream-names this ending is not uncommon. *Duinnid* in Kintail, with *Inbhir Dhuinnid*, Inverinate, at its mouth, is from G. *donn*, brown, W. *dwn*, dun, dusky. Glorat, near Lennoxtown in Stirlingshire, is for *Glóraid*, 'babbling stream,' from *glór*, voice, speech; it is primarily the name of the brook at that place; compare *Tobar na Glóire*, 'babbling well,' in Ireland. Inverdovet in Fife is Invirdubet, 1391 (RMS), from G. *dubh*, O.W. *dub*, W. *du*, black; it seems to be the Inuerdofatha, Inuerdofacta, etc., of the old chroniclers (Skene, *P.S.*). Mossat, a tributary of Don, is from the base of G. *mosach*, *musach*, nasty, W. *mws*, stale, stinking; with it goes *Musadaidh* in Stratherrick, parallel to *Treasaid*, *Treasaididh*. The river of *Gleann Lìomhaid*, Glen Livet, is doubtless from the same base as *Lìomhunn* of Glen Lyon.

(7) A special class under this ending is that of diminutives in *-nat* and *-that*, formed like O.Ir. *siurnat*, 'little sister,' from *siur*; *bónat* from *bó*, cow; *dergnat*, flea, from *dearg*, red; *tírthat*, 'agellus,' a small field, from *tír*, a district. *Conaid*, a tributary of Lyon, is 'little hound,' from the compositional form of *cú*, hound; compare *Conág*, a stream of Kintail. *Fearnaid* of Glenfernate in Athol is 'alder-water,' from *fearn*, alder. *Gleann Geunaid* in Kirkmichael parish, Perthshire, now Glen Derby in English, is Glengaisnot, 1510, -gaisnoth, 1512, -ganot, 1511, 1521, 1538, -ganacht, 1516, -genet, 1564, -gynit, 1629, -gennet, 1630 (RMS). Local tradition connects it with geese: the glen is a regular stage in the flight of wild geese from east to west. It is probably a diminutive of O.Ir. *géd*, Ir. *gé*, Sc.G. *gèadh*, a goose, meaning 'little goose, goosey,' and

hence 'goose-stream,' by a transference of meaning not uncommon. The *s* of the earlier forms might suggest *gés*, a swan, but it is probably in error for *f*, representing *dh*. Glengennet in Carrick appears to be the same—'binae Glengynnetis,' 1540 (RMS).

Aberbuthnot, now Arbuthnot in Kincardineshire, is Aberbothenoth, 1242 (Pont. Eccl. S. Andr.), Abirbuthenoth, Aberbuthenot, 1282/3 (Reg. Arbr.), etc. Here 'Bothenoth,' etc., is the name of the stream close to St. Torannan's church, and is for the genitive of *Buadhnat*, 'little one of virtue,' lit. 'little triumph, little virtue'; in other words, it was a holy stream possessed of healing power. Similarly the stream which enters the sea in front of Maol Rubha's church of Loch Carron is called *Buadhchág*, 'little one of virtue': it also was a sacred stream of power. So too in St. Kilda and in Barra there are wells called *Tobar nam Buadh*, 'the well of virtues.' We have seen how the bell of Cill Chuimein, called *am Buadhach*, 'the one of virtue,' conferred on the water of Loch Ness the virtue (*buaidh*) of curing cattle (p. 436). The Bow River, *Abhainn na Buaidhe*, in Ireland had the *buaidh* or virtue that if you drove cattle into its water on May Day, it preserved them from disease for the coming year.[1] Buadnat was, or might be, a personal name: it is given as the name of Herod's daughter.[2]

Aberluthnot, now Marykirk in Kincardineshire, is Aberluthenoth, 1242 (Pont. Eccl. S. Andr.), Abirlothenot, *c.* 1275 (Reg. Arbr.), -louthnot, 1370 (RMS), etc. The Luthnot has been confused with the Luther, but it is doubtless the stream which is adjacent to the old church and forms part of the boundary of the manse garden. It is most probably for *Luathnat*, 'little swift one.'

Glen Turret at the head of Glenroy in Brae Lochaber is in Gaelic *Gleann Turraid*; the Pass of Turret is in MacVurich *Láirc Thurraid*. This is for *Turthaid*, from G. *tur*, dry, with the diminutive suffix *-that*, meaning 'little dry

---

1 Joyce, *Irish Names of Places*, iii., *s.v. Bow*.

2 'Do tsíl Buadnaite ingine Irhuaith,' 'of the seed of Buadnat, daughter of Herod.'—*Acall.* (Stokes), l. 6250.

one,' from the fact that it shrinks in summer. Its opposite is the Irish *Buanaid*, 'little lasting one' (Joyce); compare *Buanán Cille Ruaidh*, explained as 'indeficiens rivulus Chelle Ruaid,' 'the unfailing brook of Cell Ruaidh.'[1] The Turret of Perthshire and the Turret from Mount Battock in Kincardineshire are doubtless of the same origin. We may compare *turloch*, a loch that dries up in summer, and the Welsh rivers Hepstwr, for *Hespdŵr*, 'dry water,' *Hesp Alun*, 'the dried-up Alun.'

(8) The ending *-ág* serves in Gaelic as the diminutive suffix feminine. In O.Ir. it is *-óc*, masculine; and from the O.Ir. period it is common in affectionate forms of saints' names, especially in reduced forms, *e.g.* *Mo Chóemóc* from *Cóemgin*. It is a curious fact that this suffix, which has attained so great a vogue in Gaelic, is not of Gaelic origin but borrowed from Welsh *-awc*, now *-og*, representing the early *-ācos* which appears in Gaelic as *-ach*. Stream-names with this ending are numerous, especially in the north. In their present form at least they do not belong to the oldest stratum, but, like the personal names of the same class, they may be reduced from older forms. The only example of reduction within record times that is known to me is the *Patag*, Pattack, of Badenoch, which appears as one of the bounds of the church lands of Logykenny (Laggan) in 1239 (RM) under the name of Petenachy, about 1600, Avon Pottaig (Macfarlane). This is not a compound, but a derivative of early M.Ir. *patt*, a pot, jar,[2] later *poit*, with *-an-ach* extensions. It is to be compared with the brook called Putachi, 1273 (Ant. A. & B.) in Turriff parish, and the brook called Puthachin in Keig parish, 1233/53, *ib.*, and means 'stream of pots,' *i.e.* of potlike pools; compare *Balgaidh*, *Pollaidh*. A local rime refers to it as—

'Patag dhubh bhalgach,
An aghaidh uisge Alba.'

---

[1] Plummer, *Vit. SS. Hib.*, i. p. 53.

[2] *E.g.* 'dá phaitt, hit é lána do fín,' 'two jars full of wine'—Kuno Meyer, *Expulsion of the Dessi*, 18; isna paitti, into the pots.—*Rev. Celt.*, ix. p. 48.

'Black Pattack of baglike pools,[1] that goes against the streams of Alba,' with reference to the fact that after making straight for Spey the Pattack suddenly swerves to the south-west into Loch Laggan. Another name of similar idea is that of *Gleann Ambuill*, Glen Ample on Loch Earn, from *ambuill*, a large jar; Lat. *ampulla*, an amphora; in Sc.Gaelic it seems to have meant 'vat.'[2] The reference is to the cauldron at the falls of *Aodann Ambuill*, Edinample, 'face of the vat'; compare *Allt na Dabhaich* at Ledaig and Burn of Vat in Aberdeenshire.

Some of the names in *-ág* are from names of animals or even of insects. Several Irish rivers were called *Daol*, earlier *Dóel*, 'beetle.' In the marches of the church lands of Monymusk there was a brook called 'Doeli quod sonat carbo Latine propter eius nigretudinem,' 'Doeli, which in Latin, is "coal," on account of its blackness': it really means 'beetle' (Ant. A. & B., i. p. 172). We seem to have no *Daolág*, but we have two streams called *Dorbág*, one from Lochindorb into Findhorn, the other falls into Nethy, meaning 'little tadpole' or 'minnow.' *Eunág* of Strath Oykel means 'birdie,' from *eun*, a bird. *Eunarág*, the Endrick of Glen Urquhart, is 'the snipe'; Endrick of Lennox is probably the same—Hannerch, Anerich, etc. (Reg. Lenn.); both have twisting courses, the latter especially. Conág in Kintail is 'doggie.' *Eacháig* from Loch Eck in Cowal is 'horsie'; this, like some of the others, may be a reduced form.

Some are from colours. *Fionntág* of Glen Fintag in Brae Lochaber, is from *fionn*, white, bright. So is *Fionntág* from Loch Moy, but as it appears to have been called also *Allt na Cille*, 'church burn,' *fionn* may have here its secondary meaning of 'blessed, holy.'[3] Its infall into Findhorn is

---

[1] Macbain took *balgach* here to mean 'bubbly.'—*Place-Names*, p. 243.

[2] In early M.Ir. *ampoill* is explained as *lestar gloine*, 'a large vessel of glass.'—Kuno Meyer, *Zur kelt. Wortkunde*, No. 4. It occurs in a Loch Broom song: 'Far am faighinn stuth na Tòiseachd ged a dh'òlainn dheth ambuill'; 'where I would get the stuff of Ferintosh, though I were to drink a "vat" of it.'

[3] Macfarlane, *Geog. Coll.*, ii. p. 258.

Inbhir Fhinn, Invereen. The personal name *Fintóc*, *Fionntág*, is a reduced form of *Findbarr* or other compound of *finn*, *fionn*. *Dubhág* into Nethy is 'little black one'; there is also *Allt Dubhagáin*, a tributary of Ardle, which may mean Dubucan's, later Dubhagan's, Burn. *Gormág* into Dee, Aberdeen, is 'little green one.'

Others are of miscellaneous origin. *Labharág*, 'little loud one,' and *Balbhág*, 'little dumb one,' have been mentioned. Along with them goes *Goirneág*, 'the little crier,' from *goir*, cry, of Strath Girnock on Dee in Aberdeenshire, and of *Gleann Goirneig*, Glen Girnaig in Atholl. The Garnock of Ayrshire and of Dalgarnock in Dumfriesshire show the earlier form *gair*. *Ailneág*, a tributary of Aven in Banffshire, is 'little stony one,' from *ail*, a stone, with *n* suffix. *Breunág* of *Gleann Breun* in Stratherrick is 'the little fetid one,' from *breun*, nasty, putrid; there are two sulphur wells in the glen; compare *Tobar Loibhte*, 'stinking well,' in Sleat, Skye (Forbes), also 'Stinking Water, derived from certain sulphur springs, changed to Shoshone River by Act of legislature' in the United States.[1] *Buadhchág* in Loch Carron is 'little one of virtue' (p. 446). *Bruachág* at Kinlochewe is from *bruach*, a bank, with reference to its high banks. The stream of *Gleann Caiseig*, off Glen Finglas in Perthshire, is *Caiseág*, 'the little swift one,' from *cas*, quick. *Ceatharnág* of *Inbhir Cheatharnáig*, Invercarnaig in Balquhidder, may be compared with *Allt* and *Gleann Cheatharnaich* in Duthil parish, understood to mean 'burn and glen of the kerne,' *i.e.* warrior, later freebooter, robber. *Brothág* of *Obar Bhrotháig*, Arbroath, is probably from *bruth*, heat, gen. *brotha*, whence *brothach*, boiling, also 'scabby, eruptive'; compare *Teinntidh*, 'fiery,' for the idea. *Eibhléag*, 'the little ember,' near Dornoch, is of the same type. *Fairgeág* of *Srath Fhairgeag*, Stratherrick, and *Inbhir Fairgeág*, Inverfarigaig on Loch Ness, is connected with *fairge*, the ocean, in Sutherland 'the ocean in storm.'

Airceág of *Loch Airceig*, Loch Arkaig, is somewhat uncertain. On the north side of the loch is Arcabhi,

[1] Owen Wister, *Members of the Family*, p. 27.

stressed on the last syllable, but I have not heard the Gaelic pronunciation; some distance east of it a place called Arc appears in rentals and on an estate map. Allt Arcabhi, the burn of Arcabhi, comes down steeply in a narrow gorge. These names must be taken together, and may be compared with *Loch Aircleid*, Loch Arklet, where *Aircleid* is genitive of *Aircleathad*, 'difficulty slope,' 'strait slope,' with reference to the steepness of part of the slope on the north side of the loch; similarly in Ross we have *Éigintoll*, 'difficulty hole,' the name of a corrie difficult of access. The north side of Loch Arkaig is also very steep, and *Airceág* is probably from *airc*, a strait, a difficulty, with reference primarily to the physical configuration: the river is 'the little one of the *airc*.'

(9) A small group has the ending *-agán*, found in a number of Irish personal names, such as *Dub-ucan*, etc., a purely Gaelic ending. *Luachragán*, which enters Loch Etive opposite Ardchattan, is 'the little rushy one,' from *luachair*, rushes. *Lunndragán* at Taynuilt, Argyll, is from **lunnd*, meaning probably 'a marsh,' whence the common place-name Lundie; its diminutive *lunndán* means 'a smooth grassy place, a marshy spot,' found also in place-names, *e.g. an Lunndán*, near Aberfeldy. *Lunndragán* means 'the little marshy one.' Lower down is *Lusragán*, 'the little herbaceous one,' from *lus*, with *ar* extension. Similarly formed are *Easragán*, 'the little one of waterfalls,' higher up Loch Etive, with Inveresragan at its mouth, and *Caisreagán*, 'little swift one,' a very steep stream in Appin. It is notable that all these are in the same district. *Màileagan* of *Coire Mhàileagain*, the name of two famous corries in Ross, is from *mál*, a prince, noble; this, however, may be a personal name, as we have *Cnoc Mhàileagáin*, 'Màileagan's hill,' in Skye. *Fearnagán*, a tributary of Feshie, is 'little one of alders.'

(10) The masculine diminutive ending *-án* is not uncommon, as in *Ciarán*, 'the little dusky one,' on Loch Leven (Ballachulish), *Deargán*, 'the little red one,' on Loch Creran.

CHAPTER XIV

RIVER NAMES (*continued*)

II. Some other examples of names connected with the ideas of colour and sound may be given. *Allt Éigheach*, 'the shouting burn,' in Rannoch, comes from *Coire Éigheach*. In Kirkmichael parish, Perthshire, the *Briathrachán* of Glen Brierachan is from *briathar*, a word, *briathrach*, wordy, 'the little talkative one'; from its changes of sound the people still foretell change of weather. Aultgarney, which joins the Feugh of Aberdeenshire, is from *gairim*, I cry, like *Goirneag*. *Inbhir Chagarnaigh* (? *-aidh*), Inverhaggernie in Glen Dochart, is based on *cagar*, a whisper.

The Affric of Inverness-shire is *Afraic*, from *ath* used intensively, and *breac*, dappled, a fairly common female name. Tummel is *Abhainn Teimheil*, 'river of darkness,' O.Ir. *temel*, now *teimheal*, gloom, shade; it was so called from its thickly wooded gorge. The rocky peninsula at the south-east end of Loch Tummel is *Dun Teamhalach*, 'Tummel fort,' now anglicized Duntanlich—Duncaveloch, 1473 (RMS), (read -taveloch); Duntawlich, 1630, *ib*. There are no remains of a fort: the place would be easily defended by a stockade across the neck of the peninsula. At the head of Loch Creran is another *Abhainn Teimheil*, whose glen is now Gleann Dubh, 'the black glen,' but formerly *Gleann Teimheil*—Glentendill, 1576 (Orig. Paroch.)—an example of translation. Here the name has reference to the sunless aspect of the glen. The Aven of Glen Aven in Banffshire, is in Gaelic *Ath'inn* for *Athfhinn*, 'the very bright one.' The river is noted for its clearness:—

> 'The Water of Aven runs so clear,
> It would wile a man of a hundred year.'

Tradition makes Athfhinn to have been the wife of

Fionn mac Cumhaill, who lived at Inchrory. She was drowned in Aven, which hitherto had been called *Uisge Bàn*, 'fair water.' Then said Fionn :—

> 'Chaidh mo bhean-sa a bhàthadh
> Air Uisge Bàn nan clach sleamhainn ;
> 'S o chaidh mo bhean a bhàthadh,
> Bheirmid Athfhinn air an abhainn.'

'My wife has been drowned in Aven of the slippery stones ; and since my wife has been drowned, let us name the river Aven.' Legend connected her with Clach nam Ban, 'the ladies' stone,' where was a favourite seat of Fionn with her seat beside it, overhanging the precipice of Meall Gainmhich. Since her time no one can be drowned in Aven above Ailneag. This tale is in keeping with similar Irish traditions of how rivers got their names. (Cf. p. 229, n. 2.)

Caor at Roro in Glen Lyon, from the skirts of Ben Lawers, is *caor*, a glowing mass, used in a secondary sense of a sparkling stream, also of a torrent. Its water was so highly thought of that we have the saying—

> 'Caor is Cadan is Conghlais,
> Trì uisgeachan na h-Alba.'

'Caor and Cadan (in Rannoch) and Conghlais (behind Tyndrum), the three waters of Alba.'

An important group of survivals from British is connected with W. *pefr*, radiant, beautiful. In Haddington there are two brooks called Peffer, and near Edinburgh is Peffer Mill on what is now the Braid Burn. In 1200 Gilbert earl of Strathearn granted the clerics of Inchaffray the right of fishing 'in Pefferin' ; the outlet of this stream is now Innerpeffray. In Forfarshire the Paphrie Burn falls into West Water, and south of Arbroath there is Inverpeffer where a brook enters the sea. In Aberdeenshire 'rivulus de Paforyn' or 'Peferyn' is recorded in 1247 (Ant. A. & B.) ; it is now Silverburn in Peterculter parish, which is almost a translation of the original name. In Ross the Peffery flows through Strathpeffer, *Srath-pheofhair*, and enters the sea at *Inbhir Pheofharán*, the native name of the Norse Dingwall.

In addition to the names already given from animals some more may be mentioned. The bull appears in *Obar Thairbh*, Abertarff near Fort Augustus, 'bull's confluence,' probably an old British name preserved in full, for G. *tarbh* and W. *tarw* are identical. In Galloway rivers called Tarff fall into Dee and Bladenoch. There are Tarf in Athol, and Tarf Water in Lochlee parish, Forfarshire. Polintarf Water is in Peeblesshire. Inverherive in Glen Dochart is understood to mean 'bull's confluence.' *Tarbhán*, 'little bull,' enters Loch Lyon. In most cases the reference is probably to the impetuosity and roaring of the streams. The *Gamhnán*, 'little stirk,' joins Orchy at *Inbhir Ghamhnáin*; the *Gamhnach* is in Benbecula (p. 426). *Laogh*, calf, is the stream of *Gleann Laoigh*, Glen Loy, in Lochaber and in Cowal; the latter is named in Deirdre's lament on leaving Alba. A third Glen Loy is in Carradale, Kintyre. With these may be compared *Lóig les*, 'vitulus civitatum' or 'calf of the courts,' the name of a well at Tara. There is a distinct mythological flavour about all these names. *Bran*, raven, is the name of several streams in Wales; with us it occurs in Strath Bran in Ross, Allt Bhrain in Badenoch, Bran in Closeburn parish, Dumfriesshire, and elsewhere. Here too the reference is probably mythological, and the names may be of British origin. *Allt Saidhe*, 'the bitch's burn,' in Glen Urquhart, probably refers to its having been a haunt of wolves that bred there. The *torc*, boar, does not seem to occur as a stream-name with us, though Wales has several rivers called *Twrch*. Brig o' Turk, Callander, is not an instance: the water is the Finglas, and my Gaelic authorities for that district disclaimed any Gaelic origin for Brig o' Turk.

III. Two generic terms, *dobhar* and *glais*, occur freely in composition and sometimes singly over the greater part of Scotland.

Dobhar, in Welsh *dwfr*, *dwr*, is for E.Celtic *dubron*, water. In France it gives rise to a number of names of the form Douvres, so that it must have been common in Gaulish speech; Verdouble, a stream in the south of France is for *Vernodubron*, 'alder water.' The furthest south instance in England is Dover for *Dubris*, 'at streams,' with reference to two brooks which enter the sea near Dover harbour. Our furthest north instance is Dovyr in Strathnaver in 1260 (RM). Cormac's Glossary informs us that *dobur* is a name for water, common to Gaelic and Welsh, also that (in Irish) it means 'dark' as well as 'water.' It was doubtless borrowed from Welsh, and figures very slightly either in Irish literature or in Irish stream-names. In Scotland, as in Wales, it is common in stream-names, often in more or less disguised forms. *An t-Allt Odhar*, for instance, means—probably—'the dun burn'; but when the article is not present, as often happens, the probability is that we have to do with *Allt Dobhar*, *i.e.* the original name was *dobhar*, and *allt* was added. The so-called *Allt Odhar* near Pitlochry falls into *Poll Dobhair* in Tay; *Allt Odhar* at Spean Bridge enters Spean at *Inbhir (Dh)obhair*, and so with other cases.

Eadar dà Dhobhar, 'between two waters,' becomes by assimilation *Eadarra Dhobhar*, and then by contraction *Eadra Dhobhar*; hence Edradour on the south side of Loch Tay; the obsolete Edirdovar, the site of Redcastle in Ross, seems to be for *eadar dobhair*, 'between brooks.' After *abar*, a confluence, we have the genitive case in *Abbordoboir* of the Book of Deer, now Aberdour in Buchan, and in Aberdour in Fife. In these instances *dobhar* is in stressed position and therefore in better preservation.

When qualified by a prefixed adjective or noun used as an adjective it is unstressed and sinks to *-dar*, *-dur*, or, if aspirated, to *-ar*, *-ur*, represented in anglicized forms by *-der*, *-er*, etc. *Arder*, for *Ard-dobhar*, falls into Spey. Aberarder on Deeside, Abirardoure, 1451 (Ex. Rolls), is 'confluence of high water,' and Fearder, for *féith ard-dobhair*, is 'bog' or 'bog channel' of high water. *Allt Ardobhair*, pronounced *Ardair*, comes from *Coire Ardobhair*, and enters Loch Laggan at Aberarder. High up is *Uinneag Choire Ardobhair*, 'the window of Corarder.' A third Aberarder is in Strath Nairn. Aberchirder, Aberkerdour *c.* 1204 (Reg. V. Arbr.), etc., is probably from *ciardhobhar*, 'swart water.' Auchterarder is 'upland of high-water.' Gelder in Braemar is for *gealdhobhar*, 'bright water.' Fender in Athol is *Fionndobhar*, 'bright water'; another Fender, in 1633 Findore (RMS), falls into Braan. These are the same as the Welsh *Gwenddwr* in Brecon. Morar, 'big water,' and Duror, 'hard water,' have been mentioned (p. 124), Cander in Lanarkshire, is Candouer in 1150 (Orig. Paroch.), for W. *canddwr*, 'white water.' Glaster of Glenglaster in Inverness-shire is for *glasdobhar*, 'green water' or 'wan water.' The Alder Burn, which falls down the precipitous south side of Ben Alder, is for *alldhobhar*, 'rock water,' 'precipice water,' from *all*, a steep or precipitous rock. The mountain is *Beinn Allair* (so spelled in one of the Maclagan MSS., written in 1755), now usually *Beinn Eallair*, with the palatal sound of *beinn* carried over.

The stream-name Calder is very common from Caithness southwards. Old spellings are 'ecclesia de huchtercaledouir,' 1153/65, 'ecclesia de Kaledouer,' 'ecclesia de Westir Caledoure' (Reg. Dunf.); 'ecclesia de Caledouer,' 1165/1214 'ecclesia de Kaldover Radulfi de Clerc,' 1178/88, 'ecclesia S. Cuthberti de Kaledofre . . . decima molendini mei de Kaledofre,' *c.* 1170 (Lib. Calch.). The references are to West Calder and East Calder in Lothian, both primarily stream-names. A number of the streams are well known in Gaelic as *Caladar*. The Caithness Calder is supposed to represent *Kálfadalr*, 'calf-dale,' of the Orkneyinga Saga; the fact that in Gaelic it is *Caladal* is not quite consistent with this, for *Kálfey*, 'calf-isle,' is always in Gaelic *Calbha*; we should therefore expect *Calbhadal*. Scottis-caldar and Norne-caldar appear in 1538 (RMS); the former is now *Caladal nan Gall* in Gaelic. Final *l* of *Caladal* may therefore be modern, by assimilation. At least three Calders are in Inverness-shire, one in Badenoch, one in Stratherrick, and one on Loch Oich; the confluences of the two last are *Obar Chaladair*, Abercalder; Abbircaledouer (in Stratherrick), 1238 (RM); Cawdor in Nairnshire is *Caladar*, after its burn *Allt Chaladair*. Callander on Spey is 'Calatar super aquam de Spey' in RM. In Perthshire *Caladar*, Calder, enters the west end of Loch Rannoch, and there is Achalader, 'Calder field,' in Blairgowrie. Callander near

Crieff has been mentioned (p. 106). In Aberdeenshire the Callater is a tributary of Clunie Burn, which enters Dee. In Forfarshire the Burn of Callater joins West Water. In Argyll *Ach' Chaladair*, Achallader at the head of *Loch Toilbhe*, Loch Tulla, preserves the old name of the stream in the corrie near it; close by Blaeu has Bochaletyr; there is also *Allt Chaladair* into Kinglas on Loch Etive. In Lanarkshire there are three Calders, one falls into Avon, the others into Clyde near Uddingston and Bothwell. There is also the Calder of Renfrewshire. Blaeu has Drumkalladyr near the head of Nith. This widely spread name is a survival of an early British *Caleto-dubron*, 'hard water,' and is identical with Welsh *Calettwr* of Montgomery, which being stressed on the penult becomes elsewhere *Clettwr*, a tributary of Teifi, and *Clettwr Fawr*, 'big hard-water,' in Cardigan, with contraction of the first syllable. An equivalent name in Wales is *Caledffrwd*, 'hard stream,' in Carnarvon. *Caleto-* is W. *caled*, hard, O.Ir. *calath*, later *calad*, *caladh*, hard.

The diminutive *dobhrán*, mas., is seen in *Inbhir-dhobhráin* (*dh* quite silent after *r*), Inveroran, where *Allt Orain*—for *Allt Dobhrain*—falls into *Loch Toilbhe*, Glen Orchy. *Beinn Dobhráin*, behind Tyndrum, is 'peak of the streamlet.' In the same region, near Crianlarich, is Inver-ardran, 1377 (RMS), Inverhardgowrane, an excellent spelling (with *g* for *dh*), 'confluence of the high streamlet.' On the south side of Loch Rannoch is *Camdhobhran*, Camghouran, 'bent streamlet,' to be compared with *camdubr*, (Zeuss, 136), *Camddwr*, a tributary of Towy and elsewhere in Wales.

Dobhrág, the diminutive fem., is the name of a ditch between the farms of Arabella and Shandwick in Ross-shire; it is also the name in Gaelic of the Aldourie Burn near Dores, Inverness-shire, in the local phrase 'Duras is Darus is Dobhrág,' 'Dores and Dares and Dobhrag.' There is another Alltdourie near Invercauld, and there are Findowrie, 'white streamlet,' in Careston parish, Forfar, and Baldowrie in Kettins parish, Fife. Innerourie in Strathaven, Banffshire, is in Gaelic *Inbhir (Dh)obhraidh*.

Another common stream term is *glais*, O.Ir. *glaiss*,

fem., W. *glais*, mas., a stream, rivulet. It occurs as a common noun in the notes to the Book of Armagh and often in the stream-names of Ireland and of Wales. In the old saga of the Cattle Raid of Cualnge, when Cuchulainn is sore wounded, Lug mac Ethlend, one of the pagan gods, 'cast herbs and grasses of healing into the hurts and wounds . . . so that Cuchulainn recovered in his sleep without his perceiving it at all.' The expanded version of the saga tells that the Tuatha Dé Danann took Cuchulainn to the streams and rivers (*go glassib ⁊ go aibnib*) of his own district, and they placed plants and herbs of healing on the streams and rivers, to succour Cuchulainn. Among the twenty-one streams mentioned are Findglais and Dubglais.

Glais appears uncompounded in *Allt na Glaise* in Sutherland. In Ross-shire the river *Glais* rises in *Loch Glais* and flows through *Gleann Glais*, Glen Glass, passing then through the chasm known as *an t-Allt Granda*, 'the ugly precipice.' Mr. James Fraser of Alness says of it 'in general they say the river is not sonsy, nor yet the loch from which it comes. . . . But they think the water is sanctified by bringing water to it from Loch Moire' (St. Mary's Loch).[1] In Inverness-shire the *Glais* of Strath Glass joins the Beauly river at Struy. Another Glen Glass is at the head of Euchan Water, Sanquhar. Near Craigrothie, Fife, is the Glassiehow burn, where Glassiehow looks like a part translation of 'lag na glaise.'

Fionnghlais, 'white stream,' is the equivalent of W. *Gwenlas*, in Cardigan, *Gwenlais Fach*, 'little white stream' in Glamorgan. Our best known instance is *Gleann Fionnghlais*, Glenfinlas, near Loch Vennachar. Another Finlas Water enters Loch Lomond south of Luss, and a third Finglas flows into Loch Doon, Ayrshire.

Dubhghlais, W. *Dulas* (in L. Land. *Dubleis*, *Dugleis*), 'black stream,' becomes Douglas in English. There is a Douglas Burn in Aberdeenshire. Douglas Water enters Loch Lomond north of Luss, and another, from Loch Sloy, has its confluence at *Inbhir Dhubhghlais*, Inveruglas, near

---

[1] Macfarlane's *Geog. Coll.*, i. p. 212.

Arrochar (*dh* silent after *r*). Another enters Loch Fyne on the west; there are besides Douglas Water in Lanarkshire and Douglas Muir west of Milngavie.

Another compound is *Conghlais*, O.W. *Cingleis* (L. Land.) 'dog stream.' Conglass is a tributary of Aven in Banffshire and of Orchy in Argyll. *Gleann Chonghlais* on Loch Etive and in Lochgoilhead parish, Argyll, become Glenkinglas in English. Conghlas of the former is seen in Inuir-kunglas (*c*. 1204 Reg. Dunferm.); charters are granted at Glenquhonglas, 1387, Glenqwhonglas, 1378. The form Kinglas seems to indicate a different tradition, and may have been influenced by Welsh *Cingleis*. There was a streamlet called Conglas in East Kilbride parish in Lanarkshire, 1417 (Reg. Glas.).

Foghlais means 'sub-stream,' rivulet. It occurs in *Allt Foghlais*, Loch Maree, and in *Foghlais*, Foulis, Ross-shire. Wester Fowlis in Perthshire is spelled Fowlis, *c*. 1195; Fougles, *c*. 1198; Foglais, 1208 (Chart. Inch.); this last spelling is exactly as it would be in Irish of that period. Here the brook passes the church of Fowlis, and in Easter Fowlis, in Gowrie, the church is similarly situated. In Aberdeenshire the Fowlis burn is a tributary of Don.

An old description of the marches of Keig and Monymusk in Aberdeenshire mentions a brook called Fowlesy,[1] probably a diminutive of *foghlais*. Another brook near the cemetery of Keig was called Conglassy, 1233/53 (RPSA).

IV. The British term *aber*, a confluence, is for *ad-ber* ($\sqrt{}$*bher*), 'to-bring'; in O.Welsh it is *aper*, as in the Book of Llandaf *passim*. With this we may compare Adamnan's Stagnum Aporum, already mentioned, and Apor-crossan, the old name which became corrupted into Applecross, and Abbordoboir of the Book of Deer. Another British form appears in Oper-geleu, Oper Linn Liuan, thought to represent an older *od-ber*, 'out-bring.' In Gaelic of the present day, *aber* is generally pronounced *obar* when it stands, as it usually does, at the beginning of a name; Lochaber, however, is *Loch-abar*, and Applecross was known as *a'*

---

[1] *Antiquities of Aberdeen and Banff*, i. p. 172.

Chomraich Abrach, 'the girth or sanctuary of the aber.' In Welsh, *aber* is used as a common noun with the article. In Gaelic it is never used with the article, and it does not occur in Gaelic literature or in common speech: it appears to have been displaced at an early stage in the language by the purely Gaelic term *inbhear*.

As to distribution, *aber* occurs in Dumfries and Lothian, but not in Galloway, Strathclyde, or Ayrshire. North of Forth it is common on the east side as far as Spey. There are no instances in Morayshire or Nairnshire, but several occur in Inverness-shire; the furthest north instance on the east side is Abriachan on Loch Ness, ten miles south-west of Inverness. On the west side two certain instances are Aporcrossan and Lochaber; also Abercalder, on Loch Oich, is west of the watershed. Another probable instance is Aber Isle in Loch Lomond, just off the mouth of Endrick. Outside Scotland, *aber* is frequent in Wales, though it does not seem to occur in Devon or Cornwall. It does not occur in Ireland: Irish names which in their anglified forms appear to contain it are really from *eabar*, a marsh.

As Gaelic *ceann* displaced British *pen*, so it is safe to say Gaelic *inbhear* in many place-names took the place of British *aber*. In both cases the change took place, as a rule, before the record period, but some few traces remain. Chalmers thought he had detected one in the record of King David's grant to the monastery of May of 'Inverin qui fuit Aberin.' The correct reading, however, is 'Inuerrin quae fuit Averni,' *i.e.* 'Inverin which belonged to Avernus,' the latter being a man's name; in any case the *v* is conclusive against *aber*, for its *b* is never aspirated. A genuine instance appears to be Haberberui, 1290 (Stevenson's *Documents Illustrative*), now Inverbervie in Kincardine-shire. In Perthshire the village of Abernethy is about a mile from Innernethy, on the same stream; 'Inuirnythy on the east side of the brook' goes with Abirnithy, 1189/99 (Reg. Arbr.). Near Comrie, Inver-ruchill, Innerurchill, 1596,[1] was a burgh of barony which belonged to Campbell

[1] Erskine Beveridge, *Abers and Invers of Scotland*, p. 108.

of Aberuchil; the latter name is extant, the former seems obsolete. In both these cases the double names applied, or came to apply, to different but adjacent places.

We may now take the instances which occur in the various counties.

Dumfries. Aberlosk is near the junction of a small brook with the Aberlosk burn, called lower down the Moodlaw burn, a tributary of White Esk. It is Albirbosk vel Abirbosk, 1606 (RMS); Abirlosk, 1613, *ib.*; Aberloch, 1618, *ib.*; Aberlusk, 1653 (Ret.); Aberlosik, 1661, *ib.* In the absence of really old forms we may perhaps compare with Aberlessic in Haddington. Abermilk, once the name of the parish called later Castlemilk and now called St. Mungo, is Abermelc, *c.* 1124 (Reg. Glas.). It is supposed to have been at the confluence of Milk and Annan, which is about two miles below the old Kirk of St. Mungo, but the name appears to be long obsolete. The meaning of Melc, Milk, is obscure; it may stand for an early *Malcia*, to be compared with G. *malc*, putrefy.

Haddington. Aberlessic, according to the *Fragment* of the Life of St. Kentigern, was the name of a river-mouth on the Firth of Forth, about three miles from Kepduf, now Kilduff in Athelstaneford parish; it was the point from which St. Kentigern's mother was set adrift in a coracle. The writer of the *Life* explains the name as 'Mouth of Stench' (*ostium fetoris*), from the quantity of fish cast from the boats of fishermen and left to rot on the sand. *Lessic* is an adjective, but there is no word meaning 'stench' from which it could come. It appears to be the same as the Cornish *lesic*, 'herbaceous,' 'bushy'; compare Welsh *llusog* in *Careg Llusog*, 'bushy rock,' Merioneth; also our Lossie: the meaning is 'bushy mouth.' Aberlessic is usually equated with Aberlady, nearly four miles from Kilduff, but the second parts of the names are obviously quite different, and the places are probably different also: old forms of Aberlady are Aberlauedy, *c.* 1221 (Chart. Inch.); Aberleuedi, 1214/29 (RPSA); Aberleuedy, 1336 (Bain's Cal.); Aberlefdi (Theiner); 'portum de Abirlady in sinu aque de Pepher,' 1542 (RMS). These point to a British formation from the base seen in Ir. *lobhaim*, Sc.G. *lobh*, rot, putrefy; the parallel form in Welsh and Cornish seems to be lost. It is therefore likely that Aberlady was the 'ostium fetoris'; since Aberlady is practically contemporary with Aberlessic, the latter is most probably the old name of the mouth of Gosford burn, about a mile south-east of Aberlady. Another inference of some general importance is that an *aber* is not necessarily named after the stream at or near whose mouth it is; the stream of Aberlady is the Peffer.

Abercorn is in Bede Aebbercurnig, with variants Ebber- Linlithgow.
curnig, Aebercurnei, Aeborcurnit;[1] Abercorn, 1335 (Bain's Cal.). The Old Stat. Account says, 'the church and village of Abercorn are situated upon an angular point, and from sixty to eighty feet above the level of the sea. At the point, about one hundred yards below the church, the Cornie and Midhope burns are united.' Bede's form is British, meaning 'horned confluence'; compare W. *corniog*, horned. The horn is probably the 'angular point' between the two streams whose confluence forms the *aber*; compare G. *socach*, 'snout place' often applied to such a point, and sometimes anglicized as Succoth. The name Cornie is still extant, standing probably for *cornaigh*, the later form of the gen. of *cornach*. On the Continent, *Cornācon* in Lower Pannonia, now Sotin, was situated on a sharp hornlike bend of the Danube, specially noted by Ptolemy.[2] Abercorn is the only derivative of *corn*, a horn, which has survived in Scotland, but we have many derivatives of *benn*, *beann*, a horn, peak.

Abercraig at Newport does not seem to appear on record. Fife.
Abercrombie, now the name of a village, was formerly the name of a parish, now called St. Monans; as the old parish church stood near the Dreel Burn, the *aber* was probably near it. Old forms are Abercrumbin, Abircrumbi (Reg. Dunf.); Abbercrumby, 1270 (Bain's Cal.); Abercromby, 1296, *ib.*, etc. The second part is as in *Dul-chrombaidh*,

---

[1] Plummer's *Bede*, i. p. 26.

[2] ἡ κατὰ Κορνακὸν ἐπιστροφὴ τοῦ Δανουβίου ποταμοῦ, 'the bend of the river Danube at Cornācon'; Ptol. pp. 2, 15, 1.

'haugh of bending,' on Loch Ruthven in Strathnairn, *crombadh* being from O.Ir. *cromb*, G. *crom*, W. *crwm*, bent, formed like *lombadh* in *Innis-lombaidh*, Ross-shire, from *lom*, bare. It is probably a Gaelic adaptation of a British term of similar meaning. There are also Crombie and Crombie Point in the parish of Torryburn, and that there was here a place called Abercrombie may be inferred from the record of 'the water that runs between the land of Petliuer and the land of Gelland on the one part and the land of Abercrumbin on the other,' 1227 (Reg. Dunf.).

Aberdollo in Largo parish is '*lie sched* called Abirdolo extending to 5 acres,' 1611 (RMS); Aberdolloche, 1630 (Ret.), now apparently obsolete. The second part is *dolach*, the old genitive of *dol*, a mead, now *dail*, gen. *dalach*; compare Ballindollo in Forfarshire, as against the later—and usual—Ballindalloch. Aberdour is Abirdouer, *c*. 1329 (Reg. Dunf.); Abirdowyr, 1336 (RMS), from Gaelic-British *dobor*, later, *dobhar*, W. *dwfr*, water, stream, like Aberdour in Buchan, and means simply 'mouth of streamlet.' Abertay Sands is a bank off the north-east coast of Fife, near the mouth of Tay, but I have met no proof that this is an old name.

Aberdona, north-east of Alloa, may have denoted the junction of a brook, now nameless on the map, with Black Devon; it is Aberdonie, coupled with Barnaige *alias* Aikenheid, 1655 (RMS).

Aberargie is on the Farg, south-west of Abernethy; more than a mile lower down and near the junction of Farg and Earn is Culfargie. It is Apurfeirt (? read -feirc) in the note of the grant of Abernethy made by King Nectan to St. Brigid (*P.S.* 6). We may compare the Fergie burn, a tributary of the Banffshire Aven, possibly also the Forgue burn which gives its name to Forgue parish in Aberdeenshire. As Farg seems to be nominative case, with genitive -fargie (-fh)argie, the second part may be reasonably regarded as G. *fearg* fem., gen. *feirge*, 'wrath.' It is, however, more likely that we have to do with a gaelicized form of a British name from the base seen in O.Welsh *guerg*, 'efficax,' Gaulish *Vergo-bretus*, the designation of the annually elected chief magistrate of the Aedui, who had the power of death and life, explained by Zeuss as 'iudicio efficax,' 'effective in judgment'; also probably in *Vergivios Oceanus*, the name, according to Ptolemy, of the sea to the south of Ireland; Gr. ἔργον, 'a work, deed'; Ir. *ferg*, a warrior, hero.

Aberbothrie, near Alyth, is Abirbothry, 1373 (RMS), etc. (see p. 435).

Abercairney, east of Crieff, is Abercharni, *c.* 1221 (Chart. Inch.); Abircarnyche, Abbircarnych, *c.* 1268, *ib.*; Abercarny, 1339, *ib.*, on a burn which joins the Peffray. 'Cairnie' sometimes represents 'cardnie, cardeny,' from the term seen in Kin-cardine, etc.; if it does so here, the meaning is 'copsy mouth'; compare Aberlessic. The alternative is *carnaigh*, genitive of *carnach*, 'place of cairns,' 'rough, rocky place.' Aberdalgie, a parish near Perth, gets its name from the confluence of a brooklet on which the church stands, with Earn. It is Aberdalgin, 1215/21 (Chart. Inch.); 'Waltero de Abirdalgy,' 1315/21 (RMS), etc.; from G. *dealg*, a thorn, pin; Corn. *delc*, 'monile' (necklace). The point at the junction of Earn and Tay, seven miles east of Aberdalgie, was called Rind-algros (*Rind-dealgros*), 'point of thorn-point,' now Rhynd.

Aberfeldy is in Gaelic *Obar-pheallaidh*; old forms—none are earlier than fifteenth century—sometimes have *d*, sometimes not. It is at the confluence of the Moness burn with Tay, and commemorates the name of Peallaidh, an *ùruisg*—really an ancient water-demon—of whom there is still abundant tradition. *Caisteal Pheallaidh*, 'Peallaidh's castle,' is in the Den of Moness. Peallaidh may be for **pelldae*, formed from *pell*, a hide (Cormac's Gloss.), with reference to the shaggy coat of hair with which the *ùruisg* was covered.

Aberfoyle is in Gaelic *Obar-phuill*, 'confluence of the *poll*,' *i.e.* of the sluggish stream or 'Pow.' In the Irish *Life of St. Berchan* it is *Eperpuill*.[1] Aberlednock, at

[1] See p. 225. The speculations which connect Palladius with Aberfeldy are not to be taken seriously.

Comrie, is *Obar-Liadnáig* in Gaelic; on record Abbirlednoch 1444 (Ex. Rolls); Abirladnaucht, 1541 (RMS), etc. The stream *Liadnág* comes through *Gleann Liadnáig*, Glenlednock. The name is obscure to me, but it is to be compared with *Ard-talanáig*, Ardtalnaig, and *Ard-radanáig*, on the south side of Loch Tay, in the same district, both of which appear to contain stream-names. For Abernethy, see p. 211.

Abernyte is the name of a village and parish, the former of which is situated near the confluence of two rivulets (Old Stat. Account). It is Abernuyt, 1388 (Bain's Cal.); Abirnyte, 1415 (Ex. Rolls), etc.

Aberuchil, near Comrie, is primarily the confluence of Ruchill from Glen Artney and Earn; Abbiruchil, 1461 (Ex. Rolls); Aberquhill, 1465, *ib.*; Aberurquhill, 1594 (RMS), 1636, 1662 (Ret.); in Gaelic now *Obar-rùchail.* The local explanation of Ruchil as *ruadh-thuil*, 'red flood,' is impossible phonetically, though it suits the colour of the stream in spate;[1] but the spelling Aberurquhill suggests a compound beginning with the preposition *ar*, on, with possibly G. *coille*, W. *celli*, 'wood,' as the second part; compare *Urchaill*, Orchil, 'on-wood,' 'wood-side.'

Aberruthven is the junction of Ruthven Water and Earn; p. 387. Of Abertechan near Crieff I have neither old forms nor Gaelic pronunciation. Aberturret is the confluence of Turret with Shaggie near Crieff.

Aberugle in the barony of Kinnoull is given as the name of a parish, 1696 (Ret.).

Abercairnie in Glen Ogilvie, four miles south of Glamis Kirk, is at the junction of Glamis burn with a small brook; compare Abercairney in Perthshire.

Aberlemno is Aberlevinach, 1250 (RPSA); Abirlemenach, 1275 (Reg. V. Arbr.); Abrelemnach, 1359 (Ex. Rolls); Abirlemenach, 1322 (RMS), etc. The second part is *lemnach*, later *leamhnach*, the adjective from *lemain*, *leamhain*, an elm tree, most probably a Gaelic adaptation of a British name or stream-name. The place so called is some distance from the stream Lemno.

---

[1] *Chronicles of Strathearn*, p. 168.

Arbroath is in Gaelic now *Obar-bhrotháig* ; Aberbrudoc, 1189/98 (Reg. Arbr.) ; Abirbrothoc, 1199, *ib.* ; Abberbrodoch, 1187/1203 (RM) ; Aberbrodoc, *c.* 1202 (Chart. Lind.) ; Aberbrothoc, *a.* 1226 and *c.* 1235, *ib.* ; Abberbrodoc, 1239, *ib.* ; see p. 449.

Arbirlot, near Arbroath, is at the junction of Elliot Water and Rottenraw burn ; Aberhelot, 1198 (Reg. Arbr.) ; Arbirlot, 1202/1214, *ib.* ; Abrellot, 1323, *ib.*

For Arbuthnot and Aberluthnot, see p. 446. Kincardine.

Abercattie appears to have been the junction of Cattie and Whitehouse burns in Keig parish ; it is now according to Macdonald called Whitehouse. Old spellings are Abercawtie, 1473 (Laing Chart.) ; Abercawltye (Litill and Mekill), 1543 (Ant. A. & B.); Abircathie, 1573, *ib.*; Abircattie, 1638 (Ret.).

For Aberdeen and Aberdour see pp. 211, 454. Aberdeen.

Aberdour in Buchan is Abbordoboir in the Legend of Deer, later on record Aberdouer, etc. ; identical with Aberdour in Fife.

Abergairn is the junction of Gairn with Dee, in Gaelic *Obar-gharthain* ; Abirgardin, 1468 (Misc. Spald. Club IV) ; Abirgarny, 1497 (RMS) ; Abirgardene, 1539, confirming charter of 1468 (RMS) ; Abergardyne and Glengardyne, 1639, *ib.*

Abergeldie is at the junction of a small stream with Dee ; Geldie is a common stream-name, meaning 'white water' (p. 440).

Abersnithock, in Monymusk parish, is, according to Milne, the junction of the burn of Blairdaff with Don ; the farm which used to be so called is now Braehead. It is Abersnethok, 1573 (Ant. A. & B.), etc.

Aberbrandely is a name long obsolete which occurred in Banff. the old parish of Strathavon. It is Abberbrandolthin, 1187/1203 (RM) ; later Abyrbradalum, Aberbrandly, Aberbrandely, Aberbrandali, etc., *ib.* ; Aberbrandaly, 1293/5 (Bain's Cal.), perhaps to be compared with Mandalay, *Manndalaidh*, and *Liandailidh*, in Glengarry, where *dailidh* is a diminutive of *dail*, a dale. But this is mere conjecture.

Aberchirder, the old name of the parish of Marnoch, is now the name of a village, which lies between the Arkland burn and the burn of Auchintoul, just above their confluence ; see p. 454.

Aberlour is situated, says the Old Stat. Account, at the mouth of a noisy burn ; it denotes the junction of that burn with Spey. The old spellings are Aberlower, 1226 (RM), and such ; the stream was doubtless called *Labhar*, 'the loud one,' Welsh *Llafar* (p. 432).

Loch-aber, Abercalder, and Aberarder have been mentioned (pp. 78, 455, 454).

Abyrcardon appears 1224/33 (RM), 'one davach of land at Logykenny (*i.e. Lagan Choinnigh*, 'St. Cainnech's hollow,' now Laggan), to wit, Edenlogy, and Abyrcardon, and the land in which the church of Logykenny is situated between two streams, namely Kyllene and Petenachy.' This place is mentioned nowhere else and its exact position cannot be fixed ; it may be the old name of the mouth of Pattag ; the second part is the British term found in Kincardine (p. 352), and meaning 'copse, wood.'

Abriachan, now the name of a district on Loch Ness side, is primarily the meeting of Abriachan burn with Loch Ness. It is in Gaelic *Obr-itheachán* ; on record Abirihacyn, 1239 (RM) ; later Abriach, Aberbreachy, Abbreachy, Abireachy, Abirbryacht, Aberbryach, *ib.* It is impossible to be certain of the true full form, but the record forms, controlled by the Gaelic form, suggest *Obar-bhritheachán* or *-bhrigheachán*. The stream falls very steeply, and the name is probably from O.Ir. *brí*, W. *bre*, a hill, formed like *Giúthsachán* from *giuthas*, O.Ir. *giús*, fir.

Abernethy, the confluence of Nethy and Spey, is the same as Abernethy near Perth.

Abertarff, the meeting of Tarff with Loch Ness, is *Obar-thairbh* in Gaelic, on record Abirtarf, 1208 (RM), etc., and means 'bull's confluence' ; p. 453.[1]

---

[1] Abersky, at the west end of Loch Ruthven, is on the Faragaig river, but there is no confluence ; the Gaelic is *Abairsgigh*, with stress of course on the first syllable, though Macfarlane has it Abir Esky (ii. 557). It is a compound of *abar*, puddle, and *easgach*, marshy place, in the dative *easgaigh*, meaning 'muddy marsh,' like *Pollaisgigh*, from *poll*, hollow, pool, and *easgach* the name of a part of the farm of Clashnabuiac, Novar.

The only instance is Applecross (p. 287). Ross.

V. Some stream-names not included so far may now be mentioned. Adder of Whitadder and Blackadder in Berwickshire might, in its present form, be equated or compared with *Atur*, now l'Adour, of Gascony, *Aturavus*, now l'Arroux, a tributary of the Loire, *Aturia*, now Oria, a river of Spain, and the Adour, Adur, of Cornwall and Sussex.[1] In the *Life* of St. Cuthbert, however, and in Symeon of Durham, it is *Edre*, which can hardly be other than A.S. *œdre, édre*, 'an artery, a vein, fountain, river.'

Allan, which enters Tweed east of Galashiels, appears several times in early records: 'de Aloent usque in Twedam'; 'Raburne qui cadit in Aloent'; 'a Pennango usque ad Alewent. . . . Blachaburne cadit in Alwente,' 1153/65 (Lib. Melr.); 'via que est divisa inter Weddale et Lauuedderdale usque ad Alewentisheude,' 'the road which forms the march between Wedale (Gala Valley) and Lauderdale as far as the head of Allan,' 1165/1214, *ib.* In Blaeu it is Elvand. Compare Alowent, 1238 (Cal. of Close Rolls), now Alwent in Gainford, Co. Durham. These forms may be for an early *Alo-vinda*, 'bright-stoned one'; the name is probably the same as *Afon Alwen* in Carnarvon. O.Ir. *ail*, rock, gen. *alo*, is an i-stem, later gen. *ailech*.

Allan of Strath Allen in Perthshire, is Alun in the twelfth century (Reg. Dunf.), Strathalun, 1187 (Johnston), most probably the same as the Alun of Nant Alun in Wales (L. Land.), Pont Alun in Glamorgan, and the Alan of Cornwall, representing an early *Alauna* or *Alaunos*, '? stony one.' The Ale, which is Alne, 1176 (Orig. Paroch.) is apparently the same as the Alne of Alnwick in Northumberland, which has been identified with Ptolemy's *Alaunos*. Ancrum on Ale is Alnecrum, 1296, Allyncrom, 1304 (Bain's Cal.), Alnecrumba in Latin, *c.* 1124 (Reg. Glas.), Alncromb, *c.* 1150 (Lawrie); it is situated near the spot where the Ale turns south through a right angle, so that the second part is

---

[1] McClure, *British Place-Names*, p. 114, says 'the (Sussex) river Adour has seemingly been coined from Camden's identification' (with Portus Adurni of *Notitia Dignitatum*).

W. *crwm*, G. *crom*, bent: 'bent Alne,' or if *crwm* is used as a noun, 'Alne bend.' There is another Ale in Coldingham parish. Compare the Alne of Warwickshire and Yorkshire. In Ross there is *Alanáis*, Alness, primarily the name of the site of the parish church on a rising ground close to a stony, and in parts rocky, stream: the name seems to mean 'Allan-stance'; compare Dallas. In Knockbain parish, Ross, there is *Alán nan Clach*, 'Allan of the stones.'[1]

Another Allan, of which I have no really old forms, falls into Teviot above Hawick.

Inverallan near Grantown-on-Spey is the name of a parish, but primarily the site of the old church close to the spot where a small stream enters Spey. Old forms are: Inueraldem (? read -ein), 1187/1203, Inueraldeny, Inueralien, Inueralyien, 1224/42, Inneralien, 1389, Inneralyan, *c.* 1365, Inuerelzem (? read -ein), 1451 (RM); given me in Gaelic as *Inbhir Ailein*, but the present Gaelic tradition of this district is unreliable, being often merely a reproduction of anglicized forms.

The Legend of St. Andrew relates how Regulus and his companions 'crossed the mountain range, namely Moneth, and came to a place which was called Doldencha, but now called Chondrochedalvan.' The place last-mentioned appears later as Kindrochit, 'bridge end,' otherwise Castleton of Braemar, by which name it is now known. The old name in full means 'bridge end of Alvan,' *i.e.* of the stream bridged. A later record has 'Aucatendregen on the one side of the river called Alien,' 1214/34 (RPSA), now Auchindryne, 'blackthorn field,' on the Clunie Water at Castleton. This stream, of old Alvan, later Alien, is joined about two miles above Castleton by the Callater, 'hard water,' 'rocky water,' at a place called Auchallater.

The Elvan, a tributary of Clyde, is on early record in the term Brother-alewyn, -awyn, Brothir-alewyn, Brothyr-alewyn, perhaps for *brugh ar Alewyn*, 'mansion on Alewyn'

---

[1] *Alán*, Allan in Easter Ross, is far from stony, and I have suggested comparison with Lat. *palus*, a marsh, from the root seen in Gr. πηλός, mud.—(*Place-Names of Ross and Cromarty*, p. 75).

(Reg. Neub.); in Macfarlane it is Aluan, Aluine (vol. iii. pp. 56, 132), on Blaeu's map 'Aluan or Aldvine.' Compare Penhalwyn, 1288 (Cal. of Close Rolls), now Penhallyn, Jacobstowe parish, Cornwall.

The Alvain burn passes by the Clachan of Campsie; near its source Blaeu's map has Aldvin Hill; a place near its course is called Allanhead.

The three last-mentioned, and possibly the Allan of Inverallan at Grantown, seem to be the same in form as the first-mentioned Allan.[1]

Boyne in Banffshire, Boyen, 1368, may be a repetition of the Irish Boyne, nom. *Boend*, *Boand*, gen. *Boindeo*, later *Bóinne*, dat. *Bóind*, the *Buvinda* of Ptolemy.

Bervie in Kincardineshire is identical with the Irish *Berba*, *Bearbha*, the Barrow, with change of final *-a* to *-ie* (p. 439). It is from the root found in Ir. *berbaim*, I boil, W. *brwd*, hot, fervent. The Barrow was one of the thirteen 'royal waters' (*ríguisce*) of Ireland (LU 98 a 14). *Allt Bruthain(n)*, Burn of Brown in Banffshire, is ultimately of the same origin—W. *brwd*, hot; G. *bruthainn*, sultry heat; the ending suggests an early **Brutona*; compare *Bruto-briga*, a place in Spain of old.

Inverquharity in Forfarshire is probably for *Inbhir-chàraide*, 'inflow of the couple'; two streams fall into South Esk close together at the place, one of which is now called Carity.

Dubh'isgidh on Loch Eil is for *dubh uisge*, 'black water,' *-e* becoming *-ie* (p. 439); compare Duisk, a tributary of Stinchar.

Daer, a tributary of Clyde, is probably identical with Dare of *Aberdâr*, Aberdare in Wales.

Abhainn and *Gleann Duibhailigh*, Dubh Lighe of Ordnance Survey Map, on Loch Eil, is probably genitive of *dubhail-each*, *duibheileach*, 'black rocky place,' after *gleann* and

---

[1] Hogan compares the Irish river *Ailbine*, now Delvin, with Allan, the tributary of Tweed. The names are different, but *Ailbine*, *Ailbhine*, may be connected with *ail*, a stone, rock. In *Tochmarc Emire*, 'the Wooing of Emir,' *Indber nAilbine* is described as *clochach acus cairrcech*, 'stony and rocky.'—*Celt. Zeit.* iii. p. 244.

abhainn. Similarly its neighbour *Fionnailigh*, Fionn Lighe of Ordnance Survey Map, is '(river of the) white rocky place.'

The Esks of Kincardine and Forfar are in Gaelic *Uisge*, *Easg*; compare O.Ir. *esc . i . usce*, '*esc* means water,' in Cormac's Glossary. In a note to Oengus' *Félire*, June 17, Moling is said to have leaped '*tar araili escca*,' translated by Stokes, 'over a certain water,' but it may mean 'over a certain boggy bit,' for another note says *tar lathaigh*, 'over a puddle'; it is used like our G. *féith*, 'bog channel, boggy place.' There are also Esk na Meann, 'the kids' bog,' in Corgarff, Eskemore and Eskemulloch in Glen Livet, and other instances. *Easgach*, a bog, is in Corgarff; 'the river Diveron springes out of Escaiche in the head of Glenbuickett' (Ant. A. & B., i. p. 121); Eskielawn is in Lintrathen, from *easgaigh*, 'at fen.' *Esc*, *easg* is for E.Celtic *isca*, 'water,' given by Ptolemy as the name of the rivers now called Exe in Devon and Usk in Wales; it was also the name of some streams on the Continent (Holder).

Gleann Freoin, Glen Fruin, is Glenfrone 1225, -freone 1250 (Orig. Paroch.); it seems to be connected with G. *freoine*, rage, fury (H.S.D.), of unknown origin. Balfreoin in an old list of names in Arrochar suggests that *freoin* may be a personal name in the genitive case.

Gryfe in Renfrewshire, Strathgriua, *a*. 1153 (Reg. Glas.), is obscure; compare Cairn Gryfe in Lanarkshire and 'the marsh of Lower Grife in the thanage of Aberlemenach (Aberlemno),' 1325 (RMS), also perhaps *Gleann Grìobhaidh*, Glen Grivie, at the head of Glen Affric, which may be from Ir. *grìobh*, a claw, talon, from the claw-like shape of the stream.

The river of *Gleann Ghlaoidh*, Glen Gloy in Lochaber, with *Inbhir Ghlaoidh*, Invergloy, at its mouth, is connected with G. *glaodh*, glue, O.W. *gloiu*, liquid, W. *gloyw*, shiny; the root is *glei-*, 'sticky, liquid.' This appears to be an old British name taken over into Gaelic.

Glen Garry in Inverness-shire and in North Perthshire is *Gleann Garadh*, the latter being in the vernacular *Gleann Gar*, with *-adh* dropped as usual in this region of Perthshire. *Garadh* is thus in the genitive case; the nominative is either *garadh* indeclinable, or *gar* (? *garr*), a dental stem like

O.Ir. *eirr*, gen. *eirred*, 'chariot-fighter,' *luch*, gen. *lochad*, a mouse, etc. We may compare *Uaran Garad* in Connacht, *Glenn Garad*,[1] in Keating *Gleann Gharaidh*, in Tipperary, and *Cenn Garad*, Kingarth, in Bute. *Garadh* is well known as the name of a warrior of the Fiana, but always with gen. *Garaid*, *Garaidh*. It appears therefore that here we have not to do with a personal name—though several glens are named after persons—but probably with the genitive case of *gar*, whence the diminutive *garán*, a thicket, connected with *garadh*, a den, copse.

Leith, Inverlet, *c.* 1130 (Chart. Hol.), Inverlethe, *c.* 1315 (RMS), is in Gaelic *Lìte*. It is doubtless connected with Innerleithen, Inuerlethan *c.* 1160 (Lib. Calch.), and perhaps W. *llaith*, damp, moist, *lleitho*, moisten, may be compared.

The Leader of Lauderdale is Leder in the *Life* of St. Cuthbert, 'inter Galche et Leder, 1124/53, 'de Galue usque ad Ledre,' 'sicut Leder cadit in Twedam,' 1153/65 (Lib. Melr.). It seems to be the same as Leder, a tributary of Conway in Wales. Lauder, however, is Lauuedder-dale 1165/1214, Louueder 1208 (Lib. Melr.), Lowederdale 1337 (Bain's Cal.), etc., and appears to be unconnected with Leader.

Liver, of Loch Etive and Loch Awe, is in Gaelic *Lìbhir*, pronounced *Lì'ir*. This is probably the same as the river Liber of Leinster, said to be from a personal name (LL 160 b).

The river Nevis of Glen Nevis at the foot of Ben Nevis is *Abhainn Nimheis* or *Nibheis*—*mh* can hardly be distinguished from *bh* in pronunciation after *n*, which makes the syllable nasal in any case. The name recurs in Loch Nevis on the west coast of Inverness-shire. A third instance, which may or may not be of the same origin, is Knocknevis in Minnigaff parish, with a small brook from it into Dee. In poetry we have *tar éis leoghuin Loch Nimheis*, 'after (the death of) the lion of Loch Nevis,' in a bardic poem, *c.* 1705.[2] A poem by Iain Lom mentions *Sròn Neamhais*, 'point of Nevis' (the river), but the rime shows that the

---

[1] 'Échta Guill Glinni Garad,' 'the exploits of Goll of Glen Garadh.'—LL 204 a 34.

[2] Adv. Lib. MS., LII. p. 14 b.

poet pronounced it *Sròn Nimheis* ;[1] so also in his poem on Inverlochy he makes *Nimheis* rime with *pillein*. Adopting the spelling *Nibheis*, Macbain took the name from an early *Nebestis* or *Nebesta* ; 'the root *neb* or *nebh* is connected with clouds and water, and gives us the classical idea of Nymph, root *nbh*. . . . There was a river in ancient Spain called the *Nebis*, now *Neyva*, which may also show the same root.'[2] This is probably going too far afield. If we take the spelling *Nimheis* the name may be compared with the Irish river-name *Nem*, *Neim* (*Neimh*), gen. *Neme*, one of the thirteen 'royal rivers' (LU 98 a 14). This name is formally identical with O.Ir. *nem*, *neim*, venom, gen. *neme*, and probably means 'venomous one,' the opposite of our *Fìne*. *Nimheis* (genitive), Nevis appears to be the same with addition of a suffix : nom. *Nemess*, later *Neimheas*, gen. *Neimheis* ; this would become *Nimheas*, gen. *Nimheis* in Sc.Gaelic, as *neimh*, venom, becomes *nimh*. This explains also the anglicized form Nevis, which dates from a time before *Neimheas* became *Nimheas*. A sixteenth-century bard says of Glen Nevis :—

'Gleann Nibheis, gleann na gcloch,
Gleann am bi an gart anmoch ;
Gleann fada fiadhaich, fàs,
Sluagh bradach an mhìoghnàis.'

'Glen Nevis, glen of stones, a glen where corn ripens late ; a long, wild, waste glen, with thievish folk of evil habit.'[3] Another version has :—

'Gleann ris an do chuir Dia a chùl :
Amar sgùrainn an domhain mhóir.'

'A glen on which God has turned his back : the slop-pail of the great world.'[4]

Orrin from Glen Orrin, which joins the Conon at Urray in Ross, is in Gaelic *Abhainn Oirrinn*. T. Pont notes, 'item

---

[1] Gillies, *Collection*, p. 76.
[2] *Place-Names*, p. 149.
[3] *Inverness Gaelic Society Trans.*, vol. xxvi. p. 463.
[4] *An Duanaire*, p. 46.

Glenavaryn with Avon Ferbaryn 10 myl long with Loch na Whoying 3 myl long cuming out of Ban Whoying; it entereth on the south side of Connel, a myl above the cobil whair we cum ovir.'[1] From this and from Pont's map in Blaeu it is clear that Glenavaryn is Glen Orrin and Avon Ferbaryn, which on the map is Avon Forbarin (*i.e.* Fairbairn River), is the Orrin, from *Loch na Cuinge* (Ordnance Survey Map Loch na Caoidhe), 'loch of the yoke.' In 1440 Andrew de Rose appears as perpetual vicar of Innerafferayn in the diocese of Ross.[2] A papal Bull of 1256 mentions 'the tithe sheaves of Inverferan,'[3] but a Bull of 1257, which is essentially the same as that of 1256, has Inveraferan.[4] From these records and from other considerations which need not be gone into here it is certain that the church or vicarage of Innerafferayn or Inveraferan is to be identified with the church of Urray, situated at the junction of Orrin and Conon, and that 'afferayn,' etc., represents the genitive case of *aifreann*, *oifreann*, an offering: Inverafferayn means 'the confluence of the offering,' with reference to an offering of land to the ancient church of Urray. This is therefore a notable instance of river, glen, and confluence being all named from a circumstance connected with the Church: the old name of the river was superseded.[5]

Pareot is on record in 1273 (Ant. A. & B., i. 467) as a tributary of Deveron near Turriff. It seems to be a British name, but it would be unsafe to compare it with the Parret of Somerset, which is Pedrida, 658, Pedrede, Pedret, 893 (Johnston).

Gleann Dá Ruail, Glendaruel in Cowal, is probably to be identified with *Glenn Dá Rúadh* mentioned in Deirdre's

---

[1] Macfarlane, *Geog. Coll.*, ii. p. 551.

[2] *Calendar of Papal Letters*, ix. p. 445.

[3] Theiner, *Vetera Monumenta*.

[4] Printed in Sir William Fraser's *Earls of Cromarty*.

[5] The article on Orrin in my *Place-Names of Ross and Cromarty*, p. 111, was written in ignorance of the facts given above. Inverferan of 1256 is not Dingwall but Urray. The church of Urray is styled 'the parish church of St. Madidus.'—*Cal. of Papal Letters*, ix. pp. 426, 445.

lament on leaving Alba.[1] 'Suny Magurke's lands in Knapedale and Glenarewyle in Scotland' were granted to Duugal de Gyvelestone (? Galston in Ayrshire), valet of Edward II., in 1314 (Bain's Cal.). A well-known Gaelic song makes mention of *Clachán Ghlinn Dà Ruail*, where the metre requires *Ruadhail*, which is doubtless the true form. *Ruadhail* is formed from *ruadh*, red, in the same way as *Deargail* is formed from *dearg*, red, and means 'red spot,'[2] primarily a place-name. The name thus means 'glen of two red spots'; Ruel as applied to the river is secondary and a case of transference.

Shin in Sutherland is in Gaelic *Abhainn Sin*, formally identical with *sin*, gen. of *sean*, old. It may be for an early *Senos* or *Sena*, 'old one'; compare Ptolemy's *Sēnos*, later Sinand (LU 98 a), in Latin *Sinona*, the Shannon, where *Sēnos* seems to be for *Senos*.

Spey is in Gaelic *Abhainn* (or *Uisge*) *Spé*, where *Spé* (*Spéith*) is genitive of a nom. *Spiath*, whose diminutive is *Spiathán*, Spean: Spey and Spean are sister streams, rising near each other and flowing in opposite directions. *Spiath* is to be compared with W. *yspyddad*, hawthorn, which is for an early *squij-at-*; Ir. *scé*, hawthorn, gen. pl. *sciad(h)*; Lat. *spina*, a thorn.[3] A broch near Lentran, Inverness, is called *Caisteal Spiathanaigh*, Castle Spynie; an islet in the Kyle of Sutherland opposite Rosehall is *Eilean Spiathanaigh*; compare also Spynie in Moray. Spey means 'hawthorn stream,' the British equivalent of G. *Allt na Sgitheach*, 'hawthorn burn.'

Teatle, which enters Loch Awe near Dalmally, is in Gaelic *Abhainn Teatuill*, an obscure name.

Loch Toilbhe, Loch Tulla at the head of Glen Orchy, may be from a stream-name, also obscure to me.

---

[1] *Irische Texte*, 2nd series, pt. II. p. 128; *Celtic Review*, vol. i. p. 110. The quatrain is:—

> 'Glenn Dá Rúadh / mochen gach fer dána dual
> is binn guth cúach ar cráib cruim / ar in mbinn ós Glinn Dá Rúadh.'

'Glenn Dá Rúad, dear to me each one of its native men; sweet the cuckoo's note on bending bough on the peak above Glenn Dá Rúad.'

[2] Joyce, *Irish Names of Places*, vol. ii. p. 39.

[3] Morris-Jones, *A Welsh Grammar*, p, 143.

Tennet of Glen Tennet, in Forfarshire, is probably from O.Ir. *tene*, fire, gen. *tened*, later *teineadh*, W. *tân*, fire; compare Glan Tanat in Montgomery and Afon Tanat in Denbighshire, also Restennet, of old Rostinoth, etc., in Forfarshire, and the stream-name *Teinntidh*. Tynet or Tynot, a Banffshire stream, is probably the same.

Abhainn Tréig from *Loch Tréig* in Brae Lochaber may be connected with W. *tranc*, 'end, dissolution, death,' which if taken over into Gaelic would become *tréc*, later *tréag*; *Tréig* is of course in the genitive. Loch Tréig was traditionally the haunt of ferocious water-horses (*eich uisge*).

Abhainn Teilt, the Tilt of Athol, is obscure to me; 'Erc mac Telt' is given as the ultimate ancestor of *Clann Mael-anfaidh*, 'children of the servant of storm,' the old name of the Camerons.[1] *Loch Tailt* in Ireland (Hogan) seems to be from the same term.

The Orchy in Argyll is in Gaelic now *Abhainn Urchaidh*; Glen Orchy is *Gleann Urchaidh*; in 1801 John MacGregor wrote 'Gleann-urcha nam badan,' 'Glen Orchy of wood-clumps,' and this pronunciation is still heard in part of Perthshire. In the Dean of Lismore's book it rimes once with *gái*, once with *taoibh*, twice with *dáimh* (all in Deibhidhe), and once with *cungbháil*, proving that the fifteenth-century poets made the final syllable long. Deirdre's lament on leaving Alba mentions *Glenn Urcháin*, with variants *Orchaoin*, *Archáin*, riming with *dromcháin* (modern *dromchaoin*), by which Glen Orchy is probably, though not certainly, meant.[2] It is, however, hardly credible that the *n*, if genuine, should have dropped in Gaelic, and that too in the fifteenth century: if the reference is to Glen Orchy, the poet of the lament was probably etymologizing. The name is plainly a compound beginning with *ar*, 'on, near,' appearing often as *or*, *ur* in composition, as in *Archartdan*, now *Urchardan*, Urquhart, 'on wood,' Orchill for *ar-choill*, 'on wood.' *Urcháidh* may be for an early

[1] Skene, *Celtic Scotland*, iii. p. 480.

[2] 'Glend Urcháin/ ba hed in glend dírech dromcháin,' 'Glen Urchain! that was the straight fair-ridged glen.'—*Celtic Review*, vol. i. p. 110; compare *Irische Texte*, 2nd series, pt. II. p. 128.

Are-cēt-ia, 'on-wood-stream'; Gaulish and E.British *cēto-*, wood, *Cētion*, wood-place; W. *coed*, wood. If so, it was taken over into Gaelic early.

The Gaelic terms for a confluence are *inbhear*, *comar*, *comhrág*, *comunn*.

Inbhear, originally neuter, is for *eni-beron*, 'in-bring'; it is now generally masculine, but in West Sutherland and the Southern Isles it is sometimes at least feminine, gen. *inbhearach*, *e.g. Tigh na h-Inbhearach* in Jura, but *Achadh an Inbhir* in Loch Broom and Glen Artney. It is, of course, common in Scotland, and fairly common in Ireland now; Hogan's *Onomasticon* of names in the older literature has over one hundred instances. With us it denotes the junction of two streams or the junction of a stream with the sea, and sometimes the lower part of a stream's basin before it enters the sea.

Comar, for *con-beron*, W. *cymmer*, 'confluence,' always denotes a junction of streams, *e.g.* Comar in Strathglass and at the head of Gleann Dubh, east of Loch Lomond. *Innis a' Chomair*, 'meadow of the confluence,' is near the shepherd's house at the head of Strathrusdale in Ross; here the article shows *comar* used as a common noun. A derivative is *comrach*, place of confluence, whence Comeragh or Cumeragh in Ireland (Joyce, vol. ii.), also the locative Cumry (Joyce, vol. iii.). With us Comrie in Perthshire at the junction of Ruchil, Lednock, and Earn, is in Gaelic *Cuimrigh* or *Cuimirigh*, Comry *c.* 1268, Cumry 1271, 1275, Comri 1271 (Chart. Inch.), Cumery 1275 (Theiner). Cumrie in Cairnie parish, Aberdeenshire, is Cumery, 1226 (RM). There are also in Perthshire, Comrie near the junction of Lyon and Keltney burn and Comrie in Rannoch; Comrie in Ross, at the junction of Meig and Conon, is in Gaelic *Comraigh*.

Comhrág, O.Ir. *comracc*, means 'a meeting'; it occurs seldom, and is anglicized as Conrick in Badenoch and Dumfriesshire.

Comunn, from Lat. *communio*, 'society,' is rare in the sense of 'confluence,' but I have met *Comunn nan Caochán*, 'confluence of the streamlets.'

CHAPTER XV

SOME GENERAL TERMS

CERTAIN terms of general interest may be noted in conclusion. Some of them have come up incidentally in former chapters.

Abh, gen. *abha*, is a stream, river; in O.Ir. *ab*, *aub*, *oub*. In Argyll there is *Abhainn Abha*, the river Awe, with *Bun Abha*, Bonawe, 'Awe-foot'; but Loch Awe is *Loch Obha*. There is another Loch Awe in Sutherland, south of Loch Assynt, pronounced *Loch Abha*. The variation between *a* and *o* is capricious. From *abh* is formed *abhach*, stream-place, applied to places on a stream as Obhach (pron. O'ach, almost one syllable), Avoch, in Ross-shire. *Loch Abhaich* (Avich), Argyll, means 'loch of the stream-place,' and so does Loch Oich, Inverness. The river that flows into Loch Oich is called now *Obhaich*, but that of old it was called *Abh* is proved by the fact that above it is *Uachdar Abha*, Auchterawe, *i.e.* the ground above the Awe. The loch is named after the river, as usual, and in earlier times it may have been *Loch Abha*. *Abhlaich*, Aulich, on Loch Rannoch, is on a stream, and there is Aulich burn from *Carn a' Gheoidh* in north-east Perthshire. The suffix here may be *-lach*, denoting practically 'place of,' or *th'lach*, the aspirated and reduced form of *tulach*, a hill. Avochie, Rothiemay, has the *-ie* ending found in Luncarty, etc. *Gleann Sìará*, Glen Shiara near Inveraray, is probably for *sìorabh*, 'lasting river'; compare *Sìorghlais*, 'lasting stream,' near Blair-Atholl, and the Irish *Buanaid*.

Àth, mas., a ford, occurs often with a descriptive word or phrase following. *Àth Ruadh*, Redford; *an t-Àth Leathan*, Broadford; *Àth Chuirn*, Cairnford, are examples. Ashogle on Deveron means Rye-ford. *Àth-chrathaidh*, Achray, seems to mean 'ford of shaking'; *Crathaidh*, Cray, in Glen

Shee, is to be compared with Irish *crathaidhe, creathaidhe,* a quaking bog (Joyce, vol. iii.). When the descriptive epithet is prefixed, the vowel of *àth* is shortened owing to unstressed position. At *am Bannáth,* Bonar, there was a ford between Sutherland and Ross, by which the Kyle could be crossed at low tide, and as it was the lowest place of crossing, it was called *am Bonnáth,* 'the bottom-ford'; the original *o* is kept in the English form, Bonar. *Am Bànáth,* Bona, 'white-ford,' at the foot of Loch Ness, got its name from white pebbles; there is no ford there now owing to the change caused by the construction of the Caledonian Canal. *An Damháth,* Dava, 'ox-ford' or probably rather 'stag-ford,' is on Dorback near Grantown-on-Spey. Gella on S. Esk and Galla Ford near Harperrig Reservoir, Edinburgh, both stand for *geal-àth,* 'white-ford.' Glenshanna in Dumfries is 'glen of the old-ford,' *sean-áth*; right at its foot is a ford over Megget. Similarly there is a ford over Moneynut Water at Shanna-bank, Abbey St. Bathans. *Dùn-iaráth,* Dunira, in Perthshire, is 'fort of west-ford,' on Earn. Acharacle, in Argyll, is understood to mean 'Torquil's ford.'

Àthaigh, a derivative of *àth,* is a stream-name occurring twice in Ross, anglicized Eathie; we may compare Ethie in Forfar. Similarly from E.Welsh *rit,* a ford, there is the stream-name *Ritec* in the Book of Llandaf.

In common speech *beul-àtha,* ford-mouth, is often used, but it is rather rare in place-names. An example is *Beal-athdruim,* Belladrum, 'ford-mouth ridge,' near Beauly. *Beul an Àthain,* 'mouth of the fordlet,' in Badenoch, becomes in English spelling Balnain.

E.Ir. *ailech,* a rock, a rocky place, gen. *ailche* and *ailig,* dat. *ailiuch* and *ailich,* is in modern Gaelic *eileach,* a mill lade, a water channel. St. Brendan's monastery of Ailech was on the isle now called *na h-Eileacha Naomha* (pl.). Craig Elie in Lonmay parish, Aberdeen, is probably for *Creag Eiligh,* 'rock of the stony place.' The old adjective from *ailech* is *ailchide,* stony, whence *Creag Eileachaidh* at Aviemore and in Aberlour parish, the parasitic vowel between *l* and *ch* having developed into a full vowel.

Similar are the Water of Allachy, a tributary of Tanner; Burnside of Allachy, Aberlour; Ellachie burn in the Cabrach; Inverallochy in Lonmay parish; Elchies in Knockando, Banffshire, in 1226 Elchy, 1224/42 Elechyn (RM), later Elloquhy, Alloquhy. Elcho near Perth is in 1281 Elyoch (Chart. Lind.), in Wyntoun it is Elyhok, Elchok, apparently for *ailcheach*, stony place; we may perhaps compare Elliock and Elliock burn near Sanquhar, Dumfries.

All, a cliff, crag, has in E.Ir. gen. *aille*, but with us its gen. is *alla*, as in *mac-alla*, an echo, lit. 'lad of the cliff.' Kinnell near Killin, Loch Tay, is in Gaelic *Cinn Alla*, 'at head of crag.' Kinnoull near Perth is also *Cinn Alla*, with reference to the bold front of Kinnoull Hill. With us the final *t* of the article before this masculine noun is sometimes transferred to the noun itself, *i.e. an t-all*, 'the crag,' becomes *an tall*, as in modern G. *mac-talla*, an echo. Hence *Tall a' Bheithe*, 'crag of birchwood,' on the north side of Loch Rannoch, and *Meall Tallaig*, 'mass of crag-notch' (*eag*) in Glen Urquhart.

Airbhe, eirbhe, fem., a dividing wall or boundary; E.Ir. *airbe*, 'ribs, fence.' In certain parts of Ross and Sutherland there are walls, some of which run for many miles, made of turf and stone, which give rise to a number of names involving this term.[1] *Allt na h-Airbhe*, 'burn of the wall,' occurs in Eddrachilles and in Kildonan, but the best-known example is *Allt na h-Eirbhe*, Altnaharra, near the head of Loch Naver, spelled Aldnaheirbh in Macfarlane.[2] In this case the wall is said to run from Ben Vragie near Dunrobin to the stream which enters the head of Loch Naver; I have followed its course for some miles through Dunrobin Glen. Still another *Allt na h-Eirbhe* is found on Loch Broom opposite Ullapool, coming from *Loch na h-Eirbhe*.

[1] On the name Erray (*an Eirbhe*), in Mull, J. G. Campbell remarks, 'In olden times a wall (of turf) was commonly built to separate the crop land from the hill ground, and was known as *Garadh bragh'd*, or Upper Wall. The ground above the *Garadh bragh'd* was known as the *eirbhe*.'—(*Superstitions of the Scottish Highlands*, p. 183, *n*. 2.) I have not met this use.

[2] *Geog. Coll.*, i. p. 188.

The wall, which I have seen, runs by the north side of the loch up to the hill called *Maoil na h-Eirbhe.* At *Camas na h-Eirbhe,* in Nether Lochaber, there is a similar wall. Bogharvey, 'bog or morass of the wall,' is south-east of Aberchirder, in Banffshire. In Aberdeenshire there is Burn Hervie, for *Allt na h-Eirbhe.* Balharvie, 'wall-stead,' is in the Lomond region of Fife. Ayrshire has Auchenharvie twice, near Ardrossan, with Laigh Dykes and Dykesmains near it, and in Kilwinning parish. In Kirkcudbright there seem to be a number of instances. Black Arvie and Low Arvie are in Parton parish, and on Loch Ken are Nether Ervie and Ken Ervie (*Ceann na h-Eirbhe* ?). Kinharvie, Wall-head, is in New Abbey. The two burns called Pulharrow, which enter Loch Trool and the river Ken, are probably also to be referred to *airbhe,* and so may Cornharrow. One would expect to meet this term along the course of the so-called Deil's Dyke, but I have not met any instances there.

In old Gaelic *benn* means 'a horn,' then 'a peak'; 'rhinoceros' is neatly made *srónbennach,* 'nose-horned,' in the O.Ir. glosses. With reference to hills, strictly speaking it should be applied only to such as are peaked, and as a rule it is so applied. The usage is much commoner in Scotland than in Ireland.

From *beann* comes a diminutive *binnein,* a pinnacle, often applied to hills that taper to a sharp point. Probably the finest *binnein* in Scotland is that beside Benmore in Glen Dochart. *Am Binnein* in the Trossachs was made by Scott into Ben A'an. *Creag a' Bhinnein* is in Athol.

The adjective *beannach,* 'horned,' is not uncommonly applied to lochs that have hornlike bays; *Loch Beannach* is the equivalent of *Loch Cròcach,* from *cròc,* an antler, and both are found in Sutherland and elsewhere. If *Lacus Bēnăcus* in the north of Italy is correctly explained as a Gaulish name meaning 'horned lake,' the usage goes back to very early times. Similarly from *corn-,* the other E.Celtic term for 'horn,' comes *Cornăcon,* 'horned place,' on a sharp bend of the Danube.

From *benn,* a horn, compounded with *cor,* 'a cast,' 'a

setting,' comes *bennchar*, 'horn-cast'; compare *buachar*, 'a cow-cast,' a cow's dung, *urchar*, from *air*, 'on, in front,' and *cor*, 'an on-cast,' a shot, *clethchor*, 'a wattle-setting.' This term has given many place-names in Ireland and Scotland. Joyce considered that in Ireland it is usually applied to places where there is a peaked hill or a collection of peaks, *i.e.* the 'horn' is vertical. In Scotland it is a favourite term to denote 'horns' that are horizontal, especially horned lakes and horned part of streams. The instance of *Cornācon* just mentioned proves the antiquity of this usage. Loch Vennachar is in Gaelic *Loch Bheannchair*, formerly spelled Lochbannochquhar, Lochbanquhar (Ret.); in Macfarlane it is Loch Banchar, Loch Bennachar. The diminutive is seen in *Loch Beannchaŕain* at the head of Strath Conon and also in Glen Strathfarrar. These lakes are distinguished by long taper ends. Another 'lacus de Benchoir' appears in Peterculter parish in 1247 (Ant. A. & B.).

In Strathdearn, on Findhorn a little below Garbole but on the opposite side, is *Beannchar*, Banchor, where the river forms bends that strikingly resemble a broad brow garnished with two short horns: this is the *beannchar*. Another *Beannchar* is found on Findhorn above Dulsie Bridge. Somewhat similar horn-like bends occur on Don near Corbanchory and Edinbanchory, the latter meaning 'face of the horn-cast.' At Banchory-Ternan on Dee the river forms two broad sweeping symmetrical horns. Similarly the site of the old church of Banchory-Devenick on Dee is at a bend of the river which forms two horns spreading symmetrically from a somewhat narrow forehead. Banchory on the Isla in Perthshire is at a striking collection of horn-like bends. At *Tulaich Bheannchair*, Tullybanchor, on the Earn near Comrie, the river shows similar features, and the same is true of Bannachra (in Macfarlane Banachran) in Glen Fruin, Dumbartonshire. Benchar on the Ness is now forgotten; it probably referred to the broad bend below Dochgarroch. *Beannchar* of Glen Banchor is on the river Calder, near Aviemore, but its exact position is unknown to me.

The only instance that is unconnected with water is Banchor near Kinghorn in Fife; here a notable feature is a long narrow outcrop of rock, now grass-covered, which in all probability suggested the name.

As to the form Banchory, we have seen that Banchory-Ternan was of old *Bennchar*; other places vary between Banchor and Banchory. The probable explanation is that *beannchar*, which was originally a masculine *o*-stem, came to be declined like *cathair*, gen. *cathrach*, dat. *cathraigh*, a feminine palatal stem, thus giving a dative *Beannchraigh*; this has actually happened in the case of *urchair*, a shot, gen. *urchrach*.[1]

Bealach, mas., a pass, is found from Sutherland to Galloway, and is anglicized as Balloch. When Balloch stands for *Baile-loch*, 'lochstead,' as it does, for instance, near Inverness, the stress is on the second syllable. In Sutherland there are many passes called *bealach*, with some qualifying term, as *Bealach nam Mèirleach*, 'the thieves' pass,' between Loch Merkland and Strathmore, described as 'a lower neck of ground' joining 'Binnhee' and 'Binn-dirach.' *Bealach na h-Imriche*, 'the pass of the flitting,' is between Loch Dionard and Strath Beag; this name recurs in Glenartney, Perthshire. Among the Ross-shire instances are *Bealach Collaigh*, 'pass of hazel-wood,' west of Wyvis; *Bealach nam Bròg*, 'pass of the brogues,' between Wyvis and Lochbroom, so called, it is said, because in a battle which took place there brogues were used as bucklers; *Bealach nan Corr*, 'the cranes' pass,' on the south side of Knockfarrel, Strathpeffer, also the scene of a battle; and *Bealach nam Bó*, 'pass of the kine,' by which the public road now goes to Applecross. Another pass with the same name as this last is on the south side of Loch Katrine, named in *The Lady of the Lake*. Inverness-shire has no very notable *bealach* on the mainland, but in Skye there are *Bealach na Sgàirde*, 'pass of the scree,' 'the hie way

[1] For the mistaken idea that 'bangor' denotes in Welsh a primitive type of monastery, see Prof. J. E. Lloyd, *A History of Wales*, vol. i. p. 192. The same idea has been held with regard to our 'beannchar,' but without any basis in fact.

throw that trinket of hills generally called Klammaig,' and the famous *Bealach a' Mhorghain*, 'pass of shingle,' in Trotternish, at the foot of Beinn Eadarra. Here apparently was the *dùchas* of the *Colann gun Cheann*, the headless spectre which haunted Morar. At any rate, when forced to leave Morar the thing was heard to chant woefully :—

'Is fada bhuam féin bonn Beinn Eadarra,
Is fada bhuam féin Bealach a' Mhorghain ; '

'far from me is the foot of Beinn Eadarra, far from me is Bealach a' Mhorghain.' There are naturally many passes in Skye, and most of them are called *bealach*. Taymouth, in Perthshire, is always in Gaelic *Bealach* ; the full name of the pass is *Bealach nan Laogh*, 'the calves' pass.' Taymouth Castle is *Caisteal Bhealaich*. Balloch at the foot of Loch Lomond, the ancient seat of the Earls of Lennox, is also *Bealach*. In the parish of Mortlach, in Banffshire, there are the Meikle Balloch and the Glacks of Balloch on the pass between Deveron and Glen Fiddich, and Aldavallie, for *Allt a' Bhealaigh*, at the western end of a pass from Glen Fiddich into Corryhabbie. This is the district famed as the setting of the song 'Roy's wife of Aldivalloch.' Ballochbuie, 'the yellow pass,' in Crathie, Braemar, is best known as the name of a deer forest. In Forfar there is Balloch near Alyth. In the south-west there is Ballochmyle in Ayrshire, made famous by Burns ; Balloch-duan is in the parish of Ballintrae, with which may be compared Dorus-duan in Kintail. In Barr parish there is a long and lofty pass called the Balloch, with 'the Nick of the Balloch' at its summit, and Pinvalley, for *Peighinn a' Bhealaigh*, 'pennyland of the pass,' at its north end. There are instances also in Galloway, as Balloch in Minnigaff, Kirkcudbright. In Argyll the term is applied once at least to a sea pass, for the strait between Scarba and Lunga is *Bealach a' Choin Ghlais*, 'the pass of the gray dog,' but I have not heard the legend attached to the name.

Làirig, a pass, is common in certain districts, especially in north and west Perthshire ; it is O.Ir. *laarg*, a fork. Joyce does not mention it as occurring in Irish names of places, nor is it to be found in Hogan's *Onomasticon*. In Ross-shire there are *Làirig Bhaile Dhubhthaich*, the Lairgs of Tain, a pass between the parish of Tain and Edderton; *Làirig an Lochain*, 'pass of the lochlet,' between Strathrusdale and Dibidale; and *an Làirig* between Boath and Glenglass. In Inverness-shire, *Làirig an Tùir*, 'pass of the tower,' is on the south side of Strathnairn. *Làirig Thurraid*, 'the pass of Turret,' is at the head of Glenroy; here the pass is named after its stream, as is also *Lairig Dhrù*, between Badenoch and Braemar, from the river Druie. *An Làirig Leacach* 'the pass of flagstones' 'goes between Lianachan and the head of Loch Tréig, a high pass, once well trodden. Perthshire has many 'lairigs' connecting its glens. The road from Loch Tay side to Bridge of Balgie in Glen Lyon follows *Làirig an Locháin*. Between Glen Lochay and Glen Lyon there are *Làirig Luaidhe*, 'the pass of lead,' from *Tìr-àigh* to Moar; *Làirig Bhreislich*, 'the pass of rout, or of confusion,' from Dùn-chroisg to Bridge of Balgie; and *Làirig nan Lunn*, 'the pass of the staves,' from Kenknock (*Ceannchnoc*) to Pubil. I could find no tradition as to the origin of the two names last mentioned. The chief passes between Glen Lyon and Rannoch are *Làirig Mhuice*, 'swine pass,' and *Làirig Chalabha*, both starting from Innerwick. *Làirig Mhic Bháididh*, 'MacWattie's pass,' mentioned by Duncan Macintyre in his poem on Coire Cheathaich, is at the head of that fine corrie. *Làirig Mheachdainn* is between Lochs and Rannoch. The pass between Kenmore and Glenquaich is called *Làirig Mìle Marcachd*, 'pass of the mile of riding,' because at the summit there is a long level stretch on which a rider could put his horse to speed. The older name appears to have been *Làirig Monadh-marcachd*, for in the Chronicle of Fortingall it is Larkmonemerkyth.[1] A pass leading into Glen Almond is called *Làirig Phrasgáin*, 'pass of (the) band or troop.' The pass from Lochearnhead to Glen Dochart is called on

[1] 'Crux lapidea fuit posita in Larkmonemerkyth in magno lapide qui alio nomine vocatur clachur . . . per Dougallum Johnson primo Octobris anno domini V^{d}xxix.' 'A stone cross was placed in L. on a great stone called *clachur* . . . by Dougall son of John on first October, 1529.'

the southern side *Gleann Ògail*, Glen Ogle, but on the northern side *Làirig Ìle*, of which William Ross says :—

> 'Is ge do dhìrich mi Lairc Ìle
> Tha mo spìd air falbh uam,'

showing that he makes the *i* long, correctly. Here *Ìle* is most probably the name of the stream from *Lochán na Làirce*. The pass from Inveraray to the foot of Glenfalloch is called on its Glenfalloch side *Làirig Àirnein*, where *Àirnein* is probably a personal name. In Argyll there are not many cases of *làirig*, but we have *Làirig Ghartáin* in Glen Etive, 'Gartan's pass,' and *Làirig Nodha* between Glen Noe and Glen Strae, named from the river Nodha, which gives its name to the glen also.

With a prefixed adjective we have *Fionnlairig*, 'white pass' or 'holy pass' at Killin at the head of Loch Tay; it occurs also in Moray. Doularg, 'black pass,' is on Stinchar below Barr. Fleuchlarg, 'wet pass,' is near Gatehouse on Fleet, Wigtown. Camlarg, 'crooked pass,' is near Dalmellington.

Crasg, mas., genitive *croisg* (also *creisg*), means a crossing, as over a ridge; it is widely distributed, and is specially common in Sutherland and Ross. *An Crasg*, on the road between Lairg and Altnaharra, is well known as the gate to the Reay Country. *Loch a' Chroisg*, 'loch of the crossing,' occurs both in Sutherland and in Ross, anglicized in the latter as Loch Rosque. *Cnoc a' Chroisg*, by which the public road crosses into Boath, Alness, was given me once as *Cnoc a' Chreisg*, and my informant said that this was the old pronunciation. This illustrates Cresky in Glenmoriston,[1] now *Tom-chrasgaigh*, and other old spellings. Ardnagrasg, in Ross, shows eclipsis (*Ard na gCrasg*), and is explained locally as from the old system of cross-rigs or run-rig. *Crasgág*, 'little thwart one,' was the name of the stream that runs at the foot of Kinrive Hill in Ross. It was also the name of a place in Glen Urquhart, now called Kilmartin. *An Crasg* is the name of a crossing on *Rathad nam Mèirleach* in Abernethy parish. Instances in

[1] Macfarlane's *Geog. Coll.*, ii. p. 549.

Perthshire are *Coire a' Chroisg*, Glenample, and *Dùn-chroisg*, 'fort of the crossing,' in Glen Lochay, at the mouth of *Làirig Bhreislich*. In Argyll, *Allt a' Chroisg* is on the south-east side of Loch Awe. In Aberdeenshire, Afforsk (1391, Achqwhorsk), is for *Achadh a' Chroisg*, 'field of the crossing.' Arngask, Kinross, shows eclipsis like Ardnagrasg above, and means 'division (or possibly height) of the crossings.' Glencorse in Midlothian, Glencrosk 1336, etc. (Bain's Cal.), is 'glen of the crossings,' so called from the crossings toward Penicuik, Colinton, and Balerno, which are still rights of way. There is another Glencorse in the parish of Closeburn, Dumfries, and a third, Glencrosh, in the parish of Glencairn.[1]

The parallel roads of Glenroy are *na Casan*, pronounced exactly as the modern plural of *cas*, foot.

Conair, fem., common in Ireland, is not used in our common speech, but that it was so used at one time is shown by the name *Bad na Conaire*, 'clump or spot of the path,' in Sutherland. Darnconner and Darnaconnar in Ayrshire are probably for *doire na conaire*, 'path-side copse.'

Tóchar, a causeway, a road, though unknown in our common speech, occurs in Dun-tocher, 'causeway-fort,' near the western end of the Roman Wall between Forth and Clyde, at the site of a Roman fort. Here the causeway was the ancient Roman road. Other instances are Tocher and Tocher-ford, Rayne, in Aberdeenshire; Kin-tocher, 'causeway-head,' on the Leochel burn, Aberdeen, and in Fowlis Wester, Perth; and Craigie-tocher near Turriff. It does not seem to occur within the area that is now Gaelic speaking.

Cong, means a narrow, a gullet, usually in a stream and always in connection with water; in Irish 'a narrow neck, a strait where lake or river narrows itself' (O'Donovan).

---

[1] Robert Gordon of Gordonstoun records 'Crask-Worwair, hoc est Thani vel Comitis Crux' (*Crasg a' Mhorair*, that is the Cross of the thane or earl), a stone cross, or a stone with a cross carved on it, near Dornoch (Macfarlane). I have not met this use of *crasg* elsewhere for certain.

With us it is found only in the oblique case *cuing*, still used in the common speech of Lewis. *An Cuing* is the name of a very narrow passage leading to a landing place at *Gob na Heiste*, on the coast of Duirinish, Skye. The Whing burn is near Sanquhar, Dumfries. With the suffix *-lach* is formed *conglach*, whence the locative *Cuinglich*, the name of a place on a deep narrow gorge of the river Averon in Ross-shire. *Coire Choingligh*, 'Corrie of the narrow,' is in Lochaber. Compounded with *leitir*, a hill slope, it gives *Cuingleitir*, 'gorge-slope,' at a deep narrow part of the river Etive in Lorne; compare Whinletter Pass, Basenthwaite, Cumberland. In Mull we have *Coingleitir*, anglicized into Goladair. Similar in meaning is *Cuingleathad*, Coylet, from *leathad*, a hill-slope, in Cowal. The Burn of the Cowlatt is in Knockando parish. *Cuingleum*, 'gorge-leap,' is applied to a number of places where a stream running between rocky banks narrows so that it is possible, or nearly possible, to leap across. In Stratherrick there is a *Cuingleum* on Feachlinn above Whitebridge, with *Cnoc a' Chuingleum*, Knockchoilum, and *Cill a' Chuingleum*, Killiechoilum, near it. Hence also Coylum in Badenoch, Colzium in Kilsyth parish, Stirling, and Colzium on Camilty burn, in West Calder, Midlothian. Rob Donn mentions *an Caoileum*, for *an Cuingleum*, in his district of Sutherland. Ling-a-Wing, for *Linn a' Chuing*, is on Earn, near Crieff.

Cumhang, a narrow, a defile, is found in *an Cumhang* at the lower end of Strathrory, in Ross. The Pass of Lyon is sometimes called *Cumhang Dhubhghlais*, 'the defile of the black stream.' The Pass of Leny is *Cumhang Lànaigh*, the meaning of which is not clear. *Cumhang a' Bhrannraidh* is anglicized 'the Pass of Brander' on Loch Awe. *Brannradh* in Irish means 'a trap, snare, stocks, gibbet'; Dr. Alexander Carmichael, who knew it as a word in common speech, explained it as 'an obstruction,' and instanced the sea rock between Port Mo-Luaig in Lismore and Eriska, which is called *am Brannradh*: the instruction for crossing used to be 'seachain am Brannradh,' 'avoid the B.' The Pass on Loch Awe is correctly described by Barbour as 'ane narrow place betuix a louchside and a bra'; here Robert the Bruce was beset by three men who aimed to kill him. The older form here appears to have been *brannradhan*: Coriwrannarane 1497, Correbrandrane 1614 (RMS), Coribarnderan 1686 (Ret.).

Dorn means a fist, W. *dwrn*, Gaulish *durno-* in *Durnomagos*. *Dùn Dùirn*, Dundurn on the Earn near St. Fillans, was an important stronghold in very early times: 'obsessio Dúin Duirn,' 'the siege of Dundurn,' took place in 683 (AU). Why it was named 'fort of the fist' is unknown, perhaps from its shape; compare *Dùn dà Làmh*, 'fort of two hands,' in Laggan, Inverness-shire. The local tradition given me is that the masters of the fort used to levy a tribute of handstones (*dornagan*) from the people of the district, which they used as weapons. That handstones were so used is known both from Irish sources and from Wyntoun, who records 'the Batayle Dormang,' *i.e. cath nan dornag*, fought by the Butemen and so called 'for thai thare with stanys faucht' (viii. l. 4350). The peak above the fort is *Biorán Dùin Dùirn*, 'the spike of Dundurn.' Hill of Durn, with remains of an old fortification, is near Portsoy.

From *dorn* comes *dornach*, 'pebbly,' 'a pebbly place,' the pebbles being often fairly large, like the Greek λίθος χειροπληθής, 'a stone that fills the fist.' Hence we get Dornoch in Sutherland, Dornach on Loch Nevis, Dornoch Point on Loch Eck in Cowal, Dornock in Dumfriesshire. Hence also Durno, a place and surname in Aberdeenshire. The genitive appears in Beldornie, for *Baile Dornaigh*, on Deveron, the married home of the Gaelic poetess Silis MacDonald of Keppoch. As an adjective *dornach* appears in Drumdurno and Edindurno, 'pebbly ridge,' 'pebbly face,' in Aberdeenshire, and in Baldornoch, 'pebbly stead,' near Clunie in Perthshire.

The old dative-locative was *dornaigh*, whence *an Dornaigh* (fem.), Dornie on Loch Alsh; it is found also in Loch Broom and in Sutherland and elsewhere.

Allt Dornan in Luss probably means 'burn of handstones.' Dornal Hill is in the south-east corner of Colmonell parish, Ayrshire. Eminences called Meikle Dornell and Little Dornell, with Dornell and Loch Dornell, are in

Balmaghie parish, Kirkcudbright; these may show the *-l* extension found in *Deargail*, Dargle, 'red spot,' etc.

Eileirg, pronounced *eileirig* in Perthshire, and *iolairig* north of the Grampians, appears in the Book of Deer, which records that Máel Coluim son of Máel Brighde gave the *elerc* to the old monastery of Deer.[1] This *elerc* is now Elrick in the parish of Old Deer. A still older form of the word is found in the Milan Glosses, ascribed to the first half of the ninth century, where *in erelcaib* glosses *in insidiis*, 'in ambush' (Ml. 28 c 1); we have also 'ba hi temul dugníth Saul cona muntair intleda ocus *erelca* fri Duaid,' 'it was in darkness that Saul with his people used to make snares and ambushes against David' (Ml. 30 a 3). O.Ir. *erelc* thus means 'an ambush,' and *elerc* of the Book of Deer has undergone metathesis.

The *eileirg* was a defile, natural or artificial, wider at one end than at the other, into which the deer were driven, often in hundreds, and slain as they passed through. The slaughter at the *eileirg* was the last stage in the great deer hunts which were once so common in Scotland and which survived in the north till the eighteenth century. Though the term occurs often and widely with us, it does not seem to be found in Ireland.[2]

The furthest north instances known to me are Elrick south of Loch Affric and Elrick near the east end of Loch Ruthven in Strath Nairn. As the term is easily recognized in its anglicized form, it is unnecessary to give a detailed list of instances: it is common in Perthshire, and occurs in the counties of Aberdeen, Kincardine, Forfar, and Argyll. It appears freely in Galloway, and there seems to be one instance in Roxburgh, near the head of Borthwick Water.[3]

Eileag, fem., denotes a deer-trap. Long ago I was told that the *eileag* was V-shaped, open at both ends, and that the deer were driven in at the wide end and shot as they

---

1 'Malcolum mac Moilbrigte dorat ind elerc.'

2 Joyce does not mention it; the Rev. P. Power tells me that he has not met a case of it.

3 See further *Celtic Review*, ix. p. 156; *Book of the Red Deer*, p. 88.

passed along; this would make it to be much the same as the *eileirg* or Elrick. The term, however, appears not to have been confined to a place of this sort. The Minister of Assynt in Sutherland, writing in 1795, mentions a place in his parish called 'Fe-na-hard-elig,' *i.e. Féith na h-Airdeileig*, 'bog of the high *eileag*,' which he explains as 'a track of soft boggy moor in which, in times of old, the natives gathered deer, and when entangled, they killed them.'[1] This effective, if unsportsmanlike, manner of dealing with the deer, was adopted by the hero Cuchulainn in the first century.[2] Near the place indicated by the minister there are still *Allt Eileag*, *Loch Eileag*, and *Mòinteach Eileag*, and this last is probably the 'boggy moor' which he mentions. Sage in his *Memorabilia Domestica* mentions Ellig in Kildonan. In Ross, near Amat in Kincardine parish, there is *Eileag Bad Challaidh*, 'the *eileag* of the hazel-spot,' of which tradition says that so long as the natives held it and *Cairidh Chinn-chardainn*, 'Kincardine weir,' they could not be starved into submission.[3] *Bog na h-Eileig* and *Loch na h-Eileig* in Killearnan parish, Ross, suggest methods similar to those used in Assynt. *Eileag na Baintighearna*, 'the lady's *eileag*,' on Caiplich moor east of Abriachan, appears to have been named after a Lady of Lovat described by the Minister of Wardlaw as 'a stout bold woman,' and 'a great hunter.'[4] *Sìthean na h-Eileig*, 'fairy hill of the *eileag*,' is in Gairloch. In Banffshire there are Cairn Ellick and beneath it Ellick on Chabet Water, also Torr Ellick at the head of Glen Fiddich. A poem relating to the Cairngorm district says of a hunter:—

> 'Ann an Eileag a' Chuilinn
> 'S tric a dh' fhuilich do lann;'

---

[1] *Old Stat. Account*, xvi. p. 169.

[2] *Stories from the Táin*, p. 22.

[3] So in Skye of old: 'it is an ordinary saying with the inhabitants, they can never be ruined as long as these three are to the fore. The first of which is a well in the parochin of Uig; the 2d Loughsent dulce (*Loch Sèanta*, 'sacred loch'); 3d Hebri rock, all three within nine mile circumference.'—Macfarlane, *Geog. Coll.*, ii. p. 222.

[4] *Wardlaw MS.*, p. 110.

'in the *eileag* of holly thy blade was oft bloody.'[1] My furthest south instance is *Creag Eileig* on Kerrowmore in Glen Lyon.

Eireachd, an assembly, is Ir. *oireacht*, a faction, party, earlier *airecht*, an assembly. The *eireachd* was, often at least, a court of justice, and was held in the open air at a definite place, sometimes on an eminence, as is indicated by such names as *Cnoc Chomhairle*, 'hill of counsel,' right above the ancient 'Castle' near Fincastle Post-Office, Perthshire, and *Cnoc an Eireachd* near Duntulm in Skye. In Sc.G. *eireachd* is masculine, in Irish it is feminine.

In Sutherland there is *an Eireacht* (*sic*), apparently feminine, a flat at the head of Loch More. Near Fort William is *Eireachd*, once the home of the noted soldier Allan Cameron, who raised the Cameron Highlanders, known as *Ailean an Eireachd*, 'Allan of the Eireachd.' *Loch Eireachd*, Loch Ericht, is 'loch of assemblies'; it has transmitted its name to the river Ericht, which takes its waters to *Camas Eireachd* on Loch Rannoch. *Gleann Eireachdaidh* at Struan is for *Gleann Eireachda*, Glen Erichdie, 'glen of the assembly.' The name is repeated in Blairgowrie.

Aonach has two very different meanings: (1) a place of union, an assembly, a fair or market; (2) a solitary place, a mountain top; both come from *aon*, one. From the former we have *Tigh an Aonaich*, Teaninich, near Alness; *an t-Aonach* is the name of a fine flat field on Drummond farm near Evanton, formerly the site of a market. *Blàr an Aonaich*, Blairninich, at Fodderty, was also the site of a market, as was *Aonachán* near Spean Bridge. A well-known dance tune goes to the words of a *port-a-beul* called *Gobhainn Druim an Aonaich*, 'the smith of market-ridge.' The second meaning is seen in *an t-Aonach Dearg*, 'the red mountain top,' near Ben Nevis.

Tional, Ir. *tionól*, a gathering, gives *Cnoc an Tionáil*, 'rallying hill,' in places so far apart as Inverness, where it

[1] The *Life of Sir Ewen Cameron* of Lochiel mentions a deer hunt got up for the pleasure of certain English officers, prisoners in Sir Ewen's hands, in which the officers slew the deer with their swords.

is the old name of the site of the Cameron Barracks, and Colmonell in Ayrshire, where there is Knockytinnal on the eastern side of the parish.

Comhdhail, fem., gen. *comhdhalach*, E.Ir. *comdál*, means a tryst, from *con*, together, and *dál*, a meeting. Trysts were often held at well known stones; hence *Clach na Comhdhalach*, 'trysting stone,' in Coigach, where certain brothers are said to have been used to meet of old. This name is repeated with part translation in Codilstan, 1402 (Ant. A. & B.), now Coldstone in Aberdeenshire. *Coire na Comhdhalach* is on the western side of Cairngorm.

Eochair, bank, brink, edge of river, lake, or sea, is obsolete in Scottish Gaelic. As I have not met it within the area that is now Gaelic speaking, the instances given depend on spelling and situation. A perambulation of the lands of the church of Aberchirder called 'le Yochry' or 'the Youchre'[1]. shows that they were on the left bank of Deveron, at a part where the river forms an almost rectangular 'loop.' The name appears to be obsolete. Yoker on the Clyde, below Glasgow, is another instance: its meaning is echoed by its neighbour Clydebank. Pitteuchar on Lochty in Fife was formerly spelled Pittyochar, meaning Bank-stead. With it may be compared Culteuchar on May Water, Forgandenny, the nook (*cuilt*) on the waterside.

Gaoth, older *gáeth*, *góith*, means 'a marsh'; *gáethamail* is 'paluster,' marshy; *góithlach*, a swamp; *Gáethlaige Meotecdai*, 'the Maeotic Swamps,' the Palus Maeotis or Sea of Azov. In literature we have 'longa ag imtecht ar na gáethaibh gáibtecha sin,' 'ships faring on those dangerous marshy waters.'[2] In Ireland *gaoth* is said to denote a shallow stream into which the tide flows, and which is fordable at low water,[3] and Joyce gives instances from O'Donovan of its use in the northern half of the west of Ireland. With us it seems to occur not infrequently; the difficulty is that it cannot, as a rule, be distinguished from *gaoth*, 'wind,'

---

[1] *Ant. of Aberdeen and Banff*, ii. p. 210.

[2] *Celt. Zeit.*, vi. p. 284.

[3] Joyce, *Irish Place-Names*.

except by reference to the situation. One instance is certain, Bog of Gight, Gordon Castle, in Gaelic *Bog na Gaoithe*, for a charter of *c.* 1374 mentions 'capella de le Geth' as 'apud le Geth.' These expressions show that here we have to do not with *gaoth*, wind, but with *gaoth*, a marsh. The names that follow are probably of the same origin. Edingight in the parish of Grange, Keith, means, probably, 'face (*aodann*, O.Ir. *étan*) overlooking the marsh': near it is Mosstown. The Mains of Gight, Fyvie, has Moss-side and Old Moss near it; here also are Little Gight, Mill of Gight, and Gight Castle. Balnageith near Forres lies low on the Findhorn, and Balnagight in Kincardine is on the Cowie Water. Strageath, in Gaelic *Srath-gaoithe*, now Blackford in Perthshire, was spelled of old Strogeth, Strugeth, and means 'strath of the marsh.' Guay near Dunkeld, on the Tay, appears to be another instance, and Irongath, Linlithgow (Arnegayth, 1337) is probably *Earrann-gaoithe*, 'marsh-division.' 'The midmoiss callit the Goitt' (Ant. A. & B.) was near Cullen. 'Goitmureheidis' was in Forfar (Ret.). Balgay in Inchture parish is for *baile* (*na*) *gaoithe*. In the sense mentioned by O'Donovan and Joyce, *gaoth* does not seem to occur with us.

Longphort is a compound of *long*, a ship, also in Irish a dwelling, abode, and *port*, primarily a harbour, then a mansion, also simply a place; in modern Gaelic, a ferry. Primarily a sea term meaning 'ship-station, harbour,' *longphort* came to mean (2) an encampment, in which sense it is very common in Irish literature; (3) a palace, whence *lùchairt* in Gaelic; (4) a hunting booth or sheiling. As to (4), Taylor, the Water Poet, who travelled in Scotland in 1618 and saw a hunting in Marr, mentions the 'small cottages, built on purpose to lodge in, which they call Lonquhards.' Robert Stewart, the bard of Rannoch, says of a certain hunter, 'threobh e longard,' 'he lived in a hunting booth.'[1] Longphort is common in Irish names of places, and is usually anglicized as Longford, but appears also as Lonart, while near Limerick is Athlunkart, meaning

[1] *Orain Ghaelach*, le Raibeard Stiubhard, 1802, p. 68.

'disused fort or encampment.'[1] With us it is found in a variety of forms, from Sutherland to Wigtown, sometimes on or near the coast, but more often inland. In Sutherland the form is *Laghart*. *Allt an Laghairt*, 'burn of the encampment,' flows into Loch Loyal; there is also *Allt an Laghairt* (Alltalaird, Ordnance Survey Map) on Loch Coire. *An Sean-laghart*, 'the old encampment,' is in Strathmore. In Ross there is Loch Luichart, in Gaelic *Loch Lui(n)cheart*. *Camas-longart* on Loch Long in Kintail, means 'bay of the encampments' or 'ship-stations' and as Loch Long means 'Loch of Ships' we may well have here the original meaning of ship-station. Inland is Loch Lungard, in Kintail. In Inverness-shire there are Dail an Longairt in Badenoch, and the sheils of Badenlongart in Gaick. 'Lag in Loncart' at Beaufort, Lovat, is made 'the leagger bottom' in the Wardlaw MS. (p. 68). Banff has Auchlunkart, 'field of the encampment,' in Boharm parish. In Aberdeenshire Luncarty occurs north of Turriff, probably a dative-locative plural *longartaibh*. There is also Lumphard Hill, with which may be compared the old spelling Lumphortyn (RPSA) of Luncarty in Perthshire, and Lomphart (Blaeu), now Longford in Midlothian. In Perthshire Luncarty on the Tay above Perth is famed as the scene of a battle in which the Danes were defeated in 990 A.D. Though the authority for this battle is late and traditional, the existence of this name, 'at encampments,' tends to confirm the account of a battle there. Near Finlarig on Loch Tay there was, as I am informed, *Lùchairt na Féinne*, 'the encampment of the Fiann,' a name of considerable interest. Argyll has *Cùil an Longairt*, 'nook of the fortress,' in South Kintyre; it is well inland, and is close to an old fort. *Barr an Longairt* on Loch Killisport is also near a fort. There is another *Barr an Longairt* near Castle Lachlan on Loch Fyne. In Ayrshire there is Auchinlongford in Sorn parish, a well preserved form. Drumlamford (Drumlongfuird Ret.) is

---

[1] Joyce, *Irish Names of Places*, i. Joyce explains Athlunkart as 'the ford of the fortress or encampment,' but the other meaning is the more probable.

south-east of Barrhill. Wigtown has Barlockart near Glenluce, and there is Craiglockhart near Edinburgh, Craiglikerth in old spelling, with which we may compare Loch Luichart above. It is likely that Longformacus in Berwickshire means 'the encampment or residence of Maccus'; an early spelling is Langeford Makhous (Johnston).

In composition with *fas*, a station, we get *faslongphort*, 'a firm or fixed encampment,' M.Ir. *faslongport*. In Sutherland there are *Creag*, *Allt*, and *Bruach an Fhaslaghairt* (Aslaird, Ordnance Survey Map) behind Lochmore Lodge, and *Allt* and *Loch an Fhaslaghairt* near the head of Strathmore (Aslaird of Ordnance Survey Map). The *faslongphort* appears to have been about the place where the burn, *Allt an Fhaslaghairt*, crosses the ancient track known as *Bealach nam Mèirleach*, the Thieves' Pass. Rob Donn, the Reay bard, mentions 'leathad Allt an Fhaslaghairt,' [1] and from the context it is clear that he means the burn just mentioned.

Rinn, fem., means a point, headland; O.Ir. *rind*, neut., 'cacumen, aculeus,' top, point; Welsh *rhyn*, a point, promontory. Usually it applies to a cape on sea or loch, a point formed by the sharp bend of a stream, or by the junction of two streams. *Dubhrinn*, Durine, the former name of Durness village in Sutherland, is 'black-point.' Near it is a double cape formed by a headland cleft by a narrow inlet: its two points are each called *Leithrinn*, 'half-point,' *i.e.* 'one-of-two-points,' distinguished as Beag and Mór. On a sharp bend of the Naver there is *Roinn-imhigh*, Rínavie, a compound of *rinn* and *magh*, a plain, meaning 'point-plain'; it is of an old type, representing E.Celtic *Rindo-magos*. *Rinn a' Bheithe*, Rinavéy, in Glenshee, is 'point of the birch-wood.' A sharp cape on Loch Tay, near Morenish, is called *Rinn a' Chuilg*, from *colg*, *calg*, a straight sword, a prickle; on record as Rynachulig. At the junction of Earn and Tay there is Rhynd, 'point,' of old Rindelcros, Rindalgros; a point caused by a bend of Forth near Alloa is also Rhind.

[1] *Poems*, p. 167 (Maclachlan & Stewart).

Rindrought on Ugie, in the parish of Strichen, is 'bridge-point,' or possibly 'bridge-field' (*raon*). Redkirk Point, on the Solway near Gretna, was formerly Reynpatrick, Reynepatrick, Rainpatrick, meaning 'Patrick's Point,' from St. Patrick to whom the church which stood here was apparently dedicated. The Rinns of Galloway have been mentioned. In the Rinns of Islay, however, the term is *rann*, a division: the Rinns formed one of the three divisions of Islay; a Rinns-man is *Rannach*.

Ros, as already explained, means 'a thing forthstanding, a projection,' hence a promontory, a wood; in Welsh *rhos* means a moor, heath, mountain meadow. With us it is common in the sense of 'cape,' but the other meanings occur, and it is not always easy to distinguish. Its diminutive *Rosnat* was the Irish name of Whithorn.

The county name Ross has been mentioned. Rosskeen on the Cromarty Firth is in Gaelic *Ros-cuibhne*, *-cuibhnidh*, Roskevene 1275 (Theiner), 'antler point,' which describes the promontory at Invergordon, near Rosskeen church, now called *an Rubha*, 'the point.' Rosmarkie on the Inverness Firth is *Ros-maircnidh*, named after the burn beside it. In Abernethy parish, Banffshire, there is *Ruighe dà Ros*, 'the reach of two points,' an instance of *ros* in inland names. *An Ros Muileach* is the Ross of Mull. *Ros Dubh* and near it *Ros* are on Loch Lomond. *Ros-neimhidh*, anglicized Rosneath, is 'cape of the sanctuary.' The cape on the east side of Kirkcudbright Bay was called the Ross. *Ard-rois*, Ardross, 'height of the cape,' occurs in Ross and near Elie Ness in Fife. Montrose, in Gaelic *Mon-rois*, is probably 'moor of the cape,' *mon* being the short form of *monadh*. Kinross, 'head of cape,' is so called from the site of its church at the end of a point projecting into Loch Leven.

Melrose in Gamrie, Aberdeenshire, is probably for *maol ros*, 'blunt point,' 'bare point.' Melrose on Tweed is in Bede *Mailros*, which is probably British, yet the site of Old Melrose, which is on a peninsula formed by a loop of Tweed, suggests the meaning of 'bare or blunt promontory.' *Coillerás*, 'wood-point,' occurs (1) near Lianachan in Lochaber, where there is a mountain spur clad with a thin wood called *Coille Lianachàin* ; (2) on Loch Laggan, a small cape with *Coire Choillerás*[1] behind it ; (3) near Gairlochy, where a wooded point projects into Loch Lochy ; old spellings—Killerois, 1505, and such—show the *o* of *ros*. We may compare the Irish *Ros-cailledh*, 'point of the wood,' 'with a thousand of every kind of tree therein.'[2] *Dealganros* occurs thrice at least : (1) near Inverness, anglicized Dalcross ; (2) near Comrie, at the junction of Earn and Ruchil, the site of a Roman Camp ; (3) in Atholl, at the junction of Tilt with a small stream. The first part is not the diminutive *dealgán*, 'a thorn brake,' but the compositional form of an old *delgu*, gen. *delgon*, 'thorny place,' as in '*cath Delgon i Cinn-tíre*, the battle of Delgu in Kintyre' (Tigern., A.D. 574) ; the meaning is 'thornwood point.' Similarly Rescobie in Forfarshire, of old Roscolbyn (RPSA), is 'point of thorny place,' from *sgolb*, a splinter, thorn, whence *sgolbach*, 'thorny place.' Culross in Fife is *Cuillennros* (BB 214 a 21), *Cuilendros* (Celt. Scot., ii. p. 258), 'holly point' ; compare *Cuillendros na Féinne* (Acall. l. 1939). For Cardross, see p. 353.

In 'nemus nostrum de Rosmadirdyne,' 'our wood of the Ross of Madderty,' 1278 (Chart. Inch.), *ros* means 'wood.' *Dubhras*, Dores, may mean either 'black wood' or 'black point,' but, as there is no point on Loch Ness at Dores, the former is doubtless the meaning. Durris in Kincardineshire, on a bend of the Skeoch burn near its confluence with Dee, may mean 'black point.' These and other instances of Durres, Durris, are to be compared with Irish *Dubhros*, which is similarly ambiguous. In the case of Ross near Gargunnock Station, adjacent to Woodyett and Meiklewood, the meaning is probably 'wood.'

From *ros* comes *rosach*, a wooded place, a shrubbery, with dative *rosaigh*, as in *Rosaigh Ruaidh* in Ireland (Hogan). We have Rossie Priory, formerly the Abbey of Rosinclerach, in Inchture parish, Perthshire ; Rossie and Rossie Muir

---

[1] My authority for this is the late Dr. Sinton of Dores, who was brought up near it ; we should expect *Coire Choilleráis*, which is what I have heard.

[2] *Acallamh*, l. 1463.

south-west of Montrose; Rossie near Auchtermuchty; Rossie and Craigrossie in Dunning parish, with Rossie Ochel to the east. Near Elgin is Findrassie for *Findrosaigh*, probably from E.Ir. *indross*, 'a large wood, with prosthetic *f* as in G. *fionnar*, cool, for *induar*.

A variant of *ros*, wood, is *ras*, underwood, shrubbery. Near Evanton in Ross is *Cnoc Rais*, 'shrub-hill.' With prosthetic *f* this is *fras* in *Achadh na Fraschoille*, 'field of the shrubbery,' in Brae Lochaber, and Achafraskill in Latheron parish; compare *Fid Frosmuine*, 'the wood of Frosmuine' (LL 166 a 35), where *muine* is 'copse'; also Raskill in Cavan, explained as *Raschoill*, 'underwood, brushwood' (Joyce, iii.). Forres, in Gaelic *Farais* (open *a* in both syllables), in 1189/99 Forais (RM), is probably 'little shrubbery,' from *fo*, under, 'sub,' and *ras*.

Rubha, a point, calls for no special remark except that it has sometimes displaced an older term, *e.g.* the point at Invergordon was once a *ros*; it is now *an Rubha*, in full *Rubha Aonach Breacaidh*, 'point of Breacaidh market'; Tarbat Ness was of old Ardterbert, now *Rubha Thairbeirt*.

Aird, a cape, is seen in *Dubhaird*, Duart, 'black cape'; *Fionnaird*, Finnart, 'white cape,' 'holy cape'; *an Fharaird*, anglicized as Farout Head, 'projecting cape,' in Durness. Ardrossan is 'cape of little cape,' a pleonasm.

Fas in our place-names is practically the equivalent of W. *gwas*, an abode, dwelling; O.Ir. *foss*, rest, act of residence. It seems to occur in *Lethfoss*, a place-name in the Pictish Chronicle, meaning 'half station,' *i.e.* 'one-sided station.' A *fas* is a stance, a nice level spot such as a drover would choose as a night's quarters for his charge.

Fas na Clèithe, 'station of the hurdle,' is at the head of Loch Roe in Sutherland. *Fas na Coille*, Fasnakyle, 'station of the wood,' is in Glen Affric. *Fas a' Chapuill*, Fascaple, 'horse (or mare) station,' is in Kirkhill, Inverness. *Fas na Cloiche*, 'station of the stone,' is in Appin, Argyll. *Fas-an-darroch*, 'oakwood station,' is near Dinnet, Aberdeen.

Terminally it appears in Darfash, probably for *doire fais*, 'oak copse of the station,' in Gamrie, and in Tarnash, for *tarr an fhais*, 'protuberance of the station,' in Strath Isla,

Banffshire. Shennas, for *seanfhas*, 'old station,' is on the south border of Ballantrae parish, Ayrshire. It appears also in *Altais*, Altas, 'bluff station,' Sutherland; *Dallais*, Dallas, 'meadow station,' 'plateau station'; *Alanais*, Alness, 'Allan station'; *Dubhais*, Duffus, 'black station'; *Peit-gheollais*, Pityoulish in Badenoch, 'portion of bright station' (O.Ir. *geldae*, bright); compare *Dail-gheollaidh* near Comrie; *Rathais*, Rothes, 'fortunate or lucky station' (*rath*, good fortune, grace); *Cluanais*, Clunes in Ardclach, 'meadow station'; *Fliònais*, Fleenas in Ardclach, the meaning of which is obscure to me.

In all these cases *ais* has open *a*: open *a* in this position (unstressed) arises either from an *ā* (or *ō* which has turned to *ā*) as in *clachán* for *clachān*, or as representing *a* (or sometimes *o*) of a word originally capable of bearing the stress, *e.g.* *conghlás*, a dog muzzle; *Dubhghláis*, 'black stream'; *Coillerás*, 'wood point' (*ros*).

In O.Ir. *fossad* is an adjective meaning 'firm, stable,' and a noun used in such phrases as *fossad a mullaig*, 'the very top of his head,' *fossad a étain*, 'the very front of his face.' Later *fosadh* means 'an abiding place,' as in *sa fosadh siothbhuan sin*, 'in that everlasting abode.'[1] The corresponding Welsh term is *gwastad*, meaning as an adjective 'level, smooth, steady,' and as a noun 'a level, a plain.'

With us it is *fasadh*, applied in place-names in much the same way as *fas*, with which it is connected. In Sutherland there are *Fasadh an t-Sean Chlaidh*, 'the stance of the old graveyard,' at Loch Roe, and *Cromsac* for *Cromfhasadh*, 'bent station,' in Armadale. In Ross, *am Fasadh Àluinn*, 'the lovely station,' is the site of Duncraig Castle, Loch Alsh; *am Fasadh* on Loch Maree is the site of old ironworks.[2] In Inverness-shire, *Tigh an Fhasaidh*, Teanassie near Beauly, is 'house of the station'; *Dabhach an Fhasaidh*, Dochanassie on Loch Lochy, is 'davoch of the station'; *am Fasadh Fearna*, Fassiefern on Loch Eil, is 'the alderwood

[1] John Carswell's *Liturgy*, p. 30.

[2] For other examples from Ross see *Place-Names of Ross and Cromarty*.

station'; *am Fasadh* occurs at the foot of the Wester Bunloit burn, Glen Urquhart, also near Foxhole School, Kiltarlity, and at the parish march west of Bridge of Oich. *Fasadh an Fhithich*, 'the raven's stance,' and *Fasadh nam Feannag*, 'the hoodiecrows' stance,' as also other instances, are in Glen Moriston. *Allt an Fhasaidh* with *Dail an Fhasaidh* is on the south side of the Strontian River. *Fasadh Bradáig*, 'the thievish woman's stance,' seems to be in Lochaber. In Moray there is Marcassie, 'horse station,' like *Gleann Marcfhasaidh*, Glen Marxie in Ross, Corinacy, for *coire an fhasaidh*, is in the Cabrach. Bogfossie, 'bog of the station,' in Aberdeenshire, retains the old vowel. Foss in Strath Tummel is in Gaelic *Fas*, final *-adh* being dropped as usual in this region; Braes of Foss, however, is *Bràigh Fasaidh*. Barassie, for *barr (an) fhasaidh*, 'summit of the station,' is near Troon in Ayrshire, while in Kirkcolm parish, Wigtown, there is Knocknassy for *cnoc an fhasaidh*. Fassis near Campsie is probably an English plural of *fasadh*. The term is rare in the Isles, but there are *Rubha an Fhasaidh*, 'point of the station,' in Eigg, and *na Glasfhasadhnan*, Glassans at Port Charlotte in Islay, 'the green stances.'[1]

Magh, a plain, gen. *muighe*, *moighe*, dat. *muigh*, *moigh*, is in E.Celtic *magos*, neut., stem *mages-*; W. *ma*, place, now used only terminally. *Magh* is now feminine in Irish and in several of our place-names, though our dictionaries usually make it masculine.

For *Magh Comair* and *Magh Bhard* see pp. 241, 243. Magask in Fife is 'plain of the *gasg*,' or 'of the *gasgs*'; *gasg*, whence Gask, is a tail-like point of land running out from a plateau: near Magask are the 'Tongues of Clatto.' In 1585 (RMS) it is Magas, whence Magus Moor. 'The Maw' is near Wemyss, Fife; Maw-carse in Kinross is 'plain of the carse.'

A' Mhuigh, Muie in Sutherland, is dative, 'at plain.' Moy near Strathpeffer is *a' Mhuaigh*; Moy near Tomatin

---

[1] The last two instances and several of the others I owe to the Rev. C. M. Robertson.

is *a' Mhoigh*. *Muigh Bhlàraidh*, Muieblairie in Edderton, is 'at dappled plain.'

Druim Muighe, Drummuie near Dornoch, is 'ridge of the plain.' Moy Bridge on Conon is *Drochaid a' Mhuaigh*, a curious form. Loch Moy is *Loch na Moighe*, and Mackintosh is *Tighearna na Moighe*, 'lord of Moy.' Cambus O'May, in Aberdeenshire, is for *Camas a' Mhuighe*, 'bight of the plain,' and Rothiemay is 'rath of the plain.'

Moyness near Nairn is in Gaelic *Muighnis* for *Muigh-inis*, 'plain-meadow,' like *Maginis* in Co. Down (Hogan). Here the stress is on the first part, which is specific.

In many instances *magh* forms the second part of a true compound and is unstressed, being the generic part. In this position *gh* often becomes *ch*. With compounds of this class we may compare Gaulish *Durno-magos*, 'fist plain,' 'pebble plain'; *Gabro-magos*, 'steed plain,' *Marco-magos*, 'horse plain.' The first or specific part, which is stressed, may be an adjective or a noun.

With *mor*, the compositional form of *muir*, the sea, *magh* gives *mormhoich*, 'sea-plain,' a dative-locative; W. *morfa*. This is a common name round the coasts. Lovat near Beauly is now *a' Mhormhoich* in Gaelic; Lord Lovat is *MacShimidh mór na Mormhoich*. *A' Mhormhoich Mhór* is near Tain; there are also Morvich in Kintail; Murroch, in 1609 Murvaich (RMS), near Dumbarton; Morroch in Wigtownshire near Port Patrick, and others. It occurs inland in Badenoch, applied to an upland moor; similarly W. *Morfa* occurs inland in Carmarthenshire. *Taranaich* in Moray is anglicized as Darnaway, in 1368 Ternway (RMS), also Terneway (RM). Lachlan Shaw says 'probably from *taran* or *tarnach*, thunder, because there Jupiter Taranis might have been anciently worshipped.' Apart from the speculation about Taranis, this is probably correct in substance. 'Thunder' is in Gaelic *torunn*, W. and Cornish *taran*; Adamnan gives *Tarainus*, i.e. *Tarain*, as the name of a Pictish noble; it was also the name of a Pictish king, and is to be equated with *Taranis*, the Gaulish thunder god. *Taranaich* represents an early British *Taranu-magos*, 'thunder plain.'

Rovie in Sutherland is for *ro-mhaigh*, 'excellent plain.' *Roinnmhigh* on Naver is 'point plain.' *Multamhaigh*, Multovy, near Alness, is 'wedder plain,' and *Mucamhaigh*, Muckovie, near Inverness, is 'swine plain'; the vowel after *mult*, *muc*, is the parasitic vowel, for the old stem endings in **Molto-magos*, **Mucco-magos* disappeared early. An old parish near Beauly was called Fernua or Fernway, which is identical with the M.Ir. *Fernmag*, for an early *Verno-magos*, 'alder plain.' Near Rothes and surrounded on three sides by a bend of Spey is Aikenway, in 1229, 1224/42 Agynway (RM). On the fourth side Ben Aigan or Ben Eagen rises very steeply; with this we may compare a hill in Ross called *Éiginn*, 'distress,' 'the hill difficulty.' Aikenway is 'plain of the hill Aigan,' 'difficulty plain'; at one time it was anglicized into Oakenwall. Fossoway in Kinross is Fossedmege, *c*. 1210 (Johnston), for *fossadmag*, later *fosadhmhagh*, 'firm plain'; compare E.Ir. *fossadlár*, 'a firm level,' as in *fosadhlár Fernmuighe*, 'the firm plain of Fernmoy,' corresponding to W. *gwastad-lawr*. The Welsh equivalent of *fossadmag* is *gwastadfa*, 'a level place.' Kininvie in Banffshire is for *cinn fhinnmhuighe*, 'at head of white plain.'

On the moor near the river Add in Glassary, Argyll, there is *Cnoc Albha*, Knockalava, in 1549, Knockalloway (Orig. Paroch.), 1541, Knopalway, *ib.*; near it is Dùn Alva on a rocky eminence. Here *Albha* is probably for *allmhagh*, 'rock-plain,' 'crag-plain,' an old compound not declined after *cnoc* and *dùn*. With it probably go the other places called Alva, Alloa, Alloway. Alloa on Forth is Alveth 1357 (RMS), Alwey, Aluethe, Alloway, in notes of charters (RMS, i. App. 2). North of it is Alva, in 1489 Alweth, 1508 Alloway, Alweth 1532, etc., 'torrens de Alveth,' 1597 (RMS). Alloway in Ayrshire is Auleway, 1324 (RMS). Alvah in Banffshire is Alveth 1308, 1315, Alueth 1329/32 (Ant. A. & B.), Strathalvecht (RMS, i. App. 2), Strathalva 1425 (RMS). With the spellings Alveth, etc., in these names we may compare 'terra que dicitur Kenmuckeveth in Kennochyn schyir,' *i.e.* Kennoway parish in Fife (RPSA), where 'Kenmuckeveth' is certainly for *Cenn Mucamhaigh*, 'head of swine plain.' There is also Alvie in Badenoch, *c.* 1350 Alveth (RM), 1621 Alloway, now in Gaelic *Allmhaigh*, 'at rock plain.'

Sruth, a stream, current, is often applied to sea currents, for example *Sruth na Maoile*, the current off the Mull of Kintyre ; *Sruth nam Fear Gorm*, 'the current of the Blue Men,' between Lewis and the Shiant Isles.[1] From its diminutive form are named *an Sruthán*, Struan, in Skye, and probably Struan, near Edzell, which is at the head of a burnlet. Struan in Athol is different (p. 350). A derivative of *sruth* is *sruthach*, current-place, as in Struthach ford on Deveron, corresponding in meaning to the Fords of Frew. The old locative case of *sruthach* is *sruthaigh*, whence Strowie, now Struie, at the junction of Water of May and Chapel Burn in Forgandenny parish, Perthshire. In Inverness-shire there is Struie at the junction of Farrar and Glass, and there is also *an t-Strùigh* (fem.) in Edderton, Ross, where there are streams but no confluence. *Cnoc na Strùigh*, a bold granite hill, is near it, and there is another *Cnoc na Strùigh* at the Ord of Caithness. Another derivative is *sruthair*, common in Ireland, whence Struthers and Anstruther in Fife ; Struther near Stonehouse, and Strutherhead, with Burnhead close by, in Lanarkshire ; Bellstruther Bog and Weststruther in Berwickshire. The streams are tiny in all cases. A variant of *sruthair* is *sruthail*, whence the Struthill Well of Muthil, reputed to cure cases of madness.[2] Hence also Loch Trool for *Loch an t-Sruthail.*

The ordinary terms for a well are *tobar*, *tiobar*, *fuarán*, the last being usually applied to a natural spring. *Tobar* is the common term north of Spey ; it occurs also in Banff, Aberdeen, north and west Perthshire, and Argyll. *Tiobar* is very rare north of Spey, if it occurs at all there ; south of Forth and Clyde it is the regular term.

It is unnecessary to give detailed instances of *tobar*, and I shall mention only two instances. A charter of Alexander confirming granting the lands of Burgyn, now Burgie in

---

[1] J. G. Campbell, *Superstitions of the Scottish Highlands.*
[2] Macfarlane's *Geog. Coll.*, iii. p. 91.

Moray, to the monks of Kinloss and dated 1221, mentions as part of the marches 'ad Rune Pictorum et inde usque ad Tubernacrumkel et inde per sicum usque ad Tubernafein,' 'to the Picts' field,' etc. A note on a piece of parchment stitched on to the charter has 'Tubernacrumkel, ane well with ane thrawin mowth or ane cassin well or ane crook in it. Tubernafeyne of the grett or kemppis men callit ffenis is ane well.' The former appears to be for M.Ir. *tobar an crombeóil*, 'well of the bent opening'; the latter is for *tobar na Féine*, 'well of the warrior band' (*fian*), associated doubtless with some legend of the Fiana of Fionn mac Cumhaill.

Tiobar is in E.Ir. *tipra*, gen. *tiprat*, dat. *tiprait*, and is anglicized as Tipper, Tibber, Tiber, Chipper; the genitive becomes -tibbert -tibert. Tipperty is either for dat. pl. *tipratib*, later *tiobartaibh*, or a locative of *tiobartach*, 'well-place.' Tibbermore or -muir near Perth is supposed to mean 'big well,' from a fine spring near the churchyard, but as this used to be called 'the lady well,' the meaning may be 'Mary's well,' *tiobar Moire*, like *Tobar Mhoire*, Tobermory in Mull. Other examples are Ach-tiobairt in Rannoch, Achintiobairt in Lorne, Auchentibber near Blantyre, and in Ayrshire, Auchingibbert in Cumnock, Auchintibbert in Stoneykirk, Chippermore in Mochrum, Tibbers Castle and Auchingibbert in Dumfriesshire, the Tippert Well near Penicuik.

An old term for 'well' is *fobhar*, *fabhar*, whence in Ireland St. Féchín of Fobhar or Fore. We have the diminutive in Foveran in Aberdeenshire, of which an old writer says 'very congruously did the lairds of Foveran assume its name from a fountain adjoining the castle. For Foveran, both in the British and Irish languages, signifies a fountain, or well spring; whereof there is one here of a most pure and delicate water, which delights the beholder by playing, through a hundred several pipes, as it were, a spring or tune to the dancing atoms of earth.'[1]

[1] *Ant. of Aberdeen and Banff*, vol. i. p. 367. When he says that the word is found in British, he is perhaps thinking of W. *gofer*, a rill.

Teamhair, gen. *Teamhra*, later *Teamhrach*, meaning 'an eminence of wide prospect standing by itself,' is the Irish term anglicized as Tara; there were many places of that name in Ireland besides the famous *Teamhair Breagh* in Meath. We have *Dail Teamhair* far up Glen Casley in Sutherland. Here the genitive singular is not inflected now, but that it was inflected once appears from the record 'Dailteawrache et Polteawrache in Glen Casley,' 'dale of the *Tara* and pool of the *Tara*,' 1621 (RMS). At this part of the glen there are many conical hillocks, some on the flat, but most of them on the hill side. The most conspicuous is one beside *Allt Dail Teamhair*, the burn which flows into the long pool now called *Poll Teamhair*. *Druim Teamhra*, on the Ordnance Survey Map Druim Teamhair, north of Loch Gorm in Islay, has retained the older form of the genitive.

The old term *sét*, a path, road, later *séad*, is rare in place-names, but is found in composition with *tar*, 'across, over,' (in composition *tairm, tarm*), in *Meall an Tarsáid*, 'hill of the path across,' south of Whitebridge, Stratherrick. In Aberdeenshire there are Tarset Hill in Slains and Tersets, an English plural, in Drumoak.

Dòirlinn is an isthmus, usually covered at high water and connecting an *eilean tioram*, 'dry island' (an island accessible at low tide) with the mainland or another isle. The Norse equivalent of *eilean tioram* is *örfiris-ey*, 'ebb-isle,' *i.e.* an island which at low water is joined to the mainland by a reef which is covered at high water, appearing in the west as Orasay, Oransay. Examples of a *dòirlinn* occur at Castle Tioram in Moidart, at Oransay in Loch Sunart, at the north end of Gigha, and at *Eilean Traoghaidh*, 'ebb isle,' in West Loch Tarbet, Kintyre, and elsewhere.

Tairbeart (for *tairm-bert*, 'an over-bringing') also means 'isthmus,' but differs from *dòirlinn* in being always above water. This is well seen in Gigha, where the *tairbeart* and the *dòirlinn* are close together. On the east there are Tarbat in Ross and Tarbert in Fidra Isle in the Firth of Forth, the latter being probably the tiniest extant. On the west the term is fairly common, *e.g.* Tarbert in Harris and

Tarbert at the north end of Kintyre, across which, in 1098, King Magnus caused his ship to be dragged with himself on the poop holding the tiller.

Corrán, a low cape tapering symmetrically to a point, is from *corr*, 'taper, peaked, rounded to a point,' an adjective applied to a large variety of things—arrows, spears, blades of grass, eyes, fingers, goblets, etc. Of the many promontories on the west which are so named, a typical example is the Corran of Ardgour which divides *an Linne Dhubh* from *an Linne Sheileach*.

Machair is a plain ; *a' Mhachair Ghallda* is the Lowlands generally ; *Machair Rois* is the level expanse of Easter Ross ; part of the east coast of Sutherland is *Machair Chat*. 'I was in the low country' is 'bha mi air a' mhachair' ; 'eadar machair agus monadh' is 'between the plain and the hill ground.' Most of the isles contain one or more 'machairs' ; a large farm in Colonsay is called *na Machraichean*, Machries, 'the plains.' Machrihanish in Kintyre is for *Machair Shanais*, 'plain of Sanas' ; near it is Loch Sanish, on Blaeu's map 'Loch Sannaish.' An alternative name for it—perhaps the earlier name—was *Magh Sanais* : an elegy on 'Niall Óg Mhachra Shanuis' contains the lines 'sgaoil a fhréimh ó chian fa Shanuis,' 'his root (ancestors) spread from of old about Sanas,' and 'chaochail Magh Sanuis gu mór,' 'the plain of Sanas has greatly changed.'[1] There was another Magh Sanais in Connacht. *Sanas*, mas. or fem., ordinarily means 'a whisper, a hint, a secret, a warning,' and it may mean so here, but in some instances it seems to denote some kind of plant : *gass sanais* is 'a stalk of *sanas*' ;[2] in the tale of the Battle of Ventry it is said 'we would form a druidical host around thee of the stalklets of *sanas*' (*do na geosadánaibh sanaisi*). It is difficult to say whether the meaning here is 'whispering stalks' or 'stalks of a plant called *sanas*,' and I find nothing really decisive : *Loch Sanais* may mean

[1] *Rel. Celt.*, ii. p. 408.

[2] *Revue Celtique*, xxvi. p. 11 ; LL 186 a ; 'raithnech ⁊ clocha ⁊ bolca belcheo ⁊ gasána sanais,' 'fern and stones and puffballs (?) and sprigs of *sanas*.'—Lec. 316 b. 27.

'whispering loch,' and the plain may have been named after the loch.

The district of Cushnie in Aberdeenshire, now part of Leochel-Cushnie parish, has long been noted for cold and frost; its hills were declared by Lochaber raiders of experience to be the coldest hills in Scotland.[1] The name doubtless comes from *cuisne*, ice, frost. Another Cushnie occurs in Gamrie parish, Banffshire, a little inland from Troup Head. The old name of the latter is probably found in *Ross Cuissine*, 'Cushnie Point,' off which thrice fifty ships of the Picts (*Picardaich*) were wrecked in A.D. 729 (Tigern.). Gamrie itself (the first syllable is like Eng. 'game') may be genitive of *geamhradh*, winter, with omission of a preceding term. By the sea, between Gamrie church and Troup, is Lethnot, for *Lethnocht*, 'naked-sided place,' with reference to its exposed situation; compare *Lethnocht* in Ireland (Hogan).

Duntroon, near Crinan in Argyll, is most probably the place meant in the Lay of Deirdre, in which that heroine makes mention of her jealousy of 'inghean iarla Dúna Treoin,' 'the daughter of the earl of Dún Treoin';[2] another version has *Dún Treoir*,[3] but the extant form is in favour of the former. *Treoin* may be the old genitive of *trén*, 'mighty,' used as a personal name: 'Trén's fort.' There is also Duntrune 1540 (RMS) in Forfarshire. We may compare *Ross meic Treoin* (p. 277 n. 3).

Loch Striven in Cowal is in Gaelic *Loch Sroighean(n)*, in 1525 Lochstrewin (RMS); the point at its mouth, between it and the Kyles of Bute, is *Rubha Sroighean(n)*, anglicized as 'Strone Point.' Spelling with *bh* for *gh*, we might compare the proper name *Srob-cenn*, in Keating *Sraibhgheann*,

---

[1] Cushnie was reckoned the coldest place, next to Cabrach, in the diocese of Aberdeen.—*Ant. of Aberdeen and Banff*, i. p. 593. The adjective from *cuisne* is *cuisnech*, frosty (LL 293 b 36); as a noun *cuisnech* would mean 'frosty place,' with locative *cuisnigh*.

[2] Stokes, *Irische Texte*, ii. II, p. 115, from an eighteenth century MS.

[3] *Reliquiae Celticae*, i. p. 119, from Adv. Lib. MS. xlviii., 'probably written by the middle of the eighteenth century' (Mackinnon, *Catalogue*, p. 98).

and *sraibtine*, *sroibthene*, later *sraibhthine*,[1] 'lightning,' but the name is obscure to me.

Dunoon is in Gaelic *Dùn Obhainn* and *Dùn Othainn*, in Wyntoun 'Dwnhovyn,' 'Dwnhowyn in till Cowale'; other early spellings are Dunhoven, Dunhovyn. The site of the old fort is an eminence beside a stream, and *Dùn Obhainn* is doubtless for an earlier *Dùn Obhann*, 'fort of the river.' The name recurs in Stirlingshire as Denovan, locally pronounced Dun-níven (Johnston), on the Carron, in 1510 Dunnovane (RMS), and possibly in Dunnone 1493, Dunnoyne 1494 (RMS), in Forfarshire.

Dunollie, near Oban, a fortified site of great antiquity, is mentioned five times in AU: A.D. 686—'combussit tula aman Dúin Ollaig,' a corrupt entry, meaning that somebody burned Dunollie; 698—'combustio Dúin Onlaigh, 'burning of Dunollie'; 701—'distructio Dúin Onlaigh apud Sealbach,' 'destruction of Dunollie by Selbach'; 714—'Dún Ollaigh construitur apud Selbacum,' 'Dunollie is built by Selbach'; 734—'Talorggan filius Drostani comprehensus alligatur iuxta arcem Ollaigh,' 'Talorgan son of Drostan is captured and bound near Dunollie.' The ancient tale of *Táin Bó Fráich*, 'the Driving of the Cattle of Fráech,' tells how Fráech, son of Fidach, a hero of Connacht, who was contemporary with Cuchulainn, had his cattle stolen. Some of them were taken to Cruthentuath in the north of Alba. Fráech followed, recovered his cattle, and he and Conall Cernach came from the east past *Dún Ollaich maic Briuin*, 'the fort of Ollach son of Brión,' across the sea from the east to Aird hua nEchtach in Ulster.[2] Hogan places this fort in Ulster, but it is certainly, in my opinion, Dunollie, in Scotland. As to the meaning, all that can be said is that the fort was named after a man

---

[1] O'Davoren explains *srabtine* as *srib tine*, 'a stream of fire'; *sraibtine do neim* is 'lightning from heaven.'—*Arch. f. Celt. Lexik.*, ii. p. 463.

[2] 'Co tulatar ass anair sech Dún Ollaich maic Briuin dar muir anair i nAird húa nEchtach.'—*Celt. Zeit.*, iv. p. 47; LL 252 a 48. *Briuin* is given as genitive of *Brian* in Rawl. B 502, 121 a 51, 124 b 5, 139 b 45. See Kuno Meyer in *Über die älteste Irische Dichtung*.

called Onlach or Ollach, son of Brión, of whom nothing is known.

On the coast of Gairloch and Lochbroom the old sea beach forms a series of shelving declivities called individually *am Faithir, am Faithear*, pl. *na Faithrichean*. This term recurs on Loch Ness side as *Foithear*, anglicized as Foyers (an English plural); the Laird of Foyers is *Fear Foithreach*. In the same district another form, *fothair*, occurs: *am Fothair Beag*, with an old cemetery, is behind Knockie; *Fothair Mac Clòthain* is below Dulcrag; there is also *Tollaigh an Fhothair*, 'Tollie of the *fothair*.' These places are all on terraced slopes above Loch Ness, and it can be safely taken that the term means 'a slope, a terraced declivity.'

It seems probable that it is with this term we have to do in the names that contain *fother* or *fodder*, *fether*, or *fetter*, for, so far as I can ascertain, the distinctive feature of the places so named is the presence of a slope, often or usually shelving or terraced.

Fetters in Leuchar parish, Fife, is 'Auldmuris and Wester Fotheris' in 1536, 'Wester Fethers and Auldmures,' 1588 (RMS). The feature of this neighbourhood is sand dunes which form smooth ridges running north and south, with long canal-like pools between, whence the name has probably arisen. The plural is English. Other instances on record are Fotheris 1547, 'Fotheris et Schanwall alias Tentismuris' 1575 (Schanwall is for *seanbhaile*, 'Oldtown)'; the Futheris 1577 (RMS).

Fetteresso is Fodresach (Pict. Chron.), Fethiresach (RPSA); the second part is *easach*, the adjective from *eas*, a waterfall or rapid, and I am informed that 'waterfall-slope' describes the place. The farm of Feathers 'slopes down to the Carron at the kirk of Fetteresso'; [1] Feders 1592, Fedderis 1609, Featheris 1647 (RMS). The name is the same as Fetters in Fife.

Fetternear, on the left bank of Don, is Fethirneir, 1157 (Reg. Aberd.), Fethirnere, 1163 (RM); *-near* is doubtful,

[1] Crabb Watt, *The Mearns of Old*, p. 343.

but may be for (*a*)*niar*, 'west'; compare Wester Fotheris above.

Fetterangus is Fethiranus, 1207, 1220 (Ant. A. & B.), ffether Angus (Macfarlane), and means 'Angus' *fetter*'; the old church is on a terraced slope, and Angus is the name of its patron.

Fettercairn is Fothercardine, Fettercardin, Fethircarne (RMS i); here the second part may be British *carden*, a copse.

Fothirletir, 1428 (RMS) is in Macfarlane Fetter Letter in Fyvie; it is the same as Fodderletter in Banffshire, given me as *Farrleitir* (Ardclach), and as *Foirleitir* (Cromdale) in Gaelic, in 1502 (RMS) Fothirlettir, and meaning probably 'terraced slope' (*leitir*).

Fetter-eggie in Glen Clova is 'the *fetter* of Eggie,' the name of a place further up the glen.

Forteviot appears twice in the Pictish Chronicle (PS, 8). Kenneth MacAlpin died 'in palacio Fothuir-tabaicht'; in the reign of his brother Donald certain laws were made 'in Fothiur-thabaicth.' Here, if the text can be relied on, *fothuir* is genitive sing., *fothiur* is dative sing. The second part seems to be the same as *Tobacht*, the name of a place in Ireland;[1] *Argain Rátha Tobachta*, 'the sacking of the Rath of Tobacht,' was the title of one of the Irish 'chief tales' (LL 190 a), and Hogan identifies the place with 'the moat of Ratoath in Meath.' The meaning of *tobacht*, *tabacht* is obscure to me. There is, as I am told, a terrace at Forteviot.

Dunottar appears twice in AU: A.D. 681, 'obsessio Dúin-foither'; 694 'obsessio Dúin-fother'; the Pictish Chronicle has 'opidum Fother occisum est a gentibus (PS, 9); St. Berchan's *Prophecy* has 'for brá Dúna-foiteir,' 'upon the brow of Dunottar';[2] in the same poem 'ar in luirc os Fother-dun,' means 'on the shank above Dunottar,' rather

---

[1] Failbe ó Thobucht, 'Failbe of Tobacht,' is given in Gorman's *Martyrology* at June 19.

[2] A. O. Anderson, *Early Sources*, vol. i., p. 454; Skene, *P. S.*, p. 93. The latter has *for bhrú Dúna Foitheir*, 'on the brink of D.'; *brá* is 'an eyelid, eyebrow.'

than 'above Fordoun,' [1] *Fotherdun* being a poetic inversion. In *Dun-fother* the second part is genitive pl. Whether the ancient stronghold occupied the site of the present castle, or was situated on what is now called the Black Hill, the meaning 'fort of the slopes' suits the physical features.[2]

Foderance in Kettins parish, Forfar, is now called Lintrose (Old Stat. Acc. xvii, 13); *c.* 1120 (Lib. Scon) it is Fotheros, a compound of *fother* and *ros*, 'a point, a wood,' meaning probably 'slope-wood' or 'slope-point.' Another Fotheros is described as 'near St. Andrews (Reg. Dunf. and RMS, 1450); also as adjoining the sea (Bain's Cal., ii). A third instance is *Fotharáis* in Badenoch, situated on a terrace, and anglicized rather strangely as Phoineas; here *ros* appears to mean 'a wood,' not 'a point.'

Fodderty in Ross is 'ecclesia de Fotherdino,' 1238, Fotherdyn, 1275, Fothirdy, 1350, Fetterty, 1521, Fedderty, 1561, in Gaelic *Fothraididh.*[3] In Gaelic *fother* has undergone syncope of the second syllable owing to the stress on the first syllable, *i.e.* it becomes *fothr-*; the ending is that seen in Tressady, Navity, Bar-muckety, etc. Along with it goes Fedderat in Aberdeenshire, in 1214 Fedreth; *Fedreth* stands to *Fodderty* in the same relation as *Treasaid* in Strath Tummel stands to *Treasaididh* in Sutherland.

In these instances the term occurs (1) simply, in Foyers, etc., Fetters, Feathers, Dunnottar; (2) with extensions, in Fodderty, Fedderat; (3) followed by a genitive case, in

---

[1] The following line further defines the site as *for brá tuinne,* 'upon the brow of the wave'; A. O. Anderson, *Early Sources*, vol. i., p. 398; here also Skene has *brú.* Further, Fordoun is *Fordun isin Mairne,* 'F. in the Mearns,' in LU 4 a 39. Fordoun is the G. form of British Gordon, Gordoun, 'great fort.'

[2] The Rev. J. B. Burnett of Fetteresso informs me as follows:—'The rock of Dunottar Castle is separated from the mainland by a steepish ravine towards which the ground slopes on the landward side. . . . The Black Hill is between the Castle and Stonehaven. Between this hill and the old town and harbour of Stonehaven there is a steep terraced slope. There are remains of fortifications on the Hill, and it is said to have been occupied by Cromwell's soldiers.'

[3] The *o* is close and short; the correct spelling is therefore *fothr-*, not *fodhr-*, for the latter would yield *o* long. The vernacular English is *Fotherty,* with *th* as in *bother.*

Fetterangus, -cairn, -eggie, Forteviot; (4) followed by an adjective, in Fetteresso and, probably, Fetternear; (5) prefixed adjectivally, in Fotheros (Foderance, etc.). Fodderletter. It may be compared tentatively with Ir. *fothair*, *fuithir*, meaning apparently 'a dell, hollow,' and further with W. *godir*, 'a slope, lowland,' from *go*, O.W. *guo*, 'under, *sub*,' G. *fo*, under, and *tír*, 'land.' [1]

That cupmarked stones were noted of old appears from *Clach Phollach* in Ross and Perth-shires. A remarkable cupmarked surface on the north side of Loch Tay is called *Cragan Toll*, 'the little rock of holes,' a prominent rocky knoll. Cochno in Dumbartonshire, formerly Cochynnach, Cochinach (Chart. Pasl.), is for *Cuachanach*, 'place of little cups,' with reference to cup and ring markings on certain stones there. Some of the streams called *Cuach*, Quaich, Queich, are probably named from pot-holes.

Some names have been given from resemblance to tools or instruments. *Beinn Eighe*, 'file peak,' in Gairloch, is named from its serrated outline. Crarae on Loch Fyne is in Gaelic *Carr Eighe*, 'file rock.' *Loch Gilb*, Lochgilp, is 'chisel loch.' Loch Skene is for *Loch Sgine*, 'knife loch.' *Inbhir Snàthaid*, Inversnaid, is 'confluence of the needle,' and *Allt na Snàthaid*, which forms the confluence, is 'burn of the needle.'

One of the marvels of Alba recorded in the Irish Nennius is 'a valley in Angus (Forfarshire), in which shouting is heard every Monday night; Glend Ailbe is its name, and

---

[1] Nogo rángadar gusan fothair fírdomain inar cuired an Maol da mórcoiméd, 'till they came to the very deep dell in which the hummel cow (of Flidais) had been put for safe keeping'; *Celt. Rev.*, vol. iv., p. 108. Cáilte tells how he raced 'tar gach fothair is tar gach fán,' 'over every dell and over every slope'; *Duanaire Finn*, p. 21. Suibne Geilt leaped 'ó gach fuithir 7 ó gach fáinghlenn di aroile,' 'from each dell and from each sloping glen to the other'; *Buile Suibne* (Ir. Text Soc.), p. 62. Here 'dell' seems to be the meaning intended; but O'Clery has *fuithir* . i . *fearann*, '*fuithir* means land'; the Lecan Glossary has *fothair* . i . *gort*, '*fothair* means a field.' Compare also 'Síl mBruidge etir Taulchind 7 abaind 7 Fother,' 'Sil mBruidge are between Tulchenn and the river and Fother'; *Rawl. B.*, 502, 124, b 25; here other MSS. give *Foithir* and *i Foithiur* (Hogan).

it is not known who makes the noise.'[1] Glend Ailbe, or in modern spelling Gleann Ailbhe, is difficult to identify. It may be Glen Isla, in 1233 Glenylif, 1234 Glenylefe, 1248 Glenylef (Chart. Coupar), Glenylef (Reg. Arbr.), and this again looks like Hilef mentioned in a twelfth century description of Scotland as bounding a province extending from Tay to Hilef.[2] But *Srath Ìl'*, the present form of Strath Isla in Gaelic, cannot represent *Ailbhe*.

---

[1] Ata dno glenn i nAengus 7 eigim cacha h-aidchi luain and, 7 Glend Ailbe a ainm, 7 ni feas cia dogni fuith.—*Ir. Nennius*, p. 118.

[2] See Skene, *Celtic Scotland*, i. 340; *P.S.*, p. 136.

ADDITIONAL NOTES

P. 2, *n*. For objects of the Hallstatt period see *The Early Iron Age Site at All Canning's Cross Farm, Wiltshire*, by M. E. Cunnington; Devizes, 1923.

P. 18. Smertae: Smirgoll mac Smertha rí Fomore, 'Smirgoll son of S., king of the Fomorians,' LL 18 a 19; Smirgoll mac Smertho, grandson of Tigernmas, LL 17 b 24; Tigernmas had two sons, Smretho and Smirgoll, Rawl. B 502, 137 b 21.

P. 25. Cuchulainn's peculiar position as a *deóradh*, 'advena,' outsider in the tribe, and at the same time an *urradh* or person holding land within the tribe, forms the subject of a short legal discussion given in *Eriu*, i. p. 126.

P. 53, *n*. 4. 'At Foirthe' probably refers to Aedan's fort at Aberfoyle on Forth.

P. 55. With Nith compare also Litgarth, p. 384.

P. 59. May Water appears in Inuermeth, Inuirmed (RPSA), Innermeith, Innermeth, etc. (RMS and Ex. Rolls); ? M.Ir. *méde, méide*, now *méidhe*, a neck, trunk, stump; compare *Méde ind Eoin*, 'neck of the bird,' and *Méide in Togmaill*, 'neck of the marten (? squirrel),' otherwise *Méthe n-Eoin*, etc., apparently names of fords in Meath; Windisch, *Táin Bó Cúalnge*, p. 181.

P. 64. Fir Alba: a poem in LL 18 a says that Oengus Olmuccaid broke fifty battles on the men of Alba. The prose account which precedes has 'he broke fifty battles on Cruithentuath and the Fir Bolg,' thus making these equivalent to Fir Alba.

P. 78. Raith, bracken; also *rath*: 'ro-ruad rath . i . raithnech,' 'very red is the fern, *i.e.* the bracken,' LU 11 b 25; Rawl. B 502, 103 a 15.

P. 79. Kentra Bay: 'the bay known as Cul na Croise [nook of the cross] . . . is ideal for beaching a long ship or landing from ships' boats, and is certainly the best site for this purpose from Ardnamurchan Point to Kyle Akin.'—(*Battle Site in Gorten Bay, Kentra, Ardnamurchan*; Proc. of Soc. of Antiquaries of Scotland, 1925, p. 105.)

An alternative to Kentra Bay is Camas nan Geall, a lovely little place on the south side of Ardnamurchan, where there are a monolith with cross, an ancient cemetery named *Cladh Chiaráin*, and St. Columba's Well. The name may be for *Camas na gCeall*, 'bay of the churches.' The place is described in Miss M. E. M. Donaldson's *Wanderings in the Western Highlands and Islands*. But Buarblaig near it cannot, as has been supposed, represent Muirbolc; it is Bordblege in 1610, Borbledge in 1667 (RMS); later spellings are similar (RMS and Ret.).

P. 81. Muirbolc Mār: with Adamnan, *Muirbulc Mar* is dative sg. after the preposition *in*.

P. 104. Dalmeny may be 'my Eithne's meadow'; compare Dalmahoy, Kilmeny.

P. 106. Callander on Teith is in Gaelic *Calasráid*, sometimes shortened into *Caltráid*, for *caladh-sráid*, 'shore-street, ferry-street'; the adjective *caladh*, hard, came to mean also 'firm shore, beach,' modern 'harbour'; compare 'the Hard' of Portsmouth.

P. 110. The Howe of the Mearns is known in G. as *Lag na Maoirne*.

P. 115. Cé: In the Isle of Taransay there are *Cladh Ché* and *Teampull Ché* (Carmichael, *Carmina Gadelica*, ii. 82). Here the name is that of a saint, whom Martin calls St. Keith. No woman must be buried in St. Keith's, and no man in St. Taran's (p. 300 above). Dr. Carmichael states that 'in North Uist there is a tall obelisk called *Clach Chē*—the stone of Ce.'

P. 131. Cairell:

Ardrí Alpan, art Arand,
bág barand, bruithi bairend,
tuc aníar dar srían rudach
trí chóicait curach Cairell.

'Cairell, Alba's high-king, bear of Arran, wrath of battles, who crushed rocks, brought from the west (*i.e.* from Ireland to Alba) across the roaring sea thrice fifty curachs' (reading *rian*, sea, for *srian*); LL 182 b 4.

P. 136. Coitchionn: add 'Cuthkin Eklismagirgill,' 'the common of Exmagirdle,' 1600 (Ant. A. and B., iv. 505).

P. 143. Corstorphin: compare Inchethurfin *c.* 1150 (Reg. Dunf.), near Coupar.

P. 148. Carse-regale: ? Corse, in which case compare Crossraguel.

P. 158. Rind Snóc: ? compare 'totam terram quam habui in Snoco de Berwic'; 'toftum cum crofto super Snocum de Berwic' (Lib. Melr., p. 166). Bellenden translates Boece's *angulus muli* by 'the Mulis Nuk,' *i.e.* the Mull of Galloway.

P. 164. Kilblane: 'ecclesia sancti Foylani de maiore Sowerby,' *c.* 1200; 'Sowrby et Kirkfolan,' 1282 (Lib. Dryb.).

P. 173, *n.* 2. Ailsa Craig is also often in Gaelic *Creag Ealasaid* and *Ealasaid a' Chuain*, 'E. of the sea,' or, in the old sense of *cuan*, 'of the inlet'; here *Ealasaid* is by analogy of *Ealasaid*, Elizabeth. In the Irish tale of *Buile Suibne* (Irish Text Soc.) it is called *Carraig Alastair*, 'the abode of the sea-gulls, chilly for its guests . . . sufficient is the sheer height of its side, nose to the rushing main' (p. 96).

P. 183. Urchair: *Beinn na h-Urchrach* in Ardnamurchan, 'peak of the cast,' with reference to a cast at a certain Norseman (Miss Donaldson's *Wanderings*). *Rae Urchair*, 'field of the cast,' in Ireland, is explained thus: two brothers, engaged in clearing two adjacent plains, had one billhook and one shovel which they used in turn, and each of them would fling the implements in turn to the other into *Rae Urchair*; Stokes, *Bodleian Dinnshenchas*, No. 16.

P. 190. Crossraguel, pron. Cross-ráygel (Johnston); Crosragmol, 1225, 1236, 1265, 1269 (Chart. Pasl.); Corsragmol, 1244, *ib.*; Crossragwell, 1275 (Boiamund); Crosragal, 1286 (Chart. Pasl.); Crossraguel, 1306 (Bain's Cal.); Crosragmer, 1323, Corsragmere, 1373 (RMS); Corsragmer, 1329 (Ex. Rolls); Corsragwell, 1534, etc. (RMS); Crucisregalis (gen.), 1571, 1573, etc. (RMS). There was another place of the same name in Glassford, Lanarkshire: Schelis of Courchraguell, 1539 (RMS), Scheillis of Croceragwell, 1619, Scheillis of Corsraguell, 1663 (Ret.), known later as Hiecorseknowe (*Charters of the Abbey of Crossraguel*, p. lxvii.). Riagal of Mucinis is in Oengus's *Félire* at Oct. 16; in LL 354 d he is said to have been a son of (or descended from) one of the seven daughters of Dalbrónach, one of whom was the mother of Brigit. His name, which in AU (A.D. 748) is Reguil (gen.), is from Lat. Rēgulus. The forms -ragmol in Chart. Pasl. may be explained as misreadings of the original chartulary, which is lost; the forms -ragmer are difficult to me. With -ragal, 1286, we may compare Raggal (p. 147), from *riaghal*, Lat. *regula*, a rule.

P. 191. Troon: 'terra de le Trone,' 1371; 'le Trune' 1464 (RMS); 'the Truyn' (Blaeu). In Arran Gaelic Troon is *an Truthail*, which is most likely for *an t-Sruthail*, 'the current,' either an independent form, with reference to the current at the point, or a Gaelic rendering of Troon by folk-etymology. *Sruthail*, a variant of *sruthair*, is given as fem. by O'Reilly in the sense of rinsing, streaming; in Loch Trool it appears to be masc. Troon is now 'the Trin' in the vernacular.

P. 195. Lintheamine: the *Life* says that a man whose eye had dropped out through a violation of Cadoc's monastery went about the province of Lintheamine making a display of his mutilation for money. As he would be more likely to do this in a neighbouring district than in the neighbourhood of the monastery, I conjecture that Lintheamine may be for Lleuddiniawn, the Welsh form of Lothian.

P. 199. Cassillis, Cashley: add Glen Casley and the river Casley in Sutherland, in Gaelic *Gleann Carsla, Abhainn Charsla*. T. Pont wrote 'the water of Casla,' 'Chassil river,' 'Innerchassil,' 'Glen Chassil' (Macfarlane, *Geog. Coll.*, ii. 546, 569, 570). There are remains of brochs at Achness and further up at Croick. For the phonetics compare Kirkcarswell for Kirkoswald.

P. 211. Ythan: Bede's Ythancaestir in Essex is identified with Othona, a station under the Count of the Saxon shore; and is a different name from our Ythan.

P. 220. Balmackewan: Inshewan, 'Eoghan's meadow,' occurs in Forfarshire.

P. 226. Earn: for Professor Julius Pokorny's views see *Celt. Zeit.*, vol. xv. 197 *seq.*

P. 238. Rothiemurchus, in G. *Ràta-mhurchuis*, may be 'rath of Muirgus,' 'Sea-choice' (Macbain, *Place-Names*, p. 158), or 'of Muirgéis,' 'Sea-swan.' For the form *Ràt*, see Macbain, *ib.*

P. 246. Nemthor may be a *poetic* name for Dumbarton.

P. 248. Kyltirie, G. *Càil-tiridh*, means apparently 'nook of kiln-drying.'

P. 250. Balmadity, Barmuckety: the ending is formally similar to the O. and M.Ir. abstract ending seen *e.g.* in *tirmatu*, drought, gen. *tirmatad*, dat. *tirmataid* (in modern spelling *tirmadadh*, *tirmadaidh*). It seems fairly certain that Flichity in Strath Nairn, G. *Flicheadaidh* is dat. of an old *flichetu*, wetness; similarly Mussadie in Stratherrick, G. *Musadaidh*, is from the base of *musach*, nasty, W. *mws*, stale, rank, stinking. The extension to names like Balmadity, etc., might perhaps be explained by the common tendency of Gaelic to use abstract nouns concretely, *e.g. flaith*, fem., lordship, sovereignty, came to mean also 'lord,' still remaining fem. Thus *matud*, a dog, would give *matadtu*, *madattu*, 'currishness'; *mucc*, a swine, would give *muccatu*, 'swinishness' (*muccnatu* occurs in literature), with secondary meaning 'place of dogs,' 'place of swine.' The transition would be helped by analogy of names like Tressat, Mossat, Tarvit, Turket, etc. But looseness of usage consequent on the change from British to Gaelic may have to be taken into account. Other names similarly formed are Auchtermuchty, Vchtermukethin, 1204/14 (Laing Chart.), Ochtirmokadi, 1350 (Chart. Lind.); Ochtermegatie, in Forfarshire, 1593 (RMS); Cas-conity, in Fife, from *con-*, the compositional form of *cú*, a hound; Tressady in Sutherland; Delgaty near Turriff, and Dalgety in Fife, from *delc*, *dealg*, a prickle.

P. 254. *Clach na h-Iobairt*: the Old Stat. Account (ii. 473) mentions just about this spot 'Clagh-ghil-Aindreas, or the cemetery of Andrew's disciple,' southward of which the Andermass market was held.

P. 261. Severie: I visited the Seat recently; only a fragment of it remains, situated on a knoll with a magnificent view. It was the seat of the Judge of Menteith.

P. 265. Dewar na Mais: probably for *deòradh na mèise*, 'Dewar of the portable altar,' carried about by an attendant for the celebration of mass.

Deòraidh is in the vernacular *deòr*, *e.g.* Baile an Deoir, 'the Dewar's stead,' 'Dewarton.'

P. 271. Aibind: *Dùn Éibheann* in Colonsay is not connected.

Pp. 273, 308. I am informed that *Cill Da-Bhi*, Taymouth, was not within the Castle grounds but close by, near Stix (*na Stuiceannan*). The Rev. W. A. Gillies, Kenmore, tells me that he can find no trace of the site now. It may be added that Inchadnie is in Gaelic *Innis Chailtnidh*, the *l* being quite clearly sounded by the older people. The *innis* is the peninsula formed by a bend of Tay just east of the Castle, and the old church of Inchadnie was at the apex of the bend, on the north side of the river. *Innis Chailtnidh*, 'Keltney Haugh' must have been connected in some way with Keltneyburn a mile and a half distant. The name, therefore, is not from *Aedán*, which would give *Innis Aodháin* in modern Gaelic.

P. 275. St. Bride's Camdale [*sic*]; I do not know the place.

P. 276. In Ireland there were Gleann Cainneire and the river Cainner, so called, it is said, from Cainner, daughter of Ailill and Meadhbh of Connacht, who was slain there (*Celt. Rev.*, iii. 122). Not mentioned by Hogan.

P. 279. Colman from Ard Bó: in Rawl. B 502, 141 a 45, he is styled Colman Muccaid, 'Colman the swineherd.' To judge from his genealogy he must have been a contemporary of Columba. Maedoc of Ferns, who is given in the genealogies as ninth (LL 347 h; BB 217 a; Lbr. 14 c) or tenth (Rawl. B. 502, 90 d) from Colla Uais, died in 625 (AU).

P. 284. Lethen: Lenedycothe 1238 (RM), for Leathan Dubhthaich.

P. 285. Fínán Lobhur is usually rendered 'Finan the Leper'—unnecessarily.

P. 285. Kilellan near Toward appears on record as Kyllenane 1376, Killenane 1525, 1539 (Orig. Paroch.), and therefore probably commemorates Fínan rather than Fillan. *Fínan* is from *fín*, wine. *Tollard*, Toward, means 'hole-point,' 'point of holes.' Miss Lamont of Cnoc Dubh inform me that the rock is magnesian limestone and full of holes.

P. 289. Maelruba: for the seventeenth century rites connected with his name in Wester Ross, see *Records of the Presbytery of Dingwall*, pp. 279-282 (Scottish Hist. Soc.).

P. 292. Teampull Mo-Luigh: the saint commemorated here is more probably Mo-Lua of Cell Da-Lua (Killaloe). The *Life* of Flannan records Mo-Lua's influence 'apud Orcades et usque ad insulas Gallorum,' in the Orkneys and even as far as Inse Gall,' *i.e.* the Hebrides; Plummer, *Vitae SS. Hib.*, cxxvi., *n.* 7.

P. 294. Slios an Truinnein: also *Ceas an Truinnein, Ceas a Druinnein.*

P. 295. Móenenn: in Gorman's *Martyrology* the metre shows that *Móenenn* was stressed on the first syllable, not on *-nenn.*

P. 300. Kincardine O'Neil: the last quatrain of the poem on the seven sons of Oengus (LL 354) is:—

Na manistri fuaratar / i ndénatar a ferta
is la hUi Néill Naoigiallaig / co rrath an Spirta sechta.

'The monasteries which they obtained, wherein they did their works of wonder, belong to the hUi Néill Naoigiallaig (the descendants of Niall of the Nine Hostages), with the grace of the sevenfold Spirit.' The parish of Kincardine O'Neil borders on Banchory-Ternan, the site of Torannan's monastery; its epithet O'Neil, which distinguishes it from Kincardine in the Mearns, must be explained from this quatrain: it formed part of the patrimony of Torannan's monastery. 'Onele' appears as the name of the district in 1200 (Reg. Aberd.).

P. 304. Loch Finlagan is usually called in Islay *Loch an Eilein*, 'the loch of the island.' Rev. C. M. Robertson, however, has succeeded in getting the real name from an old *seanchaidh*, who gave it as *Loch Bhìollagain* (*io* being strangely not nasal); the *bh* is of course due to eclipsis of initial *f* by *loch*, which was once neuter, *e.g. Loch nEachach*, 'Eochaidh's loch,' now Lough Neagh. The anglicized form Finlagan or Finlaggan is stressed on the second syllable, probably a fairly recent change.

P. 307. Epscop Nέm; *Thes. Pal.*, ii. 364.

Cill Mo-Naoimhin: in Iona the name MacNiven is pronounced *MacRaoi'in* for *MacNaoimhin* (Rev. R. L. Ritchie).

P. 314. Do-Chunne: To-Chunni, Sept. 3, Mart. Taml. (LL 362 c); all entries in Mart. Taml. are three days too early just here.

P. 317. *Drostén* represents an early *Drustignos.*

P. 322. Feargán (Farragon on maps) is the name of a mountain east of Schiehallion. Rev. C. M. Robertson would connect this and the names of the wells with O.Welsh *guerg*, 'efficax,' Ir. *ferg . i . laech*, '*ferg* means a warrior'; E.Celtic *Vergo-bretos*, 'effective in judgment.'

P. 323. Kilgourie: bishop Gobran is mentioned as a contemporary of Columba in the *Life of Farannan* (*Anecdota*, iii. p. 2).

P. 324. Anatswell: Annot was a female personal name and may have been so here.

Kentigern: add St. Mungo's Chapel a little to the north of the castle of Auchterarder, and St. Mungo's Well in Gleneagles (*Old Stat. Account*, iv. p. 44).

Exmagirdle is Eglesmagrill in 1211 and *passim* till 1300 (Chart. Lind.); Eglismagrill 1442, 1567 (RMS); in 1568 (RMS) it becomes by metathesis Eglismagirll, which passed into Eglismagirdill 1618, *ib.* It is 'my Grill's church'; Grillán, the diminutive of Grill, was one of the twelve who accompanied Columba from Ireland.

P. 326. Kilrennie in Fife is Kilretheni (RPSA), for *Cill-reithnigh*, the latter part being genitive or dative of *reithneach*, as in *Both-reithnich*, Borenich, for the regular G. *raithneach*, bracken (p. 120); compare Kilranny (p. 199). The Old Stat. Account (vol. i. 409) says: 'the fishermen, who have marked out the steeple of this church for a meath or mark to direct them at sea, call it St. Irnie to this day; and the estate which lies close by the church is called Irniehill; but, by the transposition of the letter *i*, Rinnie-hill.' The writer adds a tradition that 'the devotees at Anstruther, who could not see the church of Kilrenney till they travelled up the rising ground to what they called the Hill, then pulled off their bonnets, fell on their knees, crossed themselves, and prayed to St. Irnie.' He equates St. Irnie with St. Irenaeus, bishop of Lyons. There can be little doubt, however, that St. Irnie is purely fictitious, and that we have really to do with *irnaide* (*irnaidhe*), the Middle Gaelic form of Mod.G. *urnuigh*, meaning (1) a prayer, (2) an oratory, *e.g.* 'Findsige Urnaide . i . ina h-írnaide,' '(the day of) Findsech of Urney, *i.e.* in her oratory' (LL 363 h). The name arose probably from the ancient oratory; it seems likely also that the hill mentioned was called of old *Cnoc Irnaid(h)e*, 'the hill of prayer,' or 'the hill of the oratory'; in Mod.G. *Cnoc na h-Urnuighe*.

P. 327. Cadoc: Dog, Dogg, Doig, the surname of several landed families in the Kilmadock district, is for *Gille Dog*, 'St. Cadoc's servant,' as Blane is for *Gille Bláán*, Munn for *Mac Gille Mhunna*, etc. It occurs also in the Carse of Gowrie and Forfarshire. The name is now Doag, Doig, Doak. For instances see RMS, *passim*, and Macfarlane, *Geog. Coll.*, i. 338. Cadoc died probably *c.* 570. His name in full was Catmail, in modern Welsh Cadfael, which is for an early Celtic *Catu-maglos*, 'Battle-prince.'

P. 333. Syth: 'prespitero Sadb,' *Thes. Pal.*, ii. 271, where it appears to be the name of a man.

P. 335. Volocus: Skene (*Celt. Scot.*, ii. 178) says, 'Volocus is the Latin form of Faelchu,' and identifies him with Faelchu, abbot of Iona 717-724. But there is no proof that Faelchu was so latinized, and Faelchu of Iona died on April 3, while Volocus is commemorated on Jan. 29.

P. 336. Buite: Bellenden's translation of Hector Boece's History of Scotland tells how 'King Edward tuke sic displeseir aganis this Heltane, his brothir, becaus he brint the kirk of Sanct Bute, with ane thousand personis in it, that he dang him throw the body with ane swerd, afore the alter of Sanct Johne' (ed. 1821, vol. ii. p. 432). The Latin original, however, has 'Divi Machuti aedem . . . incendio corruperat,' 'he had destroyed by fire the church of St. Machutus.' Fordun and John Major mention the slaying and the reason of it, but not the name of the church (they say 'churches'). The king was Edward III., the brother was John of Eltham, Duke of Cornwall, and the church meant by Boece was not, as might be thought from Bellenden, a church of Buite, but probably Lesmahagow. (Sir Thomas Gray says that John died at Perth *de bele mort*; see A. Lang, *History of Scotland*, i. p. 253).

Hector Boece was a native of Forfarshire, where the name was not uncommon as Bois, Boys, Boyes, Boyse, Boyische, latinized, in the case of the historian and his brother Arthur as Boethius, Boetius, Boecius. As Buite is also latinized Boetius, Boecius, and as his monastery is now Monasterboice, it might be thought that the Forfarshire name is shortened from Gille Buite (see note on Cadoc, above). But it appears certain that the Scottish name is for the

Norman-French *del Bois*, for *de Bosco*, 'of the wood'; moreover Hugh Boece or Bois, Hector Boece's great-grandfather, was baron of Drisdale (Dryfesdale) in Dumfriesshire, and got land in Forfarshire by marriage.

P. 337 (4). Of Iona Bede says: 'Its monastery was long the chief (*arcem tenebat*) of almost all those of the northern Scots and all those of the Picts, and had the direction of their people' (Bk. iii. ch. 3). Bede, who died in 735, wrote this some considerable time after the communities of Iona had been expelled from Pictland across Druim Alban by king Nectan in 717 (Expulsio familiae Iae trans Dorsum Brittanniae a Nectano rege; AU). This is why he uses the past tense. As Dr. Plummer remarks, the Columban communities among the Picts must have been numerous.

P. 400. Monediwyerge: the second part is probably *díbheirge*, gen. of M.Ir. *díbergg*, brigandage, marauding, rapine.

P. 404. Ladder Hills: in Ireland, *áradh*, a ladder, is applied to a hill with ridges across; Joyce, *Irish Names*, iii. *s.v.* Arroo.

P. 412. Pitcruive may contain the gen. sing. of *crobh*, a hand; compare Dalcruive, p. 418.

P. 419. Cupar in Fife, Coupar-Angus, and Cupar-maculty near Coupar-Angus, now Couttie, are doubtless British; the Gaelic form is *Cùbar*; the Cross of Cupar is *Crois Chùbair*.

Dulshangie: the second part may be for *seangda*, from *seang*, slim.

Crombaidh may rather be for *crombda*, adjective from *crom*, bent.

P. 422. *Tom* is 'knoll' in Lewis speech. *Reangam* or *Reangum* near *Líteagán*, Litigan, Fortingal, is for *reang-thom*. O'Davoren gives *reng . i . caol*, '*reng* means slender'; as a noun *reanga* means 'the reins of the back,' and *reanga faille* is applied to a long sharp-backed rock (Dinneen).

P. 424. *Carraig*, a loan or survival from British, O.W. *carrecc*, Breton, *karrek*, is rare in our place-names. Besides Carrick in Ayrshire there are Carrick on Loch Goil and *na Carraigean*, Carrick, at the head of Glen Fincastle in Perthshire.

P. 426. Nigra Dea: compare 'alveum fluminis Sinnae, qui dicitur Bandea,' 'a channel of Shannon which is called Bandea (goddess)'; 'trans flumen Bandae,' *Thes. Pal.*, ii. 265, 269.

Damona: compare Damia and Auxēsia, goddesses of cattle and of growth respectively, among the people of Epidaurus (Herodotus, v. 82, 83).

P. 431. Gamhar: compare the Gaulish river Samara, the Somme, which may be from *sam*, summer. The two Irish rivers Samáir, now the Erne and the Morningstar, are different, in ending at least.

Add the Crosan of Applecross.

P. 432. Tanar, Tanner: compare also the Irish river *Torand*, 'thunder,' LL 17 a 19; 16 b 37.

P. 439. Lossie: compare the Luce of Wigtown, 'river of herbs'; also Lusragan, Glen Luss.

P. 455. Add Cragganester on Loch Tay, where 'ester' is for *easdobhar*, 'waterfall stream,' which describes the tumbling burn adjoining the 'craggan' on the west side.

Scotscalder is shortened to *Cal' nan Gall* in Gaelic. *Caladal* (or *Cal'*) *nan Gall* would be expected to represent Norne (Norse) Calder, as perhaps it originally did; Norne Calder is now quite obsolete. Here, as in Lothian, Calder has come to be a district name; there are East Calder, West Calder, Mid Calder, Old Calder, North Calder.

P. 456. Aldourie: *dourie* should represent *doborda*, dark, but the usage suggests that it is a reduced form of a compound that began with *dobur*, *dobhur*.

P. 470. *Isca*: E.Celtic *i* in stressed position becomes *e* in Gaelic before *a* or *o*; thus *isca* becomes O.Ir. *esc*; *viros*, O.Ir. *fer*.

P. 472. Nevis may be compared with Nemesa, now Nims, a stream of Luxemburg mentioned by Ausonius.

P. 476. Yarrow is Gierua, *c.* 1120, Gierwa, Gieruua, *c.* 1150 (Lawrie). This, like Gala, is an English name, and may be compared with Jarrow, in Bede "In Gyruum." Tweed is obscure to me.

P. 478. Add *Ath-bhreanaidh* on Lyon below Duneaves; compare *Cill Ach-breanaidh* in Strath Brora. The second part is obscure to me.

P. 485. Lairg, a Sutherland parish, is in Gaelic *Lairg*, *Luirg*, from *lorg*, a shank. Rev. C. M. Robertson prefers *lorg*, a track; but see Lurgyndaspok.

P. 491. Aonach: Ennochdhu in Kirkmichael, Perthshire, is an *t-Aonach Dubh*.

P. 499. Open *a* in unstressed position is also found in connection with the *-nt-* ending. All the Scottish examples under (6) and (7) on pp. 444-7 have open *a*.

P. 502. Alloway, etc.: *Ailmag* in Connacht is made 'Rockfield' by Kuno Meyer (*Contrib.*), but the forms *Aelmag*, *Aolmagh*, given by Hogan point to 'Lime-plain.'

P. 506. Sanas: MacSween of Gigha, etc., is styled 'eucag Sanais na sreabh séimh,' 'peacock of Sanas of gentle streams,' in a poem in the Book of the Dean of Lismore. Loch na Shanish, near Inverness, is *Loch na Seanis*, for *Sean-inse*, 'Loch of the old Haugh'; it is now usually spelled Loch na Sanais.

P. 508. Dunoon: compare *ess oyvin* for *eas abhann*, 'a waterfall on a river,' in the phonetic Gaelic of the Dean of Lismore (*Rel. Celt.*, i. 58).

Onlach mac Briuin: compare 'Columbanus nepos Briuni,' a youth in Iona in the time of Columba, and 'Molua nepos Briuni,' a brother of the community of Iona (Adamnan, *Life of Columba*, ii. 15, 30); 'Pat. M'Abhriuin' (Macdonald Gaelic Charter of 1408) for *Mac úi Bhriuin*, which may be for an older *Moccu Briuin*. It is now Brown in Argyll.

INDEX OF PLACES AND TRIBES

NOTE.—The stress accent is indicated by a full stop placed before the stressed syllable: *e.g.* Aber·arder is stressed on the third syllable, Aber-sky is stressed on the first. Early Celtic forms and most record forms are left unmarked. The early Celtic names, and names from Latin or Greek texts are in italics. Reconstructed or hypothetical forms are marked by an asterisk.

INDEX OF PERSONAL NAMES